Forest Products Laboratory, 1910–2010

Celebrating a Century of Accomplishments

Compiled and Edited by

John W. Koning, Jr.

Forest Products Laboratory, 1910–2010

Celebrating a Century of Accomplishments

Forest Products Laboratory
1910–2010
Celebrating a Century
of Accomplishments

Compiled and Edited by **John W. Koning, Jr.**

Copy Edited by **James R. Anderson**

Designed by **Tivoli C. Gough**

Photographs: **Stephen A. Schmieding and others**

References: **Julie J. Blankenburg, Mary A. Funmaker**

USDA Forest Service
Forest Products Laboratory
One Gifford Pinchot Drive
Madison, Wisconsin
53726-2398
Phone: (608) 231-9200
Internet: www.fpl.fs.fed.us

August 2010

John W. Koning, Jr., started his career with the U.S. Forest Service on the Willamette National Forest in 1957. In 1958 he was called to duty and served for three years as an officer in the U.S. Air Force. Following service, he joined the staff at the FPL, conducting research in paperboard packaging. Following retirement in 1986, he joined the Engineering Professional Development Department, College of Engineering, University of Wisconsin, working with industry in developing educational courses. He is the author of *The Scientist Looks at Research Management* (American Management Briefing); *The Manager Looks at Research Scientists; Corrugated Crossroads: A Reference Guide for the Corrugated Containers Industry;* and *Three Other R's: Recognition, Reward and Resentment.* He was a contributor to the *Handbook of Physical and Mechanical Testing of Paper and Paperboard; TAPPI Personal Career Planning Workbook;* and the Wisconsin Department of Natural Resource booklet, *10 Ways to Protect Your Woodland Property.*

Contents

Preface

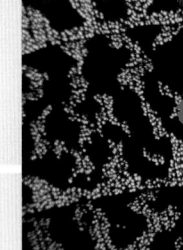

I've been asked the question many times: "Why did you devote your time and effort to compile this book?" I've often wondered the same thing, and I have no simple answer. (I do know it wasn't for the money—I worked as a volunteer, and any profits from the sale of the book go to the FPL Employees' Association and the University of Wisconsin.) Part of the answer is that after working at FPL for over 20 years and then working for the University of Wisconsin and industry, I came to view the Laboratory in a different light. I came to understand that FPL has been, and continues to be, one of the very few places that provides unbiased research to help solve difficult problems important to sustainable forest management and a diverse and fragmented industry. And I believe this information is important to the American Public, those who financially support the FPL, because hundreds of wood-derived products impact our daily lives, impact the environment, and provide employment for thousands of citizens.

When I started working on this project at least five years ago, I was focused on the 100th anniversary of the FPL and thought it would be great to have a summary of the Laboratory's major accomplishments and an update of its history. Thus, my early vision of the book had a section on many of the people who worked at the laboratory and their various activities. As the Laboratory's centennial anniversary drew closer, FPL Director Chris Risbrudt learned that I was working on this project and asked if I could complete it in time for the celebration. I said that I could but I would need help from FPL staff—which he approved. He also decided to have an official academic history prepared. Following those two decisions, I removed the historical sections covering the people and activities and ramped up my efforts, focusing mainly on the accomplishments of the FPL. (I did retain one section—"FPL People"—that lists the names of those who have worked at the FPL. Unfortunately, the list is not inclusive because of the lack of complete records. I gathered the names from official directories, phone lists, photos, and organization charts, so the list includes employees and some visiting scientists and students who worked at FPL.)

My approach in compiling the book was to first review what had already been written about FPL accomplishments, scattered in previously prepared reports, lists, and annual reports. Where possible, I matched the accomplishments from these sources to published reports. I also compiled a list of historic FPL mission statements and goals. Combining this information, I organized the book, prepared a draft of each section, and selected interesting photos emphasizing the research. I gave each section to one of the retired scientists or engineers for review and comment, asking specifically if I had missed any major accomplishments and if what was included was accurate. Then I made necessary changes and gave the section to another retiree, and so on, ending the cycle with the current staff to review and provide information on current research. This process resulted in as few as one section reviewer to as many as 11 (see "Acknowledgments"). Multiple reviews helped achieve some subject matter balance and emphasis except where unequal emphasis was intentionally included, such as the "Adhesives" section, written primarily by Bryan River; if each section were as complete, this book would be a multivolume set, not the coffee table book I envisioned.

Capturing the uniqueness of the Laboratory was a challenge, and unfortunately, no one is around to review and discuss the very early years. The Laboratory historically has been organized along general disciplines, such as chemistry, engineering, and natural sciences, with special emphasis on subjects such as preservation, sawmilling, pulp, paper, wood quality, and composite products processes. This has brought together a group of people with different research outlooks, from fundamental to applied, theoretical to experimental, esoteric to practical, and individual to team approaches.

Just as the staff represented a range of perspective, so did management—from directors who conducted hands-on research to those that were more detached—though they all worked hard to secure funding for the Laboratory and to navigate the political winds.

Out of this mix came thousands of research reports. For many of those reports, it was difficult to determine who actually had the ideas, conducted the research, or even who wrote the report. In the early days, report authorship was decided by supervisors; some felt it important to give the credit to the Laboratory, so the report was authorless. In fact, in the first seven years (1910 to 1917) almost 100 reports listed only "Forest

Products Laboratory" as author. Others carried the authorship of the supervisor, when the work was actually done by the staff. However, the vast majority of reports gave credit to the researchers in charge and his or her coworkers but generally did not include the names of the technicians or support staff.

With the diversity of the reviewers and my limited in-depth knowledge of the vast variety of subjects, I wasn't surprised that a number of reviewers were motivated to make substantial changes to draft sections. In fact, Robert White rewrote the Fire Safety section and Stan Lebow, the Preservation section. All the suggestions, corrections, and rewrites were then further revised and edited. I hope that what has emerged is reasonably accurate and understandable.

This book was designed primarily to show the breadth of FPL accomplishments. The exception to this is the "Adhesives" section. A few other subjects were also expanded to give a more in-depth—but still brief—view:

- Sustainability—Using Problematic Natural Resources, by Jerry Winandy
- Nondestructive Evaluation of Wood, by Robert Ross
- Best Opening Face & Sawmill Improvement Program, by John (Rusty) Dramm
- Overview of Composites and their Technologies, by Jerry Winandy
- Brief History of Forest Products Utilization and Marketing Assistance, by John (Rusty) Dramm
- ASTM Committee D-7: Wood, by David Green, Robert Ethington, Edward King, Bradley Shelley, and David Gromala/Forest Products Journal Vol. 54(9) (Sept. 2004)
- Development and Mill Transfer of the FPL Semichemical Pulping Process, by John McGovern/TAPPI Journal Vol. 67(6) (June 1984)
- Biological Utilization of Wood for Production of Chemicals and Foodstuff, by George Hajny/Research Paper FPL 385

Additional background for many of the accomplishments can be found in "Further Information" listed at the end of each section. Note that the number of reports listed is not a good indication of total research contribution. For example, between 1910 and 1917, more than 150 reports were prepared on research related to wood preservation (from "Reports issued by the FPL up to April 1, 1917," Sarah D. Kinney).

Results that were under-emphasized in this book were FPL's theoretical research by scientists such as Ed Kuenzi, Hank Montrey, Charles (C.B.) Norris, Mike Rosenthal, Jerry Saeman, Al Stamm, Harold Tarkow, Harry Tiemann, John Zahn, and many others. Their research does not lend itself easily to photographic representation, but in many instances it was instrumental in guiding or explaining many of the FPL accomplishments.

Finally, I have to confess that I really enjoyed putting this book together, even though others may have done a better job. The interactions with retired scientists and staff, current staff, and especially some of the new, younger scientists, was not only essential but enlightening (sure is a lot one doesn't know) and invigorating. Thanks to all who have helped make the FPL a success.

John W. Koning Jr.
FPL (retired)

PERSISTANCE

"Nothing in this world can take the place of persistence. Talent will not; nothing is more common than unsuccessful people with talent. Genius will not; unrewarded genius is almost a proverb. Education will not; the world is full of educated derelicts. Persistence and determination alone are omnipotent. The slogan 'press on' has solved and always will solve the problems of the human race."

—Calvin Coolidge

Acknowledgements

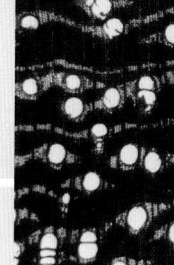

Compiling examples of the results of 100 years of research was a daunting task. I want to thank the following people for assisting me in this endeavor. (Names in *italic* are retirees.) Many suggested minor, and some major, changes to the text, provided pictures, and gave helpful guidance about what was factual and important. My apologies to those scientists whose research was not included or not given the emphasis it deserved.

Adhesives—*Al Christensen, Tony Connor*, Charles Frihart, Linda Lorenz, *Bryan River*
Anatomy—*Regis Miller*, AlexWiedenhoeft
Bending wood—Bob Ross
Biochemistry—Ken Hammel, Barbara Illman, *Kent Kirk*
Bridges—Mike Ritter, Jim Wacker
Chemicals—*Andy Baker, Dan Caulfield, Tony Connor, Bill Feist, Kent Kirk, Jim Laundrie, John Obst*, Alan Rudie, *Ed Springer, Harold Tarkow, Duane Zinkel*
Decay—Carol Clausen, Jessie Glaeser, Barbara Illman
Drying—Rick Bergman, Bob Ross, *Bill Simpson*
Economics—Rick Bergman, Dave McKeever, Ken Skog
Energy—Mark Ditenberger, *John Zerbe*
Engineering—*Russ Moody*, Bob Ross, Xiping Wang, *Ron Wolfe*
Environment—Charlie Carll, Sam Glass, Joe Murphy, *Anton TenWolde, Sam Williams*
Fasteners—Doug Rammer
Fiberboards—*Bob Geimer, Jerry Winandy, John Youngquist*
Finishes—*Bill Feist, Sam Williams*
Fire safety—Sue LeVan-Green, Robert White
Flooring and paneling—*Kent McDonald, Jerry Winandy*
FPL—Julie Blankenburg, Gordon Blum, Michael Kaspszak, *Gary Lichtenberg*, Jeff Sims, Ken Skog
Housing—Charlie Carll, Joe Murphy, Mike Ritter
Insects and marine borers—Rachel Arango, Carol Clausen, Rick Green, *Bruce Johnson*
Laminated products—*Russ Moody*, Bob Ross, *Jerry Winandy, Ron Wolfe*
Lumber grading—*Dave Green*, Dave Kretschmann, Bob Ross
Lumber manufacturing—John (Rusty) Dramm, Terry Mace, *Kent McDonald*
Machining—*Kent McDonald*
Molded products—Craig Clemons, John Hunt, *Jerry Winandy*
National defense—Bob Ross
New century—*Bruce Johnson*, Sue Paulson, Ted Wegner
Packaging—*Dunc Godshall, Bob Kurtenacker*, Alan Rudie, *Tom Urbanik*
Pallets—*Tom Urbanik, Jerry Winandy*
Paper—*John Klungness*, Alan Rudie, Ted Wegner
Particle composites—*Bob Geimer, Jerry Winandy, John Youngquist*
Plywood—*Jim Muehl*, Dave Kretschmann, *Jerry Winandy, John Youngquist*
Poles, piles, posts, and railroad ties—Dave Kretschmann, *Russ Moody, Ron Wolfe*
Preservation—Carol Clausen, *Bruce Johnson*, Stan Lebow
Pulping and bleaching—*Jim Laundrie, John Obst*, Alan Rudie, *Ed Springer*
Recycle and reuse—Bob Falk, *John Klungness*, Ted Wegner, *John Youngquist*
Statistics—James Evans
Structures—*Russ Moody*, Bob Ross, *Ron Wolfe*
Sustaining Our Natural Resources—Peter Ince, *Kent McDonald*, Ken Skog, *Jerry Winandy*
Technology Marketing Unit—Sue LeVan-Green, *Jean Livingston*
Trusses—*Russ Moody*, Bob Ross, *Ron Wolfe*
Veneer—*Jim Muehl, Jerry Winandy, John Youngquist*
Waste—Carol Clausen, Barbara Illman
Wood—*Kent McDonald*

Special thanks to Jim Anderson for making the text readable and understandable, and for helping to reconcile some of the differing recollections of past research.

Tivoli Gough deserves full credit for design and layout of the complex array of subjects and is much valued.

Julie Blankenburg and Mary Funmaker's efforts to locate publications and other resources were essential to the production of this book.

I very much appreciate the help of Steve Schmieding for taking or finding and reproducing many of the pictures taken by previous FPL photographers including *James K. Brooks, Sandra L. DeMaster, Melvin E. Diemer, Leif J. Ersland, Donovan R. Every, Robert M. Graves, Roger B. Russell, and James R. Vargo*. The following also provided photographs: Jessie Glaeser and *Hal Burdsall*—mushrooms; *Regis Miller*, Alex Wiedenhoeft, Michael Wiemann, and *Richard Kinney*—wood anatomy; *Gary Lichtenberg*—shops; and Dave Kretschmann—poles.

Thanks also to Jeff Sims, *John Bachhuber*, Pat Brumm, and *Barb Wolfe* for budget information; Sue Austin, JoAnn Benisch, Julie Blankenburg, *Mary Doran*, Cherilyn Hatfield, Sue Howards, *Mary Jane Klein*, and *Sue Parisi* for employee names; and Janet Stockhausen and Erin Bubrick's help with patents.

Special thanks to Julie Blankenburg, Gordon Blum, *John Erickson*, *Tom Hamiliton*, *Russ Moody*, Janet Stockhausen, Ted Wegner, *John Youngquist*, and *Bob Youngs* for reviewing the rough draft and offering many suggestions for improving the book.

Special thanks also to Harry Cullinan, Auburn University and Terry Gerhardt, Sonoco Corporation, for review and comments on the final draft of the book.

Support of this endeavor by Chris Risbrudt, FPL Director, is greatly appreciated.

I'm grateful for the financial support from the University of Wisconsin Madison, Office of University Relations.

To all the others who have contributed in many different ways: Thank You.

Finally, I recognize and express my gratitude for Kenneth Kruger's confidence in hiring me, Keith Kellicutt's mentoring me in my research, Dunc Godshall and Curt Peters for technical advice to me and many other FPL researchers, and the many competent technicians and other support staff that helped me. I especially appreciate the secretaries and clerk–typists who assisted me over my time at the FPL: Lorraine Moore who issued pencils to me when I first arrived at the Laboratory (things were different then), Susan Howards, Mary Douros, Cecilia Doyle, Noreen Esser, Barbara Foellmi, Delores Gust, Jane Kohlman, Mable Neugent, Eleanor Pape, Charlotte Richardson, and Judy Wegner. Special thanks to my wife for her continued patience and support.

This book is dedicated to the Forest Products Laboratory employees.

Introduction

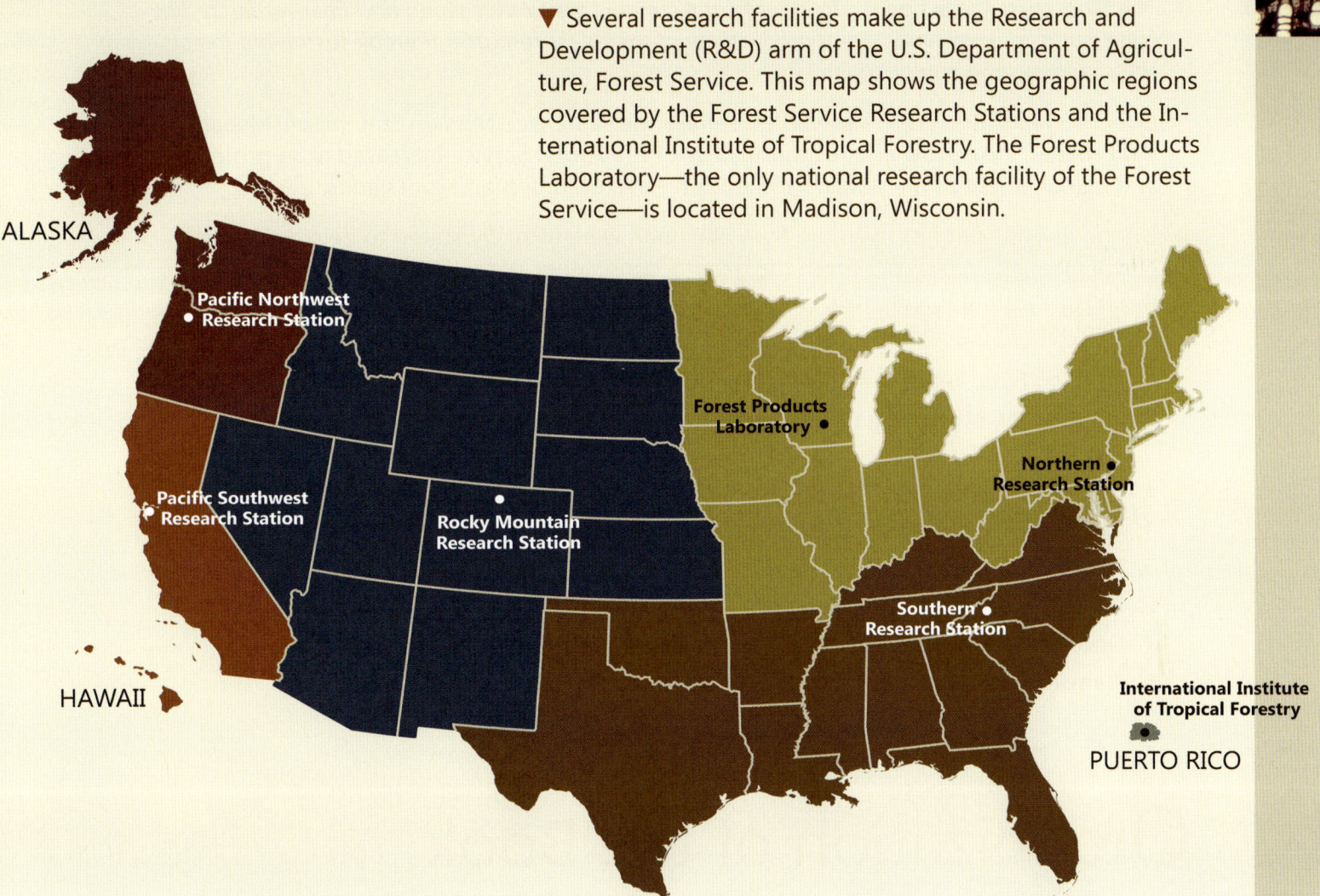

▼ Several research facilities make up the Research and Development (R&D) arm of the U.S. Department of Agriculture, Forest Service. This map shows the geographic regions covered by the Forest Service Research Stations and the International Institute of Tropical Forestry. The Forest Products Laboratory—the only national research facility of the Forest Service—is located in Madison, Wisconsin.

ALASKA

Pacific Northwest Research Station

Pacific Southwest Research Station

Rocky Mountain Research Station

Forest Products Laboratory

Northern Research Station

Southern Research Station

HAWAII

International Institute of Tropical Forestry

PUERTO RICO

This book is about many of the accomplishments of a unique institution—the U.S Department of Agriculture, Forest Service, Forest Products Laboratory (FPL). When it opened in 1910 on the University of Wisconsin campus, FPL was the only wood research facility in existence. Today, it is joined by many similar research laboratories located around the world. Many of its accomplishments have been joint efforts with universities, associations, other agencies, and industry.

Over the past 100 years, the American public has invested approximately $1,500,000,000 (current dollars) in the Forest Products Laboratory—an average of $15,000,000 per year. (For comparison, the cost to fight forest fires in 2007 alone was over $1,700,000,000.) One objective of this book is to show what FPL has accomplished with that investment.

Another objective is to give the American public an armchair tour of the FPL as an alternative to an actual visit—a much better option that you should certainly consider, if at all possible!

According to John Erickson, retired FPL Director, "The FPL has been and continues to be involved in the development of nearly all U.S. standards covering the conversion and use of wood products. These standards assure the American Public the best value and safety in the use of wood products. If you enter a house, read a paper, drive over a bridge, or travel by train or air, you can be sure that FPL research touches your life. This book documents the broad spectrum of research accomplishments that serve the public."

FPL Mission

The FPL's current mission is *"To identify and conduct innovative wood and fiber research that contributes to conservation and productivity of the forest resource, thereby sustaining forests, the economy, and quality of life."*

Over the years, the mission stayed essentially the same (see Mission Statement sidebar) but did broaden as the role and responsibilities of the Forest Service increased in its protection and management of the public's 193 million acres of forests and grasslands.

During its 100-year history, more than 264 goals were formally stated to help guide the FPL in meeting its mission. Given the general consistency of the FPL's mission statements, it is not surprising that almost every goal can be restated as part of one or more of eight general research goals:

1. Develop basic knowledge of wood and its use
2. Improve the processing of wood to increase yield and quality and reduce costs and waste
3. Extend the service life of wood products
4. Utilize the various wood species and changing quality of wood
5. Improve the quality and usefulness of wood products
6. Develop new products from wood
7. Develop basic knowledge of the biological aspects of wood
8. Transfer the research developed both nationally and internationally

Organization of the Book

The purpose of this book is to highlight many of the research accomplishments that were part of meeting these goals. To achieve this purpose, this book is organized into several sections.

The ***Introduction*** starts by painting the broader picture of *Sustaining Our Natural Resources*—seeing the variety of benefits provided by our forests helps us understand the importance of sustainable forest management. As important as forest products are, the benefits of forests go way beyond this single use. And one of the beauties of forests is that they are renewable—by practicing sustainable management, we can have both wood products and all the other benefits.

One of the threats to sustainable forests is the invasion of non-native species that crowd out native species and threaten the forest ecosystem. In his article on *Sustainability—Using Problematic Natural Resources,* Jerry Winandy briefly describes how FPL research provided a use for two invasive species, which can help in their control.

The final part of this section describes the history of FPL's physical facilities—the buildings that have housed the researchers and support staff who conducted the research necessary to meet the wood utilization challenges of the past century.

The next section describes the wood utilization ***Challenges*** that FPL research has addressed in the areas of *Decay*; the *Environment—Wind, Sun, Water*; *Fire Safety*; *Insects and Marine Borers*; *National Defense*; and *Waste*.

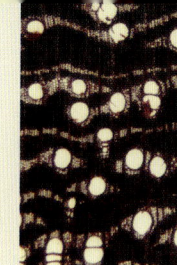

The third section describes the **Foundation Sciences** employed by FPL researchers to address the many challenges. *Anatomy*, *Biochemistry*, *Economics*, *Engineering*, and *Statistics* are among the foundation sciences applied in studying the many processes of using wood and converting it to products.

The section describing **Processes and Products** is generally ordered by "size"—from solid wood, to wood particles, to wood fiber, and finally to chemical constituents of wood. Examples from this section include research on efficient harvesting and cutting, preservative treatment (which significantly increased the useful life of railroad ties and decreased the huge drain on the forest resource early in the 20th century), new pulping processes, reducing waste, and other efficiencies in wood use.

The final sections of the book include a thumbnail **Summary** of FPL accomplishments; a glimpse into the exciting **New Century** of wood research, including the application of nanotechnology and a renewed interest in extracting chemicals from wood; a **Conclusion** reminding us that wood is a renewable gift of nature and that with sustainable management, forests can continue to provide us with wood products and clean water, wildlife, recreation, food, beauty, and solitude; and three **Related Articles** that provide more detailed insights to wood use research and development for those readers with more time and interest; and a list of the **FPL People** responsible for all these accomplishments.

Additional background for many of the accomplishments can be found in the lists of *Further Information* at the end of each section. Readers should contact the FPL for more comprehensive and up-to-date information on any subject of interest.

From Research to the Real World

The FPL staff has contributed in many ways to improving our life through the use of wood products. Many of the challenges still face us, but through research we can continue to enjoy the abundance of our forests and all its benefits, provided we practice sustainable forest management.

The principal focus of FPL's efforts is on fundamental research and development and long-term studies, rather than commercialization of specific products. Although FPL has carried some accomplishments through to commercialization, generally that has occurred through cooperative endeavors with industry.

Many of FPL's substantial contributions in providing unbiased research and design data to help establish standards, rules, and specifications in the use of wood and wood-derived products do not lend themselves to photographs. But the lack of pictures does not mean the work was less important. The results for the American Public are improved performance of wood products, safer housing, reduced costs, and help in providing a sustainable supply of wood.

Not all the accomplishments described in this book have found their way to the real world. For example, the disk separator was successful in removing contaminants from recycled paper but has yet to be successfully scaled up to production levels. Press drying of paper has not gone beyond the development stage. Automated hardwood screening and resaw for maximizing wood quality awaits improved methods of sensing wood defects. Veneer press drying is still waiting for further development.

Some ideas and concepts are ahead of their time, such as applying the mathematical equations developed for maximizing paper strength. One major delay is the lack of a rapid, online sensing system for appropriate properties. Biopulping, though successful even in large-scale demonstrations, is yet to be commercialized, partly because of the challenge

of using biological organisms in an industry geared to chemicals, the cost of installing new processing facilities, and the suggestion of an even better pulping process. Many of the innovative ideas used in the NuFrame house were hindered in their adoption by builders because of present building code requirements.

The viability of converting wood to ethanol, though technically feasible (and in fact commercialized in the 1910s), is directly affected by the price of oil.

Some accomplishments have gone to commercial scale (such as the automated waste separating process run by the City of Madison) only to be later replaced by a different approach.

Finally, many of FPL's accomplishments have been absorbed by industry, were improved by them during development, and are now commercial. An excellent example is one of the FPL's greatest contributions, a new pulping process that allowed the use of different wood species for making paperboard: neutral sulfite semichemical (NSSC) pulping allows for the conversion of many different hardwood species into usable paperboard. When developed in the 1920s, it greatly accelerated the use of corrugated fiberboard containers that now transport over 90% of the American Public's products. This research, described in the article by John McGovern (Related Articles), started a new industry, created thousands of jobs, and significantly contributed to sustainable management of forests.

A number of research areas described in this book are no longer active research areas. These include sawing, machining, pallets, wood bending, trusses, veneer, plywood, flooring, paneling, and packaging. FPL's areas of research have been started, stopped, emphasized and de-emphasized based on changing goals, budgets, personnel, advances in science, and inputs from the public, cooperators, and advisory groups.

Over its 100-year history, FPL accomplished many things. It should again be emphasized that these impressive accomplishments have resulted from significant cooperative input from universities, industry, associations, governments, and other professionals and the continuing financial support of the American Public. All contributors should take great pride in what has been jointly accomplished.

I hope you will enjoy reading about our shared accomplishments.

Top Corner Thumbnail Pictures

- Wood is a very complex material—the wood properties of each individual tree are affected by its growth rate (resulting from soil properties, weather conditions, altitude, and other factors), and each tree species is characterized by a unique cell structure. And our forest ecosystems reflect vast variety. To remind us of these complexities—and the straightforward beauty of nature—each page of this book shows the microscopic cell structure of one of many tree species or a fruiting body (mushroom) of one of numerous species of forest fungi.

Throughout this book,

- numbers in parentheses (x) are approximate dates;

- numbers in curly braces {x} are photograph numbers;

- and numbers in square brackets [x] refer to the reports in the lists of *Further Information* found at the end of each section.

Forest Products Laboratory Mission Statements

Over its 100-year history, at least 12 mission or purpose statements were formally documented for the FPL:

* 1910—To promote the greatest possible utilization of forest material with the least possible waste. (History of the U.S. FPL 1910–1963, Nelson, C.A., p. 42)

* 1922—The purpose of the Forest Service in the administration of the Forest Products Laboratory is to conserve American forests by developing the most economic methods of converting standing trees into finished products. The purpose is also to make the growing of timber more profitable by increasing the possibilities in the utilization of both used and unused species. The Laboratory seeks to develop not only new and more efficient processes, but to find ways of utilizing material which would otherwise be wasted, to find new uses for old materials and new materials for old uses. In a word, the aim is to render practical assistance to the manufacturers and users of wood and wood products and at the same time to promote forest conservation and the practice of forestry. (Forest Products Laboratory brochure, 1922)

* 1934—The business of this laboratory is to help maintain the value and marketability of forest products—to develop methods to reduce wastes and lower costs in logging and in the manufacture and use of forest products, to increase the serviceability and satisfaction of forest products to the user, and to develop new and useful products from wood. The degree to which lands not needed for agriculture may be maintained as an economic asset through forest use depends largely on the volume of useful commodities that can be profitably produced from wood. ("Purpose of the FPL Work," September 20, 1934)

* 1938—The better and more efficient and diversified utilization of forest materials. (The Forest Products Laboratory—A Brief Account of Its Work and Aims, USDA Misc. Pub. No. 306, 1938, p. 3)

* 1964—Forest Products Laboratory research is aimed at developing knowledge that is adaptable or applicable anywhere in the United States at the least possible cost. (FPL 1964 Annual Report, p. 3)

* 1967—To extend the usefulness of wood for the benefit of man." (1967 Annual Report, Foreword)

* 1970—To find those techniques, processes, and practices that will attain wood utilization goals with as little environmental or ecological disturbance as humanly possible while maintaining a flow of needed products to a growing population. ("Wood in the Service of Man, 1970–1971," p. 1)

* 1985—"Conduct research to improve the utilization of wood and fiber leading to improved management of the timber resource, meeting the needs of the United States, and contributing to the needs of the international community." (State of the Lab Forum, Dec. 1985, pg. 6)

* 1991—"To improve the use of wood through science and technology, thereby contributing to the conservation and management of the forest resource." (Strategic Plan for the FPL, Jan. 1991, pg. 3)

* 1997—"To provide the science and technology needed to maintain and extend the Nation's forest resources." (Research Progress Report from the FPL. 1997-1998. pg. 3)

* 1999—"To Provide the science and technology needed to maximize benefits received from the Nation's available timber resources; ensure consumer and environmental safety and protection in processing and use of wood-based products; and match wood use and forest management to achieve both forest health and economic health." (FPL 1999 Program. Pg. 1)

* 2002—"To identify and conduct innovative wood and fiber research that contributes to conservation and productivity of the forest resource, thereby sustaining forests, the economy, and quality of life." (FPL Strategic Plan. Jan. 2002 pg. 2)

Sustaining Our Natural Resources

◄ Sahalie Falls, located in the Willamette National Forest (Hwy 126), Oregon, helps provide clean water for the public. (2007)

A major goal of the U.S. Forest Service is to enhance the wise use of our natural resources, particularly those associated with forests. The Forest Service attempts to develop forests, manage ecosystems, and wisely use forest resources in a sustainable manner to meet the needs of humans and other life forms. [25–27, 29–31, 33, 34, 38–42]

The Forest Products Laboratory's role in meeting this goal is to enhance the wise use of the wood that is produced by forests and help ensure the forest's sustainability.

Challenge: One challenge in maintaining sustainable forests is how to remove wood from the forest in such a way as to minimize any detrimental effects, improve forest health and conditions, and enhance the many other benefits provided by forests.

FPL's Contributions: FPL research has increased our understanding of how wood grows, improved harvesting practices, determined wood's constituents and properties, and developed new products and processes that utilize species that were once considered waste materials. The research extends the wood supply by utilizing more species and smaller trees, providing uses for invasive wood species, and extending the life of wood products such as railroad ties, poles, and housing components through preservative treatments.

Results: The public has greatly benefited from adequate, low-cost wood products that enhance their daily lives while leaving millions of acres of forests to provide a wide range of additional benefits (as depicted on the following pages).

Natural Resource Benefits for the Public

▲ Lake on the Gunflint Trail, Minnesota. One of the most important benefits of forests is their contribution to clean drinking water. Watersheds around the country are a precious resource for sustaining life. "Watershed maintenance and restoration are the oldest and highest callings of the Forest Service. The agency is, and always will be, bound to them by tradition, law, and science. The national forests truly are the headwaters of the Nation." (Forest Service Chief Mike Dombeck, 1998) {91 M} [34]

◄ Hiking in the national forests is one of the great pleasures of the wilderness. "The importance of recreational use as a social force and influence must be recognized and its requirements met. Its potentialities as a service to the American people, as the basis for industry and commerce, and as the foundation of the future economic life of many communities, are definite and beyond question." (Forest Service Chief Robert Y. Stuart, 1928–1933) {44 H} [34]

◀ Forests provide homes for many species of wildlife, including this fawn. Unfortunately, overpopulation of species, such as deer, can lead to forest ecosystem damage. {44 N}

◀ Controlled grazing on the grasslands and open forest areas of the national forests helps to decrease competition among plant species. However, uncontrolled grazing can lead to destruction of forests. {44 J}

▲ Bird watching is a major activity for the public, and sustaining forests for protecting wildlife habitat is an important goal.

▲ Sustaining public access to the beauty of the mountains—such as Mt. Hood, Oregon, located in the Mt. Hood National Forest—is important to building awareness of the importance of scenic vistas. (2007)

◄ View from Indian Ridge Lookout on the Willamette National Forest. Note the regrowth of trees on the previously clearcut areas in the background, providing for a sustainable supply of clean water, animal habitat, hiking, wood, and other natural resources for the public. (2007)

◄ In addition to clean drinking water, the many streams and rivers in the forest watersheds provide outstanding fishing. "The conservation of the inland waterways of the United States constitutes, perhaps, the largest single task which now confronts our Nation." (Forest Service Chief Gifford Pinchot, 1905–1910) {44 G} [34]

▲ A major use of forests is camping and the use of off road vehicles. However, improper use of mechanical equipment can lead to severe damage to the forest, trails, and clean water.

▲ Forests provide for winter sports such as cross country skiing, snowmobiling, ice fishing, and snowshoeing. {86 I}

▲ Forests also provide the wood that is used in thousands of wood products ranging from components of housing to paper to chemicals. "The basic point of our sustainable forest management strategy is this: not only do economic stability and environmental protection go hand in hand, economic prosperity cannot occur without healthy, diverse, and productive watersheds and ecosystems." (Forest Service Chief Mike Dombeck, 1998) {20 Q} [34]

◀ An early source of chemicals from the forest was pine sap used for naval stores. Today, chemicals are primarily derived from wood chip pre-steaming in kraft pulp mills.

Another tree product is maple syrup from the sap of maple trees.

◀ Forest fungi are essential to help trees obtain nutrients from the soil. A few mushrooms, the fruiting bodies of forest fungi are edible, but most are not safe to eat. {15 N}

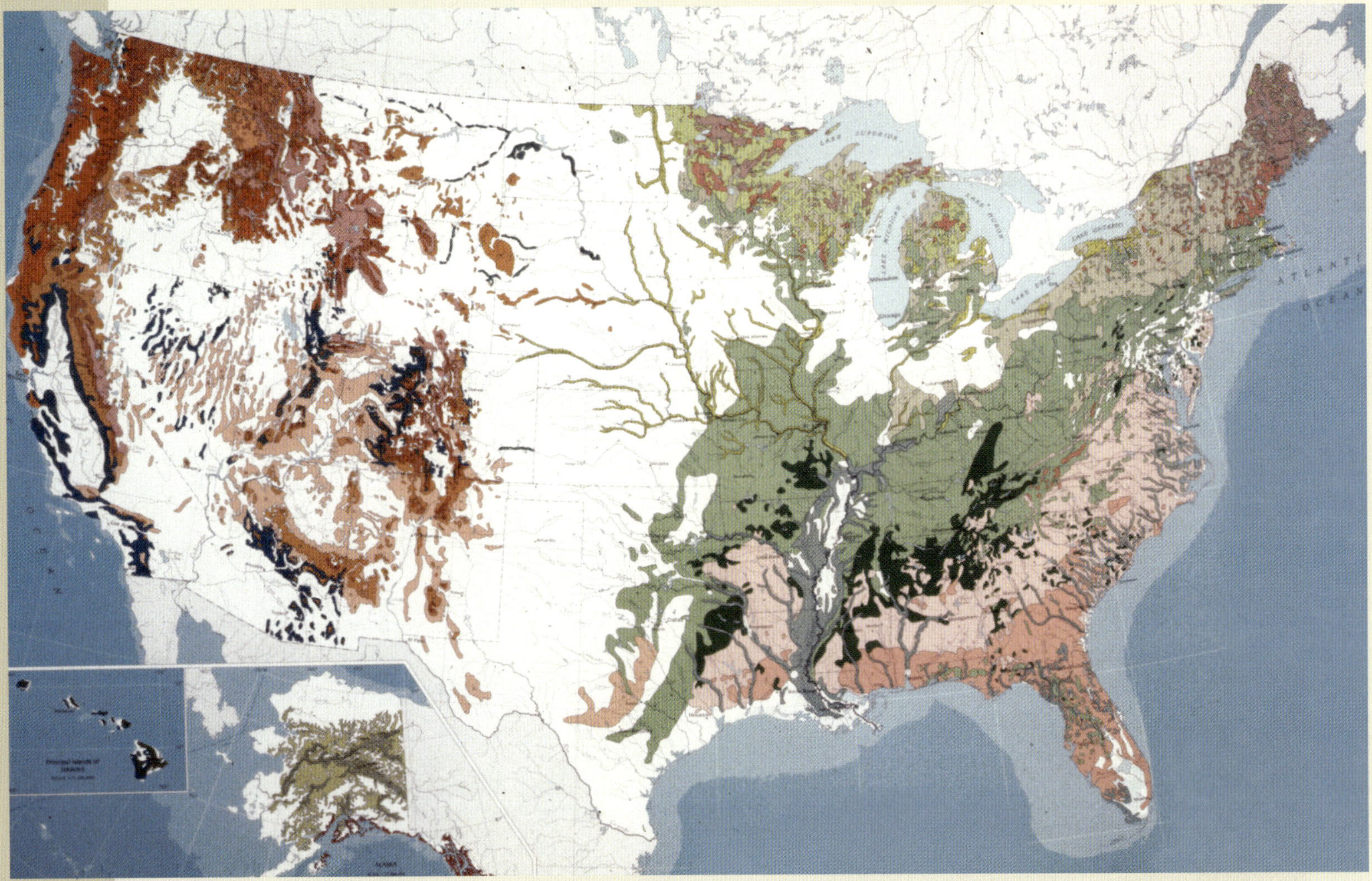

▲ Not only is use of the nation's forests complex, but the forests vary in species of softwood and hardwood trees. Each color represents a different species of tree, and all these areas have mixtures of species. This leads to the challenge of how best to manage each forest.

Types of Forests

◄ Pine plantation—Early FPL research indicated that pruned trees produce clear lumber 10 years earlier than trees subject to natural pruning. Today, most of our clear grade softwood lumber is imported from countries where pruning is practiced and labor costs are typically lower, or where economical pruning technology was developed. {22 D} [5, 7–10, 13, 14, 17, 18, 20, 21]

◄ Natural pine forest—Periodic fires have kept the growth of competing plant species under control, but in many areas, years of fire suppression have led to serious overstocking and an increased fire hazard. {115 H} [1–3, 6, 10, 11]

◄ Hardwood forest of many tree species—This diversity helps it survive natural insect and disease outbreaks. [15, 16, 19]

◄ Thinned hardwood forest to maximize growth of the best trees—Thinning allows for the removal of insect infestations and diseased trees, thus making the forest healthier. Thinning also decreases the severity of wildfires by reducing some of the dead and down material. However, thinning is economical only if a good market for small-diameter wood exists. {130 R} [4, 12]

◄ Forest of just one tree species (aspen)—Monoculture (single species) forests can facilitate harvesting, and for some species such as aspen and lodgepole pine, they occur naturally on thousands of acres. {126 G}

◄ Hardwood forest showing a variety of tree forms—One challenge is how to make usable products from poorly formed trees. {75 N} [37]

◄ Another challenge is how to efficiently make products from small trees when they are removed in a thinning operation. {110 P} [40]

◄ Some forests have trees growing in water, such as these cypress trees. This adds more complexity to proper management and generally prevents harvesting under good management practices. {21 I}

Sustainability

Each type of forest presents a unique set of challenges in efforts to conserve and sustain it. One major challenge to all types is the increasing problem of species that invade the forest and upset the ecosystem. Without control, these invasive species can negatively affect the forest (see "Sustainability Using Problematic Species"). Those "problematic species" include other plants but also include animals—and people. Many of the FPL research accomplishments presented in this book provide options for improved sustainability, such as maximizing the production of usable lumber from a log, thus extending the timber supply; maximizing our utilization of wood wastes through improving processes and recycling; and extending the life of wood products with proper selection and treatments. But much remains to be done. [25]

Utilization of Excessive Growth and Fire-Killed Timber

One paradoxical outcome of the historic efforts of the Forest Service to prevent forest fires is an overabundance of new growth and buildup of dead material in the forests. This fuel buildup has contributed to wildfires and the loss of natural resources. Both of these problems are complicated by the high costs of removing excessive new growth and dead material. One way to help decrease the serious fire situation is to develop new uses for the overstocked material and thereby offset some of the costs for removal. Unfortunately, to do nothing, as some have suggested, will just result in billions of dollars spent fighting wildfires and still leave us with a major loss in natural resources, such as clean water, wildlife, timber, and recreation. [22, 33, 39]

▲ Overstocking with young trees allows for more intense wildfires, higher suppression costs, greater damage to the forest ecosystem, and loss of natural resources.

◄ Results of a wildfire. If this forest had been thinned of the excessive number of trees, the fire may not have been as intense and would have had fewer negative effects on the ecosystem.

Utilization of Dead Trees

◄ Stand of dead Alaska yellow-cedar. (1990s)

Dead Alaska Yellow-Cedar Finds a New Life

An area of over a half-million acres of the Tongass National Forest in southeast Alaska contain an increasing percentage of dead yellow-cedar trees, which is preventing regeneration of healthy trees and may constitute a forest health problem. FPL research helped the Alaska Region of the Forest Service assess the utilization potential of this wood. Tests showed that the strength of wood from trees dead for up to 80 years was similar to that of wood from trees that had recently died and that laminated yellow-cedar lumber exceeded standard requirements for resistance to delamination, shear strength, and wood failure. Durable wood products from salvaged, but strong, yellow-cedar is possible, and that would turn what was once thought to be a wasted resource into a usable product. However, the problem of economic harvesting has yet to be solved. (1997) [28, 32]

▲ Yellow-cedar is durable and suitable for structural applications. (1990s)

▲ Stewart Lake, Mount Horeb, Wisconsin.

Sustaining Our Natural Resources

1910

The U.S. Forest Service was 5 years old.

The FPL was new.

National forests were established primarily for wood and water.

Forests were viewed by many as inexhaustible sources of wood.

"Cut and get out" was the general forestry practice.

Waste of wood was enormous by today's standards.

2010

The U.S. Forest Service is 105 years old.

The FPL is 100.

The public's 193 million acres of National Forests and Grasslands are now recognized for their multiple uses and are under management to continuously provide a broad array of benefits:

  Clean air and water

  Natural flood control

  Timber, forage, and non-wood products

  Wildlife habitat

  Endangered species recovery

  Scenic beauty

  Recreational opportunities

  Food

  Community revitalization

  Improved human health

  Carbon sequestration

Further Information

[1] "Virgin growth" and "second growth" / FPL No. 153 (1921)

[2] Shortleaf pine: the lumber-making qualities of second-growth and of virgin-growth timber / E.M. Davis / Southern lumberman Vol. 137(1770) (Dec. 15, 1929)

[3] Selective logging versus clear cutting in Shortleaf pine / R. D. Garver. Southern lumberman Vol. 141(1789) (Oct. 15, 1930)

[4] Selective logging in the northern hardwoods of the lake states / Raphael Zon, R. D. Garver / USDA, Forest Service technical bulletin No. 164 (1930)

[5] Cupping of plain-sawed lumber and checking of timbers of longleaf pine of slow, medium, and rapid growth / Benson H. Paul / Southern lumberman Vol. 140(1786) (Sept. 1, 1930)

[6] The relation of certain forest conditions to the quality and value of second-growth loblolly pine lumber / Benson H. Paul / Journal of forestry Vol. 30(1) (Jan. 1932)

[7] Pruning forest trees / Benson H. Paul / Journal of forestry Vol. 31(5) (May 1933)

[8] Knots in second-growth pine and the desirability of pruning / B. H. Paul / USDA miscellaneous publication No. 307 (June 1938)

[9] Tree pruning by annual removal of lateral buds / Benson H. Paul / Journal of forestry Vol. 44 (July 1946)

[10] Effect of crown reduction on taper and density in Longleaf pine / Ralph O. Marts / Southern lumberman Vol. 179(2249) (Dec. 15, 1949)

[11] Patterns of variation of wood structure and properties within trees of Southern yellow pine and the influence of environmental factors on these patterns / Diana M. Smith / Presented at Symposium on Wood and Wood Fiber Quality in Relation to the Southern Pulp and Paper Industry, Raleigh, N.C. (November 14-16, 1962)

[12] Products research as an aid to forest farming / J. C. Killebrew, / Forest farmer Vol. 22(11) (July 1963)

[13] Effect of tree spacing on weight yields for red pine and jack pine / Robert R. Maeglin / Journal of forestry Vol. 65(9) (Sept. 1967)

[14] Wood quality of loblolly pine after thinning / D. M. Smith / Research paper FPL 89 (1968)

[15] Relationship of black walnut wood color to soil properties and site / Neil D. Nelson, Robert R. Maeglin, Harold E. Wahlgren / Wood and fiber Vol. 1(1) (Spring 1969)

[16] Effect of nitrogen fertilizer on the growth rate and certain wood quality characteristics of sawlog size red oak, yellow-poplar, and white ash / H. L. Mitchell / Proceedings of the symposium on the effect of growth acceleration on the properties of wood, Nov 10-11, 1971. Madison, WI: Forest Products Laboratory (1972)

[17] Effect of irrigation and fertilization on wood quality of young slash pine / D. Smith / Proceedings of the symposium on the effect of growth acceleration on the properties of wood, Nov. 10-11, 1971. Madison, WI: Forest Products Laboratory (1972)

[18] Effect of mechanical stress on growth and anatomical structure of red pine: stem vibration / J. T. Quirk / Canadian journal of forest research Vol. 6 (1976)

[19] Effect of nitrogen fertilization on black walnut growth, log quality, and wood anatomy / R. R. Maeglin, H. Hallock, F. Freese, K. A. McDonald / Research paper FPL 294 (1977)

[20] Properties of wood from improved and intensively managed trees / B. A. Bendtsen / Forest products journal Vol. 28(10) (Oct 1978)

[21] Clearcut harvesting and ectomycorrhizae: survival of activity on residual roots and influence on a bordering forest stand in western Montana / A. E. Harvey, M. F. Jurgensen, M. J. Larsen / Canadian journal of forest research Vol. 10(3) (Sept. 1980)

[22] Assessing the fire hazard of structures in the wildland-urban interface / Susan L. LeVan, J. Cohen, R. Chase, J. Davis, V. Malinauskas, H. T. King / Forest Products Laboratory (1990)

[23] The importance of nutrient pulses in tropical forests / D. J. Lodge, W. H. McDowell, C. P. McSwiney / Tree Vol. 9(10) (Oct. 1994)

[24] Formation and structure of lignified plant cell wall: factors controlling lignin structure during its formation /

Noritsugu Terashima, Rajai H. Atalla / Proceedings of 8th international symposium on wood and pulping chemistry, June 6-9, 1995, Helsinki, Finland. (Helsinki: Gummerus Kirjapaino Oy) (1995)

[25] New approaches to forest sustainability / Said Abubakr, Alan Haney, Michael A. Kilgore, Betsy Daub, Kim Chapman, Eric Bloomquist./ Environmental conference & exhibit, May 5-7,1997, Minneapolis convention center, Minneapolis, MN. Atlanta, GA: Tappi Press, c1997. 477-478. Book 1 (1997)

[26] Role of wood production in ecosystem management / R. James Barbour, Kenneth E. Skog, ed. / Proceedings of the sustainable forestry working group at the IUFRO All Divisions 5 Conference; 1997 July; Pullman, Washington. FPL-GTR-100 (1997)

[27] New silvicultural practices under ecosystem management / R. James Baubour, Steven Tesch, Susan Willits, Roger Fight, Richard Gustafson, Saket Kumar, Joseph McNeel, Andrew Mason, Ken Skog / Environmental Conference & Exhibit, May 5-7, 1997, Minneapolis convention center, Minneapolis, MN. Atlanta, GA: Tappi Press, c1997 121-130. Book 1 (1997)

[28] Mechanical properties of salvaged dead yellow-cedar in southeast Alaska: Phase I / Kent A. McDonald, Paul E. Hennon, John H. Stevens, David W. Green / FPL-RP-565 (1997)

[29] Charting our future: a nation's natural resource legacy / USDA FS-630 (Oct. 1998)

[30] An update of timber certification: Potential impacts on forest management / C. Denise Ingram / Proceedings of the Society of American Foresters 1998 national convention, Traverse City, Michigan, September 19-23, 1998. Bethesda, MD: Society of American Foresters publication SAF 99-01 (1999)

[31] Ecosystem management and the use of natural resources / Marlin Johnson, James Barbour, David W. Green, Susan Willits, Michael Znerold, James D. Bliss, Sie Ling Chiang, Dale Toweill / In: Szaro, R. C.; Johnson, N.C.; Sexton, W. T.; Malk, A. J., eds. Ecological stewardship--A common reference for ecosystem management. Humans as agents of ecological change. Amsterdam, The Netherlands: Elsevier Science: Vol. 2: 558-582 (1999)

[32] Flexural properties of salvaged dead yellow-cedar from southeast Alaska /David W. Green, Paul E. Hennon, James W. Evans, John H. Stevens / Forest products journal Vol. 52(1): p. 81-88 (Jan. 2002)

[33] Using Wood-based Structural Products as Forest Management Tools to Improve Forest Health, Sustainability and Reduce Forest Fuels: A Research Program of the USDA Forest Service Under the National Fire Plan / John F. Hunt, Jerrold E. Winandy / Proceedings of the 6th Pacific Rim Bio-Based Composites Symposium. 10-13 Nov. 2002. Portland, Oregon. Oregon State University. Vol. 1: 316-322 (2002)

[34] USDA National report on sustainable forests--2003 / R. W. Guldin, H. Fred Kaiser / FS-766 (February 2004)

[35] Log sort yard economics, planning, and feasibility / John {Rusty} Dramm, Robert Govett, Ted Bilek, Gerry L. Jackson / FPL-GTR-146 (2004)

[36] Quantifying trade-offs between economic and ecological objectives in uneven-aged mixed-species forests in the southern United States / Joseph Buongiorno, Benedict Schulte, Kenneth E. Skog / FPL-GTR-145 (2004)

[37] Development of new microwave-drying and straightening technology for low-value curved timber / John R. Hunt, Hongmei Gu, Philip Walsh, Jerrold E. Winandy / FPL-RN-0296 (2005)

[38] Using wood composites as a tool for sustainable forestry / J. E. Winandy, R. Wellwood, S. Hiziroglu / Proceedings of IUFRO XXII World Forestry Congress; August 8-13, 2005; Brisbane, Australia. In: FPL-GTR-163 (2005)

[39] Small-diameter success stories II / Jean Livingston / FPL-GTR-168 (2006)

[40] Evaluation of silvicultural treatments and biomass use for reducing fire hazard in western states / K. E. Skog, R. J. Barbour, K. L. Abt, E. M. Bilek, F. Burch, R. D. Fight, R.J. Hugget, P. D. Miles, E. D. Reinhardt, W. D. Sheppard / FPL-RP-634 (2006)

[41] Use of Saltcedar and Utah juniper as fillers in wood-plastic composites / Craig Clemons, Nicole Stark / FPL-RP-641 (2007)

[42] Integrated biomass technologies: a future vision for optimally using wood and biomass / J. E. Winandy, A. W. Rudie, R. S. Williams, T. H. Wegner / Forest products journal. 58(5):10-19 (2008)

Sustainability—Using Problematic Natural Resources

As problems associated with sustaining and enhancing the world's forest and agricultural resources compete with the needs of a rapidly increasing population, the management of our land becomes a much more complex and important issue. Wood and other biofibers are renewable and can be sustainably produced. Composite wood and natural biofiber products are effective alternatives to mineral- or petrochemical-based materials, which are non-renewable and less recyclable (or not recyclable at all). We must commit ourselves to developing the fundamental and applied science and technology necessary to provide improved value, service life, and utility while meeting the needs of consumers for sustainable building materials. Engineered bio-composite technology provides a tool for resource managers because it can add value to low- or no-value fiber resources and thereby promote demand for diverse biofiber feedstocks, including small-diameter timber, fast-grown plantation timber, removals of exotic-invasive species, removals of hazardous forest fuels, and agricultural residues. At the same time, engineered wood composites serve as a tool for economic development of rural communities and provide for value-added commodity products from recycled or undervalued materials or problematic natural resources.

Challenge

We must develop tools to address resource sustainability, enhance recyclability, and minimize environmental impacts of composite processing. Then, whenever forest resource options change, or as excess waste-stream wood resources (such as discarded wood and fiber) become available, or as alternative nonwood and non-lignocellulosic materials become more economical and available, or if air and water quality regulations become more stringent, we can adapt and sustainably address each of these issues. Wood and natural biofiber–composite technologies offer an ability to meet consumer needs for products and materials that are renewable, recyclable, and carbon-neutral technologies. Such an approach promotes sustainable development and resource conservation and is better for the environment than other technologies often used today that require nonrenewable energy and materials.

FPL Contributions

This type of industrial composite processing technology can help meet the need for consumer products and construction materials from a sustainable natural resource (timber, recycled materials, and agricultural residues), giving us products made from more diverse and less desirable wood and other natural biofiber feedstocks, often from underutilized timber species of ever-changing quality. This technology development at FPL and with our cooperating partners has allowed us to use lower quality timber resources.

Results

FPL research has been focused on developing a fundamental understanding of the basic relationships between bio-based materials, processes, and composite performance properties. This understanding gives manufacturers the tools to "control the composite process on-the-fly." This allows us to accept any number of bio-based materials as input and still produce consistent value-added composite products. Most importantly, we work with forest and natural resource managers to use bio-composite technologies as a tool to help them restore damaged ecosystems and promote sustainable forest management practices. [3, 7]

Utilization of Invasive Species

A growing problem is what to do with invasive species that destroy natural ecosystems. Although efforts are being made to eradicate them, these efforts are expensive. FPL research has provided practical solutions that help offset harvest and restoration costs by developing valuable products. Following are examples of FPL research using fiber–plastic molded products for utilizing invasive juniper and saltcedar. [7]

▲ Invasive juniper in Utah. (2000s)

◄ A low-value tree of the Southwest is providing jobs thanks to FPL research. Juniper has no value for logging or pulp production and was used only for firewood and fence posts. FPL research has found that juniper works well as a fiber component in wood fiber–plastic composite products. Such composites are being manufactured into informational signs and route markers for the National Forests by a firm in New Mexico. The composite signs are considerably less expensive, more durable, and more resistant to animal damage than traditional materials. (2000s) [1, 2, 6]

▲ Another major invasive species problem is saltcedar in the southwest, which is out-competing critical indigenous species. (2000s) [7]

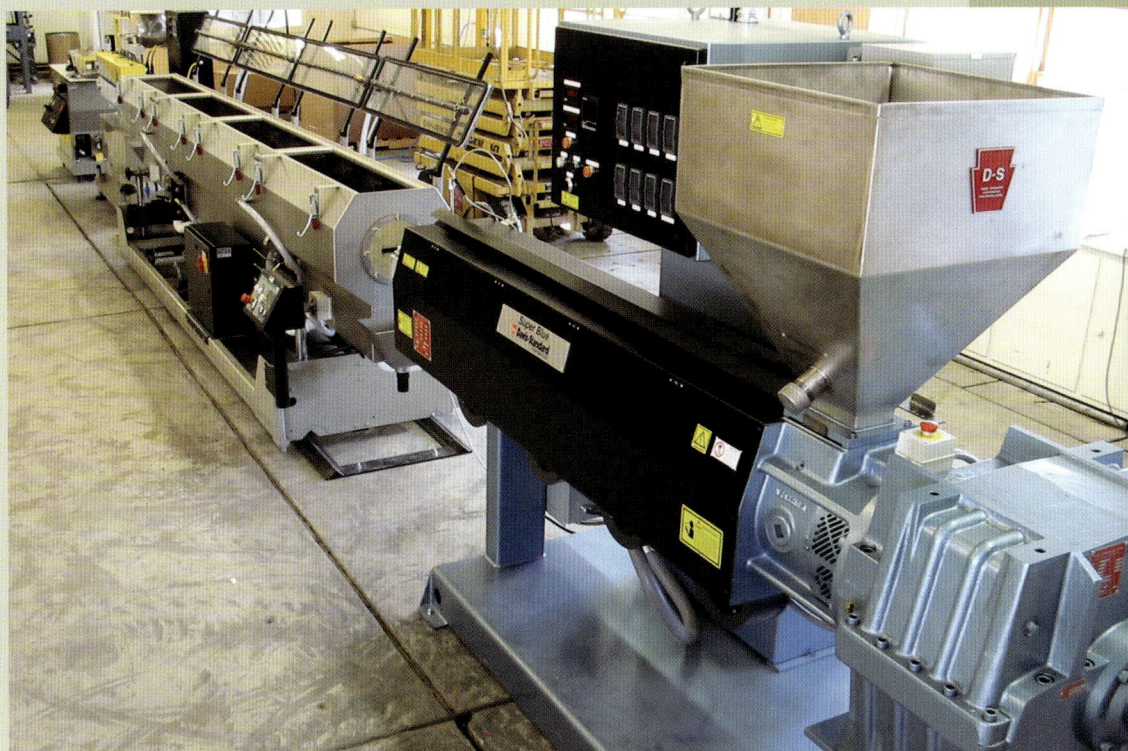

▲ FPL profiler (extruder) used to make wood–plastic boards from Utah juniper and saltcedar. (2000s)[7]

◄ Samples of saltcedar fibers used in combination with plastic. (2000s)

95% salt cedar
5% Phenolic Resin

50% salt cedar
50% recycled HDPE

95% salt cedar wood/bark mix
5% phenolic resin

50% salt cedar wood/bark mix
50% recycled HDPE

◀ Test fence used to evaluate outdoor performance of wood–plastic-composite (WPC) boards made out of juniper and saltcedar. (2000s)

Technology Transfer

USDA

United States
Department of
Agriculture

Forest Service

Forest
Products
Laboratory

General
Technical
Report
FPL–GTR–163

Using Wood Composites as a Tool for Sustainable Forestry

◀ This information is shared internationally as a way to promote global sustainability and economic development. [3]

Proceedings of 2005 IUFRO World Forestry Congress Session #090:
Using Composites as a Tool for Sustainable Forestry
Sponsor: IUFRO Div 5.05 Composites

Further Information

[1] Juniper-based signs and rangeland restoration / James H. Muehl, Andrezj Krzysik, John A. Youngquist, Theodore L. Laufenberg / Biographies & Abstracts; Forest Products Society 54th Annual Meeting: June 18-21, 2000, South Lake Tahoe, NV (2000)

[2] Wood and plastic composite material and methods for making same / Phil T. Archuletta, James H. Muehl / U.S. Patent No. 6,586,504 (July 1, 2003)

[3] Using wood composites as a tool for sustainable forestry / J. E. Winandy, R. Wellwood, S. Hiziroglu / FPL-GTR-163 (2005)

[4] Development of new microwave-drying and straightening technology for low-value curved timber / John R. Hunt, Hongmei Gu, Philip Walsh, Jerrold E. Winandy / FPL-RN-0296 (2005)

[5] Nanotechnology opportunities in residential and non-residential construction / T. H. Wegner, J. E. Winandy, M. A. Ritter / 2nd International Symposium on Nanotechnology in Construction. Held 13-16 Nov. 2005 in Bilbao, Spain. RILEM, Bagneuz, France 9p. (2005)

[6] Small-diameter success stories II / Jean Livingston / FPL-GTR-168 (April 2006)

[7] Use of Saltcedar and Utah juniper as fillers in wood-plastic composites / Craig Clemons, Nicole Stark / FPL-RP-641 (2007)

[8] Integrated biomass technologies: a future vision for optimally using wood and biomass / J. E. Winandy, A. W. Rudie, R. S. Williams, T. H. Wegner / Forest products journal Vol. 58(6) (2008)

Forest Products Laboratory

◀ This house, at 1610 Adams St., Madison, Wisconsin, was the first location for the Forest Products Laboratory. It was the temporary quarters from October 1, 1909, to April 1, 1910, during completion of the new laboratory on the University of Wisconsin campus. {M 120 172}

Original Forest Products Laboratory Building

▲ In this building (provided by the University of Wisconsin (UW) at a cost of $50,000), the Forest Products Laboratory was born in 1910. It remained here until 1932, when the present building was occupied farther west on the UW campus. This original building is now used by the UW Department of Materials Science and Metallurgy. {108 270}

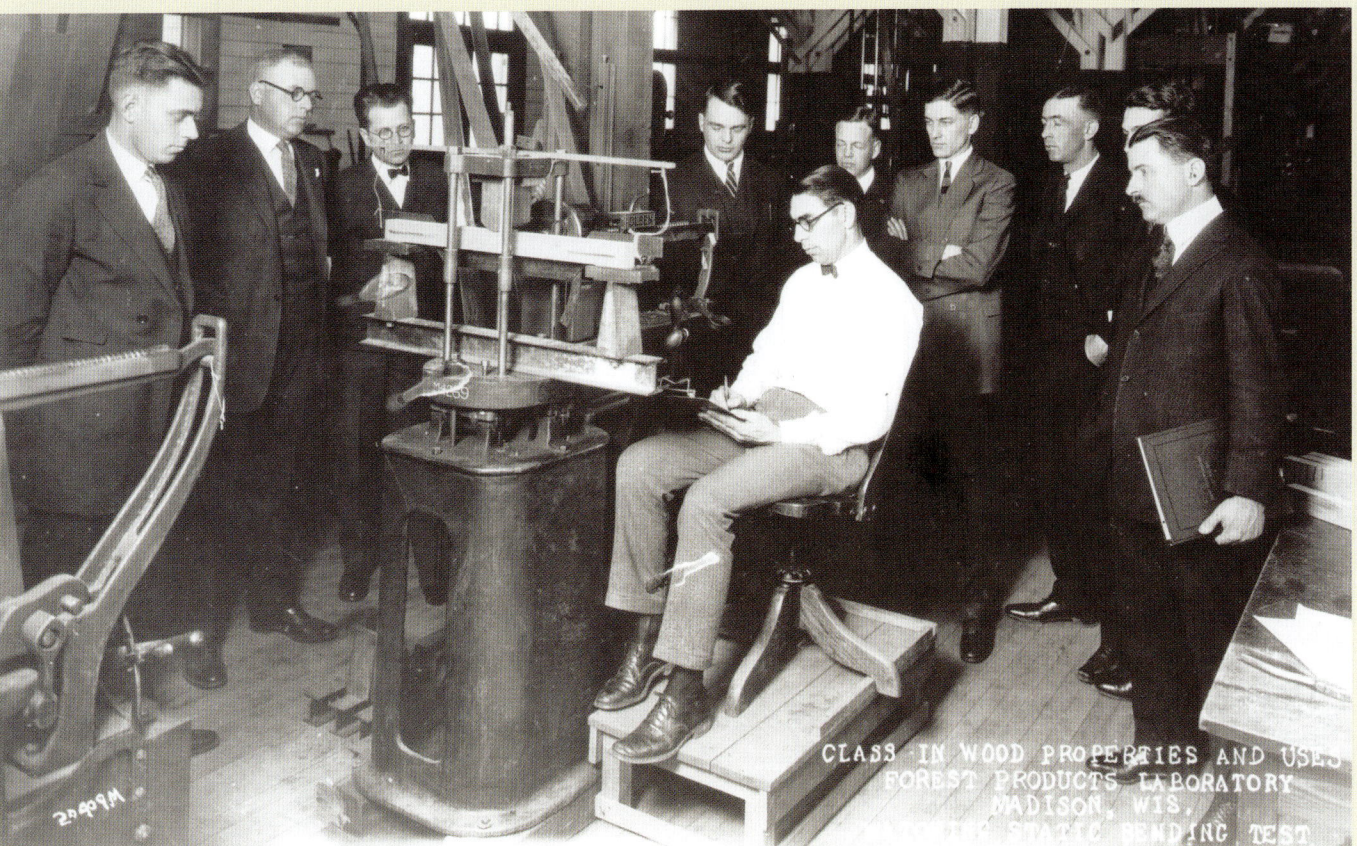

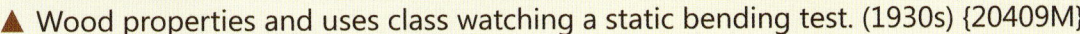

▲ Wood properties and uses class watching a static bending test. (1930s) {20409M}

◀ Red gum logs being unloaded at the FPL on the UW campus. At that time, red gum was viewed almost as a weed rather than a commercial wood because of how difficult it was to dry without exceedingly heavy degrade. The FPL, as a result of drying experiments, proved that red gum could be dried without prohibitive degrade, and it is now commercially used. (1920s) {M 85 F}

▲ In 1921, Secretary of Agriculture Henry C. Wallace inspected the USDA Forest Service, Forest Products Laboratory. *Left to right:* Carlile P. Winslow, Director of the FPL; Henry C. Wallace, Secretary, Department of Agriculture; W.B. Greeley, Forester, Forest Service; O.M. Butler, Assistant Director, FPL. {M 12429 M}

▲ Engineering test area of the FPL. Note the use of belt-driven equipment. Wood sticks on the dolly are wood specimens waiting to be tested for their strength properties. (1920s) {M123F}

Site of Present FPL Building

▲ Site of the new Forest Products Laboratory to be built in 1931–1932. This road would become Highland Ave. The railroad track is still there. {M 19122 F}

▲ Start of construction of new Forest Products Laboratory in 1931. {M 19615 F}

▲ Framework for the new Forest Products Laboratory. (1931) {M 19945 F}

Early FPL Research Emphasis Areas and Staff Contributors

Following are some early areas of FPL research emphasis and the researchers who made noteworthy contributions. (This list is based on "Reports issued by the FPL up to April 1, 1917," compiled by Sarah D. Kinney. Number of reports in each area is shown in parentheses.)

Preservation of Wood (151) George M. Hunt, Clyde H. Teesdale, Carlile P. Winslow, Howard F. Weiss, E.W. Peters

Pulp and Paper (138)
Henry E. Surface, Otto Kress, Carlile P. Winslow, Sidney D. Wells

Distillation of Wood (89)
Lee F. Hawley, Robert C. Palmer

Mechanical Tests and Properties of Wood (83)
John A. Newlin, Charles N. Betts, McGarvey Cline, Thomas R.C. Wilson

Chemical Analyses (82)
L.E. Cover, Homer Cloukey, Arlie W. Schorger

Utilization of Wood (50)
Rolf Thelen, Harry N. Knowlton

Kiln Drying (47)
Harry D. Tiemann, Charles N. Betts

Minor Chemicals, Dyes, Ethyl Alcohol, etc. (39) Frederick W. Kresssman, Rolf Thelen

Physical Properties of Wood including Identification (30) Arthur Koehler, Eloise Gerry

Naval Stores (22)
Harry N. Knowlton, Charles N. Betts

Destruction of Wood by Various Agencies (12) George M. Hunt, Ernest Bateman

Chemical Properties of Wood (11)
Ernest Bateman, Arlie W. Schorger

▲ New FPL building, fall 1932. The successful construction bid was $735,298. {M 21019 F}

▲ Aerial view of new (1932) FPL, looking northeast. Lake Mendota is in the background. Note the lack of development around the new building.

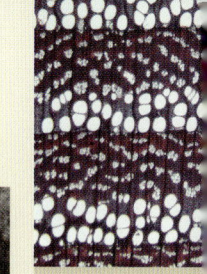

▲ Forest Products Laboratory, 1935. Note fire, packaging, and veneer laboratories to the left of the main building.

▲ Early displays in the FPL lobby. (1933) {M 22495 F}

▲ Railing around the entrance steps. Note the inlaid floor. (1933)

▲ View from the top of the entrance steps in the lobby of the new building. (1933) {M 24418 F}

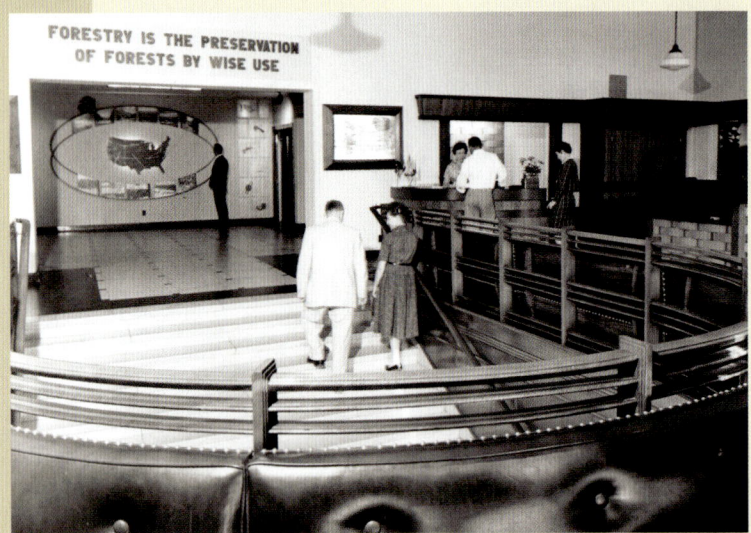

▲ Entrance view of refurbished FPL lobby. (1953)

▲ Entrance view of refurbished FPL lobby. (2008)

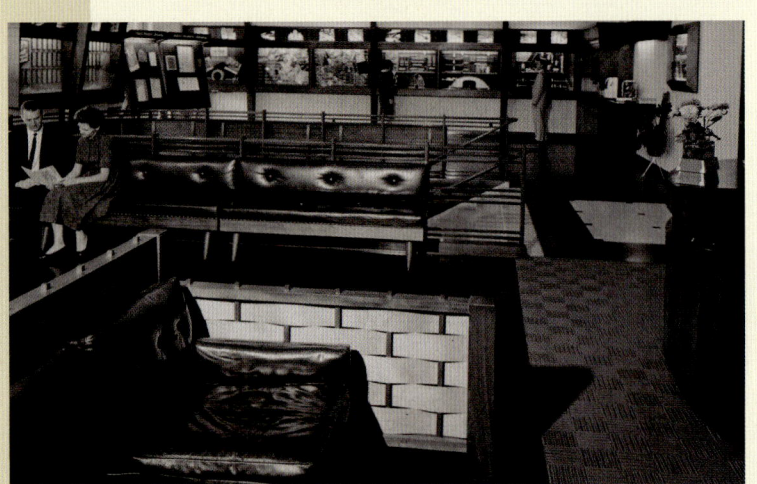

▲ Refurbished lobby with permanent display cases. (1953)

▲ Reception desk in lobby. (2008)

▲ Close up of one of the display cases. (1953)

▲ Displays in the lobby. (2008)

▲ Electrical panel for the building. (1933)

▲ Director's office with working fireplace, September 1933. In 1975, Director Robert L. Youngs, started a tradition of using his office for an all-FPL holiday open house. {M 22503 F}

▲ Incoming director John R. Erickson (left) and retiring director Robert L. Youngs (right), celebrate the 75th anniversary of FPL by pulling the cord together to make the whistle sound.

◄ FPL whistle sounding. In the early days this whistle alerted employees when to start and stop work. It was dismantled under Director Hall's tenure. It was briefly reconnected for commemorative use in celebrating the FPL's 75th anniversary.

▲ FPL in 1939. The entrance faces east. Note the experimental stress skin panel houses to the north of the main building. The two-story house (upper right) was built as a Director's residence, but actually housed FPL draftsmen.

▲ FPL in 1967. The "Director's residence" is gone and replaced with buildings housing chemical research.

FPL's fields of research and its methods of exploration include:

Structure, composition, and properties of wood—Establishing the scientific basis of wood utilization and conversion. Chemistry, the microscope, X-rays, the polarimeter applied to cellulose, lignin, pentosans, extractives, the wood fiber, fibrils, fusiform bodies, cellulose particles, and the unit cell. Aiding development of the diversified wood products—rayon silk, cellophane, wood alcohol, grain alcohol, charcoal, turpentine, rosin, molded products, tannin, and dyes.

Determination of pore structure and volume, moisture diffusion and hygroscopicity, to better methods and control of impregnating, treating, painting, gluing, etc. How growth conditions affect quality and quantity of wood grown; how to distinguish woods one from another by grain and structure.

Improving the use of wood in building and fabrication—Seasoning of wood in dry kilns to reduce shrinkage and swelling in floors, furniture, doors, trim, and framework; treatment of wood with protective chemicals to make it more resistant to decay and fire; improvement of painting; stronger and more water resistant glues and gluing. Increasing the strength of structures through tests of timbers, wall panels, built-up columns and laminated arches, nailing and bolting, plate and ring fastenings, and other types of joints. Improvement of shipping containers to reduce losses to shipper and consumer; full-scale tests of loaded boxes and crates by revolving drum and vibration table.

Improving conversion and harvesting methods—Improving practice in logging and milling to benefit the growing forest, increase future yields, raise the quality of lumber, and at the same time reduce costs. Better service of lumber through improved grading and selection. Design of instruments for quick determination of moisture, hardness, and other properties. Demonstration of better and more economical production of turpentine and rosin. How oleoresin is developed in the living tree; selection of high yield strains.

Pulp and paper—Conversion of American wood species from the log to finished paper in experimental pulping and paper making plant. Adaptation of little-used species to paper production, to make America independent of foreign pulp and pulpwood, now imported at a cost of $250,000,000 per year. New and modified processes to pulp difficult types of wood; greater economy and efficiency of pulping processes; semi-chemical pulping, refinement of fiber by rod mill; wood grinding experiments. Fiber processing, multiple-state bleaching, beating, hydration. Improvement of paper machine operation and quality of paper. Reduction of mill waste.

(From "Electric accounting machines and statistics on wood uses" / Business Machines, p. 4, Thursday, August 17, 1933)

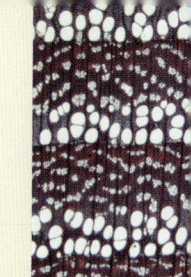

▲ FPL in the 1970s. Note the addition of the pulp and paper pilot plant, paper research, and wood chemistry buildings to the right (north). The experimental panel buildings have been tested and removed. Note the VA Hospital upper left (west).

▲ FPL in the 1990s. UW Medical Center, upper center (west); edge of Wisconsin Alumni Research Foundation, center right (north); Campus Drive, left (south).

Centennial Research Facility

◄ Site of the future FPL Centennial Research Facility (CRF) on the University of Wisconsin Campus. (2007)

◄ Start of construction of the new CRF. (2007)

◄ Construction site during winter 2007.

The Centennial Research Facility houses state-of-the-art equipment and laboratories for four major areas of research.

Wood Preservation—Microbiology and chemistry laboratories; pilot-scale pressure treatment facility

Durability—Custom-made stainless steel weathering chamber that mimics actual weather conditions, including temperature, humidity, sunlight, wind, and rain; equipment to study moisture transport properties of wood products

Engineering Mechanics—Strong floor and wall systems to perform full-scale testing of wood-frame building mock-ups and large wood structural members; open-bay floor areas accommodating large testing equipment

Composite Sciences—Compounders, extruders, injection molders, and hot presses; open-bay floor areas accommodate large equipment

Many FPL employees contributed to the planning of the new CRF. Steve Kalinosky worked with contractors to build the facility and Tom Jacobson provided safety input.

▶ Construction site early 2008.

▶ Construction site spring 2008.

▶ Construction site summer 2008. Dedicated June 2010.

Proposed FPL Bioenergy Pilot Plant

◄ The proposed Bioenergy Pilot Plant will be used for research on new methods for converting wood to usable energy such as electricity and transportation fuels.

Research Support

This section has so far documented the historical development of the FPL buildings. But the work is done by the people who work in those buildings, most of whom provide the support necessary for the scientists to conduct their research. All contribute to the accomplishments of the FPL. Following is just a sample of those who provide ongoing support.

Financial Resources Management

Jeff Sims and his group oversee the proper use of both taxpayers' research funding and cooperative funding from more than 100 cooperative agreements per year between FPL and universities, associations, institutions, industry, individual scientists, other government agencies, and non-government organizations. Throughout this book, mention is made of the "FPL partnerships." Following is just a sample of our partners from the past 6 years. Not listed, but equally important, are more than 180 industrial partners and 30 individual researchers.

Associations

American Forest & Paper Association
American Society of Civil Engineers
American Wood Protection Association
APA–The Engineered Wood Association
ASTM International
Building Materials Reuse Association
Energy and Environmental Building Association
Forest Products Society
International Society for Appraisers
National Association of Home Builders
National Frame Building Association
Oklahoma Association of Electric Cooperatives

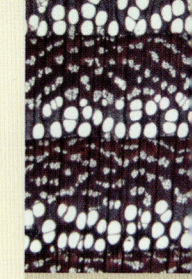

Southern Forest Products Association
Society of Wood Science and Technology
Technical Association of the Pulp and Paper
  Industry (TAPPI)
Western Governors' Association
Western Red Cedar Lumber Association

Government Agencies

Arizona State Land Department
Army–Strategic Environmental Research and
  Development Program
Caribbean National Forest
Central Oregon Intergovernmental Council
Department of Energy
Department of Veterans Affairs
Department of Homeland Security
Department of Interior—Bureau of Land
  Management
Department of Defense—Los Alamos
Federal Highway Administration
Forest Research
Forest Service National Library Program
Glacierland Resource Conservation and
  Development
Greater Flagstaff Economic Council
General Services Agency
Lane County, Oregon
Little Colorado River Plateau RC&D
National Aeronautics and Space Administration
National Academy of Science
National Agricultural Library
National Institute of Standards and Technology
National Library Program
Naval Historical Center
Naval Research Laboratory
NE Utilization and Marketing Council
Nevada Division of Forestry
Piute County, Utah
Pacific Northwest National Laboratory
Regents of California/DOE
Southwest Oregon RC&D Council
Town of Red River, New Mexico
U.S. Army
U.S. Army Corps of Engineers
U.S. Army, Natick
U.S. Customs & Border Protection
U.S. Department of Treasury
U.S. Postal Service
U.S. Department of Agriculture AID, APHIS,
  ARS, CSREES, FAS, NRCS
USDA–Foreign Ag Service
USDA, DA, OHCM (Employment–Student
  Opportunities and Internships)
Veterans Administration
Veterans Affairs
Wheeler County Development Board

Wisconsin Department of Agriculture & Trade
Wisconsin Department of Natural Resources

Universities

American Institute of Biological Sciences
Aspen Institute
Athena Institute International
Auburn University
Bejing Forest University
Capital Normal University (China)
China Ne Forestry University
Chinese Academy of Forestry
CIB CSIC Campus de La Ciudad Universitaria
  de Madrid
Clark Atlanta University
College of Menominee Nation
Colorado State University
Cornell University
Emory University
Florida A&M University
Georgia Institute of Technology
Haywood Community College
Helsinki University of Technology
Institute for Scientific Research
Institute for Agriculture and Trade Policy
Iowa State University
Institute of Science and Paper Technology
Louisiana State University
Louisiana Technology University
Louisiana State University Agricultural Center
Michigan State University
Michigan Technological University
Mississippi State University
Mt. Wachusett Community College
Nanjing Forestry University
North Carolina State
North Carolina University
Northeast Forestry University (China)
Northern Arizona University
Ohio University
Penn State University
Purdue University
Research Foundation of SUNY
Seoul National University
State University of New York
University of Alabama
University of Arizona
University of British Columbia
University of California-Davis
University of California–Berkeley
University of Canterbury
University of Colorado
University of Florida
University of Georgia
University of Guelph
University of Hamburg

University of Idaho
University of Illinois–Champaign
University of Maine
University of Melbourne
University of Minnesota
University of Minnesota–Duluth
University of Montana
University of North Dakota
University of Rhode Island
University of Sao Paulo
University of Tennessee
University of Washington
University of Wisconsin–Madison
University of Wisconsin–Platteville
University of Wisconsin–Stevens Point
University of Wisconsin & Biopulping
   International
Virginia Polytechnic Institute and State
   University
Virginia Tech
Washington State University
Western Michigan University
Yale University

Other

Aldo Leopold Foundation
American Lumber Standard Committee
American Process Energy Recovery
Canadian Lumber Standards Accreditation Board
Canadian Wood Council
Cardo DBA Lorentzen-Wettre
Cellulose Sciences International
Civil Engineering Research Foundation
Consumers Union
Coquille Tribe of Oregon
Ecole Superieure Du Bois
Edeniq
Elk Regional Health System
Elsevier
European Science FDTN
Fibre-Gen/UMD
Fleet Engineers
FPInnovations–Forintek
Greater Flagstaff Forests Partnership/RMRS/
   PNW/SRS
Haywood Habitat for Humanity
Holland Colors Americas
Holzforschung
Imporatadora Dominicana Maderas
Index Ventures Mgt SA
Intermountain Research
Iowa State University Research Foundation
Kane Area School District
Kootenai Business Park Industrial District

Lake Tahoe Unified School District
Major League Baseball
MBI
Montana Community Development Center
Mt. Adams Resource Stewards
National Museums Liverpool
Poet-Science and Technology
S. Appalachian Council
St. Maries Joint School District #41
The Reuse People
The Silver Research Consortium
US Endowment for Forestry & Communities
Watershed Research & Training Center
Wisconsin Cranberry Board, Inc.
Wood Products Council

In addition to contract administration the
group deals with the FPL budget, finance,
and purchasing.

▲ Mike Breezee, Sue Davis, and Rod Silva purchasing equipment and supplies in support of
a research contract.

Research Facilities Engineering

Providing for an excellent, safe research physical plant is the responsibility of the Research Facilities Engineering Group. Mike Kaspszak's group of engineers and trades workers not only maintain buildings, provide services, and oversee contracts, but also directly participate in scientific studies by constructing unique research equipment and preparing test specimens.

Carpenter Shop

◀ Lee Crigger and Ryan Jones observe Robert Foss preparing to saw molded fiberboard into test specimens for evaluation. (2008)

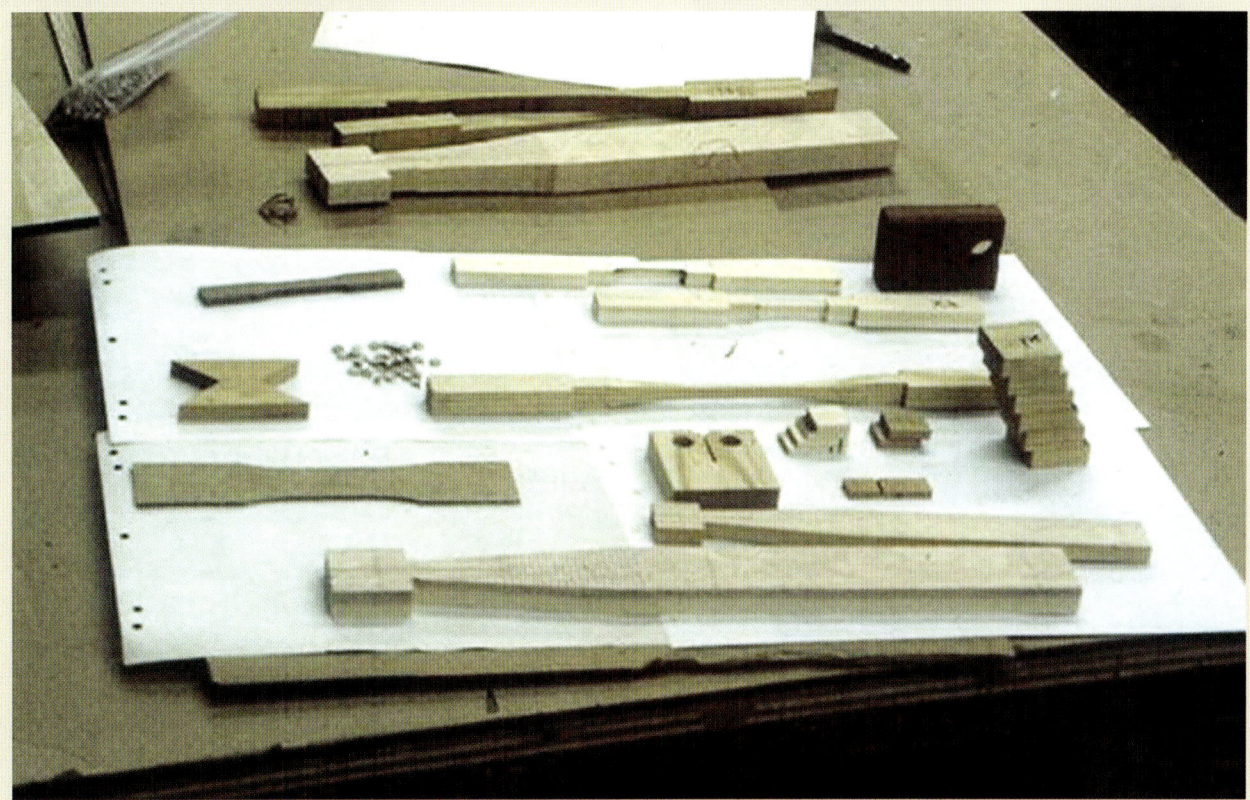

▲ Examples of types of test specimens prepared by the carpenter shop. (2008)

◄ Joe Destree (left) and Dave Eustice in the FPL metal working shop. (2008)

◄ Chester Filipowicz welding steel for a new refiner installation.

► Mike Grambsch checking electronics on the test floor.

◄ Will Kinney and Geoff Severin working on the electrical grid.

Examples of Metal Working Shop Support of Research

▲ Randy Wruck and Chester Filipowicz setting up the CD-300 preheater and discharge plug screw for an experimental pulp refiner. Wiring for this complex equipment was done by Geoff Severin and Will Kinney. This is a typical example of professional facility support for the FPL research program. (2008)

▲ A special conditioning chamber designed and built by Chester Filipowicz, John Considine, Carl Houtman, and Will Irwin for the U.S. Postal Service recycling project. It was built with many recycled parts from unused equipment. Geoff Severin and Will Kenney handled the wiring; Bob Ramos and Dante Kilgore configured the HVAC equipment; and Bob Foss insulated the unit. Brad Hotle tested the unit and found it operated successfully. Building this equipment, especially using recycled parts, required many custom modifications—made possible only by having the in-house talent. (2006)

Communications

Research accomplishments are of little use if no one knows about them. Many groups are involved in spreading the knowledge: the FPL Office of Communication; the Forest Service National Library and the Forest Service Patent Program, located at the FPL; State and Private Forestry (see Technology Marketing Unit section); the State Utilization Forester; and University of Wisconsin Extension staff. Many methods are used.

FPL Office of Communications

The FPL has always made an effort to get its research information out. Lon Yeary and his staff use many different forms of outreach, but generally research reporting starts with written reports of research results. Technical editors and writers help clarify the information.

Research reports are published in peer-reviewed technical journals and association proceedings and as FPL reports.

▲ Henry Spelter working with Madelon Wise on one of his economic reports.

Methods used by the FPL to communicate its research results include the following:

Classes
Committees
Conferences
Cooperative research
Demonstrations
Displays
E-mails
Films
General publications
Internet
News releases
One-on-one discussions
Patents
Phone calls
Photographs
Posters
Radio
School Experiments
Site visits
Special Programs
Symposia
Technical Reports
Tours
Television
University Theses
Videos

Communications Methods

Classes—When the laboratory was established at the University of Wisconsin, the staff participated in teaching some of the classes on campus. Later the laboratory held classes in the FPL, such as drying and packaging courses. At present, the scientists are again lecturing at the University of Wisconsin.

Committees—A major method of communicating research results is through FPL staff serving on standards-writing committees such as ASTM International (ASTM); American Society of Heating, Refrigerating and Air-conditioning Engineers (ASHRAE); and many government agencies.

Conferences—Presentation of research results at technical conferences provides another method of transferring research findings. The FPL also hosts conferences at which two-way interaction can occur—not only transferring information to outside participants but also listening to them about new challenges they face in using wood.

Cooperative Research—For those who desire to do joint research with the FPL, cooperative research agreements can help ensure that results will be transferred (see Financial Resources Management for lists of recent partners).

Demonstrations—During FPL open houses, active demonstrations, such as the testing of a beam or papermaking, are set up for the public. Large scale demonstrations, such as the National PATH house at FPL or the Structural Insulated Panel House that was shown at the mall in Washington, D.C., are also used to convey new ideas and research results. Demonstrations are accompanied by FPL staff members who explain details of the project (see Housing section).

Displays—Static and dynamic displays are prepared by FPL staff and used throughout the facility to inform visitors of the FPL's research. Some of these displays are portable and can be used as part of a meeting or conference. Two of the most recognized, movable displays are Smokey Bear (wildfire prevention) and Woodsy Owl (pollution prevention).

E-Mail—E-mail is a growing method of communication with other scientists and the public.

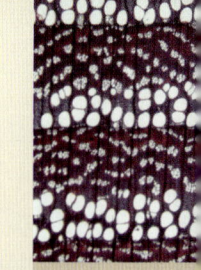

▲ Smokey Bear with Mike Dombeck (retired Forest Service Chief) and Woodsy Owl with Tom Hamilton (retired FPL Director) at the 90th Anniversary of the FPL. Smokey and Woodsy are great symbols in support of forest management and FPL research.

▲ Judy Patenaude at the FPL reception desk providing FPL information to Bonnie Johnson and Suki Croan.

Films—Films are still made to document experiments but are being replaced with other electronic methods for data storage and playback.

General Publications—FPL staff editors and writers like Jim Anderson, Jaina Roth, James Spartz, and Rebecca Wallace work with designers Karen Nelson and Tivoli Gough to prepare general publications of research results in language that is better understood by the general public as some of the research results are very technical. Examples include "TechLine," "NewsLine," e-newsletter, and newspaper articles.

Internet—Raj Lal develops and maintains the FPL Web site to provide public access to a wide variety of specialized technical and "how to" information.

News Releases—Breaking news is provided to the media through news releases prepared by the staff.

One-on-One Discussions—One-on-one visits between staff and the public provide opportunities to share knowledge, which many times has solved specific problems. FPL staff answer thousands of phone calls, e-mails, and letters requesting information.

Patents—Janet Stockhausen assists scientists in preparing patents that are a growing method of technology transfer. The passage of the Technology Transfer Act of 1986 increased incentives for organizations and scientists to consider patenting new ideas in order to give licensees protection for high-risk investments in developing ideas.

Phone Calls—The FPL is served by Sandy Morgan, Anna Bales, and other communication staff as receptionists who answer the phone or greet visitors and try to help answer their questions or refer them to someone else who can.

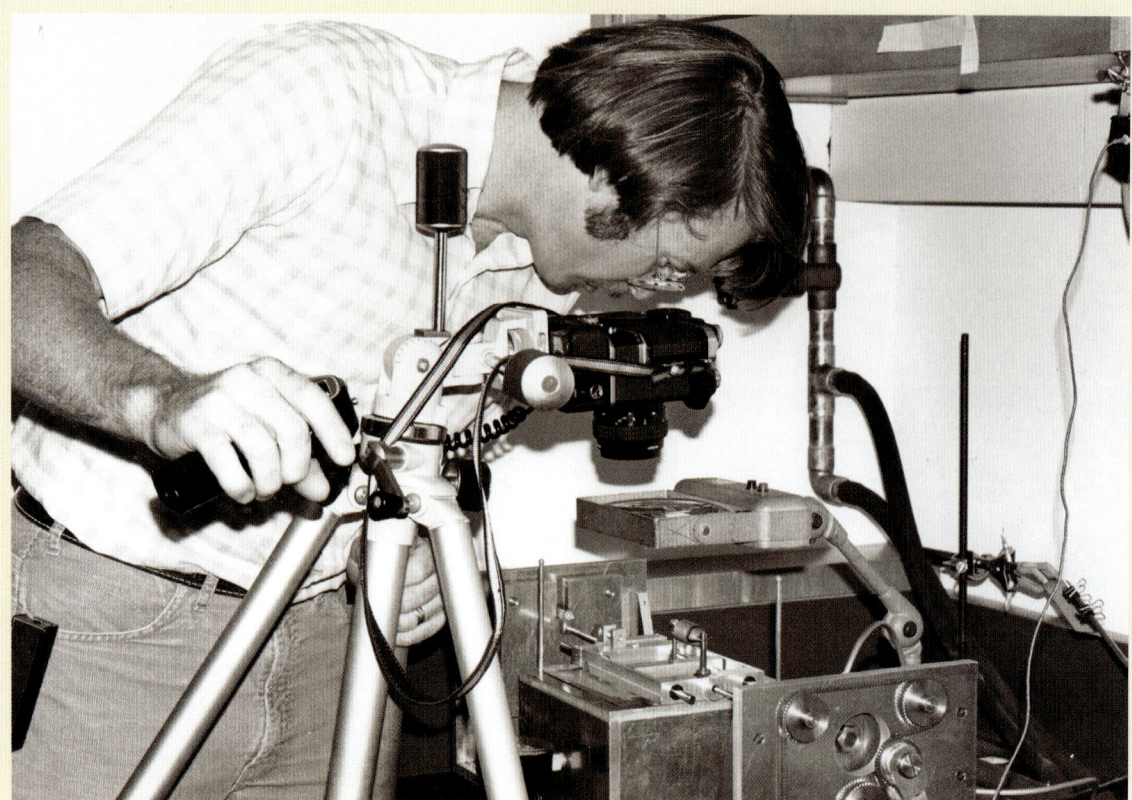

◄ Steve Schmieding capturing a close-up image of a test device. (1990s)

Photographs—Pictures, from high-speed photography to shots of tiny mechanisms, continue to be a major method of showing results.

Posters—Posters are prepared by Karen Nelson and Tivoli Gough for FPL staff; posters are used in hall displays or at conference poster sessions.

Radio—Talk radio is another source of oral communication. Call-in formats are a direct means of meeting the needs of the public.

School Experiments—An approach to communicate with students is by providing wood-related experiments that teachers can use in the classroom. This was facilitated by Al Foulger in his Classroom demonstrations of wood properties / A. N. Foulger / U.S. Forest Service, Forest Products Laboratory (1969).

Site Visits—In some instances, FPL staff will make a site visit to assess a situation and if possible make constructive suggestions for improvements. One major program that involved site visits was the Sawmill Improvement Program (see Lumber Manufacturing section).

Special Programs—Technical information and limited grant money for special programs are available through the Technical Marketing Unit of State and Private Forestry (see Technical Marketing Unit section).

Symposia—On some occasions, the FPL has hosted symposia in cooperation with the Forest Products Society, TAPPI, or other organizations.

Technical Reports—The primary method of transferring research information is through technical reports. Many of these reports are available to the public on the FPL web site (www.fpl.fs.fed.us). Some reports are available in printed form from the FPL. Some requests are handled by James Godfrey, Dulcie Dobson, and Donald Kiefer of the Office of Communication; others are handled by Julie Blankenburg and her staff, Shelley Bishop, Mary Funmaker, William Paine, and David Smith, of the National Forest Service Library located at the FPL.

Tours—The Laboratory also conducts tours of the facilities for the public and for special groups. Tours are the best way to grasp the breadth and depth of the FPL facilities and research program.

Television—On special occasions, television coverage conveys information to the public. An episode of "This Old House" TV program featuring the FPL aired in 1993.

University Theses—University students working at the FPL on advanced degrees provide useful information in their research and report this in thesis form. Over the years, hundreds of theses have been prepared.

Video—Video tapes, CDs, and DVDs have replaced film and are used to document research and transfer the information in more interesting and helpful ways.

▲ Sue Paulson explaining FPL's research to a group of school children and special guest, Woodsey Owl (Rebecca Wallace).

Librarian Julie Blankenburg describes the historical development of information access at FPL:

The Information Services section contributes to the FPL research program by supporting the researchers in various ways. Vast improvements were made over time in the tools used for finding information published by the Laboratory. In the 1970s, Sue Parisi entered all the Laboratory publications onto punch cards. She was then able to sort by author, title, and keyword in context (KWIC) to produce printed indexes. In the early 1980s, Roger Scharmer took that information and it was uploaded into a FAMULUS database so that it could be searched online.

As more and more data was moved onto computers, the FPL library decided that these items needed to be upgraded and converted into MARC electronic records, and a five-year conversion project commenced. This massive project put the data that was in the FAMULUS files into Machine Readable Cataloging (MARC) format that could then be utilized in electronic library catalogs. Subject headings and publication information were added. It was funded by the National Hardwood Research Center so that they could have access to the research that had been done at the Forest Products Laboratory. This data was eventually merged into the FS INFO database and was also used as the base data to form the electronic FPL Library catalog. In 1999, another large retrospective conversion project began. It was to convert the FPL library card catalog into electronic format. The tail end of this project continues today. Over 90% of the FPL library catalog has been converted.

These projects were done to make it easier to locate older FPL-authored publications. As full-text articles started appearing online, it became apparent that when the public was looking for Forest Service Research information, they had to go to seven different websites. To solve this problem, the Forest Service research stations got together to cooperatively operate the Treesearch database. Treesearch allowed all the research station publications to be searched together at one time. It was also designed to allow Google to index all the publications listed so that the public could find Forest Service research publications through Google. The FPL Office of Communications staff and the FPL Library were heavily involved in the formative stages of Treesearch and continue to add new items to the database as they come online. This has created greater visibility for FPL publications than our own website could provide.

The formation of the FPL website was a project of the FPL State and Private Forestry unit, which started it as a demonstration in 1995. Two years later in 1997, it was transferred to the Information Services section of FPL, where it continues today. FPL was one of the first Forest Service units to have a website of its own. Over time, it has proved to be a hit with users who can directly download our publications without having to request them from our publication distribution office.

An example from the late 1970s of how the Information Services section (now Office of Communications) supported the Laboratory research program was the "truss framing" program. The truss frame idea—using 2 by 4 lumber instead of large beams as framing

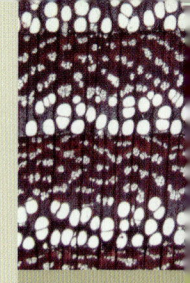

◀ Mary Funmaker and Julie Blankenburg double-checking information in the last 10% of the old card catalog that has not been computerized.

members in houses—was developed at the Forest Products Laboratory. To get this information out to designers and builders, a mass mailing was done that involved stuffing over 1,000 envelopes with 5 to 10 publications each. The Information Services section edited and printed the publications, coordinated the mailing addresses, stuffed the envelopes, and mailed the items out. These days the same thing might be done by posting the documents to a webpage on the topic and sending out a news release.

◀ Library staff: From the left standing Florence B. Usher, Florence R. Steffes, Dorothy Webber, Eleanor C. Lunde, Sella Webb. Seated is Mary Adams. (1960s)

Computational Support

◀ During WW II, computation of research results was handled by groups of employees using slide rules and mechanical calculators. Most of this work is now handled by computers. (1944) {M 56955)

Secretarial Support

◀ In early years, employees as part of a steno pool provided typing and secretarial support to researchers. (1944) {M 56955)

Program Support Assistants

◀ Today, Sue Howards and other program support assistants provide typing support but also handle budgeting, personnel documents, travel planning, seminar scheduling, and many other requests made by the researchers in multiple research projects to help them accomplish their research.

External Methods in Support of Research

Forest Products Society

In 1947, the creation of the Forest Products Research Society grew out of a conference of university, industry, and state and federal forest products researchers held at the FPL.

The Society's mission is to foster innovation and research in the environmentally sound processing and use of wood and fiber resources by disseminating information and providing forums for networking and the exchange of knowledge.

In 1992, the name was changed to Forest Products Society to reflect a larger scope of interest.

A complete history and activities of the Forest Products Society can be found on their Web site (www.forestprod.org).

International Cooperation

Although the FPL was the first national laboratory dedicated to research on wood, it has been emulated by countries around the world. Because of the importance of wood to the people of the world, sharing our research to help others in its wise use helps us all. Thus, FPL is an active participant in the International Union of Forest Research Organizations (IUFRO) and other international scientific and technical societies. This two-way interaction provides for advancing the latest developments in forest products science, wise wood use, and emerging global challenges.

In September 1963, the FPL hosted two international organizations devoted to forest products research and its technological applications: Forest Products Section 41, of the IUFRO, and the Fifth Conference on Wood Technology of the United Nations Food and Agriculture Organization (FAO), which met under joint sponsorship of the U.S. Department of Agriculture and the U.S. State Department. Sixty delegates from 32 foreign nations participated in one or both meetings.

Throughout its history the FPL has hosted visiting scientists from all parts of the world. This international interaction has greatly benefitted the FPL staff, helping them understand the importance of FPL research and its impact on other countries.

Clark C. Heritage Memorial Series on Structural Uses of Wood

Clark Heritage was an early leader in pulp and paper research at FPL and later Director of Wood Products for Weyerhaeuser Co. His legacy funded development of the Heritage Series that FPL played a major role in producing.

The series consists of four volumes:

Volume I, *Wood: Its Structure and Properties*, 1981, discusses in nine modules basic insights to the structure, treatment, and properties of wood.

Volume II, *Wood as a Structural Material*, 1982, describes both old and new wood products and their uses as structural materials in eight modules.

Volume III, *Adhesive Bonding of Wood and other Structural Materials*, 1983, covers the rapidly expanding application of adhesive-bonded wood products and structures through nine modules.

Volume IV, *Wood: Engineering Design Concepts*, 1987, includes eight modules that provide information and data on design of major structural members.

This series was developed to help engineering and materials science professors prepare courses on wood's utility in construction and design.

Forest Products Laboratory Directors

▲ McGarvey Cline
1910–1912

▲ Howard F. Weiss
1912–1917

▲ Carlile P. Winslow
1917–1946

▲ George M. Hunt
1946–1951

▲ J. Alfred Hall
1951–1959

▲ Edward G. Locke
1959–1966

▲ Herbert O. Fleischer
1967–1975

▲ Robert L. Youngs
1975–1985

▲ John R. Erickson
1985–1993

▲ Thomas E. Hamilton
1993–2001

▲ Christopher D. Risbrudt
2001–

Budget of FPL in Dollars

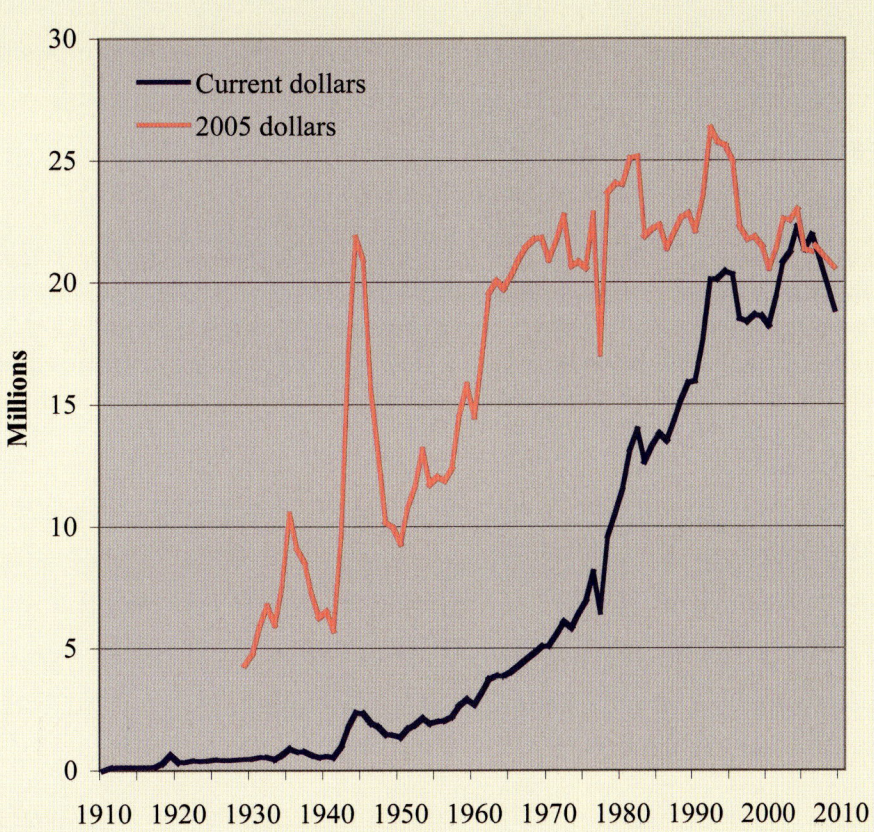

Number of FPL Employees over Time

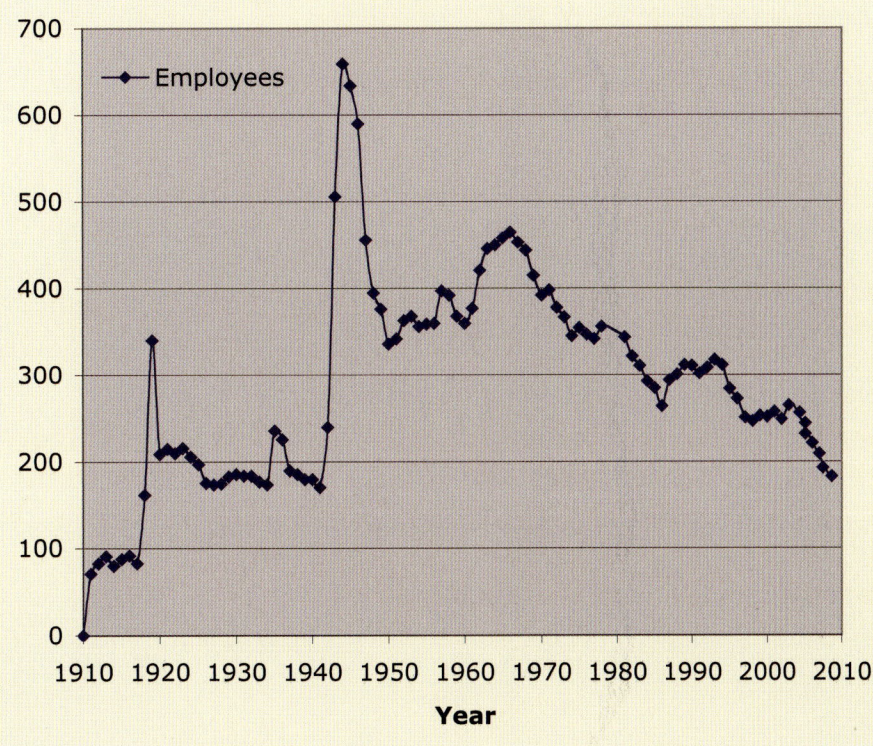

Facilities of the Forest Products Laboratory

1909 A house at 1610 Adams Street near the University of Wisconsin campus becomes temporary quarters for FPL.

1910 All staff and machinery move into the new Forest Products Laboratory at 1509 University Avenue on land donated by the UW.

1917–1919 World War I research space in part or all of 10 UW buildings. Much of the space was used for airplane propeller research.

1932 Post–WW I crowding is relieved when a new 188,000-square-foot building is constructed off Walnut Street at the west end of the campus on land donated by the UW.

1934 Packaging research gains space with construction of a building supported by laminated arches.

1935 Composite products research moves to a new building constructed with experimental siding to test finishes. Fire test laboratory was built.

1943–1945 Expanded World War II research finds space on three floors of a downtown building at 120 S. Fairchild Street.

1956 The Laboratory establishes a rural exposure site near Madison to evaluate wood products.

1960 First addition to Fire Test Laboratory built.

1967 The innovative new Pulp & Paper Pilot Plant and its complementary Fiber Research Building increase FPL space by 90,000 square feet.

1969 The new Chemistry Research Building completes the trio with another 32,000 square feet.

1974 The acquisition of several small buildings from the nearby Wisconsin Alumni Research Foundation provides more space for engineering research.

1982 Started weatherization and modernization of the 1932 building including energy-efficient windows and improved utility systems.

2001 Completion of research–demonstration house east of FPL.

2004 Second addition to the Fire Test Laboratory built.

2007 Groundbreaking for FPL's new Centennial Research Facility.

2010 Dedication of the Centennial Research Facility.

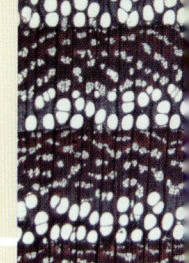

Further Information

The Forest Products Laboratory: A decennial record / Howard F. Weiss, et al. / FPL (1921)

Electric accounting machines and statistics on wood uses / Business Machines, pg. 4, Thursday, (August 17, 1933)

Some accomplishments of the Forest Products Laboratory / FPL / (February 1, 1937)

The Forest Products Laboratory: A brief account of its work and aims / Forest Service, U. S. Department of Agriculture / Miscellaneous Publication No. 306 (1938)

Toward a wiser use of wood / Forest Products Laboratory Employees / (1950)

Forest products research: How it has paid and how it can continue to pay / FPL / (November 1, 1953)

The Forest Products Laboratory: Some accomplishments, work, and trends / FPL / DO 148 (April 1954)

A golden anniversary record / FPL / (1960)

Forest products research on the march / John C. Killebrew / (June 1963)

At your service... The Forest Products Laboratory / Frank and Merle Sinclair / Exclusively Yours (April 10, 1967)

History of the U. S. Forest Products Laboratory (1910–1963) / Charles A. Nelson / FPL (1971)

Products of American forests / F. J. Champion / USDA, Forest Service, Miscellaneous Publication No. 861 (1973)

Founding the Forest Products Laboratory—An eyewitness account / Lorraine J. (Mark) Markwardt / FPL Chips (1984)

Age of wood: progress through wood research / FPL / (June 1985)

Progress through wood research: The Forest Products Laboratory: Past, present, and future / FPL 75th anniversary / (December 1985)

The Forest Products Laboratory—Giving its work away / Wood & Wood Products Centennial 1896–1996 / (1996)

Challenges

Decay

Mushrooms are the fruiting bodies of decay fungi. Many are edible, but some are deadly—picking wild mushrooms must be done with great care. The thousands of forest fungi have one thing in common: they are organisms that live off—and in the process, decay—other material. In the case of Shiitake, the fruiting body or mushroom is edible and a commercial product. However, in many instances fungi lead to unwanted decay and wood failure.

◄ Shiitake mushrooms growing on wood logs. These edible mushrooms are a favorite food for many. [8]

Damage Because of Fungi

Although fungus is the organism that causes the decay, it cannot survive if the proper conditions of temperature and moisture are absent. Because outside temperature is difficult to control, the best practical method for controlling or eliminating decay is to decrease moisture, which many times results from leaks, poor construction, improper landscaping, or lack of proper maintenance. Preservative-treated wood products should be used when wood is expected to periodically be wet, such as in contact with soil or when used outdoors. [31]

Challenge: Billions of dollars are spent annually to replace wood products destroyed by wood decay fungi. A large percentage of this loss is incurred by consumers, either directly by home maintenance or indirectly by increases in the cost of services and products. Decay caused by brown rot fungi is by far the most common and most serious kind of deterioration because it can cause rapid structural failures. One-tenth of the timber harvested annually in the United States goes to replace wood products that have decayed while in use. Identifying and understanding the growth characteristics and role of fungi in decay of wood and forest diseases are complex.

FPL's Contributions: The Forest Products Laboratory houses the largest collection of forest fungi—more than 5,000 species—which is shared with other researchers around the world. It has classified numerous species and advanced our knowledge of how fungi influence a forest's health, destroy wood, and can be used in beneficial ways. FPL researchers have designed an immunodiagnostic wood decay test that is a rapid, inexpensive, and accurate way to detect decay before it has caused significant harm to the wood. [37] Scientists have refined techniques of DNA sequencing for fungal identification and evolutionary analysis (see Preservation section).

Results: Understanding the fundamental properties of fungi helps in finding ways to control fungi and direct them in positive applications. Having access to and understanding the properties of fungi provide fundamental data for research on biological processing of wood, such as biopulping, hazardous waste site cleanup, protection of historically important structures, and more environmentally acceptable ways to protect wood from decay. [14, 17, 20, 33, 40, 42]

▲ Moisture problems resulted in fungal decay and mold growth in the siding of this home. {77 C} [27]

▲ Removing the siding shows the extensive damage of fungal growth and mold in the walls of a home. [27]

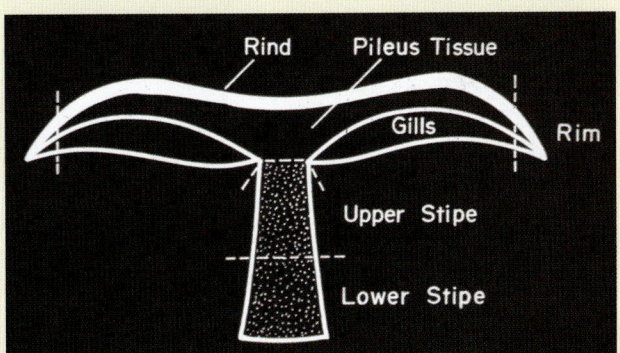

◀ Diagram of the cross section of a mushroom. The mushroom is the fruiting body of a fungus. {77I}

◀ Your Wood Can Last for Centuries report by Rodney C. DeGroot (USDA Forest Service) (1976).

▲ Frances Lombard studying a fungus using a light microscope. {M 147 000-11} [2, 3, 13]

Center for Forest Mycology Research

In 1971, the Forest Mycology Research staff were transferred from Washington, D.C., to the FPL and called the Center for Forest Mycology Research. The move brought to FPL one of the world's largest collections of wood-rotting fungi. The staff continue to define geographical limits of species, identify causes of decay and tree diseases, and provide a source of decay organisms for other researchers. In 2007, administration of the Center was transferred to the Forest Service Northern Research Station, but Jessie Glaeser and staff remain located at FPL. [10, 21, 32, 41]

▲ Kyah Norton, John Haight, and Mark Banik sampling a spruce tree for fungi in Alaska

▲ Jean Lodge examining forest fungi in New Hampshire. These fungi will be added to the extensive collection for identification, classification, and research. [34]

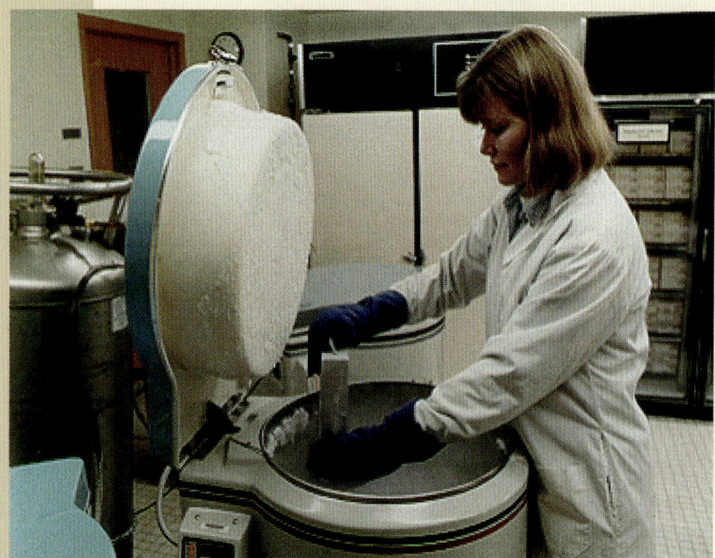

▲ Rita Rentmeester loading fungal cultures into liquid nitrogen for long-term storage.

▲ Hal Burdsall in the herbarium, which contains thousands of dried samples of forest fungi used in studies on forest mycology and bioconversion of wood. [15]

▲ Karen Nakasone identifying a species of fungus. {118 I} [6, 16, 26]

Advances in Understanding the Mechanism of Degradation

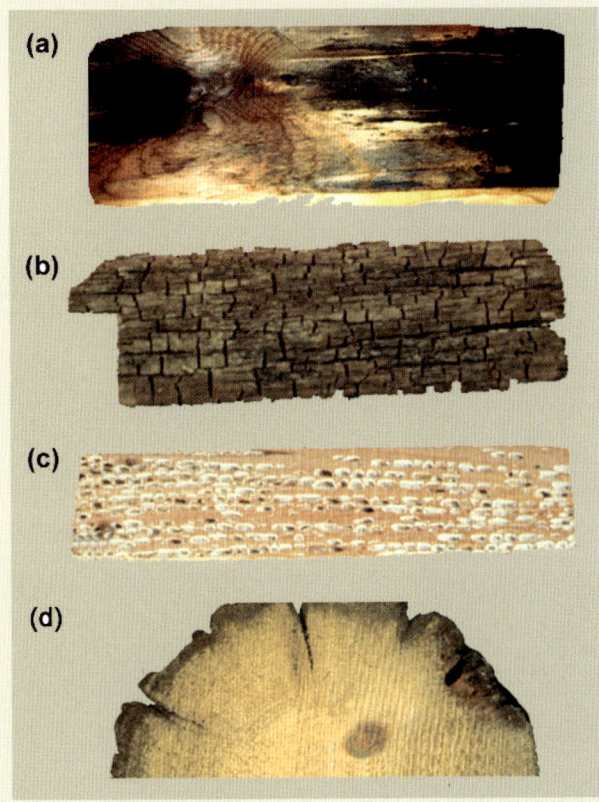

(a)

(b)

(c)

(d)

◄ Representative samples of four common types of fungal growth on wood: (a) mold discoloration, (b) brown-rot, (c) white-rot, (d) soft rot.

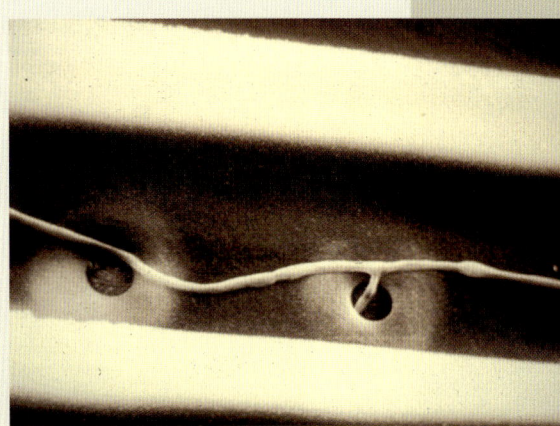

▲ Fungal hypha growing inside a wood fiber and through a border pit. {46 I} [4, 29]

Brown-rot fungi infect wood by producing enzymes and degradative agents, such as acids, to help them digest the wood. The acids help the fungi tolerate or inactivate some chemical wood preservatives, making it difficult to control decay. FPL researchers have studied how these acids help the fungi tolerate preservatives in order to develop better ways to protect wood from decay. [7, 19, 22, 39]

◀ Samples of brown-rot decay. {59 B} [19, 22, 35]

Enzymes from Poria placenta Isolated

The FPL staff have isolated and identified wood-degrading enzymes from the brown-rot fungus, *Poria placenta*. This discovery will play a vital role in development of new, non-toxic wood decay methods and may help convert wood waste into energy sources or livestock feeds. It was also used for the development of immunodiagnostic tests for early detection of decay (see Preservation section). [5, 7, 18, 19, 22, 24]

◀ Carol Clausen conducting research to isolate wood-degrading enzymes.

Synchrotron (Particle Accelerator) Applications in Wood Deterioration

Several non-intrusive synchrotron techniques are being used to detect and study wood decay. Microtomography was used to characterize loss of wood structural integrity. The techniques are providing information about molecular structures and compositions in the heterogeneous matrix of wood. [25, 38]

Shiitake Mushrooms

In the late 1970s, Gary Leatham completed a doctoral thesis based on his research on growth of the Japanese Islands variety of Shiitake. Continuing his research at FPL, in 1982 he published "Cultivation of shiitake, the Japanese forest mushroom, on logs: a potential industry for the United States." This paper and a number of public seminars stirred a lot of interest and helped make commercial cultivation possible in the United States. It was also discovered that the edible Shiitake mushrooms (*Lentinus edodes*) grow faster on shredded oak residue than on logs. This finding was a spin-off of long-term basic research toward selective degradation of lignin for biopulping. By producing Shiitake, low-quality hardwoods can be converted directly to a high-value food. [8, 28]

Shiitake growing is a budding industry in the United States, creating jobs and utilizing wood that would otherwise be wasted.

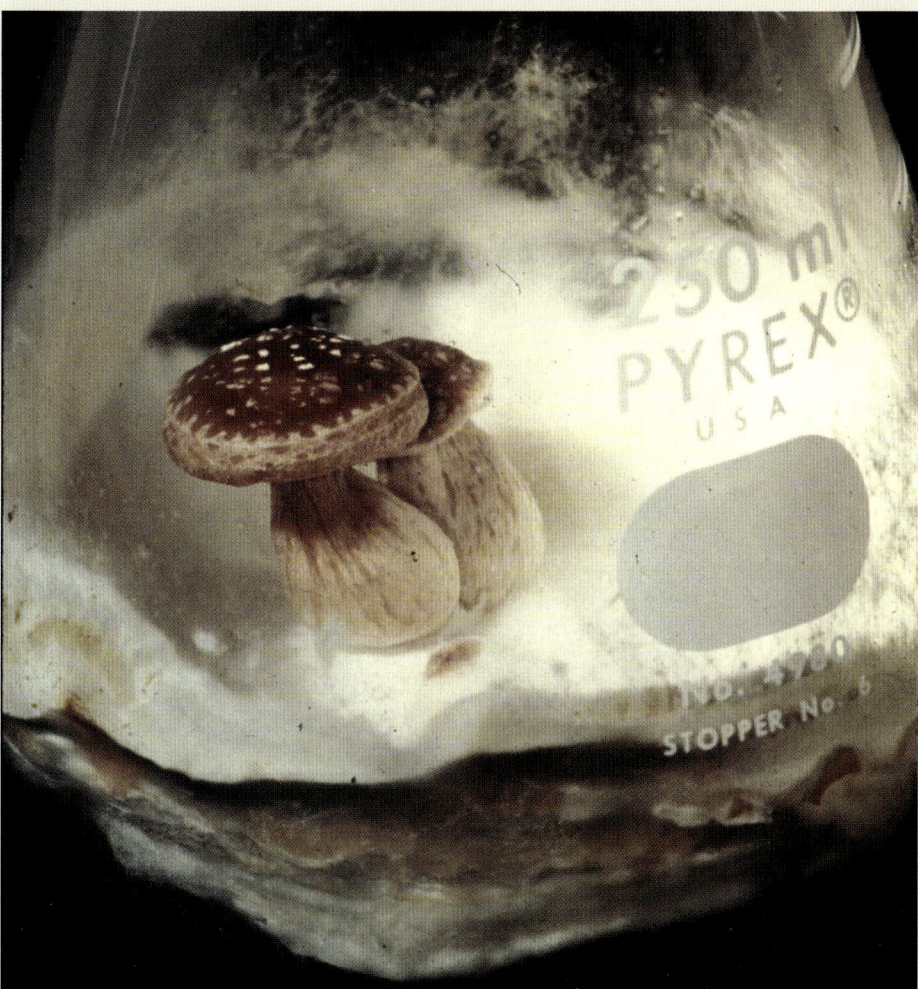

◀ Shiitake mushroom growing on a synthetic growth medium. Experiments such as this led to a better understanding of the growth characteristics of the mushroom. [9]

▲ Jane Hasselkus checking growth of the shiitake mushroom fungus in a flask. This research helped in understanding the conditions favorable to mushroom growth. {M 86 0170-4} [12]

◀ Chittra Mishra, a research associate from the National Chemistry Laboratory in Poona, India, researching solid-wood bioconversion using shiitake mushrooms, an edible delicacy grown on oak logs in Asia (1987). {M 87 0009-5}

▲ These beautiful shiitake mushrooms are growing on sawdust rather than logs. This research led to the creation of an alternative method of commercially growing these mushrooms. {122 D}

1910

Little classification of fungi or understanding of the mechanism of decay.

2010

Thousands of forest fungi are classified.

Advances in understanding the mechanism of decay have improved our ability to protect wood from decay.

Expanded commercial use of an edible mushroom that can be grown on wood that would otherwise be wasted.

Further Information

[1] Evaluating wood preservatives by soil-block tests. Pt. 1–2 / Catherine G. Duncan and C. Audrey Richards / American Wood-Preservers' Association. Proceedings of the 46th annual meeting. Washington, DC: AWPA, (1950)

[2] Wood attacking capacities and physiology of soft-rot fungi / Catherine G. Duncan / FPL No. 2173 (1960)

[3] Fungi associated with principal decays in wood products in the United States / Catherine G. Duncan, Frances F. Lombard / Research paper WO 4 (1965)

[4] Penetration and degradation of cell walls in oaks infected with Ceratocystis fagacearum / I.B. Sachs / Phyto-pathology Vol. 60(9) (Sept. 1970)

[5] Unique polysaccharide- and glycoside-degrading enzyme complex from the wood-decay fungus Poria placenta / K.E. Wolter, T.L. Highley, F.J. Evans / Biochemical and biophysical research communications Vol. 97(4) (Dec. 31, 1980)

[6] New species, Phlebia brevispora, a cause of internal decay in utility poles / K.K. Nakasone / Mycologia Vol. 73(5) (Sept.–Oct. 1981)

[7] Is extracellular hydrogen peroxide involved in cellulose degradation by brown-rot fungi? / T.L. Highley / Material und organismen Vol. 17(3) (1982)

[8] Cultivation of shiitake, the Japanese forest mushroom, on logs: a potential industry for the United States / G.F. Leatham / Forest products journal Vol. 32(8) (Aug. 1982)

[9] Chemically defined medium for the fruiting of Lentinus edodes / G.F. Leatham / Mycologia Vol. 75(5) (1983)

[10] Type studies and nomenclature considerations in the genus Sparassis / H.H. Burdsall / Mycotaxon Vol. 31(1) (Jan.–Mar. 1988)

[11] Lignin degradation by Phanerochaete chrysosporium / T. Kent Kirk / ISI Atlas of science. Biochemistry Vol. 1(1) (1988)

[12] Biology and physiology of shiitake mushroom cultivation / G.F. Leatham, T.J. Leonard / Shiitake mushrooms: Proceedings of a national symposium and trade show, 1989 May 3–5, St. Paul, MN. St. Paul, MN: University of Minnesota (1989)

[13] Taxonomy and nomenclature of Phellinus weirii in North America / Michael J. Larsen, Frances F. Lombard / Proceedings of the seventh international conference on root and butt rots: Vernon and Victoria, British Columbia, Canada, August 9–16, 1988. [S.l.]: International Union of Forestry Research Organisations, Working Party S2.06.01; Victoria, B.C.: Forestry Canada, Pacific Forestry Centre (1989)

[14] Fate of pentachlorophenol (PCP) in sterile soils inoculated with the white-rot basidiomycete Phanerochaete chrysosporium: mineralization, volatilization and depletion of PCP / Richard T. Lamar, John A. Glaser, T. Kent Kirk / Soil biology and biochemistry. Vol. 22(4) (1990)

[15] Taxonomic mycology: concerns about the present; optimism for the future / H.H. Burdsall, Jr. / Mycologia 82:1-8 (1990)

[16] Cultural studies and identification of wood-inhabiting Corticiaceae and selected Hymenomycetes from North America / K.K. Nakasone / Mycological memoir 15:1-412 (1990)

[17] Laboratory studies on control of sapstain and mold on unseasoned wood by bacteria / Terry Highley, Riana Benko, Suki Croan / Stockholm, Sweden: IRG Secretariat, (1991)

[18] Characterization of monoclonal antibodies to wood-derived [beta]-1,4-xylanase of Postia placenta and their application to detection of incipient decay / C.A. Clausen, F. Green, T.L. Highley / Wood science and technology Vol. 27 (1993)

[19] Research on biodeterioration of wood: 1987–1992. I, Decay mechanisms and biocontrol / Terry L. Highley, C.A. Clausen, S.C. Croan, F. Green, B.L. Illman, J.A. Micales / FPL–RP–529 (1994)

[20] Field evaluation of the remediation of soils contaminated with wood-preserving chemicals using lignin-degrading fungi / R.T. Lamar, J.A. Glaser / Bioremediation of chlorinated and polycyclic aromatic hydrocarbon compounds. Boca Raton, Fla. Lewis Publishers (1994)

[21] Identification of Armillaria species from Wisconsin and adjacent areas / Mark T. Banik, Jennifer A. Paul, Harold H. Burdsall, Jr. / Mycologia Vol. 87(5) (1995)

[22] Mechanism of brown-rot decay: paradigm or paradox / Frederick Green III, Terry L. Highley / International biodeterioration & biodegradation Vol. 39(2–3):113–124 (1997)

[23] Control of wood decay by Trichoderma (Gliocladium) virens—I. Antigonistic Properties / Terry L Highley / Material. und Org. 31(2):79–89 (1997)

[24] Immunological detection of wood decay fungi: an overview of techniques developed from 1986 to the present / Carol A. Clausen / International biodeterioration & biodegradation Vol. 39(2–3) (1997)

[25] Nondestructive elemental analysis of wood biodeterioration using electron paramagnetic resonance and synchrotron X-ray fluorescence / B.L. Illman & S. Bajt / International biodeterioration & biodegradation Vol. 39(2–3) (1997)

[26] Ribosomal DNA internal transcribed spacer sequences do not support the species status of Ampelomyces quisqualis, a hyperparasite of powdery mildew fungi / Levente Kiss, Karen K. Nakasone / Current genetics Vol. 33: 362–367 (1998)

[27] Decay of wood and wood-based products above ground in buildings / Charles G. Carll, Terry L. Highley / Journal of testing and evaluation 27(2):150–158 (1999)

[28] Bioconversion of wood wastes into gourmet and medicinal mushrooms / Suki C. Croan / In: Environmental aspects. Sec. 5. Document IRG/WP 99-50129 (1999)

[29] Fluorescent labels, confocal microscopy, and quantitative image analysis in study of fungal biology / Russell N. Spear, Daniel Cullen, John H. Andrews / In: Conn, P. Michael, ed. Confocal microscopy. Vol. 307. San Diego, CA: Academic Press: 607–623 (1999)

[30] Evaluation of white-rot fungal growth on southern yellow pine wood chips pretreated with blue-stain fungi / Suki C. Croan / In: Biology. Sec. 1. Document IRG/WP 00-10349 (2000)

[31] Recognize, remove, and remediate mold and mildew / Carol A. Clausen / Proceedings of the 2nd annual conference on durability and disaster mitigation in wood-frame housing: Nov. 6–8, 2000, Madison, WI Forest Products Society, p. 231–234 (2000)

[32] The genus Laetiporus in North America / H.H. Burdsall, Jr., M.T. Banik / Harvard papers in botany 6:43–55 (2001)

[33] A new fungicide combination for prevention of kiln brown stain of white pine / Elmer L. Schmidt, Terry L. Highley, Michael H. Freeman / Forest products journal Vol. 52(11/12):51–52 (2002)

[34] Basidiomycetes of the Greater Antilles project / D.J. Lodge, T.J. Baroni, S.A. Cantrell / In: R. Watling, J.C. Frankland, A.M. Ainsworth, S. Isaac, C. Robinson, (eds). Tropical mycology, Volume I Macromycetes. New York: CABI Publishing. pp. 45–60 (2002)

[35] Relationships between mechanical properties, weight loss, and chemical composition of wood during incipient brown-rot decay / Simon F. Curling, Carol A. Clausen, Jerrold E. Winandy / Forest products journal Vol. 52(7/8):34–39 (2002)

[36] Experimental method to quantify progressive stages of decay of wood by basidiomycete fungi / Simon F. Curling, Carol A. Clausen, Jerrold E. Winandy / International biodeterioration & biodegradation Vol. 49:13–19 (2002)

[37] Evaluating wood-based composites for incipient fungal decay with the immunodiagnostic wood decay test / C.A. Clausen, L. Haughton, C. Murphy / FPL–GTR–142 (2003)

[38] Synchrotron applications in wood preservation and deterioration / Barbara L. Illman / Wood deterioration and preservation: advances in our changing world. Washington, DC : American Chemical Society, ACS symposium series 845: p. 337–345 (2003)

[39] Oxalic acid overproduction by copper-tolerant brown-rot basidiomycetes on southern yellow pine treated with copper-based preservatives / Carol A. Clausen, Frederick Green / International biodeterioration & biodegradation. Vol. 51 (2003)

[40] Long-term efficacy of wood dip-treated with multicomponent biocides / Carol A. Clausen, Vina W. Yang / In: The international research group on wood protection, Section 3, Wood Protecting Chemicals. Paper prepared for the 36th Annual Meeting, Bangalore, India, April 24–28, 2005. IRG/WP 05-30379: 7 p (2005)

[41] Species diversity of polyporoid and corticioid fungi in northern hardwood forests with differing management histories / D.L. Lindner, H.H. Burdsall, Jr., G.R. Stanosz / Mycologia 98:195–217 (2006)

[42] Protecting wood from mold, decay, and termites with multi-component biocide systems / Carol A. Clausen, Vina Yang / International biodeterioration & biodegradation. Vol. 59 (2007)

Environment—Wind, Sun, Water

◀ Damage resulting from a hurricane. (2000s)

Structures are occasionally exposed to severe weather such as thunderstorms, hurricanes, or tornadoes. These events cause abnormally high forces (loads) on structures. During extreme events, abnormally high wind loads may cause building destruction and threaten the lives and safety of occupants. Buildings that are not destroyed may be penetrated by wind-borne objects—likewise a life-safety threat. FPL scientists have evaluated building structural performance during natural disasters and worked to refine building practices to limit the prevalence or severity of weather-related damage. [12, 14, 19, 20, 23, 24]

Earthquakes usually do not pose a life-safety threat to the occupants of wood-framed buildings, because wood-framed buildings can typically endure substantial seismic motion without collapse, even if they sustain structural damage. Likewise, flooding does not usually result in collapse of wood-framed buildings, unless the floodwaters move rapidly, in which case the building may be destroyed by removal from its foundation. Despite the ability of wood-framed buildings to commonly withstand flooding and earthquakes without collapse, they can nonetheless suffer structural damage from these events. [12, 19, 24, 27, 35, 43, 49]

Some water-induced damage can be expected to accompany weather events such as hurricanes. Water-induced damage can also occur in the absence of extreme environmental events, if conditions are conducive to moisture accumulation in the building.

Wood exposed outdoors (building exteriors, wood decks, posts, and fencing) experiences a variety of stressing factors. Understanding the factors, and limiting their influence through good design, material selection, finishing, and maintenance, can prolong the service life of exterior wood.

Challenge: Research challenges are (1) to provide information that permits design of robust structures capable of withstanding extreme weather events and anticipated exposure to various states of moisture (vapor, liquid, or solid) and (2) to provide information that results in longer service life for wood and wood products used in exterior applications.

FPL's Contributions: FPL helped develop instruments and techniques to measure effects of wind, sun, and water and the interactions of these factors. FPL scientists played a significant role in development of wood moisture meters and in the understanding of

◀ Example of damage caused by an earthquake—the foundation is separated from the main structure.

wood–moisture relations. FPL staff have provided technical advice in the development of building codes. They have developed recommendations on installation and operation of residential heating and air-conditioning systems. FPL staff have developed improved design of wood homes and other wood structures that help decrease the negative effects of the environment. FPL has studied wood–paint interactions and developed a stain for protecting wood.

Results: Wood buildings are more rationally designed with regard to safety and durability challenges posed by both extreme events and more typical in-service conditions. Recommendations for maintenance and operation have been made available to building owners.

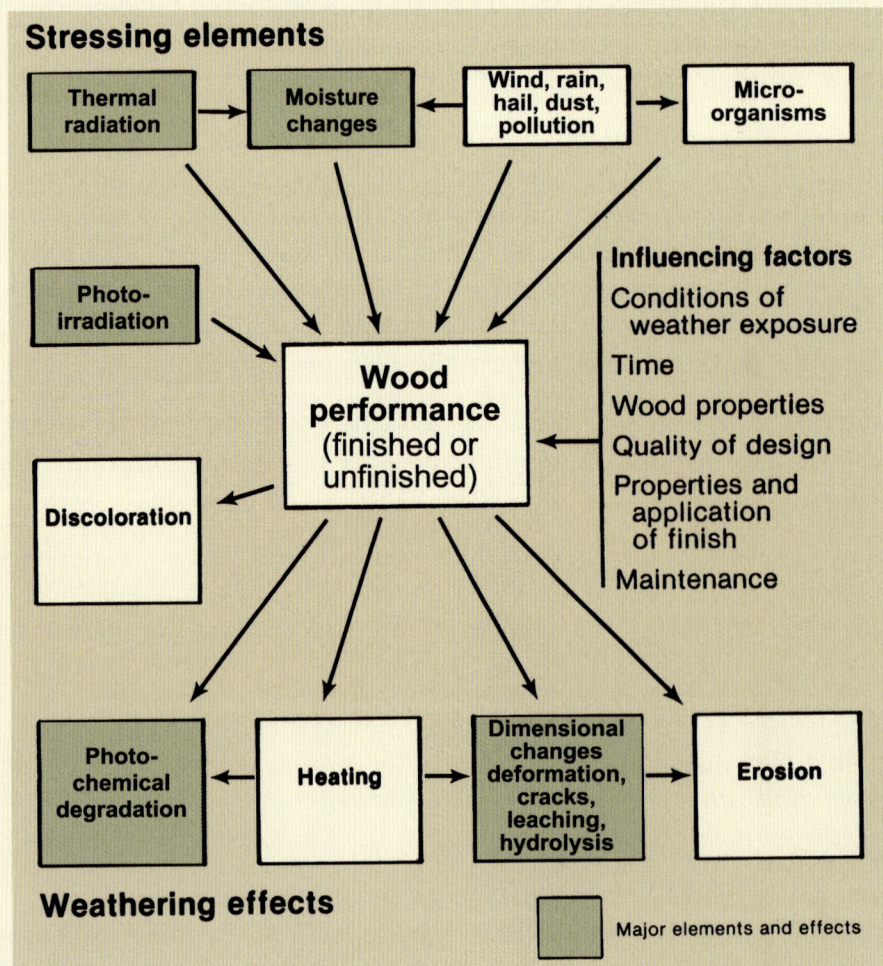

◀ Schematic illustration of factors affecting wood–finish performance and resultant weathering effects. (1988) {ML 885 561} [17, 28, 46, 60, 65]

Wind Effects [12, 14, 19, 20, 23, 25, 52, 63]

The effects of high winds on a wood structure are complex. To improve wood structure design, it is necessary to understand this complexity and how the structure reacts to wind forces. The FPL, in cooperation with the Department of Housing and Urban Development and the Department of Interior, is collecting actual field data to help understand the effects of wind. [88]

◀ Wood structure built at the Naval Live Oaks Area of the Gulf Islands National Seashore, Gulf Breeze, Florida. (2000s)

◀ The roof of the experimental building is instrumented with 78 pressure sensors and 68 load cells that measure the effect of wind on the roof. All the sensors feed data to an onsite computer that collects the information and transfers it to a computer at the Forest Products Laboratory in Madison, Wisconsin. (2000s)

Severe Weather Safe Rooms

With the occurrence of hundreds of tornadoes and a number of hurricanes each year in the United States, interest is growing in safe rooms that will increase the chances of surviving windborne missiles that may impact a house during violent weather. FPL researchers are evaluating different materials, such as laminated logs, to determine if they can provide the level of protection established by Federal Emergency Management Agency standards for safe rooms. [86]

◀ Experimental safe room wall made from laminated wood sections and oriented strandboard, prepared for missile impact test. Ann Bartuska, Deputy Chief for Research, Forest Service, is being briefed on the safe room research by Joe Murphy. (2000s)

▲ FPL's air-pressurized missile launcher used to propel materials—such as a 15-lb, No. 2 Southern Pine 2 by 4—at 100 miles per hour into experimental walls to simulate tornado debris. (2000s)

◀ Zhiyong Cai observing the penetration of a 2 by 4 in a safe room material sample.

Sun

Wood exposed to the elements deteriorates over time because of the complex action of solar radiation, moisture, wind, and airborne particles. [16, 31, 69, 70, 71]

Through design, selection of wood species, and treatments (such as paints, stain, or preservative coatings), the erosion of wood can be slowed (see Finishing and Housing sections).

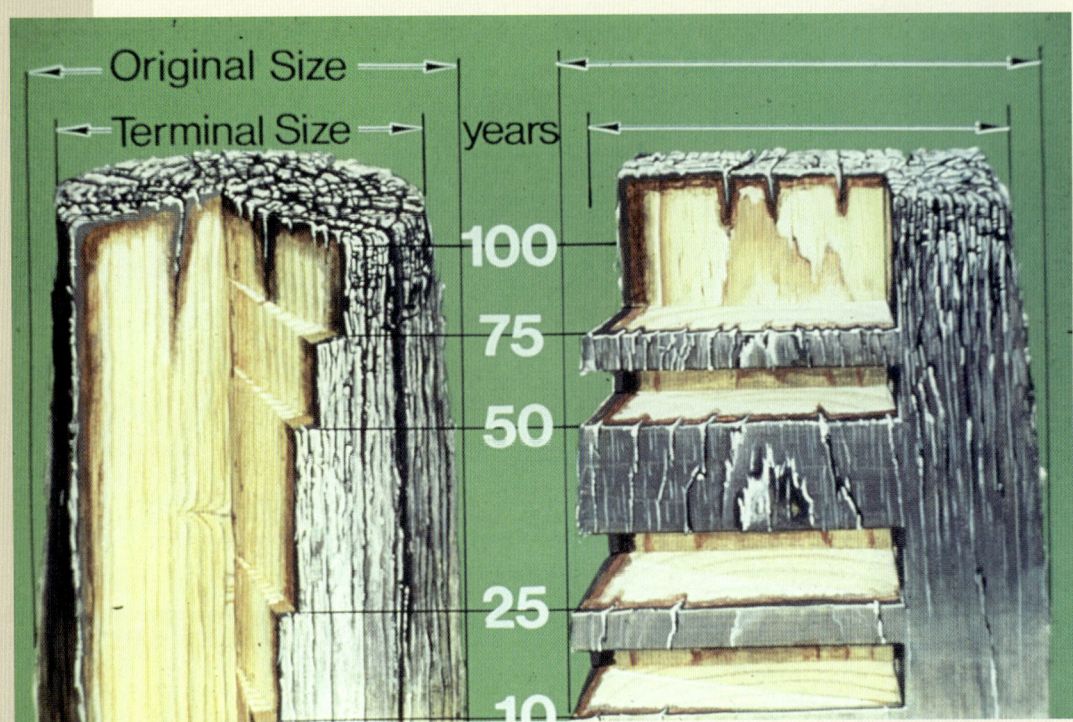

◀ Artists rendition of the weathering process of round and square timbers.(Approximately 1/4 inch of wood eroded per 100 years.) Cutaway shows that interior wood below the surface is relatively unchanged. (1970s) {M 146 221}

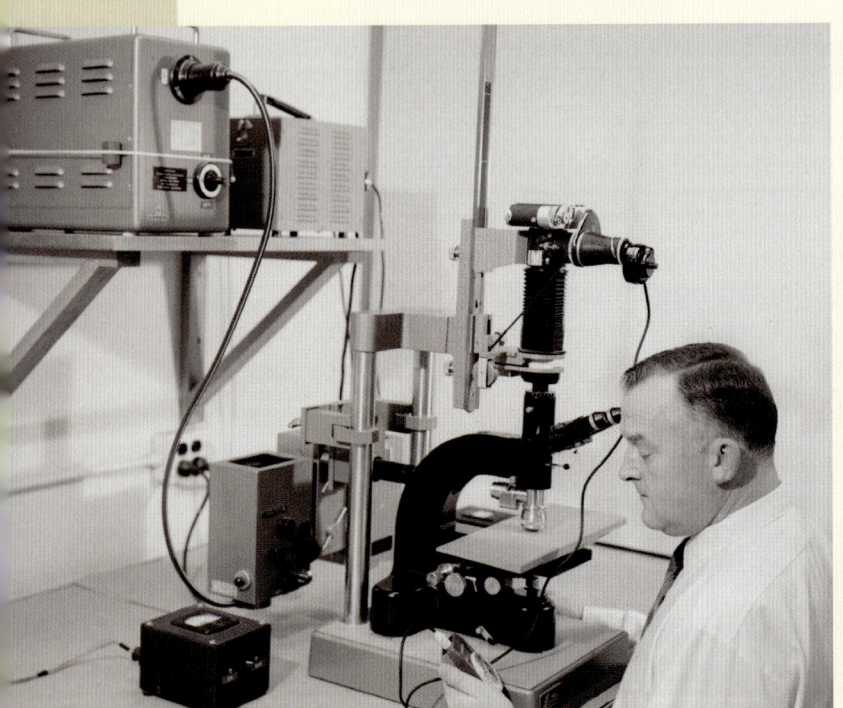

▲ Victor Miniutti investigating wood surface deterioration using an optical microscope (1960s) {M 136 316}

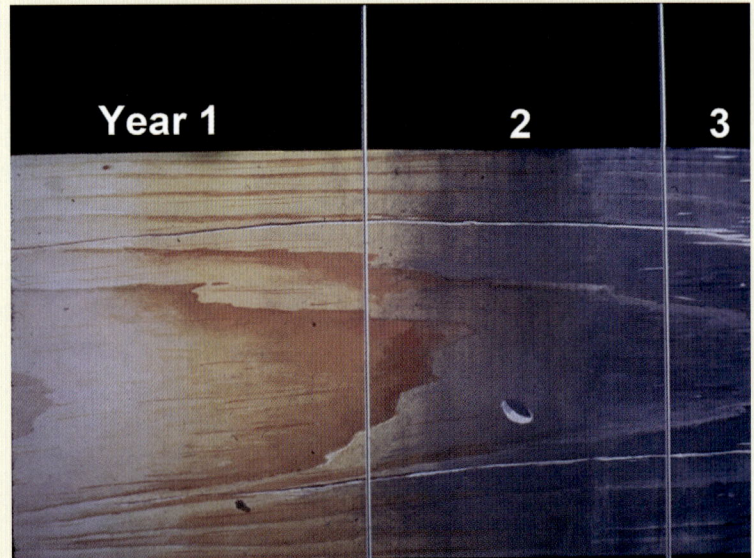

▲ Monochromatic rendition of color changes and surface wood changes during the outdoor weathering process for a typical softwood. (1970s) {M 146 222} [31]

Moisture in Structures

FPL has a long history of theoretical and experimental research on moisture and its movement in buildings. The presence and movement of moisture in the liquid state (for example, flooding and rain) are usually obvious, and moisture in this form has long been recognized for its potential to cause damage. The presence and movement of moisture as vapor is more insidious. Substantial moisture accumulation at various locations within a building can occur by movement of water vapor. The accumulation often occurs at hidden locations (for example, within walls); accumulation usually occurs seasonally, with dissipation occurring during other seasons. Such seasonal moisture accumulation can result in paint failures, growth of mold, and propagation of decay fungi, which can lead to structural problems. The challenge is to keep the moisture content of wood below 20% so the wood does not decay. [5, 11, 13, 18, 22, 32, 33, 37, 38, 47, 48, 50, 53, 54, 58, 61, 72, 81, 83–85]

Moisture Management

As used in this section, the term "moisture management" relates to building construction and operation practices that avoid or limit moisture accumulation in building assemblies or systems. The two general concepts involved in moisture management are (a) inhibiting moisture movement into assemblies or systems and (b) promoting moisture dissipation from assemblies or systems. The approaches taken by FPL research in moisture management include laboratory experiments, analytical studies, field experiments, and development of theoretical models. [58, 66, 67, 84, 85]

FPL's involvement in this field has also yielded guides for practitioners. [34, 36]

◀ Extensive fungal staining of sheathing resulting from seasonal moisture accumulation driven by vapor migration. Mold and decay fungi were present. Siding was removed to view the sheathing. The siding, incidentally, showed only minor evidence of moisture stress; the moisture accumulation was thus essentially hidden. [66]

Interior Vapor Barriers

FPL was an early proponent of the use of interior vapor barriers (referred to as "retarders" since the 1970s) in walls of insulated wood-frame buildings in cold climates. The prescription for use of interior vapor barriers was incorporated into the *Property Standards and Minimum Construction Requirements for Dwellings* issued by the Federal Housing Authority in 1942. As model building codes were developed after WW II, they incorporated the prescriptive requirement for inclusion of an interior vapor barrier in heating climates. Interior vapor retarders will be largely ineffective if there are moderate degrees of air movement through

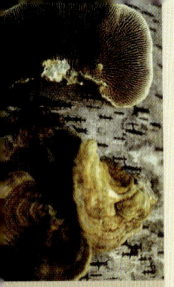

the construction, inasmuch as air movement can carry significant quantities of water vapor. Sheet vapor retarders can, however, also block air movement, provided that they are not compromised by holes, tears, gaps, or unsealed seams. [7, 9, 30, 33, 54, 56, 63, 85]

◄ Samuel Glass and Anton TenWolde working with a device for accurately measuring vapor transmission through building materials. (2000s) [83]

Measuring Moisture

An important issue in addressing moisture is the ability to measure wood moisture content. The electrical properties of wood allow its moisture content to be identified without having to resort to the labor-intensive (and usually destructive) oven-drying method. FPL played an important role, directly and indirectly, in development of electric wood moisture meters.

By 1927, Al Stamm had quantified the relationship between moisture content and electrical resistivity for nine American species. On September 6, 1932, U.S. Patent No. 1,875,359 for a "Moisture Indicating Instrument for Wood" (known as the "blinker meter") was awarded to Chauncey Suits and Mathew Dunlap, and "dedicated to the free use of the Government and the Public." With the concept of an electric moisture meter proven, commercial manufacturers would go on to develop, by the late 1940s, readily portable electric moisture meters. The post-war commercial meters allowed for direct numeric reading of wood moisture content. Many of these meters used a similar electrode configuration. FPL researchers published moisture content versus resistance relationships applicable to this (largely standardized) electrode configuration, for various wood species in an FPL report entitled "Electric Moisture Meters for Wood," dated 1947. The report, which found widespread use, has been periodically updated and remains the most commonly referenced guide for use of moisture meters. [1, 4, 6, 8, 10, 26]

Moisture Content of Interior Woodwork

Moisture content of millwork that is used in buildings becomes approximately equal to the moisture content determined by average relative humidity where the wood is used. Early FPL studies (1930s) determined the average moisture contents of interior wood found around the country to help guide the industry. However, these studies were conducted before air conditioning. Properly sized air conditioners tend to keep interior wood drier during the summer. [2, 3, 84]

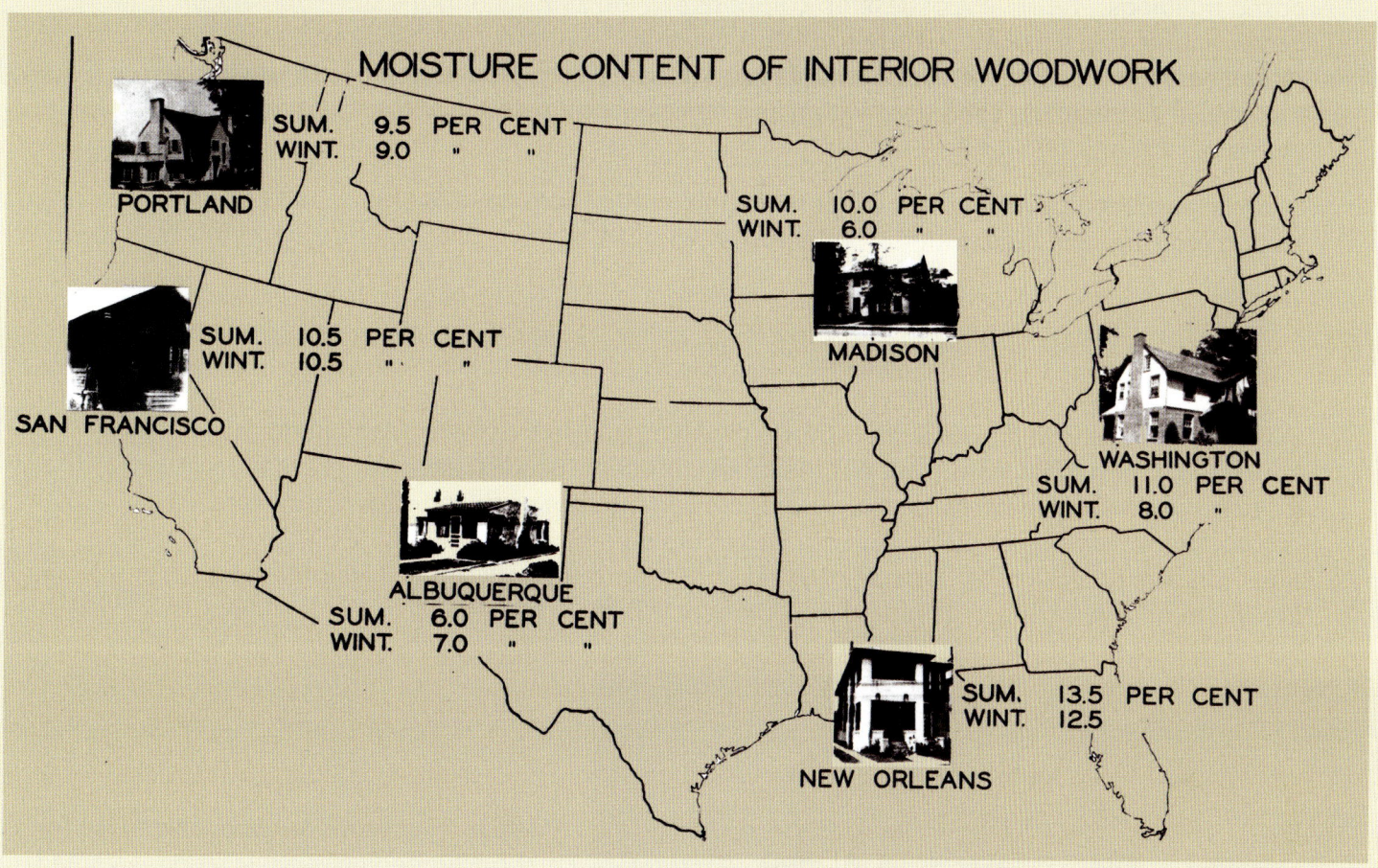

MOISTURE CONTENT OF INTERIOR WOODWORK

PORTLAND
SUM.   9.5 PER CENT
WINT. 9.0   "   "

SAN FRANCISCO
SUM.   10.5 PER CENT
WINT. 10.5   "   "

MADISON
SUM.   10.0 PER CENT
WINT. 6.0   "   "

WASHINGTON
SUM.   11.0 PER CENT
WINT. 8.0   "

ALBUQUERQUE
SUM.   6.0 PER CENT
WINT. 7.0   "   "

NEW ORLEANS
SUM.   13.5 PER CENT
WINT. 12.5

▲ Typical moisture content of interior woodwork in various areas of the United States. (1930s) {M 19319 F}

A moisture content sensor, fabricated of wood, was developed at FPL by John Duff in 1966. The sensor is small and therefore is useful for unobtrusively monitoring conditions in structures. The sensor relies on the well-recognized relationship between electrical resistance of wood and its moisture content. The sensor can also be used for monitoring relative humidity. It is relatively robust compared with commercially manufactured thin-film polymeric sensors for measurement of relative humidity, remaining functional when exposed to wet conditions. [1, 15, 26, 59]

Predicting Moisture Conditions

The most commonly used calculation method for predicting moisture conditions in insulated assemblies, such as walls, during mid-winter, had its origin at FPL. The method is often called the dewpoint method but is officially known by the American Society of Heating, Refrigerating and Air-Conditioning Engineers (ASHRAE) as the profile method. The method is reasonably easy to execute, involving a modest number of calculations that can be performed by calculator or computer spreadsheet. The method has undergone continual refinement and remains in widespread use as a rapid evaluation tool for specified or proposed buildings. The advent of desktop computers in the 1980s led to the development of iterative time-step simulation models, which allow analyses under changing conditions. FPL researchers demonstrated that the advantage of the time-step models, relative to the profile method, is in predicting conditions within assemblies during seasonally changing environmental conditions. [5, 57, 68, 72]

Models have been developed at FPL for predicting indoor humidity conditions and for predicting moisture and temperature conditions in attics [54, 61]

Ice

In climates with snowfall, roof icing can be a problem, and serious water damage to the structure can result. Roof ventilation is the most commonly used strategy for prevention of ice dams. A comprehensive review of studies concerning roof ventilation by various research organizations throughout North America was completed by Anton TenWolde (FPL) and Bill Rose (University of Illinois, Champaign–Urbana). In that review, guidelines for limiting the likelihood or severity of ice damming are outlined. [73, 74]

▼ Weather conditions conducive to ice formation, limited ceiling insulation, and inadequate air leakage control have resulted in formation of ice dams on this home. Control of heat and moisture flows is an ongoing area of research at FPL.

▲ Heat loss into the attic from air leaks, ducts, flues, or insufficient insulation can lead to ice dams even when the attic is vented. Staining can occur because of ice dams and lack of water control.

Moisture Engineering

FPL scientists play an important role in current efforts to develop moisture control as an engineering discipline. They have lead roles in the development of design tools, criteria, and standards, or in their revision. [51, 55, 81, 87]

Acid Rain

Acidic rain, which can include snow, fog, and other atmospheric depositions, are suspected of causing more rapid deterioration or weathering of many materials. Researchers at the FPL, in cooperation with the Atmospheric Research and Exposure Assessment Laboratory of the Environmental Protection Agency, Raleigh, North Carolina, investigated the incremental effects of acids on the weathering of wood and painted wood. FPL researchers found that brief exposure of wood to acid during accelerated weathering increased the deterioration rate. The rate was affected by the strength and type of acid. The more rapid weathering will shorten the service life of unpainted wood.

◀ Sam Williams checking wood specimens after exposure to simulated acid rain. (1980s) [39–42]

Sealants

Sealants are commonly used to "weather-proof" (inhibit rainwater intrusion into) building walls. Based on frequent questions about the performance or nonperformance of sealants, the FPL is conducting fundamental studies into the performance of sealants. [64, 75–80, 82]

◀ This sealant test apparatus applies cyclic fatigue to specimens. The movement is driven by moisture-induced dimensional change of wood or thermally induced dimensional change of metal. (2000s)

1910

Little scientific understanding of environmental effects, such as moisture, temperature, or solar radiation, on wood existed.

2010

Research helps explain the effects of environmental factors on the use of wood, thus providing more durable and safe structures.

Further Information

[1] The electrical resistance of wood as a measure of its moisture content / A.J. Stamm / Industrial and Engineering Chemistry Vol. 19 (1927)

[2] Dimension changes in millwork due to varying atmospheric conditions / J.S. Mathewson / FPL No. 958 (1930)

[3] Drying millwork to meet local moisture and humidity conditions / Edward C. Peck / Wood working industries Vol. 10(3) (Sept. 1931)

[4] The "blinker": a new instrument for rapid moisture-content determinations in wood / C.G. Suits, M.E. Dunlap / Southern lumberman Vol. 140(1782) (July 1, 1930)

[5] Condensation in walls and attics / L.V. Teesdale / FPL No. 1157 (1937)

[6] Electrical moisture meters for wood / M.E. Dunlap / Wood products 43(4) (April 1938)

[7] Comparative resistance to vapor transmission of various building materials / L.V. Teesdale / ASHVE Transactions, Vol. 49 (American Society of Heating and Ventilating Engineers) (1943)

[8] Electrical moisture meters for wood / M.E. Dunlap, E.R. Bell / FPL No. 1660 (1947)

[9] Water-vapor permeability of building papers and other sheet materials / E.R. Bell, M.G. Seidl, N.T. Krueger / Heating, Piping and Air Conditioning (American Society of Heating and Ventilating Engineers) (December 1950)

[10] Electrical moisture meters for wood / W.L. James / FPL No. 1660 (1954)

[11] Condensation and decay prevention under basementless houses / C.S. Moses / FPL No. 2010 (1954)

[12] Observations of damage to houses by high winds, waves, and floods and some construction precautions / R.F. Luxford, Walton R. Smith / FPL No. 2095 (1957)

[13] Condensation problems in modern buildings / L.V. Teesdale / FPL No. 1196 (1939) (reviewed and re-issued, 1941, 1953, 1959)

[14] Houses can resist hurricanes / L.O. Anderson, Walton R. Smith / Research paper FPL 33 (1965)

[15] A probe for accurate determination of moisture content of wood products in use / John E. Duff / FPL–RN–0142 (1966)

[16] Microscopic observations of ultraviolet irradiated and weathered softwood surfaces and clear coatings / Miniutti, V.P. / Research paper FPL 74 (1967)

[17] Water repellents and water-repellent preservatives / William C. Feist, Edward A. Mraz / Research note FPL 0124 (1968)

[18] Moisture distribution in wood-frame walls in winter / J.E. Duff / Forest products journal Vol. 18(1) (1968)

[19] Diaphragm action of diagonally sheathed wood panels / D.V. Doyle / Research note FPL 0205 (1969)

[20] Performance of wood construction in hurricanes / B. Bohannan / Proceedings of the tornado conference, Madison: University of Wisconsin (1970)

[21] Lower housing costs through improved design and wood utilization / J.A. Liska / Journal of forestry Vol. 68(7) (Jul 1970)

[22] Condensation problems: their prevention and solution / L.O. Anderson / Research paper FPL 132 (1972)

[23] Wood structures can resist hurricanes / G.E. Sherwood / Civil engineering Vol. 42(9) (Sept. 1972)

[24] Performance of wood construction in disaster areas / J.A. Liska / American Society of Civil Engineers. Journal of the structural division Vol. 99, No. ST12 (Dec 1973)

[25] A conventional house challenges simulated forces of nature / Roger L. Tuomi, William J. McCutcheon / Forest products journal Vol. 25(6) (June 1975)

[26] Electric moisture meters for wood / W.L. James / General technical report FPL 6 (1975)

[27] Vibrational loading of mechanically fastened wood joints / T.L. Wilkinson / Research paper FPL 274 (1976)

[28] Review of the Housing and Urban Development minimum property standards as they relate to protection of wood in use. Forest Products Laboratory. Committee for review of HUD minimum property standards, Chair Rodney C. De Groot / Committee members include several U.S. Forest Service staff (1977)

[29] Lightweight truss-framed house for safety and energy efficiency / Roger L. Tuomi / Agriculture engineering Vol. 58(5) (May 1977)

[30] Paint as a vapor barrier for walls of older homes / G.E. Sherwood / Research paper FPL 319 (1978)

[31] Comparison of outdoor and accelerated weathering of unprotected softwoods / W.C. Feist, E.A. Mraz / Forest products journal Vol. 28(3) (Mar. 1978)

[32] Moisture movement and control in light-frame structures / G.E. Sherwood, A. Tenwold / Forest products journal Vol. 32(10) (Oct 1982)

[33] Condensation potential in high thermal performance walls—cold winter climate / G.E. Sherwood / Research paper FPL 433 (1983)

[34] Controlling moisture in houses / A. Tenwolde / Solar age Vol. 9(1) (Jan 1984)

[35] Low-rise buildings subjected to seismic, wind, and snow loads / L.A. Soltis / Journal of structural engineering Vol. 110, ST4 (Apr 1984)

{36} Condensation problems in your house: prevention and solution / L.O. Anderson, G.E. Sherwood / Agriculture information bulletin No. 373 (1974)

[37] Steady-state one-dimensional water vapor movement by diffusion and convection in a multilayered wall / A. TenWolde / ASHRAE transactions Vol. 91(1) (1985)

[38] Moisture movement in walls in a warm humid climate / A. TenWolde, H. Mei / Proceedings U.S. Dept. Energy thermal envelopes III conference (1986)

[39] Acid effects on accelerated wood weathering / R.S. Williams / Forest products journal Vol. 37(2) (Feb 1987)

[40] Effects of dilute acid on the accelerated weathering of wood / R.S. Williams / Journal of the Air Pollution Control Association Vol. 38(2):148–151 (1988)

[41] Effects of acid rain on painted wood surfaces: effect of the substrate. Part I. / R.S. Williams / American paint and coatings journal Vol. 72(30):37–43 (1988)

[42] Effects of acid rain on painted wood surfaces: effect of the substrate. Part II / R.S. Williams / American paint and coatings journal Vol. 72(31):37–45 (1988)

[43] Seismic behavior of low-rise wood-framed buildings / R.H. Falk, L.A. Soltis / Shock and vibration digest Vol. 20(12) (Dec. 1988)

[44] A mathematical model for indoor humidity in homes during winter / A. TenWolde / Proceedings symposium on air infiltration, ventilation and moisture transfer. National Institute of Building Science Conference (1988)

[45] Wood-frame house construction / O.C. Heyer, L.O. Anderson / USDA Agriculture handbook 73 (1955) Latest revision of text published in (1989)

[46] Chronicle of 65 years of wood finishing research at the Forest Products Laboratory / Thomas M. Gorman, William C. Feist / FPL–GTR–60 (1989)

[47] Moisture in building envelopes / Stephen L. Quarles, Anton TenWolde / Executive summaries: 43rd annual meeting, June, Reno, Nevada. Madison, Wis. Forest Products Research Society (1988) (i.e.1989)

[48] Moisture transfer through materials and systems in buildings / AntonTenWolde / Water vapor transmission through building materials and systems. Philadelphia, PA. American Society for Testing and Materials (1989)

[49] Seismic performance of low-rise wood buildings / Lawrence A. Soltis, Robert H. Falk / Shock and vibration digest Vol. 24(12) (1992)

[50] Effect of cavity ventilation on moisture in walls and roofs / A. TenWolde, C. Carll / Proceedings of the ASHRAE/DOE/BTECC conference; 1992 December 7–10; Clearwater Beach, FL. Atlanta, GA: American Society of Heating, Refrigerating and Air-Conditioning Engineers (1992)

[51] Design tools / Anton TenWolde / Moisture control in buildings. Philadelphia, PA : American Society for Testing and Materials. ASTM manual series MNL 18 (1993)

[52] Wind resistance of conventional light-frame buildings / R.W. Wolfe, R.M. Riba, M. Triche / In Proceedings Hurricanes of 1992: Andew and Iniki one year later. American Society of Civil Engineers (1994)

[53] Moisture control in crawl spaces / William B. Rose, Anton Ten Wolde / Wood design focus Vol. 5(4) (1994)

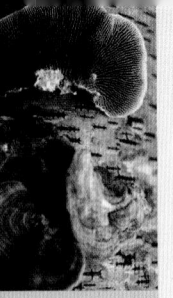

[54] Ventilation, humidity and condensation in manufactured houses during winter / A. TenWolde / ASHRAE transactions (1994)

[55] Moisture control strategies for the building envelope / Anton TenWolde, William B. Rose / Wood design focus Vol. 5(4) (1994)

[56] Airflows and moisture conditions in walls of manufactured homes / A. TenWolde, C. Carll, V. Malinauskas / ASTM Special technology publication No. 1255 (1995)

[57] Moisture accumulation in walls: comparison of field and computer-predicted data / A. TenWolde and C. Carll / Proceedings U.S. Dept. Energy thermal envelopes VI conference (1995)

[58] Manufactured housing walls that provide satisfactory moisture performance in all climates / Douglas M. Burch, Christopher A. Saunders, Anton TenWolde / Gaithersburg, MD: National Institute of Standards and Technology, Building and Fire Research Laboratory, NISTIR 5558 (1995)

[59] Accuracy of wood resistance sensors for measurement of humidity / Charles Carll, Anton TenWolde / Journal of testing and evaluation Vol. 24(3) (May 1996)

[60] Finishes for exterior wood: selection, application, and maintenance: a comprehensive guide to the painting/staining and maintenance of homes, decks, log structures, and more / R. Sam Williams, Mark T. Knaebe, William C. Feist / Madison, WI: Forest Products Society (1996)

[61] FPL roof temperature and moisture model: description and verification / Anton TenWolde / FPL–RP–561 (1997)

[62] Standards for structural wood products and their use in the United States / David W. Green, Roland Hernandez / Wood design focus Vol. 9(3) (1998)

[63] Air pressures in wood frame walls / A. TenWolde, C. Carll, V. Malinauskas / Proceedings U.S. Dept. Energy thermal envelopes VII conference (1998)

[64] Rehabilitation of wood-frame houses / A. TenWolde, R.C. De Groot, W.C. Feist, G.E. Sherwood / USDA Agriculture handbook No. 704 (1998)

[65] Wood Handbook: wood as an engineering material / FPL–GTR–113 (1999)

[66] Mold and decay in TriState homes / Anton TenWolde / Proceedings of the 2nd annual conference on durability and disaster mitigation in wood-frame housing (Nov. 6–8, 2000) Madison, WI: Forest Products Society (2000)

[67] Performance of back-primed and factory-finished hardboard lap siding in southern Florida / C. Carll, M. Knaebe, V. Malinauskas, P. Sotos, A. TenWolde / FPL–RP–581 (2000)

[68] Manual analysis tools / A. TenWolde / Chapter 7 of ASTM MNL 40: Moisture analysis and condensation control in building envelopes. American Society for Testing and Materials (2001)

[69] Erosion rates of wood during natural weathering. Part I. Effects of grain angle and surface texture / R. Sam Williams, Mark T. Knaebe, Peter G. Sotos, William C. Feist / Wood Fiber Sci. 33(1): 31–42 (2001)

[70] Erosion rates of wood during natural weathering. Part II. Earlywood and latewood erosion rates / R. Sam Williams, Mark T. Knaebe, William C. Feist / Wood fiber science Vol. 33(1): 43-49 (2001)

[71] Erosion rates of wood during natural weathering. Part III. effect of exposure angle on erosion rate / R. Sam Williams, Mark T. Knaebe, James W. Evans, William C. Feist / Wood Fiber Science Vol. 33(1): 50–57 (2001)

[72] Interior moisture design loads for residences / Anton TenWolde, Iain S. Walker / Performance of exterior envelopes of whole buildings VIII integration of building envelopes. Atlanta, GA: American Society of Heating, Refrigerating and Air-Conditioning Engineers (2001)

[73] Venting of attics and cathedral ceilings / William B. Rose, Anton TenWolde / ASHRAE journal Vol. 44(2) (Oct. 2002) (ASHRAE is the American Society of Heating, Refrigeration and Air Conditioning Engineers) (2002)

[74] Attic and crawlspace ventilation: implications for homes located in the urban-wildland interface / Stephen L. Quarles, Anton TenWolde / Proceedings from the woodframe housing durability and disaster Issues conference: October 4-6, 2004, Las Vegas, Nevada, USA. Madison, WI: Forest Products Society (2004)

[75] Studies on the effect of movement during the cure on the mechanical properties of sealant / C.C. White, D.L. Hunston, R.S. Williams / Proceeding of the 27th annual meeting of the Adhesion Society, Wilmington NC (Feb. 2004)

[76] The effect of movement during cure of a silicone sealant / C.C. White, D.L. Hunston, R.S. Williams / Proceedings of the ANTEC conference from Society of Plastics Engineering, Chicago, Il, (May 2004)

[77] Development of test apparatus for service life prediction of sealant formulations and evaluation of data from 18 months of outdoor exposure / R.S. Williams, S. Lacher, C. Halpin / In: Proceedings of the fourth inter-

national wood coatings congress, 25–27 October, The Hague, The Netherlands. Paper 16. The Paint Research Association, Teddington, Middlesex, UK. (2004)

[78] Evaluating weather factors and material response during outdoor exposure to determine accelerated test protocols for predicting service life / R.S. Williams, L. Lacher, C. Halpin, C. White / In Martin JW, Ryntz RA, Dickie RA. Service Life Prediction, Challenging the Status Quo, Federation of Societies for Coatings Technology. Proceedings: third international symposium on service life prediction, Feb 1–6, 2004 Sedona AZ (2005)

[79] Development of a powered outdoor sealant fatigue test apparatus / L. Lacher, R.S. Williams, C. Halpin / In Martin JW, Ryntz RA, Dickie RA. Service Life Prediction, Challenging the Status Quo, Federation of Societies for Coatings Technology. Proceedings: third international symposium on service life prediction, Feb 1–6, 2004 Sedona AZ (2005)

[80] Development and testing of a hybrid in-situ testing device for sealant / C.C. White, N. Embree, C. Buch, R.S. Williams / Review of scientific instrumentation 76, 045111 (2005)

[81] Proposed new standard 160, design criteria for moisture control in buildings / Released as a public review draft, Sept. 2006 / American Society of Heating, Refrigerating and Air-Conditioning Engineers (2006)

[82] The ins and outs of caulking / Charles Carll / FPL–GTR–169 (2006)

[83] Measurements of moisture transport in wood-based materials under isothermal and nonisothermal conditions / S.V. Glass / Proceedings U.S. Dept. Energy thermal envelopes X conference (2007)

[84] Review of in-service moisture and temperature conditions in wood-frame buildings / Samuel V. Glass, Anton TenWolde / FPL–GTR–174 (September 2007)

[85] Moisture performance of a contemporary wood-frame house operated at design indoor humidity levels / C. Carll, A. TenWolde, R. Munson / Proceedings U.S. Dept. Energy thermal envelopes X conference (2007)

[86] Constructing severe weather safe rooms using small-diameter timber / Joe Murphy / Research in progress RIP–EML–008 (2008)

[87] Standard guide for limiting water-induced damage to buildings. Designation E241-08. ASTM International (2008)

[88] Evaluation of wind pressure measurement techniques for test structures / Joe Murphy / Research in Progress RIP–EML (2008)

Fire Safety

◄ This burning test structure is at the point of flashover—the sudden spread of flames over an area when it becomes heated to the flashpoint. Results of these tests indicated that sandwich panels provided structural integrity for various lengths of time depending on the facing material used for the panels. (1975) {19 L} [33]

Fire safety research addresses the potential contribution of wood products to the growth of a fire, the ability of structural wood elements to withstand a fire, and the chemical treatment of wood products to reduce their flammability. [7]

Challenge: How do we ensure that traditional and new innovative wood products do not adversely contribute to loss of life and property in fires?

FPL's Contributions: FPL helped quantify the fire performance of wood products and contributed to the development of treatments of wood that reduced their flammability. This included research that helped define the fundamentals of fire behavior and efforts to develop methodologies for fire testing of wood and composite materials to ensure proper measurements of relevant performance characteristics. Demonstration of satisfactory fire performance is needed to meet requirements of the building codes. More recently, contributions have been toward data and models required for fire safety engineering of forest products in a performance-based building code environment.

Results: Maximum use of traditional and innovative wood products in safer structures.

Fire Safety Research: Fire research began at the FPL about two years after it was established in 1910. However, it was short lived due to the priorities of WW I. Research was started again in 1927, with some assistance and cooperation from the National Lumber Manufacturers Association. At that time, the emphasis was on performance of wood under fire conditions by defining three general characteristics: (1) ignition, (2) fire spread, and (3) resistance to penetration by fire.

Fire-retardant treatments of wood products have been an important component of the fire

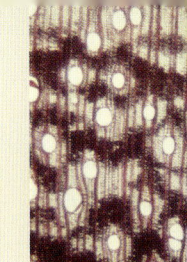

research program throughout its history. In efforts to develop effective fire-retardant treatments, methodologies of fire testing have been a key component of the FPL fire research program.

In a very early effort (1912–1915), R.E. Prince used a furnace apparatus to expose samples of untreated and fire-retardant-treated specimens to constant temperature conditions and recorded the times for ignition. Chemicals evaluated in early work by R.E. Prince in the 1910s included sodium carbonate, sodium bicarbonate, oxalic acid, borax, ammonium chloride, and ammonium sulfate plus ammonium phosphate. Ammonium salts and borax were found most effective. [1]

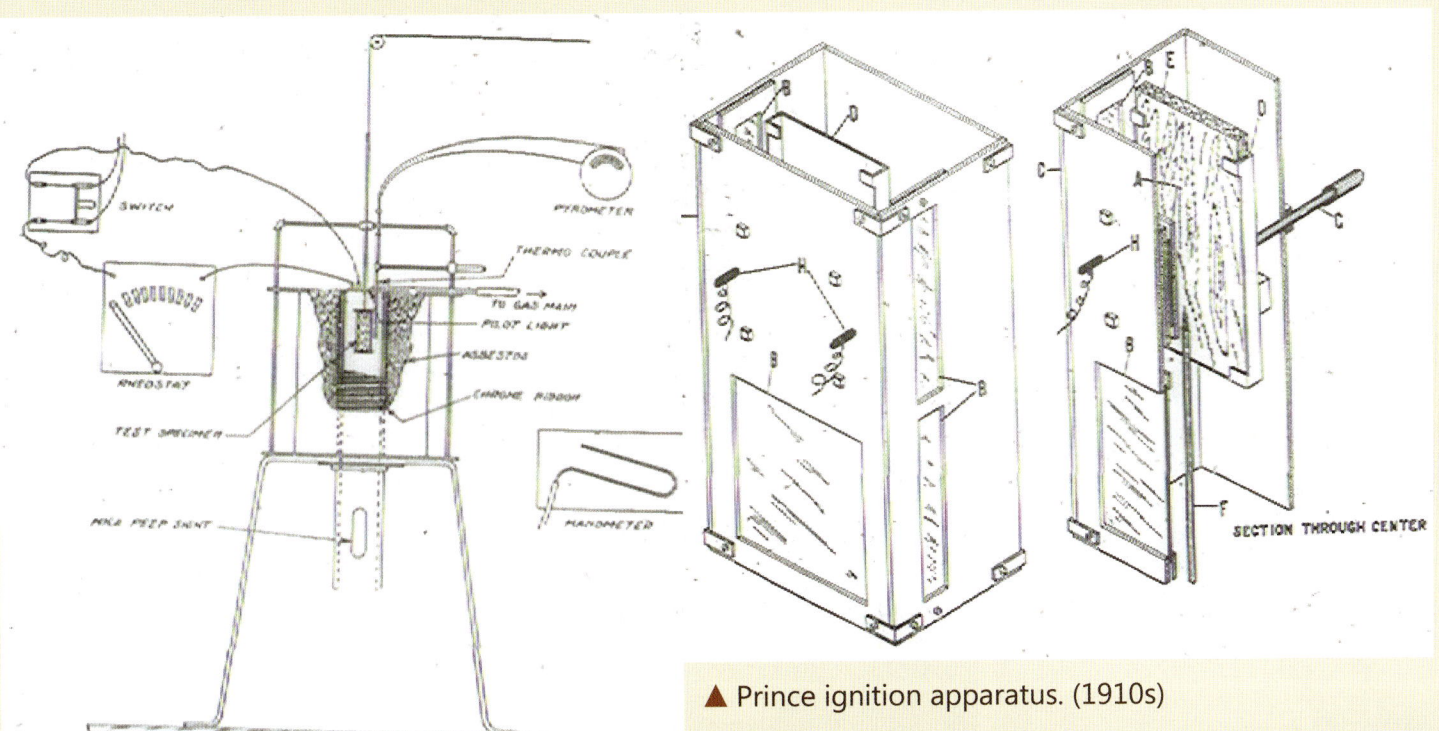

▲ Prince ignition apparatus. (1910s)

Development of the fire tube apparatus in the 1920s by Mathew Dunlap provided a simple methodology to evaluate the effectiveness of fire-retardant treatment of wood products. The methodology uses the residual mass of the sample as an indicator of flame retardancy. The method became an ASTM standard (E-69) in 1946. [3]

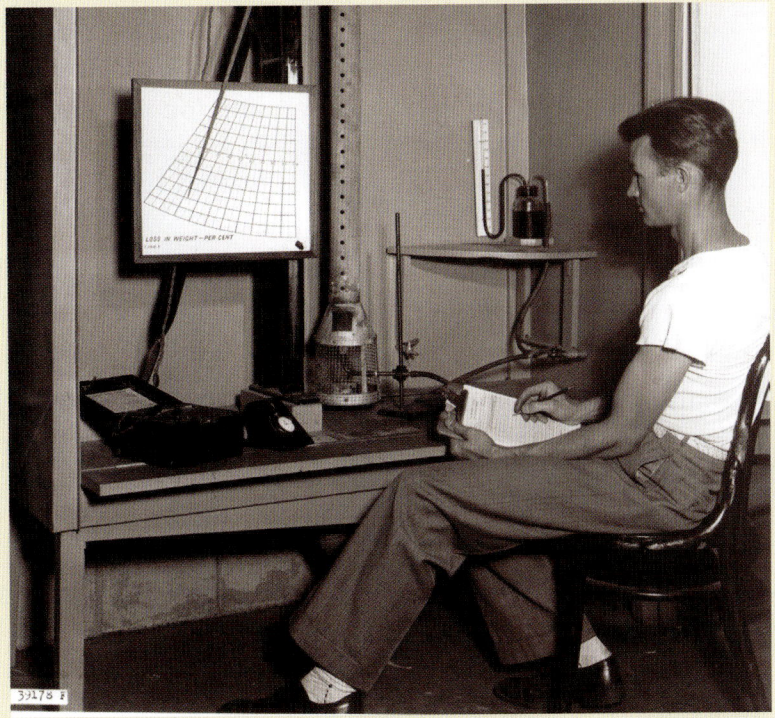

◄ Technician observing weight loss of a wood specimen as it is exposed to the FPL-developed fire tube test. (1940s) {M 39178}

Use of Chemicals in Forest Fire Control

In the 1930s, Thomas Truax investigated the use of chemicals in the control and suppression of forest fires. FPL tests and field experience indicated that "boric acid showed appreciable superiority over water. At 5% concentration, which was about the maximum obtainable with boric acid, its effectiveness compared favorably with that of phosphoric acid or of the mono- and di- basic ammonium phosphates at similar concentration. Tests were also made on combinations of two chemicals in the same solution but the results indicated no advantage over the single chemicals at the same total concentration." [4]

Fire Retardants and Fire Bombs

At 1/2 minute

At 5 minutes

At 42 minutes

▲ An attic section in which the rafters, roof-boards, inside of end-wall, and top ply of flooring were impregnated with a moderate degree of fire retardant and exposed to a 5.25-lb magnesium bomb. The treatment completely stopped the spread of fire on exposed surfaces, but the untreated subfloor was ignited by the excessive heat transmitted through the flooring. (1940s) {46351 F}

Fire-Retardant Treatments [29, 36, 37, 47, 66]

FPL investigated about 130 treatments of single or combinations of inorganic chemicals. Combinations of chemicals were used to obtain the best performance in both fire and other performance properties, such as corrosion, leaching, gluing, finishing, and costs. Such chemicals included ammonium sulfate, mono- and di-ammonium phosphates, ammonium chloride, zinc chloride, borax, and boric acid. The phosphates were identified as the most effective. These chemicals were used in the first generation of commercial fire-retardant-treated wood in the United States.

As part of its continuing efforts to evaluate new fire-retardant treatments and coatings, FPL investigated horizontal furnaces for measuring flame spread. This led to the development of the 8-ft tunnel test in the 1950s as a smaller scale alternative to the regulatory 25-ft tunnel test. The 8-ft tunnel method was standardized as ASTM E 286 in 1965. [8]

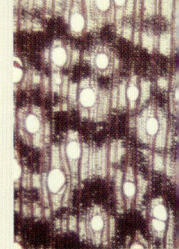

▲ Original horizontal tunnel test apparatus. (1950s)

◄ Ronald Knispel removing a test sample of fire-retardanttreated wood shingles from the FPL-developed 8-foot tunnel test. (1970s)

Panel Products

In addition to the 8-ft tunnel test, a corner wall test was also used to evaluate surface flammability of panel products.

Some other projects involved fire-retardant treatments for reducing flame spread and smoldering characteristics of cellulosic insulation. Fires involving self-heating, smoldering, and flames continue to be an area of research interest. [32, 34]

◄ A corner wall test with a wood crib as the ignition source. (1970s) [30]

▲ Standard "cigarette" test for evaluating the smoldering characteristics of treated cellulosic insulation. (1980s)

Fire Fundamentals

Gaining a fundamental understanding of thermal degradation and combustion of wood products has been a key component of FPL efforts to measure the potential contribution of wood products to a fire and facilitate the development of new fire-retardant-treated wood products.

In the 1960s, considerable fundamental research was conducted by Fredrick Browne, Walter Tang, John Brenden, and Frank Beall on thermal degradation (or pyrolysis) of wood using thermogravimetric and differential thermal analysis. [9, 10, 15, 16]

This early research provided a foundation of knowledge that is used today to develop better computer models for predicting the contribution of wood products to a fire.

◄ Walter Tang operating the thermogravimetric apparatus. (1960s)

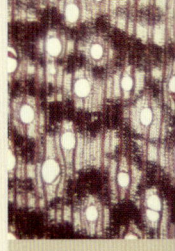

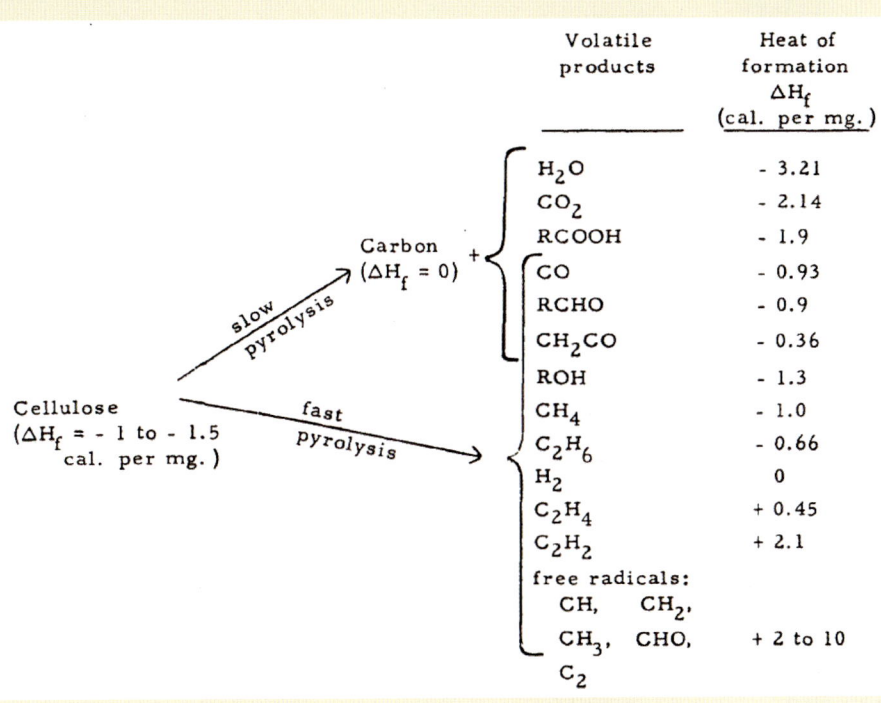

Volatile products	Heat of formation ΔH_f (cal. per mg.)
H_2O	- 3.21
CO_2	- 2.14
RCOOH	- 1.9
CO	- 0.93
RCHO	- 0.9
CH_2CO	- 0.36
ROH	- 1.3
CH_4	- 1.0
C_2H_6	- 0.66
H_2	0
C_2H_4	+ 0.45
C_2H_2	+ 2.1
free radicals: CH, CH_2, CH_3, CHO, C_2	+ 2 to 10

◀ Chemical reaction products to pyrolysis of wood. (1958) [5]

The 1958 publication by Fredrick Browne, "Theories of the Combustion of Wood and Its Control—A Survey of the Literature," remains much in demand by those conducting research in this area. [5]

Rate of Heat Release (RHR) Measurement

As an early promoter of the use of heat release rate as a measure of relative flammability, John Brenden developed in the 1960s the FPL apparatus to measure heat released by a burning material. FPL continued to obtained new equipment as better technologies were developed to measure heat release rates. [23, 26, 28]

In the 1980s, the original RHR apparatus was replaced with an Ohio State University RHR apparatus. [39, 40, 43]

In the 1990s, the Ohio State University RHR was replaced with a cone calorimeter apparatus, which had been developed at the National Bureau of Standards (now National Institute for Standards and Technology). The cone calorimeter is used in current investigations into fire-retardant treatments for composite materials and fundamental research on the fire behavior of wood. [58, 65]

▼ Cone calorimeter (1990s)

▲ Original heat release apparatus at FPL. (1960s)

▶ Anne Fuller and Mark Dietenberger use the cone calorimeter.

In addition to their use in evaluating the effectiveness of fire-retardant treatments, test methods for rate of heat release were critical in the development of models to predict flame spread behavior of wood and times for flashover in the standard room-corner test. [58]

Full-scale testing is often necessary to fully understand the effect of materials on a fire. FPL constructed a dwelling room to examine the effect of furnishings on fire development.

◀ Flashover in a standard room-corner test in the 1990s.

◀ Dwelling room facility used to conduct corner wall tests. A wood crib was used to start a room fire to study flame growth and spreading. Tests like these also assisted in evaluating fire retardants. (1940s) [6]

Full-scale testing of wood products continued with the construction of a standard room-corner test facility. Ignition source is provided by a propane burner in the corner, and the test material is installed on three walls and the ceiling. As in the cone calorimeter, heat release rate is measured using oxygen consumption methodology.

The trend towards performance-based codes creates new demands for improved material properties and computer fire models. This demand is difficult to meet for wood-based materials because of their complex pyrolysis and combustion processes. The cone calorimeter, modified to obtain additional data, has been the primary test method for deriving properties of surface ignition, fuel mass loss, and heat release rates of wood volatiles.

Investigating the effect of fire-retardant chemicals on the non-fire properties of wood products has been a key component of FPL research from the beginning. Efforts included the effects of fire retardants on adhesive quality, strength, and corrosion. Such efforts have included the development of new test methods to address the concerns. [29]

To investigate the ability of an exterior fire-retardant treatment to retain its effectiveness, methodologies for accelerated weathering of samples prior to fire testing were investigated. The efforts resulted in a weathering chamber and test protocol that was incorporated in ASTM D 2898 as an alternative Method B. [20, 22]

The weathering chamber was used in various investigations of fire retardant treatments for wood shingles. [20, 38]

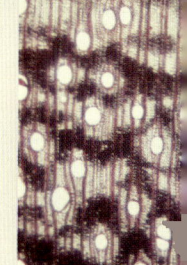

◀ Donald Fisher checking the FPL weathering chamber, which included UV exposure. (1970s) {132}

▼ Carlton Holmes and Ronald Knispel studying the effect of fire brands placed on wood shingles. (1960s) {M 130 918}

▲ Fire brands burning on fire-retardant-treated wood shingles. {57 J}

Thermal Degradation of Fire-Retardant-Treated Plywood

In the 1980s some fire-retardant treatments began causing failure of plywood roof sheathing in service. A major research effort was to understand and evaluate the ability of fire-retardant-treated wood to retain its structural integrity. The test protocols that were developed were incorporated in a series of ASTM standards that were implemented to ensure that fire-retardant-treated plywood could continue to be used as roof sheathing. [41, 42, 44, 46–49, 60–62]

Structural Integrity during a Fire

To ensure that occupants have time to escape and to limit the extent of property damage, it is important that structural members of a building maintain their structural integrity during a fire. For wood, a critical factor in the fire resistance of a structural member is charring rate. The standard ASTM E 119 defines fire exposure in testing of structural elements for fire resistance. [27, 35, 59]

◄ Illustration of the charring of wood when exposed to the standard fire exposure defined in ASTM E 119. {13 L}

In addition to early efforts to measure times for fire penetration and charring rate, large solid wood walls were tested in the large vertical furnace

Much of the work on charring rate of wood was conducted using the FPL small vertical furnace.

◄ Solid wood wall with section removed to show amount of charring. (1990s)

► Como Caldwell preparing to light the small vertical furnace.

► Small vertical furnace installed in the 2004 addition to the fire research laboratory

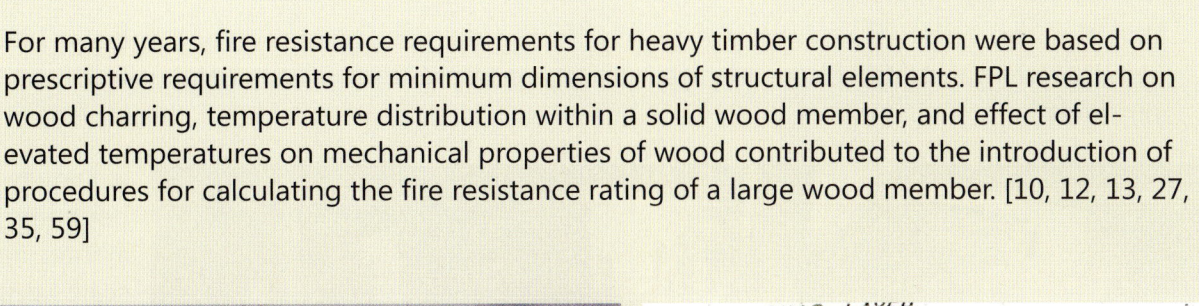

For many years, fire resistance requirements for heavy timber construction were based on prescriptive requirements for minimum dimensions of structural elements. FPL research on wood charring, temperature distribution within a solid wood member, and effect of elevated temperatures on mechanical properties of wood contributed to the introduction of procedures for calculating the fire resistance rating of a large wood member. [10, 12, 13, 27, 35, 59]

▲ Illustration of charred glued-laminated member.

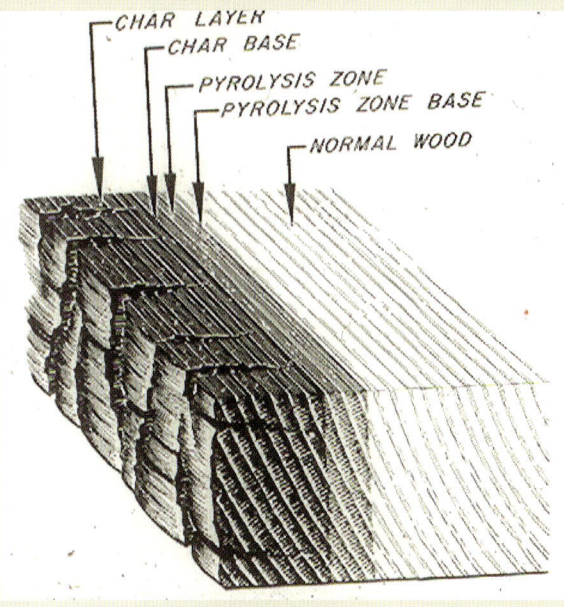

▲ Char zones in wood.

Fire resistance of light-frame construction has also been a large part of the FPL fire research program. This has included research on fire performance of innovative products for structural applications. Work in the 1970s involved sandwich panels. Tests in the standard wall furnace were followed by tests of actual structures with loaded walls and more realistic fire exposures. Research illustrated the need to add gypsum board or another protective membrane to the load-bearing sandwich panels to ensure adequate fire resistance.

◀ Wall test conducted in the large vertical furnace showing wall panel at point of fire burn through. Information of this nature is used in building designs to help ensure time for people to exit a burning structure and help contain the fire for ease of suppression. {52 K}

With the acquisition of an intermediate-scale horizontal furnace/tension apparatus in 1988, research was conducted on metal-plate-connected trusses and various composite lumber products.

◀ Dimension lumber being tested in intermediate-scale horizontal furnace/tension apparatus. (1980s)

◀ Large wood timber being fire tested in the intermediate-scale furnace while loaded in tension.

The large vertical furnace was also used to evaluate the fire resistance of wood doors. [21]

◀ Solid-core wood flush door being tested in large vertical furnace. (1970s) {19 J}

Results of Fire Safety Research

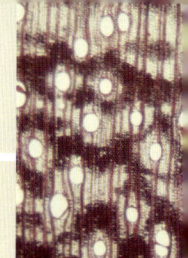

Fire–Preservation Treatment: Determining that treatment with fire-retarding ammonium salts inhibits, rather than favors, decay of wood made it unnecessary for the treating industry to make restricted use of these fire-retardant materials.

Building Codes: Many building codes are now giving wood construction a better fire rating because of FPL studies on the fire endurance of wood, flame-spread characteristics, fundamental studies of combustion, and treatments to improve fire resistance.

Fire Retardant for Wood: Ammonia-base liquid fertilizers are effective fire retardants for wood. Used in place of ammonia salts, savings of 30% to 40% in chemical and treatment costs can be realized. More economical and improved fire-retardant treatments for wood in building construction contribute greatly to enhanced safety and reduction of costs.

Improved Fire Safety of Wood Products: Greater fire safety levels can be attained through the use of chemical borate inhibitors, treating wood building materials with sodium dichromate to reduce smoke generated under certain flaming and non-flaming conditions, and using the developed test equipment to measure rate of heat release.

Improved Fire-Resistance Models: Fire-resistance models were improved for engineered wood products including trusses, I-Joists, and composite lumber products.

Adhesives: Fire performance of adhesives was characterized for structural applications.

Large Timbers: Research results helped characterize the improvement obtained by directly applying protection to large timbers.

Post-Fire Assessment: Methods were assessed for evaluating structural wood members for damage from exposure to elevated temperatures following a fire.

Flammability: The relative flammabilities of new and innovative forest products were evaluated.

Fire Growth: Existing fire growth models were modified by developing and validating specialized computer algorithms.

Leach-Resistant Fire-Retardant Treatments for Wood Shingles: Of 35 treatments studied, four showed excellent performance, both in leach resistance and in fire-retardant performance after leaching exposure, and are suitable for use on wood shingles and shakes. All acceptable treatments involve pressure impregnation of the chemical solutions into the wood and must be applied prior to installation of the shingles and shakes. Overall results of these studies indicated that treated wood shingles can meet a Class B fire rating but not Class A.

Fire-Retardant Chemicals: New and innovative forest products (including wood–plastic composites and oriented strandboards) were treated with fire-retardant chemicals and evaluated for relative flammability

Test Procedure: Procedures were developed and validated to predict the performance of forest products in the ASTM E 84 tunnel test from data obtained with smaller-scale test methods.

Ornamental Plants: Relative flammability of ornamental plants and invasive species was assessed.

Fire Propagation: Models for assessing likely fire propagation within a structure were developed.

Exterior Materials: The fire performance of new materials used for decking and other exterior applications was evaluated.

Structure Protection: Temporary measures that can be used to protect a structure were investigted.

◀ Wood burns but steel loses its strength when heated. Undated picture shows nailed wood beam holding up steel beams. Wood beams, though burning, help give people time to get out of the structure and help protect firefighters from building collapse. (19 A)

1910

Limited knowledge of the effect of fire on structural behavior of wood

2010

Based on fire research, developed and standardized tests to evaluate and help understand fire behavior and its effects on wood

Evaluated chemical fire retardants

Improved fire safety in structures

Modified building codes

Further Information

[1] Tests on the inflammability of untreated wood and of wood treated with fire-retarding compounds / R.E. Prince / National Fire Protection Association. Proceedings (1915)

[2] Making wood fire resistant with paint / FPL Technical note No. 106 (1920)

[3] A new test for measuring the fire resistance of wood / T.R. Truax, C.A. Harrison / In: ASTM proceedings Vol. 29 (II): 973–989 (1929)

[4] The use of chemicals in forest fire control / T.R.Truax / U.S. Forest Service Fire Control Notes 3(4):1–3 (Oct. 1939)

[5] Theories of the combustion of wood and its control: a survey of the literature / F.L. Browne / FPL No. 2136 (1958)

[6] Experimental dwelling-room fires / H.D. Bruce / FPL No. 1941 (1959)

[7] Fire research and results at the Forest Products Laboratory / T.R. Truax / FPL No. 1999 (Oct. 1959)

[8] Surface flammability as determined by the FPL 8-foot tunnel methods / C.C. Peters, H.W. Eickner / FPL No. 2257 (1962)

[9] Basic research on the pyrolysis and combustion of wood / Herbert W. Eickner / Forest products journal Vol. 12(4) (Apr. 1962)

[10] Effect of flame retardants on pyrolysis and combustion of [alpha]-cellulose / Walter K. Tang, Wayne K. Neill / Journal of Polymer Science: Part C, Polymer Symposia No. 6 (1964)

[11] Effect of wall linings on fire performance within a partially ventilated corridor / E.L. Schaffer / Research paper FPL 49 (1965)

[12] Review of information related to the charring rate of wood / E.L. Schaffer / FPL–RN–0145 (1966)

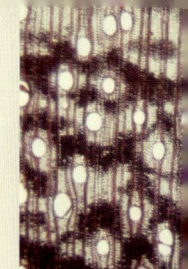

[13] Charring rate of selected woods: transverse to grain / E.L. Schaffer / Research paper FPL 69 (1967)

[14] Calorific values of the volatile pyrolysis products of wood / J.J. Brenden / Combustion and flame Vol. 11(5) (Oct. 1967)

[15] Effect of Inorganic salts on pyrolysis of wood, cellulose, and lignin determined by differential thermal analysis / W.K. Tang / FPL–RP–82 (1968)

[16] Thermogravimetric analysis of wood, lignin and hemicelluloses / F.C. Beall / Wood and fiber Vol. 1(3) (Fall 1969)

[17] Effect of fire-retardant treatment on bending strength of wood / C.C. Gerhards / FPL–RP–145) (1970)

[18] Differential calorimetric analysis of wood and wood components / F. C. Beall / Wood science and technology Vol. 5 (1971)

[19] Usefulness of a new method for measuring smoke yield from wood species and panel products / J.J. Brenden / Forest products journal Vol. 21(12) (Dec. 1971)

[20] Methods of evaluating fire-retardant treatments for wood shingles / C.A. Holmes / Forest products journal Vol. 22(3) (Mar. 1972)

[21] Fire resistance of "solid-core" wood flush doors / H.W. Eickner / Forest products journal Vol. 23(4) (Apr. 1973)

[22] Correlation of ASTM exposure tests for evaluating durability of fire-retardant treatment of wood / C.A. Holmes / Research paper FPL 194 (1973)

[23] An apparatus developed to measure rate of heat release from building materials / John J. Brenden / Research paper FPL 217 (1973)

[24] Flammability of selected wood products under Motor Vehicle Safety Standard / C.A. Holmes / Journal of fire and flammability Vol. 4 (July 1973)

[25] Fire endurance of wood-frame and sandwich wall panels / H.W. Eickner / Journal of fire & flammability Vol. 6 (2): 155–190 (Apr 1975)

[26] Wood-base building materials: rate of heat release / John J. Brenden / Journal of fire and flammability Vol. 6(3) (July 1975)

[27] State of structural timber fire endurance / Erwin Schaffer / Wood and fiber Vol. 9(2) (Summer 1977)

[28] Measurements of heat release rates on wood products and an assembly / John J. Brenden / Research paper FPL 281 (1977)

[29] Effect of fire-retardant treatments on performance properties of wood / Carlton A. Holmes / Wood technology: chemical aspects / Irving S. Goldstein, editor. Washington, D.C.: American Chemical Society, ACS symposium series No. 43 (1977)

[30] Room corner-wall fire tests of some structural sandwich panels and components / Carlton A. Holmes / Journal of fire and flammability Vol. 9 (Oct.1978)

[31] Fire performance of structural flakeboard from forest residue / Carlton A. Holmes, Herbert W. Eickner, John J. Brenden, Robert H. White / FPL–RP–315 (1979)

[32] Smoldering initiation in cellulosics under prolonged low-level heating / E.L. Schaffer / Fire technology Vol. 16(1) (Feb 1980)

[33] Fire development and wall endurance in sandwich and wood-frame structures / Carlton A. Holmes, Herbert W. Eickner, John J. Brenden, Curtis C. Peters, Robert H. White / Research paper FPL 364 (1980)

[34] Efficient application of boron fire retardant to cellulosic loose-fill insulation / T.H. Wegner, C.A. Holmes / Thermal insulation, materials, and systems for energy conservation in the '80s. Philadelphia, PA: American Society for Testing and Materials, ASTM STP; 789 (1983)

[35] Structural fire design: wood / E .L. Schaffer / Research paper FPL 450 (1984)

[36] Chemistry of fire retardancy / S.L. Levan / Chemistry of solid wood / Roger M. Rowell, ed. Washington: American Chemical Society, Advances in chemistry series 207 (1984)

[37] Performance of fire retardants and fire-resistive coatings on wood / S.L. Levan / Presented at The Fire Retardant Chemicals Association And The National Paint And Coatings Association fall meeting Pinehurst, NC, Oct 27–30, 1985 Flame retardant coatings: problems and opportunities. Lancaster, PA: Fire Retardant Chemicals Association (1985)

[38] Effectiveness of fire-retardant treatments for shingles after 10 years of outdoor weathering / Susan L. LeVan, Carlton A. Holmes / Research paper FPL 474 (1986)

[39] Heat release measurement of wood products using the Ohio State University apparatus / Hao C. Tran / Proceedings of the 13th international conference on fire safety: January 11–15, Millbrae, California. Sunnyvale, Calif.: Product Safety Corporation (1988)

[40] Rates of heat and smoke release of wood in an Ohio State University calorimeter / Hao C. Tran / Fire and materials Vol. 12 (1988)

[41] Choosing and applying fire-retardant-treated plywood and lumber for roof designs / Susan LeVan, Mary Collet / FPL–GTR–62 (1989)

[42] Effects of fire retardant treatments on wood strength: a review / Susan L. LeVan, Jerrold E. Winandy / Wood and fiber science Vol. 22(1) (Jan. 1990)

[43] Modifications to an Ohio State University apparatus and comparison with cone calorimeter results / H.C. Tran / Heat and mass transfer in fires: presented at AIAA/ASME thermophysics and heat transfer conference, June 18–20, 1990, Seattle, Washington. New York, N.Y.: American Society of Mechanical Engineers (1990)

[44] Thermal degradation of fire-retardant-treated plywood: development and evaluation of a test protocol / Jerrold E. Winandy, S.L. LeVan, R.J. Ross, S.P. Hoffman, C.R. McIntyre / Research paper FPL 501 (1991)

[45] Wall and corner fire tests on selected wood products / Hao C. Tran, Marc L. Janssens / Journal of fire sciences Vol. 9 (Mar./Apr. 1991)

[46] In-place evaluation of fire-retardant-treated plywood / Robert J. Ross, John Cooper, Zhitong Wang. Proceedings of the 8th International nondestructive testing of wood symposium; 1991 September 23–25, Vancouver, WA. Pullman, WA: Washington State University (1992)

[47] Combined effects of fire-retardant treatments and extended exposure to elevated temperature / Jerrold E. Winandy, Susan L. LeVan, Robert J. Ross / American Wood-Preservers' Association. Proceedings of the 87th annual meeting Woodstock, MD: AWPA (1992)

[48] Effects of fire retardant treatments after 18 months of exposure at 150° F (66° C) / Jerrold E. Winandy / FPL–RN–0264 (1995)

[49] Mechanical properties of fire-retardant-treated plywood after cyclic temperature exposure / Susan L. LeVan, Jong Man Kim, Robert J. Nagel, James W. Evans / Forest products journal 46(5): 64–71. (1996)

[50] Flammability of christmas trees and other vegetation / Robert H. White, Denise DeMars, Mark Bishop / In: Proceedings, 24th international conference on fire safety; July 21–24; Columbus, Ohio. Sissonville, WV: Product Safety Corporation: 99–110. (1997)

[51] Modeling the char behavior of structural timber / P.W.C Lau, I. Van Zeeland, R. White / In: Proceedings of the 5th international conference of fire and materials '98; 1998 February 23–24; San Antonio, TX. London, United Kingdom: Interscience Communications Ltd.: 123–135. (1998)

[52] Comparison of test protocols for standard room/corner tests / Robert H. White, M.A. Dietenberger, H. Tran, O. Grexa, L. Richardson, K. Sumathipala, M. Janssens / Fire and materials '98: 5th international conference, February 1998, San Antonio, TX, USA. London: Interscience Communications (1998)

[53] Flammability of native and ornamental plants for homes in WUI (Wildland/urban interface) / Robert H. White, David R. Weise, Susan Frommer / Biographies & Abstracts, durability and disaster mitigation in wood frame housing: Nov. 1–2, 1999, Madison, WI. Madison, WI: Forest Products Society (1999)

[54] Wildland/urban interface fire research at the USDA Forest Service, Forest Products Laboratory: past, present, and future / Robert H. White / Proceedings of the international conference on fire safety Vol. 30, (2000)

[55] Fire performance of hardwood species / Robert H. White / In: Forests and society: the role of research. Proceedings, 11th IUFRO World Congress; 2000 August 7–12; Kuala Lumpur, Malaysia (2000)

[56] Fire retardancy of wood treated with inorganic salts after exposure to elevated temperatures / Robert H. White, Melissa D. Hill / Proceedings of the 2nd annual conference on durability and disaster mitigation in wood-frame housing: Nov. 6–8, 2000 Madison, WI: Madison, WI : Forest Products Society, 2000: p. 243 (2000)

[57] Update for combustion properties of wood components / Mark Dietenberger / In: Proceedings, 7th international conference, fire and materials 2001 conference; 2001 January 22–24; San Francisco, CA. London, UK: Interscience Communications: 159–171 (2001)

[58] Reaction-to-fire testing and modeling for wood products / Mark A. Dietenberger, Robert H. White / Twelfth annual BCC conference on flame retardancy: May 21–23, 2001, Stamford, CT. Norwalk, CT: Business Communications Co. p. 54–69 (2001)

[59] Analytical methods for determining fire resistance of timber members / Robert H. White / SFPE handbook of fire protection engineering. 3rd ed. Quincy, Mass.: National Fire Protection Association; Bethesda, MD: Society of Fire Protection Engineers p. 257–273. (2002)

[60] Evaluation of a boron-nitrogen, phosphate-free fire-retardant treatment. Part I. Testing of Douglas-Fir plywood per ASTM Standard D 5516-96 / Jerrold E. Winandy, Michael J. Richards / Journal of testing evaluation Vol. 31(2):133–139 (2003)

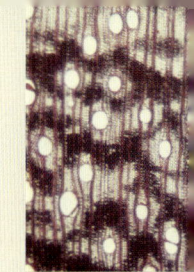

[61] Evaluation of a boron-nitrogen, phosphate-free fire-retardant treatment. Part II. Testing of small clear specimens per ASTM Standard D 5664-95, Methods A and B / Jerrold E. Winandy. Douglas Herdman / Journal of testing evaluation Vol. 31(2): 140–147. (2003)

[62] Evaluation of a boron-nitrogen, phosphate-free fire-retardant treatment. Part III. Evaluation of Full-Size 2 by 4 Lumber per ASTM Standard D 5664-95 Method C / Jerrold E. Winandy, William McNamara / Journal of testing evaluation Vol. 31(2): 148–153. (2003)

[63] Mechanical properties of gypsum board at elevated temperatures / S.M. Cramer, R.H. White / Fire and materials 2003: 8th international conference, January 2003, San Francisco, CA, USA. London : Interscience Communications Limited p. 33–42 (2003)

[64] Fire resistance of engineered wood rim board products / Robert H. White / FPL–RP–610 (2003)

[65] Cone calorimeter evaluation of wood products / Robert H. White, Mark A. Dietenberger / Proceedings of the conference on recent advances in flame retardancy of polymeric materials: volume XV, applications, research and industrial development, markets, held in Stamford, Connecticut, June 6–9, 2004. Norwalk, CT: Business Communications Co., p. 331–342 (2004)

[66] Chemical mechanism of fire retardance of boric acid on wood / Qingwen Wang, Jian Li, Jerrold E. Winandy / Wood Science and Technology Vol. 38: 375–389 (2004)

[67] Use of the cone calorimeter to detect seasonal differences in selected combustion characteristics of ornamental vegetation / David R. Weise. R.H. White / International journal of wildland fire Vol. 14: 321–328. (2005)

[68] Fire resistance of structural composite lumber products / Robert H. White / FPL–RP–633 (April 2006)

[69] Fire performance of oriented strandboard / Robert H. White, Jerrold E. Winandy / Proceedings of the conference on recent advances in flame retardancy of polymeric materials: volume XVII, Applications, research and industrial development, markets. Norwalk, CT: BCC Research, (2006)

[70] Fire resistance of structural composite lumber products / Robert H. White / FPL–RP–631 (2006)

[71] Flammability tests for regulation of building and construction materials / K. Sumathipala, R. H. White / Flammability testing of materials used in construction, transport and mining. Cambridge, England: Woodhead Publishing; Boca Raton: CRC Press, c2006. Woodhead Publishing in materials: p. 217–230 (2006)

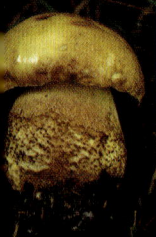

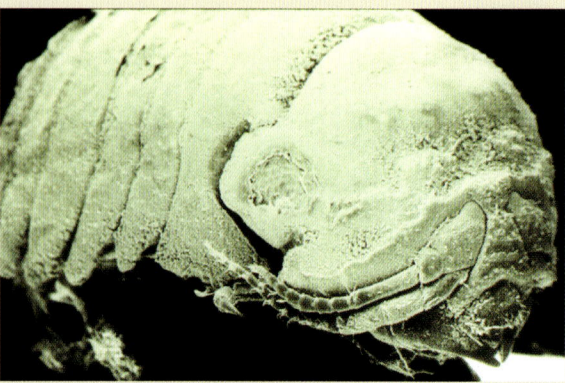

◀ Termite enjoying a tasty piece of wood. {16 J}

▼ Pill Bug, a type of marine borer.

Insects, such as termites, invade more than 600,000 homes and cause over $1,500,000,000 in damage annually.

Marine borers, small invertebrate sea animals that burrow into wood for food and shelter, destroy several hundred million dollars worth of waterfront structures each year.

Challenge: A major challenge is how to extend the life of wood products exposed to insects and marine borers.

FPL's Contributions: Research has focused on methods of attracting insects, especially termites, to a toxic compound that they will take back to their colony, thus destroying the infestation. Researchers found a compound in brown-rot fungi that attracts termites and coupled it with a species-specific insecticide to help control them.

Understanding the different types of marine borers has led to the development of specific types of effective preservative treatments.

Results: Extended service life of buildings, pilings, and other wood structures, which has decreased costs of damage, repair, and replacement.

Termites

The attractant found in brown-rot-decayed wood is now known to be the same chemical compound that termites use to mark trails.

◀ Glenn Esenther probes the termite-damaged porch of a building. Experiments with a poison lure were conducted in an attempt to protect the building and exterminate the insects. (1960s) {M 133 347 -7} [1]

▲ *Reticulitermes flavipes* is a species of termite that has taken up residence in Wisconsin and is apparently surviving cold winters and summer dry spells. Upper left, Glenn Esenther inspects experimental stakes set out around a house to test their protective value; the stakes were treated with an insecticide and an attractant he and Raymond Beal (Forest Service Southern Station) developed. Lower left, Glenn Esenther collects termites from a stake of untreated wood. Right, enlarged portion of stake shows termites scurrying into cavities they bored in wood. (1960s) [4]

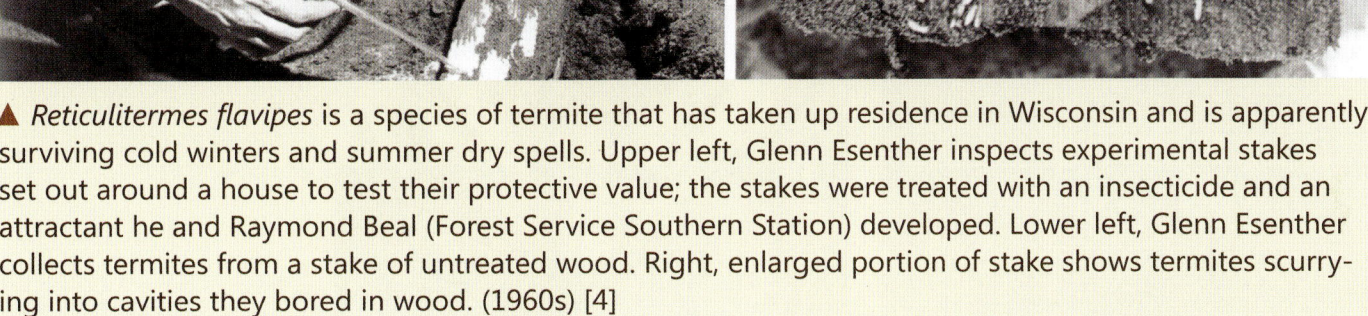

▲ Comparison of a termite and an ant. {79 B}

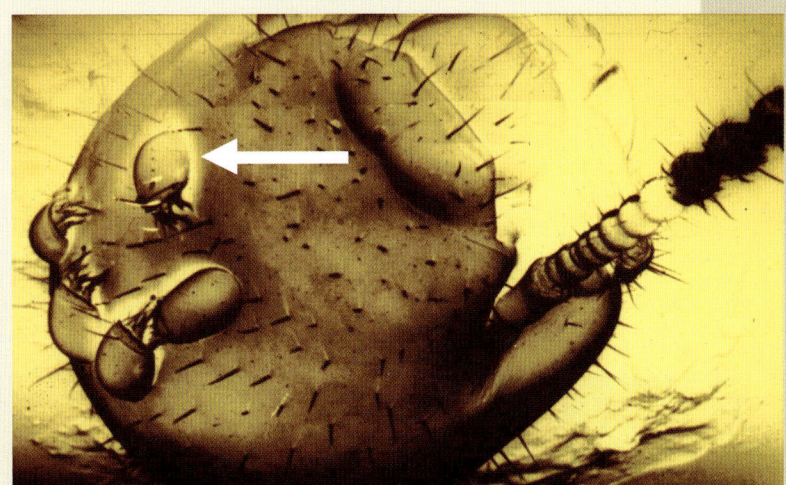

▲ Electron microscopic view of mites on the head of a termite. {I G}

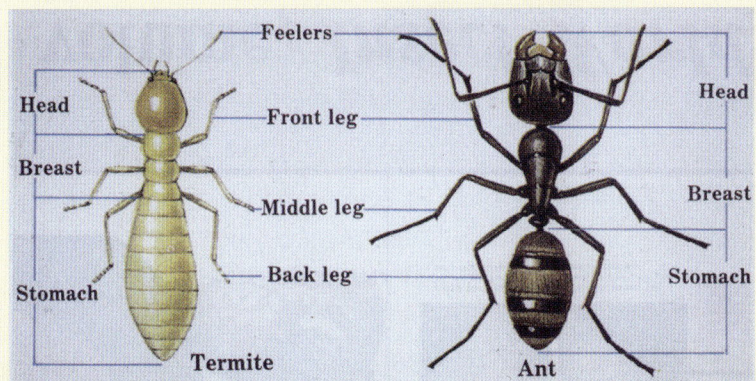

◄ Glenn Esenther and Raymond Beal found that a chemical given off by a brown rot fungus will attract termites. From this they developed a system using small wood blocks treated with the fungus and a small amount of insecticide to help protect buildings from termites. The system uses much less insecticide than conventional termite prevention measures and may be more effective because it destroys the termite colony. (1960s) {M 125 394} [6]

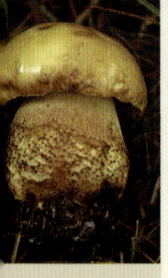

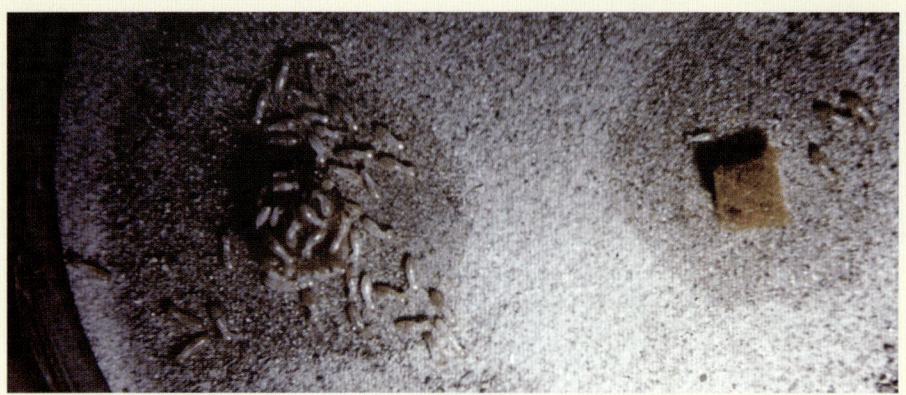

◄ Experiment demonstrating the lure of the brown rot compound. (1960s) {M 118 548}

◄ Insecticide-treated corrugated fiberboard trap. (1960s) {M 127 769}

◄ The corrugated fiberboard trap was eaten by termites. (1960s) {M 127 770}

FPL scientists investigated and perfected the mixing of dieldrin and concrete in a mixture that is fatal to termites in contact with the dried concrete. This preparation is excellent for termite protective construction in foundations and is not poisonous to humans or animals. (1960s) [2]

◄ Corrugated fiberboard being eaten by termites. Corrugated fiberboard is one of the termite's preferred foods. (2000s)

▼ Rachel Arango identifying insect specimens from indoor infestations. (2000s)

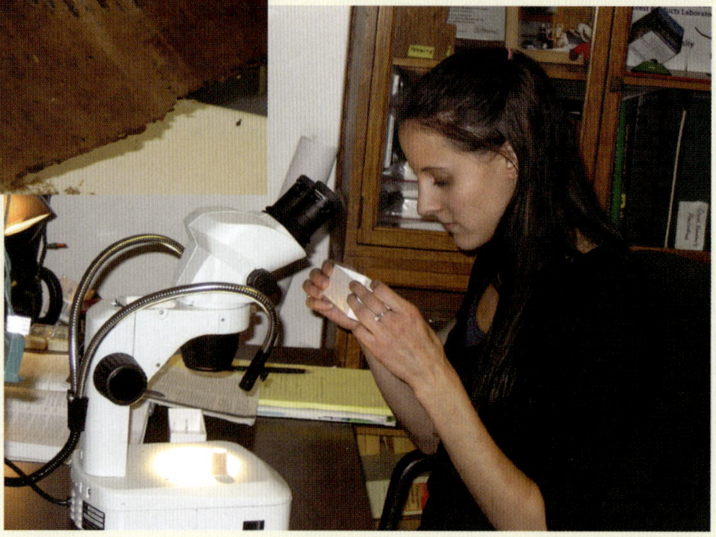

▲ Rachel Arango and Frederick Green extracting termites from corrugated fiberboard traps for a termite experiment. (2000s)

Environmentally friendly termiticide: A slow-acting termite bait that uses the compound N-hydroxynaphthalimide (NHA) as the toxicant was jointly developed by FPL and Agriculture Research Service researchers. Unlike most termiticides, NHA contains no heavy metals, which decreases its effect on the environment. The compound was combined with a cellulose matrix to attract termites, especially the highly destructive Formosan subterranean termites, which tend to avoid existing termite baits. (U.S. Patent No. 6,691,453 B1) [21]

Marine Borers

For more than a century, marine borers have caused much damage to even preservative-treated waterfront structures such as piers, bulkheads, seawalls, and fender pilings.

◀ Damage caused by crustacean marine borers. *Liminoria* (gribble) on the left and *Sphaeroma* (pill bug) on the right. (1990s) {78 C} [13]

▼ Pill bugs (*Sphaeroma terebrans*) burrowing in untreated pine. (1990s)

◀ CCA-treated southern pine pile extensively damaged by pill bugs during 9 years of exposure in brackish water at Tarpon Springs, Florida. (1990s)

▲ Extensive shipworm damage of treated marine piling resulting from inadequate preservative retention. (1990s)

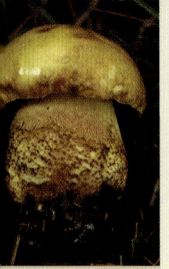

▼ Shipworms (*Bankia gouldi*) have riddled the interior of this 2 by 4, lining their tunnels with white calcium carbonate as they burrow. Several of the worms are shown. Note the rounded bivalve shells at their "head," which they use to rasp the wood. (1990s)

▲ Shipworms destroyed this untreated pine 2 by 4 in about 5 months in Chesapeake Bay. The only external evidence of the borers is the tiny white protrusion (near the tip of the pen. The interior of this 2 by 4 is shown in the next photo. (1990s)

Pine specimen treated with CCA was still okay after 7 years.

Marine Borer Research

FPL's research helped our understanding of the distribution and preservative tolerance of marine borers around the country and helped define the hazard. Many different organisms can have a negative effect on wood structures, and different preservative treatments provide different levels of protection. [9, 14, 15]

FPL research on treatment problems has explained early failures of marine timbers and led to solutions. [8, 12]

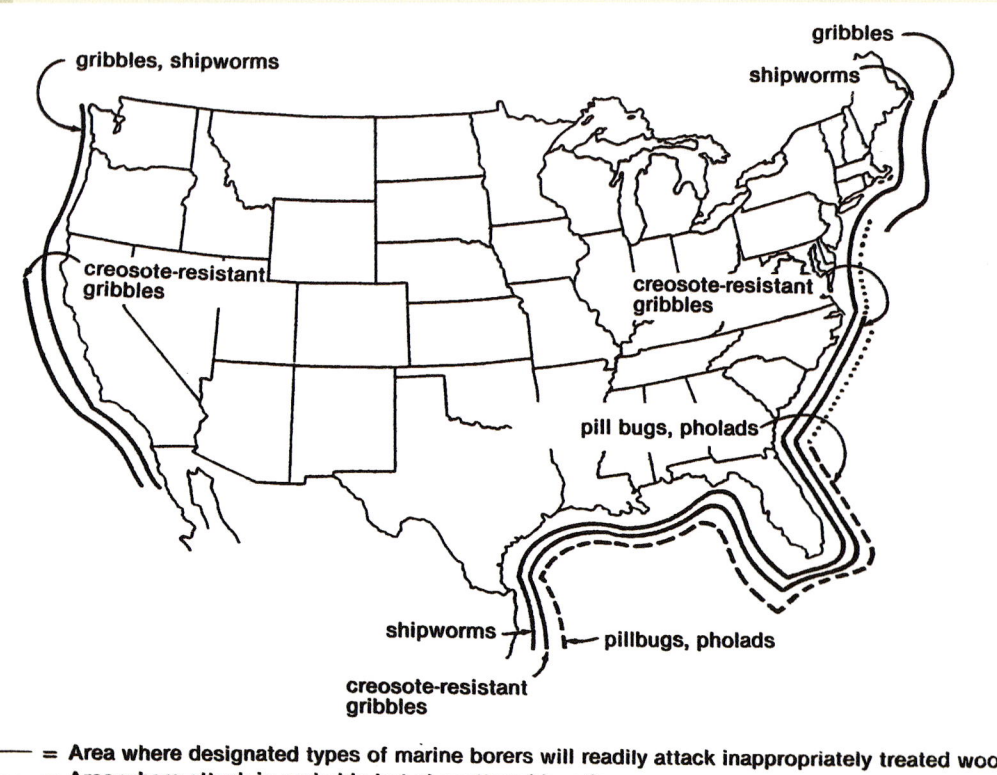

◀ Map of the United States showing the location of marine borer threats. (1987) {M 875 498}

gribbles, shipworms

gribbles

shipworms

creosote-resistant gribbles

creosote-resistant gribbles

pill bugs, pholads

shipworms — pillbugs, pholads

creosote-resistant gribbles

—— = Area where designated types of marine borers will readily attack inappropriately treated wood.
- - - - = Area where attack is probable but at scattered locations.
······· = Transition areas where problems with designated borers are unlikely but could occur.

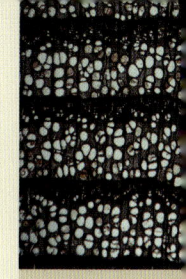

Research on conventional and unconventional preservatives has defined what preservatives, and combinations thereof, are most effective in protecting against marine borers. [10, 11, 13]

Borer damage to wooden marine structures can be prevented, but it requires knowing what types of borers are present at a particular site, selecting the proper preservative, and obtaining sufficient depth of preservative penetration and retention to protect the structure from natural surface erosion. [14]

Preservative treatments for lumber and timber exposed to marine borers (1990s) [14]

| | Effectiveness of preservative treatments[a] | | |
Marine borer type	Creosote	Water-borne preservative (CCA, ACA, ACZA)[b]	Dual treatment[c]
Shipworms (teredinids)	+	+	+
Gribbles (*Limnoria* spp.)			
L. quadripunctata or *lignorum*	+	+	+
L. tripunctata	Not recommended	+	+
Pholads	+	Not recommended	+
Pill bugs (*Sphaeroma*)	Questionable	Not recommended	+

[a]See American Wood-Protection Association UI–08 for recommended types of preservative treatments, penetration, and retention specifications for marine applications.
[b]CCA, chromated copper arsenate; ACA, ammoniacal copper arsenate; ACZA, ammoniacal copper zinc arsenate.
[c]Water-borne preservative followed by creosote.

Risk Assessment of the Importation of Pests into the United States

A major research program to identify pests that could invade the United States and mitigation of these potential threats was undertaken by a team of Forest Service scientists that included Harold Burdsall of the FPL. Working with colleagues in other countries, the team identified potential pests, evaluated their potential threat to the United States, and explored methods of mitigating the threat. An example of such a potential threat developing into a serious problem was the importation of the Asian longhorn beetle on wood used in pallets. With no known natural enemy in the United States, the beetle destroyed thousands of trees until massive clear-cutting in infested areas appears to have contained it. The gypsy moth is presently spreading from east to west across the United States, and the emerald ash borer threatens to wipe out ash trees across the country. This team of researchers is assessing the potential threat of other pests and working on ways to mitigate the threat. [16–18, 20].

With the increase in international trade this research is critical to the future viability of the trees in the United States.

1910

Little understanding of the life cycles of pests that destroy wood.

Few control methods or treatments existed.

2010

Understanding of the life cycles of some wood-destroying pests greatly improved.

Effective lures, treatments, and methods of control developed.

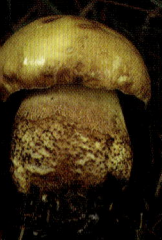

Further Information

[1] Termite attractant from fungus-infected wood / G.R. Esenther, T.C. Allen, J.E. Casida, R.D. Shenefelt / Science Vol. 134(3471) (July 7, 1961)

[2] Toxicity of Dieldrin-concrete mixtures to termites / T.C. Allen, G.R. Esenther, E.P Lichtenstein / Journal of economic entomology 57(1) (1963)

[3] Effectiveness following kiln-drying of insecticides applied to green lumber to control Lyctus powder-post beetle attack / Glenn R. Esenther / Forest products journal Vol. 14(10) (Oct. 1964)

[4] Termites in Wisconsin / G.R. Esenther / Ann. Entomology Society of America 62(6) (1969)

[5] Nutritive supplement method to evaluate resistance of natural or preservative-treated wood to subterranean termites / Glenn R. Esenther / Journal of economic entomology Vol. 70(3) (June 1977)

[6] Termite control: decayed wood bait / G.R. Esenther, R.H. Beal / Sociobiology 4(2) (1979)

[7] Efficacy of avermectin-silica carrier dust formulations to suppress Reticulitermes flavipes (Kollar) in simulated gallery injection treatments / Glenn R. Esenther / Forest Products Laboratory (1985)

[8] Protection of timber bulkheads from marine borers / B.R. Johnson / Timber bulkheads: proceedings, Geotechnical Engineering Division of the American Society of Civil Engineers: Atlantic City, N.J., Apr. 29,1987 / Geotech. Spec. Pub. No. 7. New York, American Society of Civil Engineers (1987)

[9] Sphaeroma terebrans Bate: a note on distribution and preservative tolerance in Florida coastal waters / B.R. Johnson, E.D. Estevez, S.A. Rice / Eighteenth meeting of the International Research Group on Wood Preservation (Working Group IV: Marine Wood Preservation): Honey Harbour, Ont., Can., Document No. IRG/WP/4135 (May 17–22, 1987)

[10] Resistance of chemically modified wood to marine borers / Bruce R. Johnson, Roger M. Rowell / Material und organismen Vol. 23(2) (1988)

[11] Comparison of preservative treatments in marine exposure of small wood panels / Bruce R. Johnson, David I. Gutzmer / FPL–RN–0258 (1990)

[12] Durability of heartwood in treated southern pine bulkheads / Bruce Johnson, Roy Jackson / Forest products journal Vol. 40(7/8) (July/Aug. 1990)

[13] Protection of wood from marine borers / Bruce R. Johnson, Stanley A. Rice / Proceedings. 1990 National Marine Research Conference. Wickford, RI: International Marina Institute (1990)

[14] Marine Borers / B.R. Johnson / FPL Techline (1990s)

[15] Relative tolerance of CCA by larvae and adults of the common shipworm, Bankia gouldi / Bruce R. Johnson, Stan T. Lebow / Material und organismen Vol. 30(1) (1996)

[16] Pest risk assessment of the importation into the United States of unprocessed Pinus and Abies logs from Mexico / Borys M. Tkacz, Harold H. Burdsall, Jr., Gregg A. DeNitto, Andris Eglitis, James B. Hanson, John T. Kliejunas, William E. Wallner, Joseph G. O'Brien, Eric L. Smith / FPL–GTR–104 (1998)

[17] Heating times for round and rectangular cross-sections of wood in steam / William T. Simpson / FPL–GTR–130. (2001)

[18] Heat sterilization time of ponderosa pine and Douglas-Fir boards and square timbers / William T. Simpson, Xiping Wang, Steve Verrill / FPL–RP–607 (2003)

[19] Decay and termite resistance of medium density fiberboard (MDF) made from different wood species / Nami S. Kartal, Frederick Green III / International Biodeterioration & Biodegradation 51: 29–35. (2003)

[20] Pest risk assessment of the importation into the United States of unprocessed logs and chips of eighteen Eucalypt species from Australia / John T. Kliejunas, Harold H. Burdsall, Jr. Gregg A. DeNitto, Andris Eglitis, Dennis A. Haugen, Michael I. Harverty, Jessie A. Micales, Borys M. Tkacz, Mark R. Powell / FPL–GTR–137 (2003)

[21] Naphthalenic compounds as termite bait toxicants / Maria Guadalupe Rojas, Juan A. Morales-Ramos, Frederick Green III / U.S. Patent No. 6,691,453 B1 (Feb 17, 2004)

[22] Resistance of borax-copper treated wood in above ground exposure to attack by Formosan subterranean termites / Stan Lebow, Bessie Woodward, Douglas Crawford, William Abbott / FPL–RN–0295. (2005)

[23] Naural durability of tropical and native woods against termite damage by Reticulitermes flavipes (Kollar) / Rachel A. Arango, Frederick Green III, Kristina Hintz, Patricia K. Lebow, Regis B. Miller / International Biodeterioration & Biodegradation 57 (2006)

National Defense

◀ Hand shaping airplane propellers at FPL during World War I. {M 14156 F}

Documenting the complete accomplishments and contributions of the FPL to national defense would require several volumes. This section provides a brief overview of some these accomplishments in the context of the crucial role wood plays in war. For example, "The demand for forest products during World War II was truly insatiable. Wood in the form of lumber, plywood, paper, plastics, and other materials appeared in countless war uses. Some 25,000 trainer aircraft and gliders were made of wood and plywood." [25] "Wood was used in great quantities to build fighting ships including minesweepers, submarine chasers, PT boats, and even battleships, not to mention the swarms of landing craft so important for amphibious invasion. Each minesweeper and submarine chaser contained enough timber to build ten average houses while the famous PT boat—constructed of spruce keels, mahogany planking, and plywood hulls—used 28,000 board feet of wood. The decking for the average battleship consumed 200,000 board feet of lumber and in the construction of the "Liberty" ship, nearly 700,000 board feet of lumber were used in shipway, staging, and scaffolding." [49]

"Three hundred thousand prefabricated dwelling units—together with vast numbers of conventional construction—were built largely of wood and plywood to house the multitudes of war workers that thronged the production centers of the nation, while many thousands of other wood structures were erected at military encampments around the country and abroad. Government statistics reveal that it required 1,400 board feet of lumber to house each fighting man, 300 feet to send him overseas, and 50 feet per month to keep him supplied." [20] "A military officer reported in 1944 that 61,547 tons of lumber were needed to land 100,000 men on a typical Pacific island." [21] "To restore the port of Naples, Italy, for temporary use by the allies required 50,000,000 board feet of lumber—equivalent of 2,000 train carloads." [19]

"Even more surprising was the tremendous quantity of lumber required for packaging and materials of war. When it is realized that over 700,000 different military items had to be shipped overseas for the North African campaign—most of these items being shipped in boxes, crates, or paper cartons—the enormity of the packaging problem becomes clear. Lumber requirements for boxing and crating increased steadily from 1942 to 1944; in 1944

nearly 17 billion board feet of lumber—nearly half the total lumber consumption—was consumed in domestic and military packaging." [20] "The amount of lumber required to package certain military items was staggering indeed. For example, each 105-millimeter howitzer took 711 board feet, each 40-millimeter Bofors anti-aircraft gun required 1,040 feet, while each giant bomber shipped overseas consumed 5,000 board feet of blocking and crating lumber." [49]

"Chemical utilization of wood accounted for additional consumption. The nationwide rationing program required ration books for every man, woman, and child in the United States. The manufacture of cellulose compounds for explosives, plastics, and other products, and spinning great quantities of rayon for textiles, and the like, ate voraciously into the nation's pulpwood supplies." [49] "Under Secretary of War Robert P. Patterson expressed the vital military role of wood when he declared in 1943 that, 'lumber comes close to the heart of the whole war problem. There are 1,200 different items of military and naval equipment that can use lumber and each day we find new and important ways to use wood in our weapons'." [15]

Challenge: With the tremendous demand for wood in national defense, decreasing the amount of wood used and improving the performance of wood in its many uses are critically important.

FPL's Contributions: Through improved design and use of new wood species, FPL researchers decreased the weight of aircraft, boxes, and crates; improved the drying processes of wood; and developed a process for making ethanol from wood. This research was transferred to the military by conducting courses, developing test methods, and preparing specifications, manuals, and handbooks.

Results: Significantly contributed to the defense of the country by extending the wood supply, providing safer aircraft, and reducing damage to food and equipment when shipped around the world.

World War I (1910s)

National defense activities of FPL during WW I and after can be divided into four general research areas: aircraft, packaging, drying, and chemistry. [1, 3, 8, 11, 49]

Aircraft

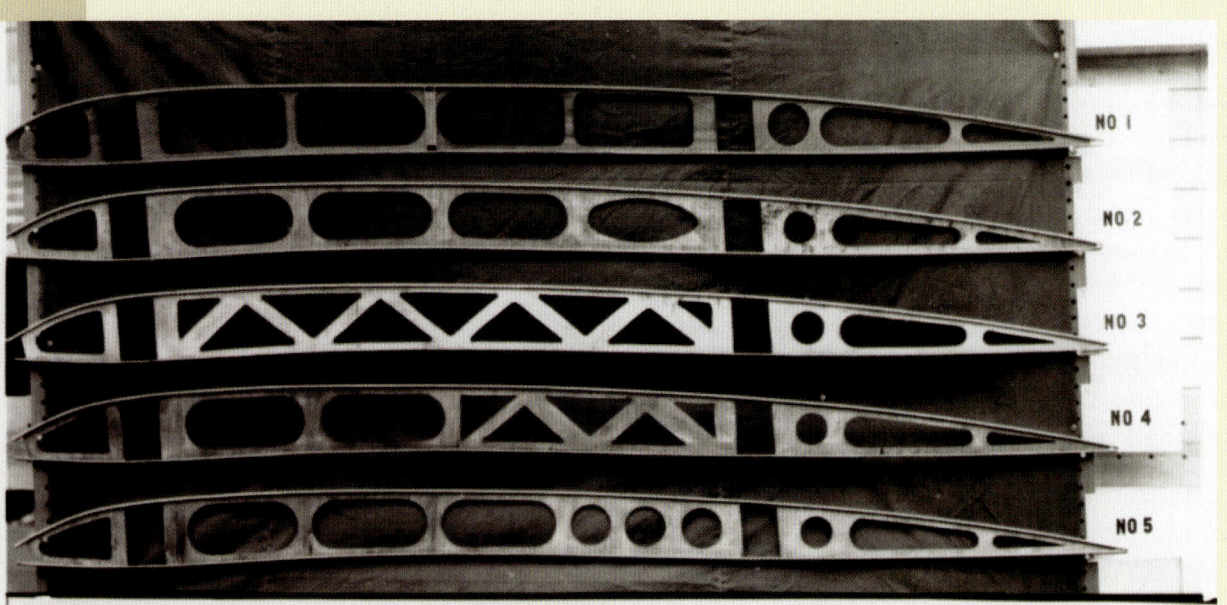

◀ Different designs of wing ribs that were evaluated at the FPL to determine their strength. The top rib (No. 1) was made of birch and yellow poplar. The other four were made of combinations of maple, yellow poplar, Spanish cedar, or basswood—all were lighter than No. 1, but stronger. {6733M}

The use of wood in aircraft has been studied from the beginning of flight. The Wright brothers experimented with various designs, and as the industry grew and demands for new and higher performance planes increased, improving the use of wood was essential. The FPL contribution was focused on proper selection and drying of spruce, acceptable use of substitute species, and engineering design and testing of aircraft parts such as ribs, struts, and propellers. FPL's research on wood propellers, aircraft design, wood manufacturing processes, and drying substantially contributed to the proper use of wood in aircraft. [49]

RECEIVING YARD — SPRUCE SUBSTITUTES

AIRPLANE PARTS LABORATORY

MAIN LABORATORY

AIRPLANE WOODS LABORATORY

EXPERIMENTAL PROPELLER FACTORY

WAR QUARTERS OF THE FOREST PRODUCTS LABORATORY

▲ Expansion of FPL activities during WW I using buildings on the University of Wisconsin campus, demonstrating the close cooperation and support of the University. Left to right: FPL log yard, UW Education building, UW FPL building, UW Agriculture Engineering building, UW Soils building. Much of the extra space was needed for propeller production. {6900M}

◄ Some of the staff involved in propeller research. (1916)

◄ FPL research on propeller blades that resist warp, twist, and unbalancing with changes in humidity. [7, 49]

▲ Manufacture of propellers. Ten propellers were produced per week by workers on a three-shift-per-day schedule.

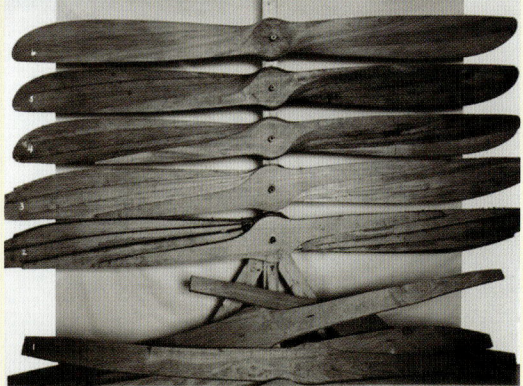

▲ Experimental propellers produced from seven species of wood.

Packaging

WW I military operations required safe and efficient shipping of materiel throughout the world, and packaging research at the FPL greatly assisted in this effort. Building on previous research on mechanical strength of wood, engineers were able to design more efficient containers, thus reducing weight and volume, yet providing adequate product protection. [4, 5, 9, 49]

Box Construction

FPL scientists studied the usefulness of species other than white pine, the preferred species, for use in wood boxes. Also studied was the optimum number of nails and kinds of nails to maximize box performance.

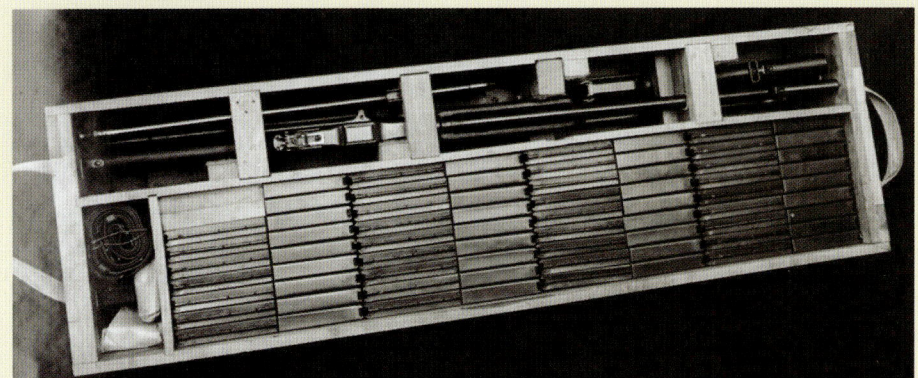

◀ Proposed box for Browning automatic machine rifles and equipment. {6956M}

Container Testing

▲ Class observing the use of the revolving drum test, developed by the FPL to simulate rough handling of containers in transport. Based on this test, weak points of the container were determined and the container was redesigned. {M 936 F}

Crates

▲ Wood crates showing the three-way corner and cross bracing. Crates of the types shown, with either block reinforcement or diagonal braces nailed to the sides of edge members, will withstand approximately twice as much load in diagonal compression as crates with diagonal braces nailed only to the edges of the frame members. {M 3345 F}

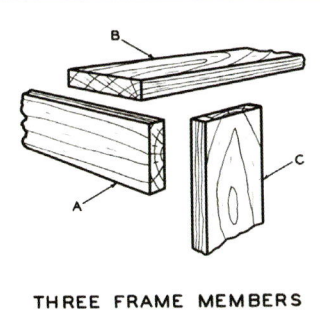
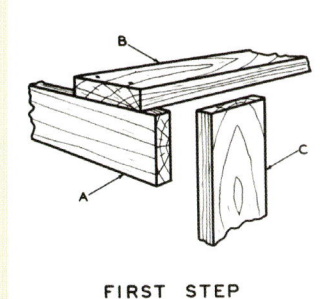
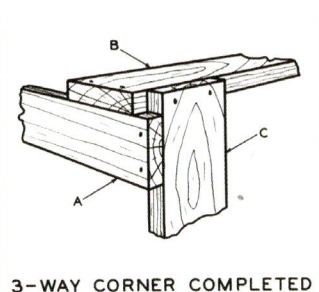

THREE FRAME MEMBERS   FIRST STEP   3–WAY CORNER COMPLETED

◀ Diagram of a three-way corner for maximizing the strength of the crate. Simple diagrams like this were used in manuals supplied to the military for field use. {M 52249 F}

FPL researchers designed many basic sheathed and unsheathed crates for the military services that have been adopted by commercial shippers of heavy industrial products. In most instances, adoption of the engineered designs has resulted in greater strength and rigidity, lower weight, small cubic displacement, and use of less material than for the previous designs. [10]

This early research for the military provided the basis for improving performance of commercial packaging.

Wood Drying

Proper drying of wood was vital to reducing wood waste and improving performance. [2] By adopting FPL dry kiln technology, drying waste was decreased from 60% to less than 5%. Rigid requirements for drying wood for military use taught much about wood drying that aided later commercial drying. [49]

◄ Improper drying of wood parts used in military vehicles, wagons, cannon carriages, gun stocks, etc., led to large rejections of wood and tremendous waste. In one instance, 60,000 walnut gun stocks were lost because of improper drying. FPL explored the use of salt treatments and developed faster and better drying schedules for drying wood, thus saving wood and money. {Library of Congress, Prints & Photographs Division}

Chemistry

Alcohol

FPL helped meet the demand for alcohol through research on increasing the yield of ethyl alcohol from dilute acid hydrolysis and fermentation of wood. FPL scientists determined the yields of ethyl alcohol that could be obtained by dilute acid hydrolysis and fermentation from 24 species and the optimum cooking conditions for a single-cook process. This information led to the successful operation of two commercial plants during the war. Research was also conducted on gunpowder and charcoal. [49]

▲ To meet the need for an alternative to cotton linters for manufacturing nitrocellulose, FPL developed several types of nitrated wood pulp cellulose for use by the artillery. [6] {Library of Congress, Prints & Photographs Division}

◄ FPL studied charcoal for absorbing chlorine gas used in chemical warfare. Coconut charcoal was found to be the most effective, followed by charcoal from hydrolyzed wood waste. Common wood charcoal was ineffective. [49] {Library of Congress, Prints & Photographs Division}

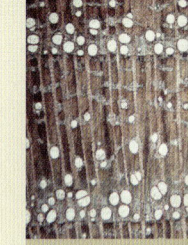

Batteries

FPL developed a low-cost, general-purpose plastic molding powder from wood waste. This type of material, which had exceptional acid resistance, was used for certain parts of storage batteries going to the Army and Navy. [49]

World War II (1940s)

In WW II, just as in WW I, the FPL was called on to assist in the defense of the nation. Emphasis was again on aircraft and packaging, with an added emphasis on development of new combinations of materials to ease the shortage of steel. [14, 49]

Aircraft

The use of wood was noteworthy in its application to the mosquito bomber and gliders of WW II. Working closely with industry, universities, and associations, major advances were made in adhesives, proper wood usage through design, and significant reductions in weight, volume, and waste. [12, 13, 16–18, 29–31]

Compreg

FPL developed a resin-treated, laminated, compressed wood (compreg) of greatly improved moisture resistance and dimensional stability. This material was produced for use in propellers, drill jigs, antenna masts, and other products. [23, 49]

◀ Assembly of veneers built up for pressing into graduated-density compreg wood—25 plies at the high-density left end and 11 plies at the low-density right end. {M 38341 F} [49]

◀ Pressed graduated-density compreg wood. {M 38343 F}

Propellers

Researchers at FPL enhanced a method of molding compreg propeller blades to finished shape, thereby eliminating the carving of hard, compressed blanks. [23]

◀ Samples of compreg wood: (left to right), section through airplane tail wheel, specimen with density graduated longitudinally, model propeller, and scarfed piece to show finish inherent through thickness of material. {M 42965 F}

▶ Products: (left to right) strip of papreg-faced aircraft carrier decking, three commercial compreg propellers, and airplane radio antennae mast. {M 55319 F}

Impreg

FPL researchers developed a resin-treated, laminated wood (impreg), superior to normal wood in dimensional stability, hardness, electrical resistance, and other properties. Two firms produced this material for housings for electrical control equipment and for certain aircraft parts. [34, 49]

Auto Die Models: The suitability of impreg for producing dimensionally stable die models for automobile manufacturing operations was also investigated. Two major auto manufacturers adopted the use of impreg for models that required the ultimate in accuracy and stability. [34]

Plywood

FPL provided technical advice for emergency plywood orders by Great Britain. [49]

◄ Beechcraft AT-10 trainers. Fuselage, wings, and tail assemblies were made of plywood. FPL was active in the development of water-resistant plywood that could be molded. {M 48495 F} [49]

◄ Molded plywood parts for military aircraft.

◄ America's first transport plane to be built almost entirely of wood. The Curtiss (C-76) Caravan shown here was flown for the first time at St. Louis, Missouri, January 5, 1943. {Credit: Curtiss-Wright Corporation} [49]

The FPL's contribution to the C-76 was summed up by G.A. Page, chief engineer of the Curtiss-Wright division at St. Louis, Missouri: "It [the Design Handbook prepared by FPL] has expedited and facilitated our work in connection with the design of the C-76 airplane to a degree that is hard to estimate." [49]

Papreg

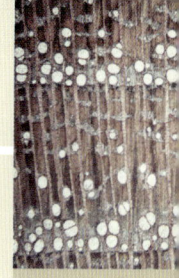

◄ B-24 Liberator bomber. A paper-based laminated plastic (called papreg) developed at FPL was used in gun turrets, seats, shields, and ammunition boxes on the bomber. This material was produced by 12 companies. After the war the concept of papreg was successfully commercialized. [49]

Honeycomb Core

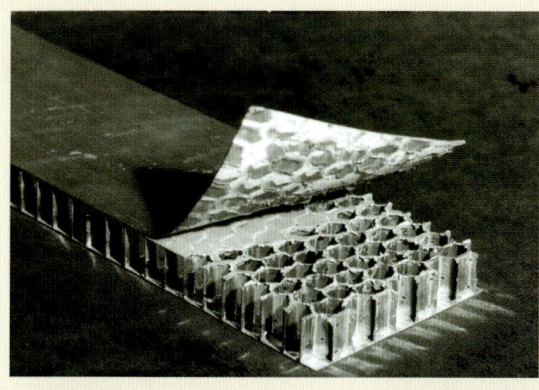

◄ A large contribution was the application of engineering design principles, developed by FPL staff from research on paper honeycomb core materials, to aluminum honeycomb cores that are used in modern aircraft. Also, development of the "FPL etch" led to an acceptable method of bonding aluminum. [26–28, 32, 35]

▼ ANC–FPL–Aircraft Industry Advisory Meeting held at FPL on January 31, 1949. This committee was instrumental in the development of modern aircraft. {M 80342} [29–31]

Packaging

The problem of shipping vast quantities of materials to support the troops was a major challenge. Fortunately, research on packaging had continued following WW I, and some of it was summarized in handbooks such as "Wooden Box and Crate Construction," by Clyde A. Plaskett. However, fighting on two fronts and requiring more and better shipping containers resulted in the FPL increasing its packaging research effort substantially. [49, 51]

◄ Damage of shipped good because of inadequate packaging, handling, loading, or transport environment. {M 50938 F}

◀ Damage to a wood container because of poor design and/or mishandling in transportation. {M 44847 F}

▼ FPL packaging laboratory during WW II. {M 61549 F} Thorwald (TAC) Carlson headed up the FPL packaging program during WW II. [49]

▲ Some containers were large. {M 55743 F}

▼ Crate being lifted for a drop test to determine its impact resistance. {M 43328 F}

◀ Sheathed crate with a superimposed dead load to simulate loading that could occur in a warehouse or in a ship. {M 47926 F}

▼ A complete tractor in its crate ready for shipment. {M 50023 F}

▶ Sample of block and bracing of an engine in a wood crate. This requires knowledge of the product, the potential shipping environment, properties of the crate material, and how to properly block and brace the item. {M 39957 F}

Reusable Containers

FPL developed over 100 reusable container specifications, resulting in decreased damage and waste. [49]

◀ Display of some smaller items that FPL helped develop packaging for or evaluated for the military. {M 60790 F}

Packaging Training

◀ One of many classes held at the FPL to train packaging personnel for the military under the direction of George A. Garratt, Dean of the Yale University School of Forestry. He was on leave to be Chief of the Division of Technical Service Training at the FPL. More than 16,000 personnel attended 300 courses at FPL or in the field. {M 50026 F} [49]

In 1951 FPL developed the Packaging Training Course for Rossford Ordinance Depot. It evolved into the Joint Military Packaging Training Center.

Packaging Materials

"The Laboratory technicians of Materiel Containers concerned themselves with testing packaging materials and designing more efficient containers. This work included tests of boxes and crates of solid wood, plywood, veneer, and special fiberboards. Specialty materials for containers, such as laminated papers, plastics, and paper-wood combinations were tested as to suitability for packaging, together with greaseproof and waterproof papers used as wraps or liners, cushioning materials ranging from foam rubber and excelsior to asbestos fibers, and plastic coatings that would be sprayed over machinery for moisture protection. There was little drama in testing of ordnance wrapping materials, but the results were vitally important to the overall packaging program. From 1941 to 1945, actual tests were made of 1,500 samples of greaseproof wrappings, 500 samples of waterproof papers, 350 samples of adhesives used for closing case liners, 325 samples of adhesives sealing fiberboard boxes, and 325 samples of pressure-sensitive tapes." [25]

Lumber Handbook

"The Army Corps of Engineers, which has general charge of wood products procurement for the Department of Defense, requested the FPL to prepare a simplified Lumber and Allied Products Handbook as a guide for the many Army, Navy, and Air Force officials engaged in procurement. The basic objective of the Handbook was to show these officials how to requisition the right sizes and qualities of material needed for a given purpose." [33]

During 1943, conservative estimates of lumber savings were $37,500,000. [49]

Manuals, Specifications, Guides

"Concurrently with the work on individual packaging instructions, the Division of Materiel Containers cooperated with the Ordnance Department and other agencies in the preparation of general publication on packaging. Among these were 'U.S. Army Specification 100-14A, Army-Navy Specification for Packaging and Packing for Overseas Shipment;' 'TM 38-305, General Instructions for Corrosion Preventive Processing and Packaging;' and 'TM9-2854, Instruction Guide, Ordnance Packaging and Shipping (Posts, Camps, and Stations).' In all, nine manuals, 37 specifications, 1,560 packaging instructions, and numerous guides and directives on packaging were prepared entirely or in large part at the Laboratory." [24]

To keep the far-flung packaging people informed of new publications and developments, the (FPL) Division of Materiel Containers issued a weekly bulletin called the "Sound Box." [49]

Alcohol Research

FPL continued research on alcohol, which included assisting in the building of a pilot plant in Springfield, Oregon. Jerry Saeman and other FPL staff members traveled to Europe after the war to gather information on production and use of alcohol in Europe.

Ships

FPL provided the department of the Navy with a manual entitled "Wood: A Manual for Its Use in Wooden Vessels." This manual summarized information relating to design, construction, repair, and maintenance of wood ships, boats, and landing craft and was used by the Navy and boat builders in design and construction of wood vessels. [22]

Laminated Ship Timbers

In cooperation with the Navy Bureau of Ships, the War Production Board, and commercial manufacturers, FPL developed a technique for production of durable, laminated ship timbers that were used for the fabrication of keels, stems, frames, and other parts of Navy vessels. Commercial manufacturers expanded use of this laminating technique and now supply laminated members for general construction. [37]

Boat Panels

FPL scientists invented a formable panel of impreg/plywood faces and cellulose acetate cores that can be used in construction of boats. Boats of considerable size can be constructed with these panels, which are strong and light. [39]

Korean Conflict (1950s)

◀ FPL studied new adhesive systems to withstand temperature extremes in high-altitude flights and also continued engineering analysis and design specifications. [36] {U.S. Air Force file photo}

Packaging

Pallets

For the Air Force, 16 types of lightweight pallets were evaluated by simulated rough-handling tests. The test became part of a new specification for expendable pallets. Part of the evaluation was made using the 14-foot drum test that was developed by the FPL. {123 435} [40]

Fasteners

"The dynamic loading capabilities of nails, lag screws, and bolts were examined with a pendulum-impact type apparatus to establish requirements for fastening blocking and bracing in crates to prevent shifting of contents during transit. Blocking and bracing are also used to restrain movement of containers and heavy equipment shipped on railroad cars and other cargo carriers. Special electronic equipment was devised to observe, measure, and record reactions to dynamic forces imposed on the fasteners." [40, 46, 48]

Wood Crate Design Manual

◀ Results of countless hours of research on crates were summarized in the "Wood Crate Design Manual" (1964) [44]

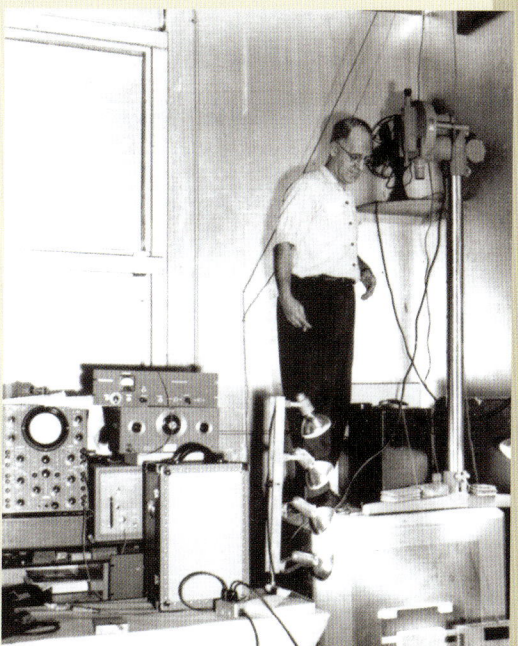

Cushioning Design Handbook

"A handbook on packaging cushioning design, the outgrowth of years of research in this field, was prepared for the Air Force." "This document provided the Department of Defense and other government agencies, as well as industry, with means of applying sound engineering principles to problems of cushioning for a wide range of machinery, equipment, and instruments in need of such protection during transit." [42, 45]

▲ Roger Russell setting up a high-speed camera for recording the swinging pendulum test that was used to determine the cushioning properties of various materials. FPL provided the methodology, collected data, and prepared a cushioning design manual so package designers could select the most efficient cushioning material to protect the product. (1960s)

Test Procedures

"An extensive job of standardizing packaging test procedures was done for the Navy Bureau of Naval Weapons. More than 1,400 test procedures have been extracted for review from some 335 government packaging specifications: each is referred to the most qualified member of the FPL staff for review and improvement. After military coordination and acceptance, they are included in Federal Standard No. 101." [40]

Packaging Cost Manual

FPL developed a Packaging Cost Manual to aid Air Force Packaging and procurement personnel. "In one 10-month period, Air Force savings on procurement contracts reportedly amounted to $2,243,000." [40]

Chemical Resistance of Wood

FPL evaluated treatments and treatment methods for enhancing performance of wood products used in pallets, consolidators, and skids. [54, 58]

Wood Preservation

The potential of a dual preservative treatment for protecting piling from attack by limnoria and shipworms, marine wood destroyers that are common in warm-water harbors, was demonstrated in experiments. These wood-boring crustaceans extensively damage port facilities. After 7 years, piling specimens given a two-stage (dual) treatment with copper and arsenic salts (by double-diffusion) followed by a pressure treatment of creosote remained free of attack. (One specimen remained exposed for another 33 years without significant damage.) Control specimens treated only with creosote were seriously attacked within 44 months (see Wood Preservation section). [47, 50, 53]

Wood Stabilization

Stabilization of wood gun stocks using polyethylene glycol (PEG) improved accuracy. This FPL research led to broader use of PEG for stabilizing wood carvings and large sections of wood, such as table tops, made from cross sections of trees. [41]

◀ Inspecting chemically stabilized target rifle stock at Marine Corps headquarters at Camp Perry during 1961 National Matches. Left to right are Major Robert Dawson, in charge of Marine Corps Rifle Team, Harold Mitchell of the FPL, Marine Major General August Larson (Director of Personnel), and Frederick Wieseman, Deputy Chief of Staff, U.S. Marine Corps. {ZM 120 149}

Desert Storm (1990s to 2000s)

Recycled Lumber

FPL scientists worked cooperatively with the U.S. Army to recycle and reuse more than 4,700 cubic meters of lumber and timber. The U.S. army estimates they have 250 million board feet of lumber and timber in WW II buildings that are slated for disposal. Reuse of this material in new construction is limited by a lack of appropriate science-based grading rules and engineering design values. Consequently, much of it ends up as broken-up waste in landfills. FPL researchers are developing a new grading system to ensure that residual properties of this lumber and timber will meet performance requirements of many applications. Using recycled lumber and timbers in new construction conserves existing forests and encourages the most efficient use of harvested materials. This research will help to recycle a portion of the estimated three-trillion board feet of lumber that has gone into buildings over the past 100 years (see Recycle section). [56, 57]

◄ Recycling lumber and timbers. (2000s)

Advanced Material Evaluation

► Devices used to provide test protocols that simulate actual in-service loading conditions to determine the mechanical behavior of advanced materials. (2000s) [59]

Connector Systems

◄ Evaluating connector systems in mine countermeasure vessels.

Mine Countermeasure Vessels

FPL scientists enhanced protection of mine countermeasure vessel's laminated wood engine mounts from explosive impacts through use of engineering analysis and redesign of connector systems. (2000s) [59]

Wood Propellers

"Nearly 90 years after the first research project on wood propellers, FPL investigated the effects of dry heat on wooden propellers for the Shadow® 200 unmanned aerial reconnaissance vehicle being used by U.S. forces in the Middle East. FPL combined time-tested manual examination with computer-aided analysis." [59]

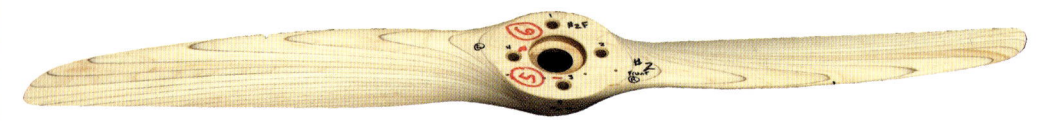

▲ Wooden propeller from Shadow® 200 unmanned aerial reconnaissance vehicle. (2000s)

▲ Mark Joyal taking measurements on a propeller sample.

▲ Launching of the Shadow® 200 unmanned aerial reconnaissance vehicle. (2000s)

This overview of the FPL's support of national defense supports a major conclusion —"that the existence of a national laboratory with the technical skills, knowledge, and equipment necessary to solve the problems and meet the challenges of the United States military in critical times has saved the nation time, valuable resources, and lives. Had some of the major problems that were brought to FPL not been addressed effectively, much greater quantities of wood products would have been consumed by the military, with a concomitant drain on our forests. Further, military equipment would not have performed as well, resulting in greater losses. ... Wood resources still play an important role in military operations, from pallets and other shipping materials, to buildings and transportation structures, to weapons components. Wood has been useful to mankind for thousands of years, in peacetime and war, and it appears this will remain true for decades to come. If our forests are managed wisely, and if we maintain and continue to build our intellectual capacity to meet the challenges of evolving human needs and changing wood characteristics, this amazing material that is wood will serve the nation well for years to come." [59]

1910

United States not at war.

Little information available on effective use of wood by the military.

2010

Provided research, problem solving, and training in the use of wood to defend the country in WW I, WW II, Korean, Vietnam, and Iraq wars. Savings in the billions of dollars and reduction in loss of life resulted.

[1] Osage-Orange dye for uniforms / Textile colorist Vol. 36:424, (1914)

[2] Kiln drying and the war / H.D. Tiemann / Southern lumberman Vol. 79(1049) (Oct. 20, 1915)

[3] War time use of forest products / Southern lumberman Vol. 52(1099) (Oct. 14, 1916)

[4] Relative efficiency of wooden partitions and corrugated fiberboard partitions in hand grenade boxes / D.L. Quinn / FPL (1918)

[5] Metal strapping on wooden boxes / FPL Technical notes: No. B3 and B6 (1919)

[6] Nitrating of wood pulp cellulose / S.D. Wells, V.P. Edwards / Paper Vol. 23 (Feb. 12, 1919)

[7] Research problems in airplane propeller manufacture / A.L. Heim / FPL (1919)

[8] War work at FPL / C.W. Gould / Timberman Vol. 20(10) (Aug. 1919)

[9] Wooden box and crate construction / FPL / National Association of Box Manufacturers, Chicago, Illinois (1921)

[10] Principles of box and crate construction / C.A. Plaskett / USDA, Forest Service Technical bulletin No. 171 (1930)

[11] Some accomplishments of the Forest Products Laboratory / FPL (February 1, 1937)

[12] Resume of recent developments in wood, plywood, and conversion products of interest in aircraft construction / L.J. Markwardt / FPL (1938)

[13] Manual for the inspection of aircraft wood and glue for the United States Navy / prepared by Forest Products Laboratory, Forest Service, Department of Agriculture. 1st ed. Washington, DC: U.S. Navy Department, Bureau of Aeronautics (1928) (2nd edition 1941)

[14] Wood goes to war / C.P. Winslow / FPL No. 1426, (1942)

[15] Forest Service information digest (December 23, 1942)

[16] ANC handbook on the design of wood aircraft structures. Supplement to the ANC handbook on the design of wood aircraft structures. Issued by Army–Navy–Civil Committee on Aircraft Design under the supervision of the Aeronautical Board. No. 1 published Dec. 1942; No. 2 published Feb. 1943; No. 3 published (April 1943)

[17] Wood aircraft fabrication manual / Prepared by Forest Products Laboratory / Issued by the Aeronautical Board. Supplement No.1 (Washington, D.C.) (1943)

[18] Repair of wood aircraft structures / Published under the authority of the Commanding General, Army Air Forces, the Chief of the Bureau of Aeronautics, and the Air Council of the United Kingdom, and approved by the Aeronautical Board, (Engineering handbook series for aircraft repair) (1944)

[19] American Forests Vol. 50(5):210) (1944)

[20] American Forests Vol. 50(7):327) (1944)

[21] American Forests Vol. 50(11):514) (1944)

[22] Wood: a manual for its use in wooden vessels / prepared by Forest Products Laboratory, Forest Service, U.S. Dept. of Agriculture, in cooperation with the Research and Standards Branch, Bureau of Ships, Navy Dept. (Washington: U.S. Govt. Printing Office) (1945)

[23] Repair of wood and compreg propellers and test clubs / Published under joint authority of the Commanding General, Army Air Forces, and the Chief of the Bureau of Aeronautics, and approved by the Aeronautical Board, Engineering handbook series for aircraft repair II, 48 p (1946)

[24] House of wood magic / Erle Kauffman / American Forests Vol. 52(1):40 (1946)

[25] House of wood magic / Erle Kauffman / American Forests Vol. 52(2):75, 90 (1946)

[26] Fatigue of sandwich constructions for aircraft: aluminum face and paper honeycomb core sandwich material tested in shear / Fred Werren / FPL No. 1559A (1947)

[27] Evaluation of several adhesives and processes for bonding sandwich constructions of aluminum facings on paper honeycomb core / H.W. Eickner / National Advisory Committee for Aeronautics technical note 2106 (1950)

[28] Short-column compressive strength of sandwich constructions as affected by the size of the cells of honeycomb-core materials / C.B. Norris, W.J. Kommers / FPL No. 1817 (1950)

[29] Design of wood aircraft structures / Prepared by Forest Products Laboratory, Forest Service, U.S. Dept. of Agriculture and by ANC-23 Panel on Sandwich Construction for Aircraft, Subcommittee on Air Force-Navy-Civil Aircraft Design Criteria, Aircraft Committee, Munitions Board / ANC-18 2nd ed. (1951)

[30] Wood aircraft inspection and fabrication / Prepared by Forest Products Laboratory, Forest Service, U.S. Dept. of Agriculture and by ANC-19 Panel on Sandwich Construction for Aircraft, Subcommittee on Air Force-Navy-Civil Aircraft Design Criteria, Aircraft Committee, Munitions Board ANC-19 (1951)

[31] Sandwich construction for aircraft / Prepared by Forest Products Laboratory, Forest Service, U.S. Dept. of Agriculture and by ANC-23 Panel on Sandwich Construction for Aircraft, Subcommittee on Air Force–Navy–Civil Aircraft Design Criteria, Aircraft Committee, Munitions Board, Ed. (1951)

[32] Adhesive bonding properties of various metals as affected by chemical and anodizing treatments of the surfaces / H.W. Eickner / FPL No. 1842 (1954)

[33] The Forest Products Laboratory: some accomplishments, work, and trends / FPL / DO 148 (April 1954)

[34] The application and properties of impreg / Ray M. Seborg / Ames forester Vol. 42 (1955)

[35] Mechanical properties of aluminum honeycomb cores / Edward W. Kuenzi / FPL No. 1849 (1955)

[36] A study of the deterioration of adhesives in metal bonds at high temperatures / J.M. Black, R.F. Blomquist / Technical report WADC; 55–330 (1955)

[37] Laminated, bolted, and solid keels for 50-foot Navy motor launch compared for strength / R.F. Luxford, R.H. Krone / FPL No. 1625 (1956)

[38] A comparison of the buckling strength of thin-walled cylindrical and barrel-shaped plywood shells / Edward W. Kuenzi / FPL No. 1323 (1956)

[39] Molded composite for boat hulls / B.G. Heebink, G.H. Stevens, E.W. Kuenzi. / FPL BuS–8 (1960)

[40] 1961 Annual Report of the FPL, pp. 15,16 (1961)

[41] Stabilized wood gunstocks in Marine Corps marksmanship competition / Major Robert E. Dawson, Major Edward G. Usher, Jr., Harold L. Mitchell / FPL No. 2245 (1962)

[42] 1962 Annual Report of the FPL, pg 18 (1962)

[43] 1963 Annual Report of the FPL, pg 9 (1963)

[44] Wood Crate Design Manual / USDA Agricultural handbook 252 (1964)

[45] Package cushioning design handbook / R. K. Stern, C. W. Roe / Department of Defense (1964)

[46] Performance of container fasteners subjected to static and dynamic withdrawal / R.S Kurtenacker / Research paper FPL 29 (1965)

[47] 1966 Annual Report of the FPL, pg 12 (1966)

[48] Performance of nailed cleats in blocking and bracing applications / R.S. Kurtenacker / FPL–RN–0200 (1968)

[49] History of the U.S. Forest Products Laboratory (1910–1963) / Charles A. Nelson / (Washington, D.C.) Govt. Printing Office (1971)

[50] Single- and dual-treated panels in a semi-tropical harbor: preservative and retention variables and performance / B.R. Johnson, L.R. Gjovik, H.Roth / American Wood-Preservers' Association. Proceedings of the 69th annual meeting Washington, DC: AWPA (1973)

[51] Packaging perspective, 1910–1985 / J.W. Koning, Jr., J.F. Laundrie / FPL–GTR–51 (1985)

[52] Forest Products Laboratory and the U. S. Navy join forces / R. E. Dudgeon / Forest products journal Vol. 36(6) (June 1986)

[53] Resistance of chemically modified wood to marine borers / Bruce R. Johnson, Roger M. Rowell / Material und organismen Vol. 23(2) (1988)

[54] Projected costs for treating wood pallets to impart chemical resistance / George B. Harpole, Rodney C. De Groot, Charles B. Vick / Forest products journal Vol. 41(4) (Apr. 1991)

[55] Structurally durable epoxy bonds to aircraft woods / C.B. Vick, E. Arnold Okkonen / Forest products journal 47(3): 71–77 (1997)

[56] Evaluation of lumber recycled from an industrial military building / Robert H. Falk, David Green, Scott C. Lantz / Forest products journal Vol. 49(5) (May 1999)

[57] Engineering evaluation of 55-year-old timber columns recycled from an industrial military building / Robert H. Falk, David Green, Douglas Rammer, Scott F. Lantz, / Forest products Journal Vol. 50(4): 71–76 (Apr. 2000)

[58] Chemical modification of wood: a short review / Roger M. Rowell / Wood material science and engineering. Vol. 1 (2006)

[59] Supporting the nation's armed forces with valuable wood research for 90 years / C.R. Risbrudt, R.J. Ross, J.J. Blankenburg, C.A. Nelson / Forest products journal Vol. 57(1/2) (January/February 2007)

Waste

◄ Uncontrolled forest fires are a major loss of wood fiber, wildlife, water quality, and recreation. {44 J}

Wood Waste [6]

Wood waste comes in many forms:

- Natural events such as uncontrolled fires, ice storms, tornados, hurricanes, insects, and diseases
- Forest remnants (slash), such as tops, limbs, and broken trees, from harvesting operations
- Manufacturing residues such as slabs, edgings, sawdust, and wood that is stained or of poor quality
- Demolition of wood structures either by natural forces or by redevelopment
- Urban forestry such as tree trimming, tree replacement, and shrub control
- Recycled wood products such as used railroad ties, pallets, and shipping materials

Challenge: How to significantly decrease wood waste or improve its utilization.

FPL's Contributions: FPL scientists developed new products from waste material and improved wood-using processes to help decrease wood waste.

Results: The results of this research have led to improved forest management, created new companies, extended the timber supply, improved forest aesthetics, and decreased negative environmental impacts.

► Remains of a forest fire may contain usable fiber if it is recovered before insects have a chance to destroy what is left. {21 M}

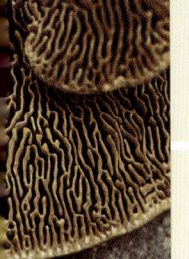

Forest Fires

According to the National Interagency Fire Center, the 2006 fire season was one of the worst on record in terms of number of fires and acres burned—96,385 fires affected 9,873,429 acres in a year that was 125% more destructive than the 10-year average. These fires in 2006 cost the federal government (American Public) over $1.5 billion to fight, a figure that does not include costs to local and state governments or the value of the lost wildlife, usable timber, clean water, or recreational opportunities.

Harvesting Waste

In the process of removing trees from the forest, a substantial amount of waste wood (slash) is created. An estimated 48,000,000 dry tons of residue have now built up in this form. This build-up of forest slash poses a real fire danger if not handled properly. [4]

◄ Forest waste is commonly referred to as logging waste or slash.

◄ One method of reducing the fire danger is to decrease the slash by controlled burning. However, this results in a loss of fiber and adds to air pollution. {34 P}

◄ Forest slash can contain large trees that are filled with rot and thus not usable for lumber. However, the remaining sound wood can be used for composite products, pulp and paper, and energy. A major concern is how to extract this type of material economically. [2]

Manufacturing Waste

▲ Slabs of wood created in the sawing of logs.

▼ Sawdust from the sawmill.

▲ Tepee burner at night. {43 J}

When a tree is sawn for lumber, a considerable amount of waste, in the form of slabs and edgings, is produced. FPL has conducted research to increase the efficiency of manufacturing processes and the utilization of waste that is produced. The results of this research, such as "Best Opening Face" discussed in the Lumber Manufacturing section, have greatly improved sawmill efficiency. Other FPL research has shown that the remaining waste can now be processed and used in composite products, such as flakeboard, particleboard, fiberboard, and pulp and paper production. [1]

Tepee burners were used to get rid of sawmill wastes, such as sawdust, before research provided economic uses for the waste. Now sawdust is used in a number of products, including animal bedding, soil amendments, energy, pulp, chemicals, and particleboard. {35 L} [3]

Wood Waste from Demolition of Structures

◄ Dilapidated buildings are another source of wood, but the cost to recover usable material is high.

◀ Building demolition is a source of wood fiber, but it is usually contaminated with many other materials, making recycling and reuse difficult and expensive. Some progress is being made by selective removal of usable material prior to demolition. Approximately 11,700,000 dry tons that are not recycled or burned end up in landfills. {35 B} [8]

Urban Forestry Waste

Urban forestry efforts, such as tree trimming, tree replacement, and shrub control, are another and growing source of wood waste. Much of this wood ends up in landfills, and in some areas it is destroyed in burn piles.

Approximately 1,700,000 tons are estimated to be available. This material can be utilized, but one unique challenge to this source of wood is separating the metal that is imbedded in some trees by people who hang items by nails and bolts. [8]

◀ Waste wood from trees that are trimmed and removed in the urban setting. {35 A}

Recycled Wood Products Waste

After wood has served its useful life as railroad ties, pallets, shipping materials, etc., these various products become another source of wood waste. Much of this material ends up in landfills, but as a result of research, more of it is being recycled. Grinding used pallets, and adding coloring to make attractive wood mulch, is a good example of research and technology developed and applied by industry to recycle the wood into another product. Other untreated wood can also be used for this relatively new product. However, treated wood, such as railroad ties, poses another challenge.

Used Railroad Ties

Used railroad ties are a special disposal problem because of the wood preservatives used to extend their life as ties. One outlet is in landscaping, where strength of the ties isn't as critical, but the treatment still provides for longer service life in an alternative use.

Research has indicated that another possibility is chipping and reconstitution of used ties into new railroad ties. [5]

Another approach is to remove the chemical treatment or render it harmless to the environment by methods such as mycoremediation.

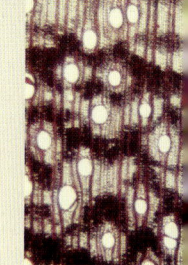

▶ Used railroad ties. {52 C}

Mycoremediation

Research is being conducted on the use of certain biological organisms to remove chemicals used to preserve the wood so that the wood can be recycled. This new field of mycoremediation uses fungi to remediate chemically treated wood, such as CCA-treated railroad ties. [7, 9–13]

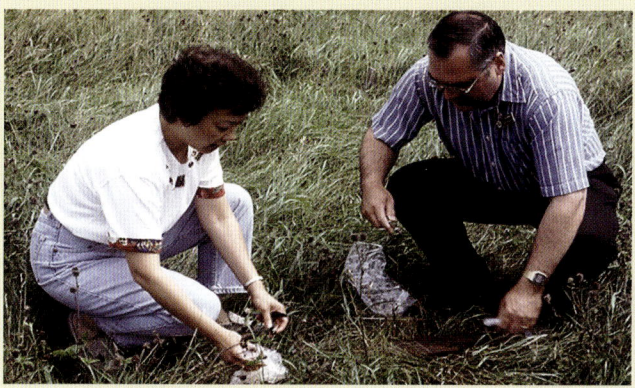

◀ Vina Yang and Les Ferge recovering 40-year-old wood decay test stakes from FPL field site. Unique fungi isolated from wood stakes, along with hundreds of fungi from the collection in the FPL Mycology Center, were part of a mycoremediation research project on biodegradation of waste wood treated with toxic chemicals. The research was awarded seven patents, including a patent for remediation of ACQ-, cresote-, pentachlorophenol-, and CCA-treated waste wood. (2000s)

1910

Large losses of wood fiber in all phases of wood utilization.

2010

Significant reduction in waste through wise use, new methods of utilization, and new products.

Futher Information

[1] Uses for slabs, edgings, and trims / FPL / FPL–RN–038 (1964)

[2] Uses for forest residues / L.H. Reineke / FPL–RN–092 (1965)

[3] Uses for sawdust, shavings, and waste chips / J.M. Harkin / Research note FPL–RN–0208 (Supersedes 1964, Uses for sawdust and shavings / FPL Research note 047) (1969)

[4] Volume of wood, bark, and needles after clearcutting a lodgepole pine stand / A.N. Foulger / Journal of forestry Vol. 71(2) (Feb. 1973)

[5] Feasibility of producing reconstituted railroad ties on a commercial scale / Robert L. Geimer / Research paper FPL 411 (1982)

[6] Resource potential of solid wood waste in the United States / David B. McKeever / In: Proceedings, The use of recycled wood and paper in building applications. Proceedings No. 7286. Madison, WI: Forest Products Society: 13–20. (1996)

[7] Bioprocessing preservative-treated waste wood / Barbara L. Illman, Vina W. Yang and Les Ferge / IRG/WP; 00-50145, Stockholm, Sweden: IRG Secretariat (2000)

[8] Successful approaches to recycling urban wood waste / Solid Waste Association of North America / FPL–GTR–133 (2002)

[9] Fungal degradation and bioremediation system for ACQ-treated wood / Barbara L. Illman, Vina W. Yang, Leslie A. Ferge / US patent No. 6,387,691 (May 14, 2002)

[10] Fungal degradation and bioremediation system for creosote-treated wood / Barbara L. Illman, Vina W. Yang, Leslie A. Ferge / US Patent No. 6,387,689 (May 14, 2002) and US Patent No. 6,664,102 (Dec. 16, 2003)

[11] Fungal degradation and bioremediation system for pentachlorophenol-treated wood / Barbara L. Illman, Vina W. Yang, Leslie A. Ferge / US Patent No. 6,383,800 (May 7, 2002) and US Patent No. 6,727,087 (April 27, 2004)

[12] Fungal degradation and bioremediation system for CCA-treated wood / Barbara L. Illman, Vina W. Yang, Leslie A. Ferge / US Patent No. 6,495,134 (Dec. 17, 2002) and US Patent No. 6,972,169 (Dec. 6, 2005)

[13] Bioremediation of treated wood with fungi / Barbara L. Illman and Vina W. Yang / Environmental impacts of treated wood. Boca Raton, FL: CRC/Taylor & Francis, p. 413–426 (2006)

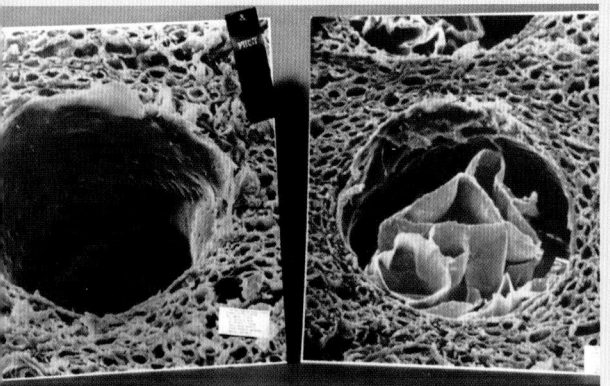 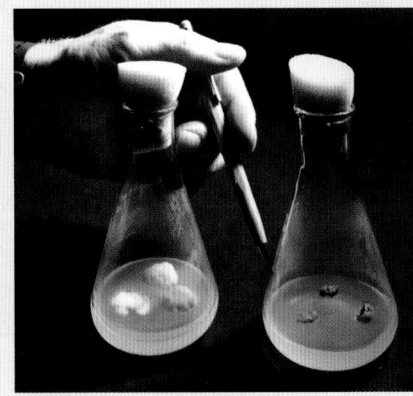 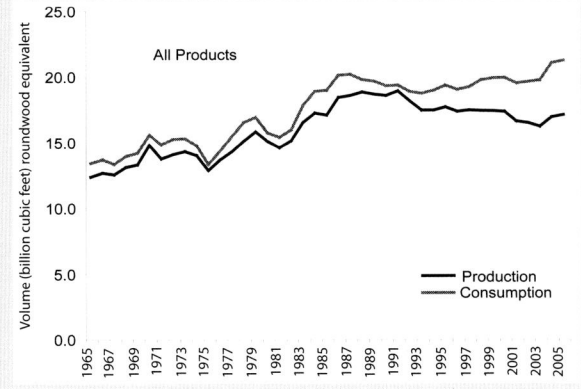

Foundation Sciences

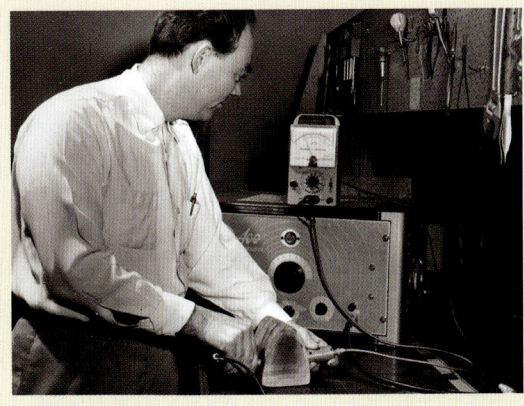

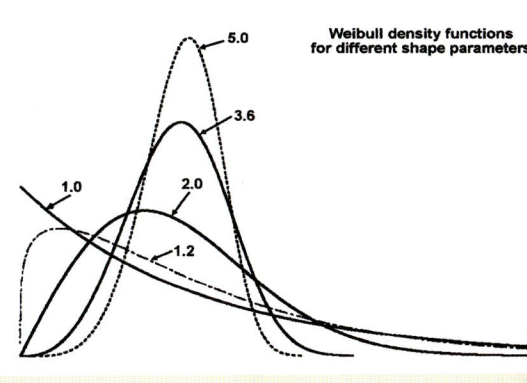

Weibull density functions
for different shape parameters

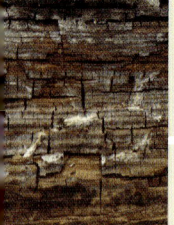

Anatomy

◄ Light microscope and cross section of a wood log. {138 B}

Forests supply thousands of different tree species, each with unique characteristics and properties. The first step toward the wise use of wood is to identify it.

Challenge: How to identify and understand the thousands of different wood species.

One method of identifying wood species is through wood anatomy, which deals with the structure of wood. Understanding wood anatomy is essential to understanding the many properties of wood. It also helps greatly in appreciating the huge diversity that exists between tree species. The primary tool for studying the anatomy of wood is the light microscope, and the images observed open an unusual world of colors, shapes, and patterns.

Not only does the study of anatomy provide a basic understanding of the cellular make-up of wood, but it also provides a means of identifying various species from samples of wood.

FPL's Contributions: FPL scientists identify wood samples for other scientists, the public, and other organizations. It is conservatively estimated that FPL has identified more than 225,000 such samples since 1910. They have developed guidebooks for wood characteristics, assisted in computerizing the process of wood identification, and assisted in forensic cases. They also established a web site for the "Center for Wood Anatomy Research" to provide information to the public.

Results: Research in anatomy provides fundamental information on the characteristics of wood leading to identification and classification of various wood species. This research unit provides valuable support in forensic investigations, assists the public in wood identification, and provides a basis for understanding tree characteristics.

Gross Anatomy of a Tree

Bark
Sapwood
Heartwood

Pith

Earlywood

Latewood

◄ Gross anatomy of a softwood tree. Starting at the center of the log is the pith, then heartwood, sapwood, and finally the bark. The earlywood (grown in the spring of the year) is usually made up of thinner-walled cells, and the latewood (grown during the summer) is composed of thicker-walled cells. {113 B} [13, 23, 24]

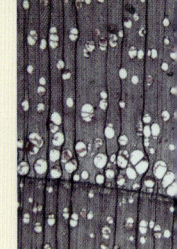

◄ Eloise Gerry, the first female scientist at FPL (1910–1954), conducted research in wood anatomy. {99 I} [1]

She also conducted pioneering research in naval stores, leading to more productive methods for harvesting tree sap for turpentine with less damage to the trees. [4]

Microscopic View of the Cross Section of Slippery Elm

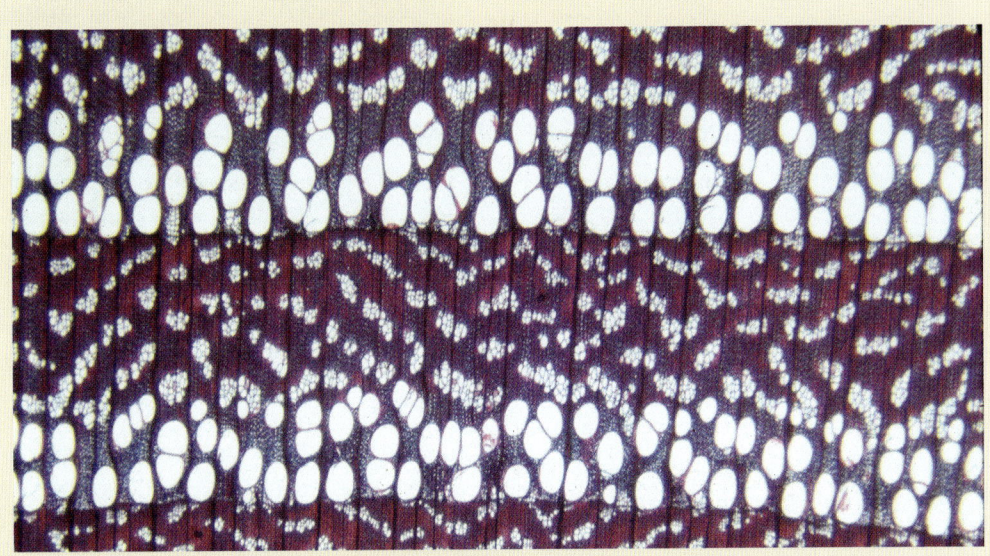

◄ Cross-section of slippery elm showing two growth rings of the tree. The large pores (vessels) were produced in the spring of the year, and the smaller pores were produced later in the season. {122 Q}

Three-Dimensional View of Wood Anatomy

► Three-dimensional rendition of the complex cell structure of softwood.

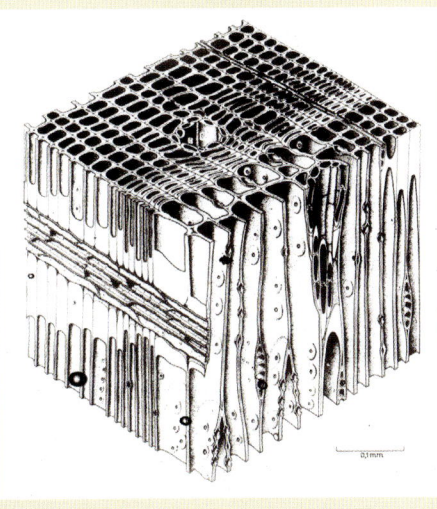

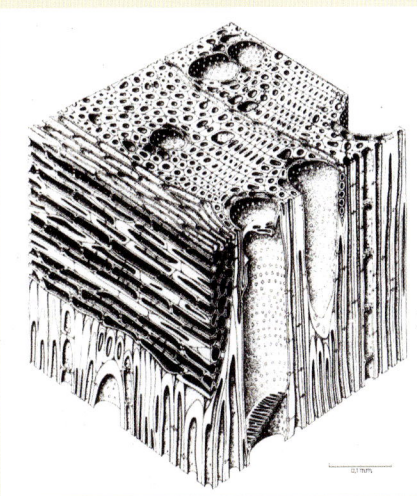

◄ Three-dimensional rendition of the complex cell structure of hardwood.

Tyloses

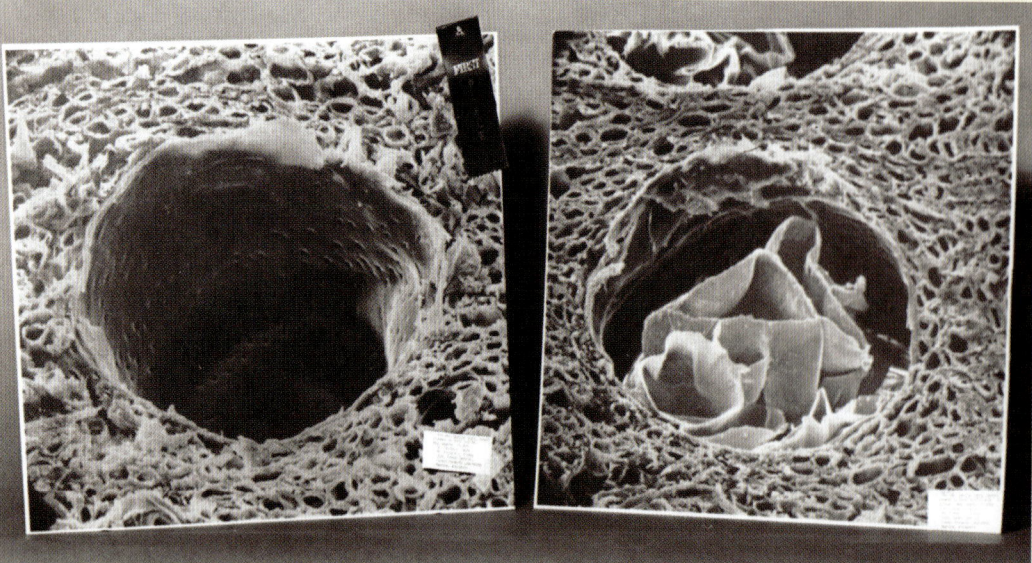

◄ Scanning electron microscope cross sections of red oak (left) and white oak (right). The material shown in the white oak vessel is tyloses, a substance that affects the movement of liquids in white oak. Lidija Murmanis used a transmission electron microscope to study the development of tyloses. (1970s) [9]

Eloise Gerry did pioneering research on tyloses. [1]

Vessel-Ray Pit Membranes

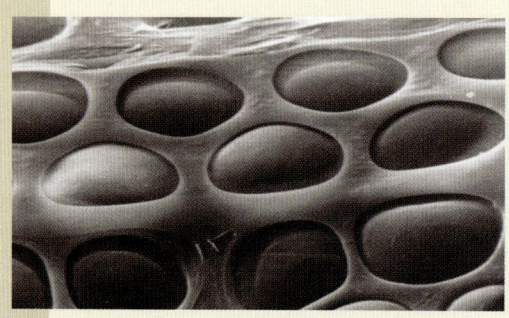 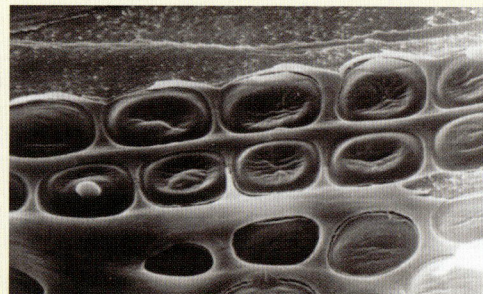 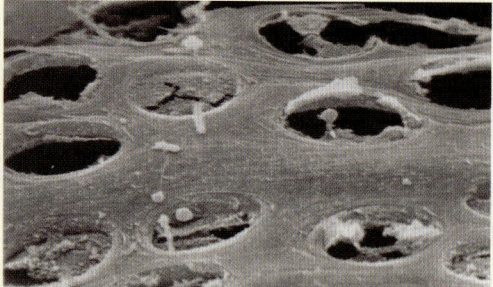

▲ Differences in vessel-ray pit membranes for normal sapwood (left), normally dying sapwood (center), and bacterially infected sapwood (right), which is commonly called wetwood. (1980s) [14]

Bordered Pits

The study of the openings between wood components, such as bordered pits in softwoods, allows for comparing species in a way that can lead to practical results. For example, results of comparing coast and inland Douglas-fir indicated that the bordered pits in the cell walls are quite similar, thus suggesting that both should respond to preservative treatment in a similar manner. [3]

Measuring Cell Wall Thickness

▶ Diana Smith measuring cell wall thickness of southern pine. (1960s) {M 130 625} [6]

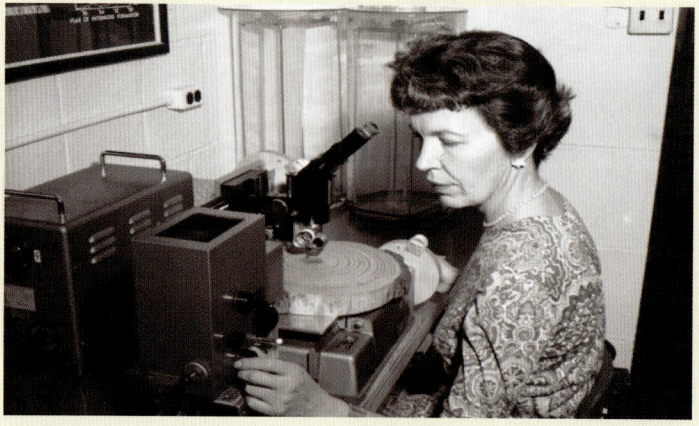

Studying the Dimensions and Microfibril Angle of Cell Walls

◄ Charlotte Hiller extended the research on wood quality by studying the dimensions of cells and microfibril angle of cell walls. (1960s) [7]

Methods of Wood Identification

◄ Arthur Koehler at his laboratory bench. Koehler and his staff developed methods for identifying native woods and some commercially important foreign ones. (1930s) {M 30022F}

The results of this early work were very helpful in the identification of woods used for railroad ties and mine timbers. [2]

Hand Lens

◄ B. Francis Kukachka ("Kuky") using a hand lens to examine a wood specimen unearthed during paleontological excavations in Oklahoma mammoth beds. Estimated to be 10,100 years old by radioactive C14 technique, the wood crumbled badly (pieces at right) until treated with polyethylene glycol. Wood structure was characteristic of elm. (1960s)

Kuky extended wood identification research to tropical woods. {M 123 118} [8]

Light Microscope

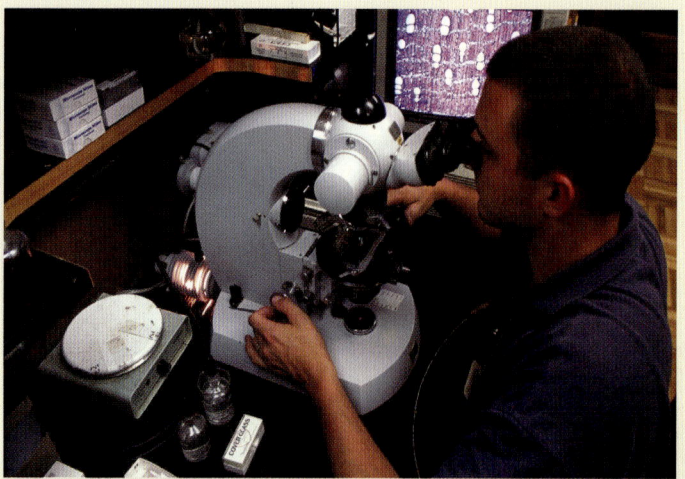

◄ Alex Wiedenhoeft using the basic tools of the trade in the study of wood anatomy—light microscope, slides, and a microtome knife. (2008) {127 B}

Relying just on the microscopic view of the wood sample for accurate identification does have definite limits. This is particularly important when dealing with expensive antiques or forensics. [25]

◄ Microtome knife (placed in wood block to protect the sharp blade) used by Donna Christensen and others to make a clean cross-section cut of a wood specimen to reveal the cell structure as an aid to wood identification and to make thin sections for observation in the light microscope. {16 N}

◄ Robert Koeppen operating a special device used to sharpen microtome knives for cutting wood sections for hand lens or microscopic examination. (1960s) {M 123 122}

Surface Fluorescence

Normal Light

Ultraviolet Light

▲ Using ultraviolet light to help identify certain wood species by their surface fluorescence. {24 L & 24 M}

Wood Collection

This xylarium has a collection of 17,000 species of wood representing 85% of the world's woody species. The FPL acquired the 55,000-specimen J. Record Memorial Wood Collection from Yale University in 1969. In 1970, a substantial wood collection from the Field Museum in Chicago was added to the Madison collection, bringing FPL's total to more than 100,000 specimens. The combined array is the largest in the world and continues to grow. Samples from this collection are used for anatomical research and for wood identification. [20]

▶ Regis Miller comparing unknown sample to known sample in the wood collection. (1990s) {121 G}]

Herbarium

◀ Robert Koeppen and "Kuky" Kukachka cross-checking species identification with the FPL herbarium collection. The herbarium collection is now located at the University of Wisconsin. (1960s) {M 126 049} [12]

Seed Cones

▶ Alex Wiedenhoeft examining a pine cone. All parts of a tree are useful in making wood identification. (2000s) [25]

Computerized Wood Identification System

In cooperation with the International Association of Wood Anatomists (IAWA) and an international committee of leading wood anatomists, the Center for Wood Anatomy Research at FPL helped develop the Standard List of Characters Suitable for Computerized Hardwood Identification and adopted a system of computer programs for wood identification. These early interactive programs allowed the user to select characteristics of wood in any order, use quantitative measurements, ask the computer for useful characteristics, and make a positive identification. [11] The latest computer program was developed by North Carolina State University. Regis Miller serves on the IAWA committee dealing with the list of microscopic features for hardwood and softwood identification. Alex Wiedenhoeft serves on the IAWA committee dealing with the list of microscopic features for softwoods.

WW II

The war years demanded an emphasis on foreign wood supplies as the nation looked for substitutes for cork, rubber, and sponge and for new sources of wood for container production. Emphasis after the war swung back to a concern for domestic supplies when it was realized that some commercial timber was "second-growth" with different characteristics from our virgin forests. Wise use of this "new" wood called for an expanded understanding of the structure of many common woods.

▲ Regis Miller developing an early computerized wood identification system. (1980s) {125 Q}

Simple Test Separates White from Red Oak

FPL developed a quick commercial test to distinguish white oak from red oak using a chemical solution. Spraying a 10% solution of sodium nitrite on the ends of oak logs will differentiate red from white oak by the color that results after applying the solution. To verify the accuracy of the method, 10,000 oak logs among 17 white oak and 18 red oak species at 30 sawmills throughout the eastern United States were tested. The tests fulfilled requirements by the European Economic Community Commission to prevent importation of oak-wilt-prone red oak wood into the European market area. Prior to this, all logs had to be fumigated, a costly, time-consuming process. [16]

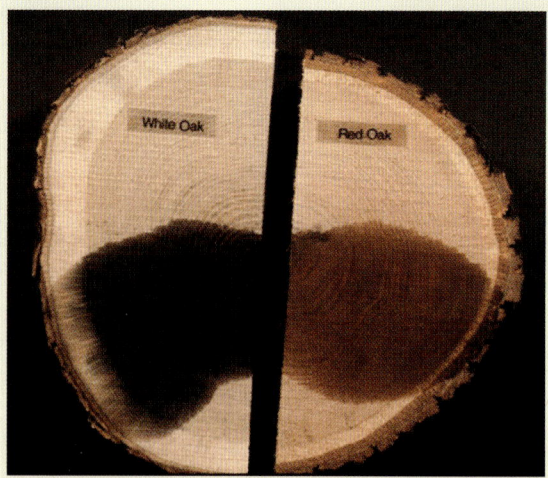

▲ Sodium nitrite test: Dark purple to black on species in white oak group, orange on species in red oak group. (1980s)

Tropical Woods

◀ Martin Chudnoff holding a copy of
Tropical Timbers of the World. (1984)
{M 150 124-5}

Comprehensive Reference Book on Tropical Timbers of the World

Tropical Timbers of the World, by Martin Chudnoff, provides university, industrial, and other research facilities a reference book that describes tree and timber characteristics of 370 species found in three main areas of the world: Tropical America, Africa, and Southeast Asia and Oceania. The book not only describes the various species but also has comparative tables of numerous properties. Trade names, common names, and scientific names are listed. [15]

Tropical Hardwoods

Progress has been made toward facilitating expanded use of tropical hardwoods to augment domestic timber supplies. Commercially important wood properties of more than 100 tropical species were reported at an international symposium in 2001. [21]

Endangered Tropical Species

In an effort to protect endangered tropical species, an Identification Guide for tropical woods controlled under the Convention on International Trade on Endangered Species (CITES) of wild fauna and flora was prepared by Regis Miller and Alex Wiedenhoeft of FPL and Marie-José Ribeyron (Coordinadora nacional adjunta, Inspecciones, Dirección de la aplicación de la ley, Ministerio del Madio Ambiente de Canadá). [22]

Hardwoods and Softwoods of North America

Harry Alden prepared a set of two books on hardwoods and softwoods of North America. Descriptions include scientific name, trade name, distribution, tree characteristics, wood characteristics (such as general, weight, mechanical properties, drying, shrinkage, working properties, durability, preservation, uses, and toxicity). Information for these books was provided by David Green, David Kretschmann, Kent McDonald, and Susan LeVan–Green of FPL; John "Rusty" Dramm of State and Private Forestry; Scott Leavengood and James Reeb of Oregon State University; and Lisa Johnson, Southern Pine Inspection Bureau. [18, 19]

Technology Transfer

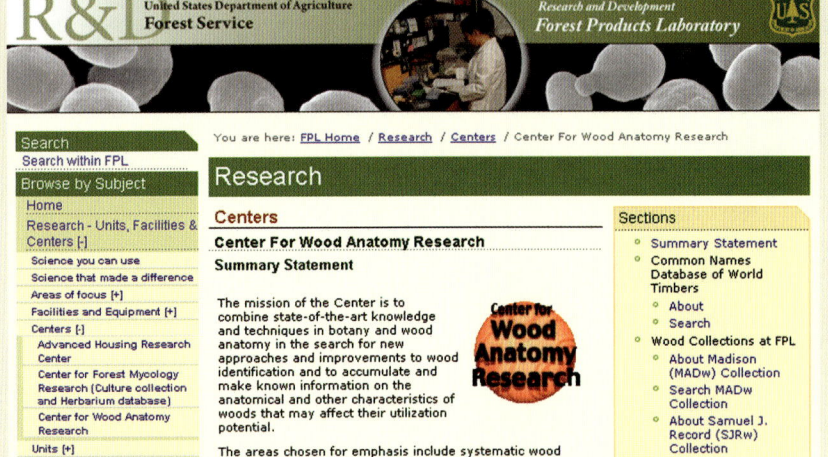

◄ Information on anatomy and properties of wood is available on the "Center for Wood Anatomy Research" web site at http://www.fpl.fs.fed.us/research/centers/woodanatomy/index.php

Forensics

Another use of wood anatomy is in comparative analysis to help identify items used in the commission of crimes or involved in accidents.

Lindbergh Kidnapping Case

In 1932, the son of world-famous aviator Charles Lindbergh and Anne Morrow Lindbergh was kidnapped from his second-floor bedroom. A wood ladder was involved in the case, and Arthur Koehler was asked to study it. Koehler identified, through microscopic techniques, all the species of wood employed in the homemade ladder left behind at the scene of the kidnapping. By studying the nail, tool, and machine markings in the wood, it was possible to trace the North Carolina pine uprights in the ladder to one particular planing mill by means of the number of knives used in the face and edge cutters and the rate at which the material passed through the mill. From this planing mill, Koehler then traced a shipment of lumber to a New York lumber dealer for whom the kidnapping suspect had worked. It was from this dealer the kidnapper had purchased lumber during the interval when the shipment was in stock. At the trial Koehler proved that one board used in the ladder had been taken from the attic floor of the kidnapper's home.

This was the first major appearance of wood identification as formal evidence in a major trial. (1934)

◄ Arthur Koehler inspecting a portion of the ladder used in the kidnapping. {M 23345 F}

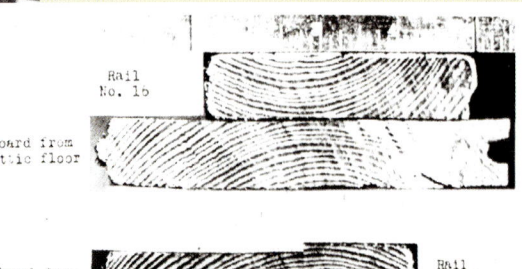

◄ Match-up of anatomical characteristics of a board used in the ladder and lumber taken from the attic floor.

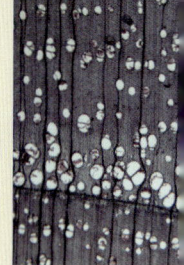

▲ Ladder leaning against the house (left) and three sections of the disassembled ladder (right) used in the kidnapping.

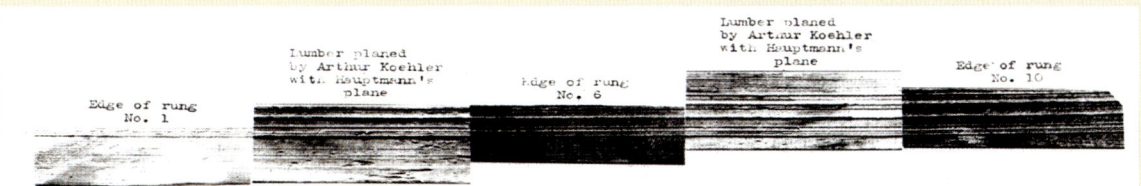

▲ Unique pattern of ridges left by nicks on both freshly planed pine and the kidnapping ladder were identified as being left by a plane from the accused kidnapper's home.

Red Paint Chip Case

In the "Case of the Red Paint Chip," a murder had taken place in Georgia. The victim was found in the woods covered with debris, including chairs, rugs, cushions, and two red painted posts. This area was used by hunters, who apparently brought the items and left them there. The red posts, however, were thought to belong to the suspect, the husband of the victim, because (1) the victim's son said that the red posts were stored in the house, (2) the son had seen them in his father's pickup truck, and (3) the husband admitted at one time that the red posts were in his pickup. In the victim's home, under the stairs where the red posts had apparently been stored, a red paint chip was found on a piece of yellow pine (*Pinus* sp.) 2 by 4 lumber. This red paint was compared with the red paint on the posts, and a good match was noted in gross structure and chemical analysis. The forensic scientist in charge noted that a few fibers were clinging to the back of the red paint.

Regis Miller was asked to examine these fibers to determine whether they were conifer fibers from the yellow pine lumber or fibers from the red painted posts, which were identified as from a species in the white oak group (*Quercus* spp.). The ray cells did not have any cross-field pitting and no ray tracheids were evident, suggesting that they were hardwood ray cells and not rays from some conifer. The fibers were of two types: pointed thick-wall fibers and long, broad, thin-walled fibers. Fibers of two distinct types do not occur in conifers, but they do occur in some hardwoods. Miller concluded that the fibers were

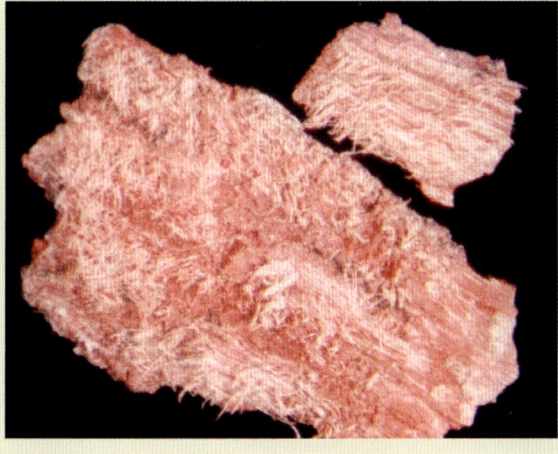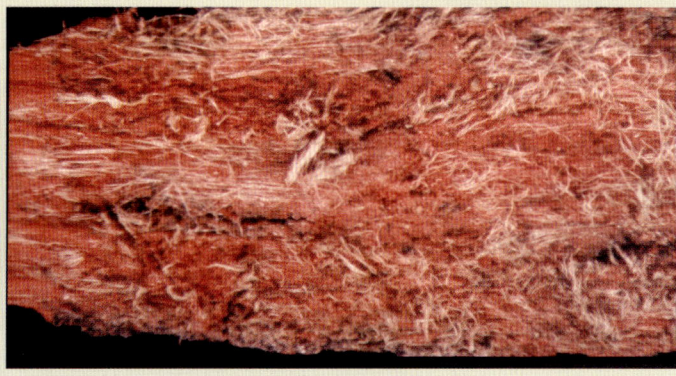

▲ Sample and magnified view of the paint chip found on a pine 2 by 4 in the basement of the accused murderer's home. (1990s)

libriform fibers and vasicentric tracheids that are typical of oak. Under high magnification and polarized light the fibers from the red paint chip appeared to have large slit-like structures that initially were thought to be pits. Upon closer scrutiny, however, these slit-like structures were found to be openings caused by decay fungi. This decay pattern was found both in the fibers clinging to the red paint chip and the white oak posts. Not only were the decay patterns the same, but also they came from the same type of fungus. This case used both identification and comparative anatomy to show that the red paint chip found on the stairs did come from the red painted posts found at the scene of the crime. [17]

"Analysis of the fibers attached to the paint chip indicated that the paint was from some hardwood posts that were found with the victim's body and had been stored in the basement of the accused murderers home, and not from the pine 2×4. The paint was evidently chipped off when the posts were moved from the basement and used to help cover the victim's body." [17]

Explosives from Central America

"Sometimes an exact identification is not necessary, as in the case of the explosives from Central America. The wood filler or sawdust used in bombs confiscated in Central America was examined to see if the bombs were made outside Central America. The theory was that wherever the bombs were made, local sawdust would be used. The fragments were so small that they could not be sectioned. Regis Miller examined the fragments under a light microscope, looking for anything that could be recognized as a species of north temperate origin. He was unable to identify the species of any fragments, but concluded that the wood filler was probably tropical in origin. He was able to make that statement because the fragments did not have spiral thickenings, a feature common to many north temperate species. In addition almost all the fragments were hardwoods. This is expected of tropical sawdust because most trees in the tropics are hardwoods; in the north temperate regions, softwoods are much more common. The only softwood fragment closely resembled a species of Podocarpus, one of the few softwoods confined to the tropical regions. He also found a fragment from a monocot (perhaps a palm), several fragments that contained a large amount of prismatic crystals in axial parenchyma cells, and fragments with small parenchyma cells or chambered cells. The features are fairly common in some tropical species, but they are rare in north temperate species." [17] The conclusion was that the evidence was consistent with material of Central American Origin.

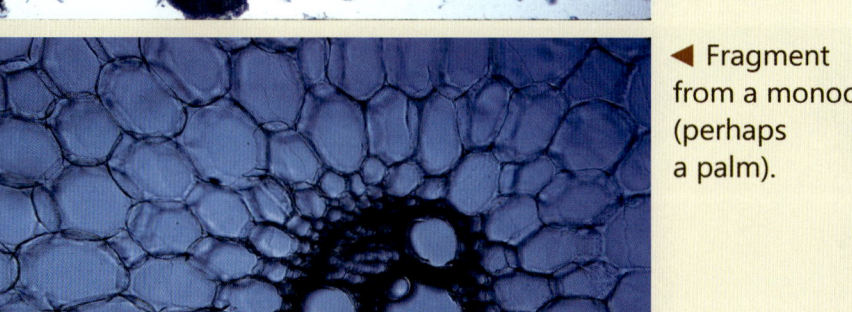

◀ Small fragments of the sawdust.

▼ Sample of wood filler or sawdust used in bombs.

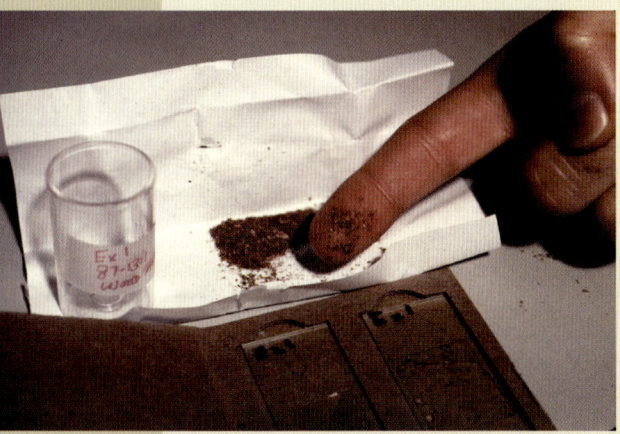

◀ Fragment from a monocot (perhaps a palm).

Breaking and Entering

"In breaking and entering cases, robberies, murders, and other types of crimes, wood fragments can become trace evidence. Doors, windows, walls, ceilings, or other wooden objects at the scene of the crime are broken or used as weapons. In the process, wood fragments get attached or linked to the perpetrators, or victims. These fragments can often be identified and linked to standards from the crime scene. In one case, sweepings from the suspects revealed several very small wood fragments. One was thin-sectioned by hand, and the other two fragments were so small that were place directly on a slide and examined under the light microscope. One piece was a species in the white pine group (*Pinus* spp.), another a species of maple (*Acer* spp.), and the third a species of basswood (*Tilia* spp.). All three pieces were part of a ceiling that the suspects broke through to gain access." (1990s) [17]

▶ Samples of wood from the broken ceiling.

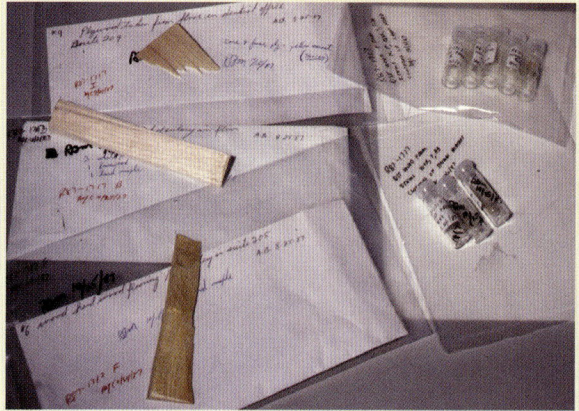

◀ Sweepings from the suspects.

▶ Wood sweeping from the suspect.

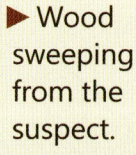

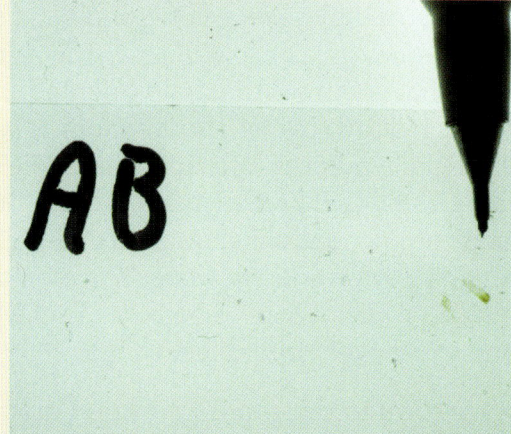

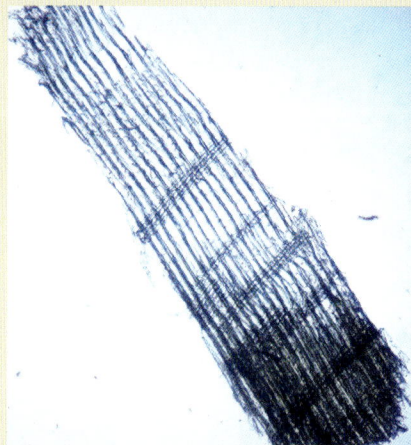

◀ Magnified sweeping, which was identified as the same species of wood as one of the samples of wood from the ceiling. In some instances larger fragments can actually be replaced in the piece of wood where they originally came from to show a good physical match.

Accident Investigations

Though wood identification can be critical to helping solve crimes or explain accidents, the material at the scene must be properly collected, documented, and stored. [26] For example, in the case of a plane crash, it was important that the wood materials involved in the case were provided.

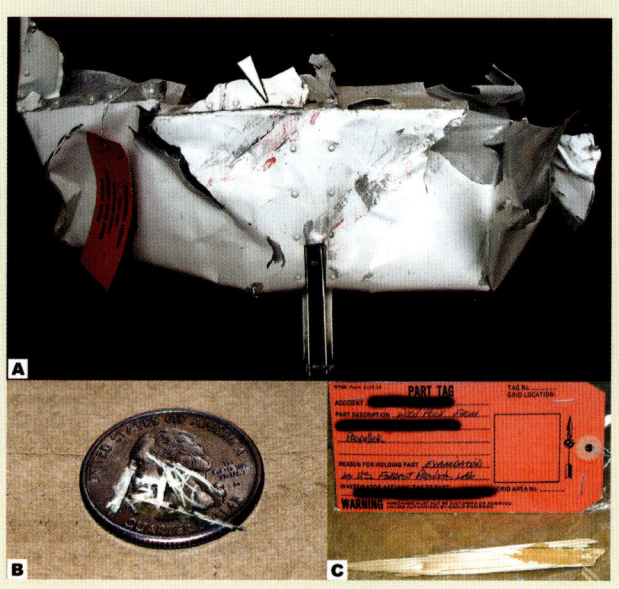

◀ Wood evidence from a mid-air collision between two airplanes. The portion of the plane submitted (A) held small pieces of wood evidence (B) under a metal flap (arrowhead in A). A piece of the wooden propeller of another plane (C) provides a reference sample. (2006)

Baseball Bats

If you followed baseball in 2008, you know that almost all major television sports outlets peppered their nightly baseball coverage with images of broken bats flying towards fans, pitchers, or ducking infielders. Broken bats have always been a part of the game, but over the past few years the frequency and intensity of breaks seemed to be on the rise. In fact, from July through September 2008, more than 2,232 bats broke during Major League games. And to Major League Baseball, this is a serious issue with potentially harmful consequences. Much of the discussion over "why" this was happening centered on the use of maple bats. Most of the more dramatic multiple piece breaks that were shown on television were from maple bats.

▲ Broken bat during a baseball game.

Dave Kretschmann, FPL scientist, was asked to head a interdisciplinary team of experts to look at what could be done to make wooden bats safer for players and fans alike. In his mind, nothing about the structure of maple should have made it seven times more likely to break than ash. "There are structural differences between maple and ash and over 60% of the bats that major leaguers use are made from sugar maple; so you would expect there to be a difference in failure rates, but not to the degree that was observed" said Kretschmann.

Still, until the research was done, he could not be absolutely sure. Starting in July 2008, MLB started collecting broken bats and shipping them to Madison, where Kretschmann and his partners from Timberco Inc., Harvard University, and University of Massachusetts-Lowell began measuring and analyzing each specimen to begin looking for patterns. "What we found was that the majority of multiple piece failures were caused by severe slope of grain. Essentially, the grain pattern for maple is more difficult to discern than ash, making it harder for manufacturers to grade it properly, and also process it so that you have the grain oriented properly. We also confirmed that the old adage we all learned in Little League about swinging with the 'trademark up' could be put to good use for reducing the number of maple bat failures, by putting the trademark in a different place for maple bats," said Kretschmann. After months of work, Kretschmann and his team sent their initial recommendations to Major League Baseball's Safety and Health Advisory Committee. The Committee and the Major League Baseball Players Association announced on December 9, 2008, at their winter meetings in Las Vegas that they had adopted the nine recommendations, which went into effect for the 2009 season.

As a result of the team's work, all bats now must conform to slope-of-grain wood grading requirements that apply to the 2/3 length of the billet that will constitute the handle and taper regions of the bat. All manufacturers must identify and grade the handle end prior to production of the bat to ensure that its slope of grain satisfies the grading requirement. All manufacturers must place an ink dot on the tangential face of the handle of sugar maple and yellow birch bats before finishing. Placing an ink dot enables a person to easily view the slope of grain of the wood. Also, the orientation of the hitting surface on sugar maple and maple bats should be rotated 90° (one quarter turn of the bat). The edge grain in maple that is currently used as the hitting surface is the weaker of the two choices. To facilitate such a change in the hitting surface, manufacturers must change their placement of logos by 90°. Finally, handles of sugar maple and yellow birch bats must be natural or clear finish to allow for inspection of the slope of grain in the handles. Also as a result of the teams work, a third-party inspection system was implemented where manufacturers would be visited on a regular basis by MLB or its designated representatives to audit each company's manufacturing processes and record keeping with respect to bat traceability.

1910

Limited information on wood anatomy, identification, or cell structure of U.S. woods existed.

2010

All commercial U.S. wood species identified and anatomy determined. Information is used in research, forensics, and service to the public.

Further Information

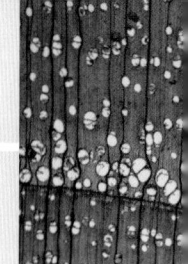

[1] Tyloses: a study of their occurrence and practical significance in some American woods / Eloise Gerry / Journal of agricultural research 1 (Mar. 25, 1914)

[2] Guidebook for the identification of woods used for ties and timbers / Arthur Koehler / United States, Dept. of Agriculture (1917)

[3] Bordered pits in Douglas fir: a study of the position of the torus in mountain and lowland specimens in relation to creosote penetration / Gertrude J. Griffin / Journal of forestry Vol. 17(7) (Nov. 1919)

[4] Oleoresin production: a microscopic study of the effects produced on the woody tissues of southern pines by different methods of turpentining / Eloise Gerry / USDA bulletin No. 1064 (1922)

[5] Identification of coniferous woods / B. Francis Kudachka / Tappi Vol. 43(11) (Nov. 1960)

[6] Microscopic Methods for Determining Cross-Sectional Cell Dimensions / D. M. Smith / Research paper FPL 79 (1967)

[7] Comparison of dimensions and fibril angles of loblolly pine tracheids formed in wet or dry growth seasons / Charlotte Hiller, Rodney S. Brown / American journal of botany Vol. 54(4) (Apr. 1967)

[8] Properties of imported tropical woods / B. Francis Kukachka / FPL Research paper 125 (1970)

[9] Formation of tyloses in felled *Quercus rubra* L. / L.L. Murmanis / Wood science and technology Vol. 9(1) (1975)

[10] Sectioning refractory wood for anatomical studies / B. F. Kukachka / Microscopica acta 80:301–307 (1978)

[11] Wood identification via computer / Regis B. Miller / IAWA bulletin n.s. Vol. 1(4) (1980)

[12] Wood anatomy of the Neotropical Sapotaceae: I through XXXVIII / B.F. Kukachka / Forest Service Research Papers 325–331, 349–354, 358–363, 371–374, 395–398, 416–426. (1978–1982)

[13] Wood: colors and kinds / Dave Green / USDA Agriculture handbook No. 101 (original publication, written by D.A. Zischke, B.F. Kukachka, 1956) (1981)

[14] Bacterial oak: drying problems / J.C. Ward, D.A Groom / Forest products journal Vol. 33(10) (Oct 1983)

[15] Tropical Timbers of the World / Martin Chudnof / USDA Agriculture handbook No. 607 (1984)

[16] Identifying white oak logs with sodium nitrite / R.B. Miller, J.T. Quirk, D.J. Christensen / Forest products journal Vol. 35(2) (Feb. 1985)

[17] Identification of wood fragments in trace evidence / Regis B. Miller / In: Proceedings of the international symposium on the forensic aspects of trace evidence: June 24–28; Quantico, VA U.S. Dept. of Justice Federal Bureau of investigation; 91–111. (1994)

[18] Hardwoods of North America / Harry A. Alden / FPL–GTR–83 (1995)

[19] Softwoods of North America / Harry A. Alden / FPL–GTR–102 (1997)

[20] Xylaria at the Forest Products Laboratory past, present, and future / Regis B. Miller / Wood to survive Vol 25, Freddy Maes–Hans Beeckman, editors (1999)

[21] Major Timber Trees of Guyana Wood Anatomy / Regis B. Miller, Pierre Detienne / Tropenbos Series 20, Tropenbos International, Wageningen, the Netherlands (2001)

[22] CITES identification guide—tropical woods: guide to the identification of tropical woods controlled under the convention on international trade in endangered species of wild fauna and flora / Regis B. Miller, Alex Wiedenhoeft, Marie-José Ribeyron / Distributed by the Government of Canada Depository Services Program, Environment Canada (2002)

[23] Wood anatomy / Regis B. Miller / McGraw-Hill encyclopedia of science & technology. 9th ed. New York: McGraw-Hill (2002)

[24] Structure and function of wood / Alex C. Wiedenhoeft, Regis B. Miller / Handbook of wood chemistry and wood composites. Boca Raton, Fla.: CRC Press (2005)

[25] The limits of scientific wood identification / Alex C. Wiedenhoeft / Professional appraisers information exchange (March 2006)

[26] Wood Evidence: proper collection, documentation, and storage of wood evidence from a crime scene / Alex C. Wiedenhoeft / Evidence technology magazine Vol. 4(3):28 (May–June 2006)

Biochemistry

Scheme depicting the lignin-degrading system of *Phanerochaete chrysosporium*. FPL scientists described most of the details of the process. (1997) [38]

Biochemistry deals with the chemical compounds and processes occurring in organisms and the chemical characteristics and reactions of a particular living system or biological substance (such as chlorophyll). [12, 17]

Challenge: Learn how microorganisms break down wood and how that knowledge might be used in industrial processes. Biological processes are inherently more environmentally friendly than chemical ones.

FPL's Contributions: Research made many seminal, fundamental discoveries about how fungi and their enzymes break down lignin, cellulose, and the hemicelluloses—the building blocks of wood.

Results: Our understanding of these processes, which are fundamental to life on earth, was greatly improved, and a number of practical applications of that understanding were—and are being—investigated at FPL and around the world.

◀ Ken Hammel and Kent Kirk (recipient of the 1985 Marcus Wallenberg Prize) discuss the chemistry of ligninase action on a pollutant molecule (a PAH, or polyaromatic hydrocarbon), which it degrades. (1986) (M 860 167-3) [22]

Lignin-Degrading Enzyme

Lignin is the second most abundant renewable organic compound on earth. Most of it is in wood, where it surrounds the most abundant compound, cellulose. The lignin barrier must be breached before cellulose-degrading microorganisms can get at the cellulose. Thus, lignin degradation, a role played by fungi, is actually key to the earth's carbon cycle. Until the work at FPL, very little was known about how fungi break down lignin. Chemical breakdown and removal of lignin from wood frees the cellulose for paper and other fiber products. FPL scientists also closely examined the possibility of using fungi, rather than chemicals, to pulp wood. Results showed that certain fungi can modify enough lignin in wood chips so that it can be pulped (mechanically) with much lower energy requirements. [15, 19–21, 24, 28, 35, 36, 48]

Biopulping

In an effort to decrease the energy used in pulping wood, FPL scientists explored the use of fungi to modify the lignin so that wood can be more easily pulped. Laboratory experiments using different fungi and growing conditions led to the selection of fungal strains that preferentially modify lignin. FPL scientists took the study of "biopulping" from a laboratory scale to a 50-ton scale, clearly demonstrating that it works on an industrial scale. Biopulping saves energy, improves mechanical pulp properties, and expands the usefulness of mechanical pulps. Studies at the University of Wisconsin indicated that biopulping is economical—and that was before the huge rise in oil prices in 2008. [25, 31, 42, 43, 45]

▲ Close-up of biopulping: fungal mycelium colonizing the wood chips. The fungus preferentially attacks the lignin in the wood chips.

▲ Fifty-ton wood chip pile experiment to evaluate the use of a fungus to modify the lignin in wood chips and decrease the energy required to grind the chips to make wood pulp for paper. (biopulping). The experiment was successful. Energy reduction was approximately 30%, but the process has not yet become commercial. (1990s) {101 K}

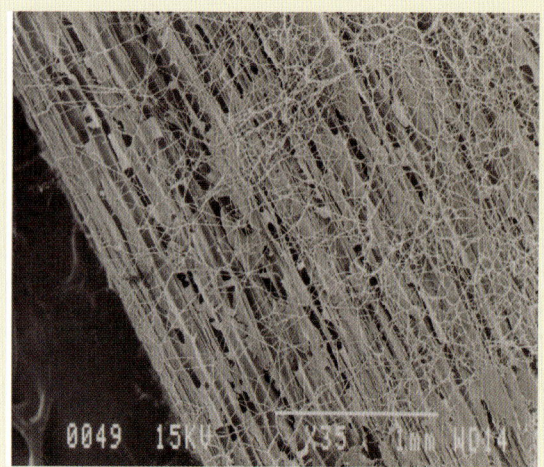

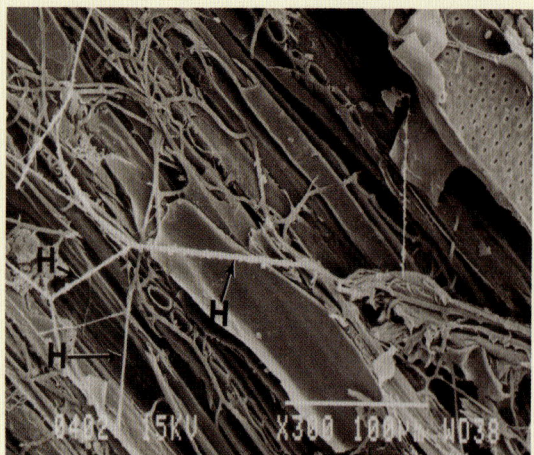

◄ Close-up of fungal mycelium growing on a wood chip and penetrating the cell walls. The fungus is destroyed in the subsequent pulping processes. (1988) {M 880 149 1 & 2}

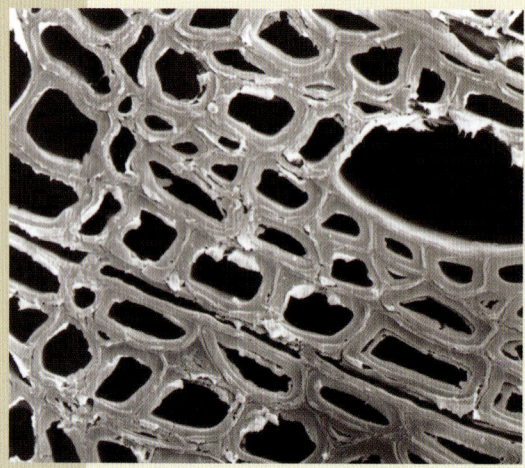

◄ Cross section of sound wood chip showing normal cell walls. (1988) {M 880 154 A}

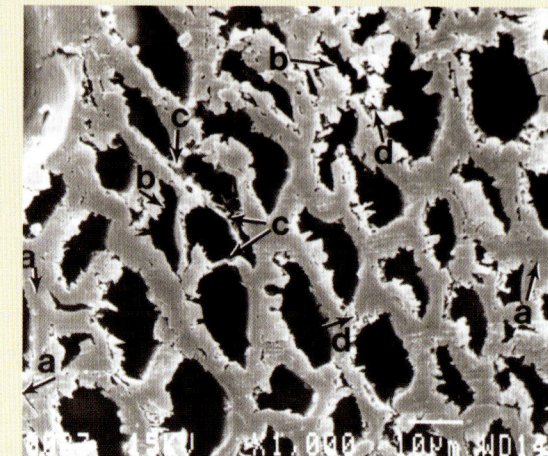

◄ Cross section of wood chip showing the effect of the fungus on cell walls. (1988) {M 880 154 B}

Oxalic Acid Pulping

An unexpected spin-off of biopulping research was the discovery by collaborator Robert Blanchette of the University of Minnesota that the best biopulping fungi produce copious quantities of oxalic acid as they degrade lignin. This led Masood Akhtar, University of Wisconsin, and FPL staff to investigate whether oxalic acid, which is inexpensive, could be used directly in pulping. The answer: yes, in much the same way as the fungi themselves—to decrease energy for mechanical pulping. This work was recently extended at FPL to a volatile derivative of oxalic acid, which as a gas can be introduced easily into wood where heat and the natural acids in wood convert it back to oxalic acid. The process currently is under study by industry. (2000s) [46, 53, 54, 56]

Wood-Decay Fungi Can Be Used to Clean Up Contaminated Water and Soil

FPL researchers showed that the lignin-degrading enzyme system is so non-specific that it also degrades pollutants.

Working with researchers from North Carolina State University, FPL scientists showed that lignin-degrading fungi can remove the dark color from pulp mill wastewater. The dark color comes from lignin during chemical pulp bleaching. Most of the research was with the fungus *Phanerochaete chrysosporium*. Called FPL/NCSU MyCoR (Mycelial Color Removal), the process underwent pilot-scale testing and offered the paper industry a new method of effluent treatment. This process was not commercialized because the bleaching process was changed and less emphasis was placed on color removal. (1982) [18]

◄ Two flasks, one containing colored pulp mill effluent and the other one containing clear liquid that results from the action of a fungus that degrades lignin. (1980s) {83 P}

◄ Scale-up to a bench model rotating bioreactor containing a fungus that decolorizes pulp mill effluent. (1980s) {109 L}

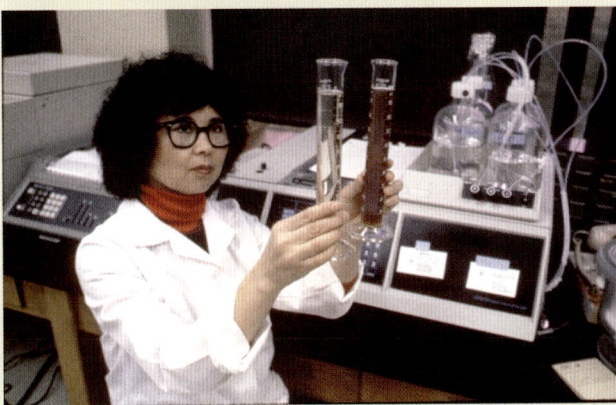

◄ Suki Croan checks lab samples to determine decolorization effect of wood decay fungus. (1980s) {M 150 678-6}

Decontaminating Soil

◄ Richard Lamar moistening samples of contaminated soil after they were inoculated with the lignin-degrading fungus *Phanerochaete chrysosporium*. Such bioremediation of contaminated soils using fungi was shown to work and has been used commercially. (1990s) {113 J} [26, 27, 29, 30]

DNA of the Fungus Phanerochaete chrysosporium

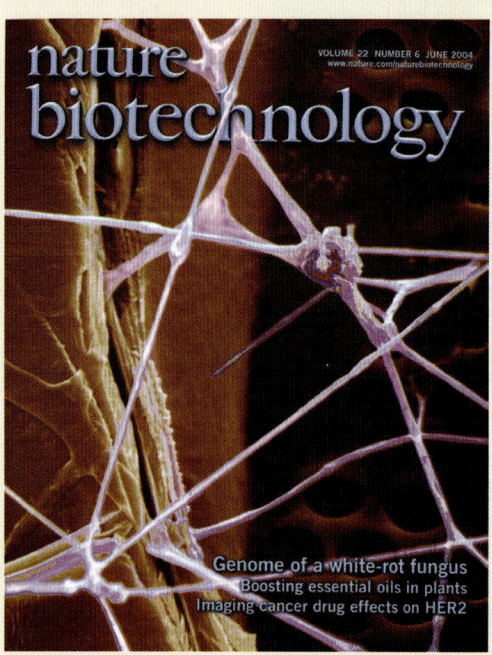

◀ Hyphae of the lignin-degrading fungus, *Phanerochaete chrysosporium*. A joint effort of Dan Cullen, FPL, and scientists from seven other organizations resulted in the genome sequencing of the lignocellulose-degrading fungus *Phanerochaete chrysosporium* strain RP78. This genome sequence is the first for the higher fungi that are so central to the earth's carbon cycle. This fungus is closely related to plant pathogens. This genome provides a high-quality draft sequence of a basidiomycete, a major fungal phylum that includes important plant and animal pathogens. (2004) [48, 49]

Natural Chlorination of Organic Material

It is common knowledge that compounds containing chlorine occur often in the environment as industrial pollutants, but is less well known that they are also produced naturally. Research by a team of FPL scientists has shown that forest fungi chlorinate lignin and are thus likely contributors to the production of chlorinated compounds in soils and plant litter. The natural occurrence of these compounds probably accounts, in part, for the fact that useful pollutant-degrading microorganisms have evolved over eons and are already available to clean up the results of environmental spills that we humans accidentally cause. [52]

Types of Wood Decay

White rots are caused by lignin-degrading fungi such as *Phanerochaete chrysosporium*. [41]

Recent FPL research has shown that brown-rot fungi generate a very powerful inorganic oxidant that attacks all the components of wood, leaving a strongly modified lignin residue. In the coniferous forests where brown-rot fungi predominate, this residue goes on to become humus, which is essential for forest soils to maintain their nutrient- and water-holding capacity. Brown-rot fungi are also major contributors to the failure of wood in service. FPL's ongoing elucidation of their decay mechanisms thus provides information that may lead to the development of new antifungal wood preservatives. [50]

◀ These pieces of rotting wood illustrate the three kinds of rot caused by fungi. A is "white-rot"; B is a specialized kind of white-rot that leaves cellulose-filled pockets; C is "brown-rot"; and D is "soft-rot." FPL scientists have greatly improved the understanding of the chemistry and biochemistry of these types of rot. [8, 15, 28, 39, 40, 47, 50]

Ethanol

The FPL plays a major role in converting wood to ethanol. A brief history of this involvement is outlined in the Energy section, and further information on the Madison Process for making alcohol from wood is found in the Chemicals section.

The FPL research covered here relates to the major effort to find useful organisms that will convert the xylose that is extracted from wood hemicellulose to ethanol.

One of the main building blocks of hardwoods, grasses, and crop plants is the five-carbon sugar xylose. When these plant materials are broken down with acid, about 1/3 of the resulting sugars is xylose. (The other 2/3 is cellulose-derived glucose.) Thus, any process that involves fermentation of these sugars must work with xylose as well as with glucose. FPL scientist Tom Jeffries has pioneered research on xylose fermentation, discovering yeasts that do it and much about how they do it. Tom was issued U.S. Patent No. 4,663,284 (May 5, 1987). Tom, along with Haiying Ni and Jose M. Laplaza, were issued U.S. Patent No. 7,285,403 (Oct. 23, 2007) on Xylose-Fermenting Recombinant Yeast Strains [1, 2, 17, 32–34, 55]

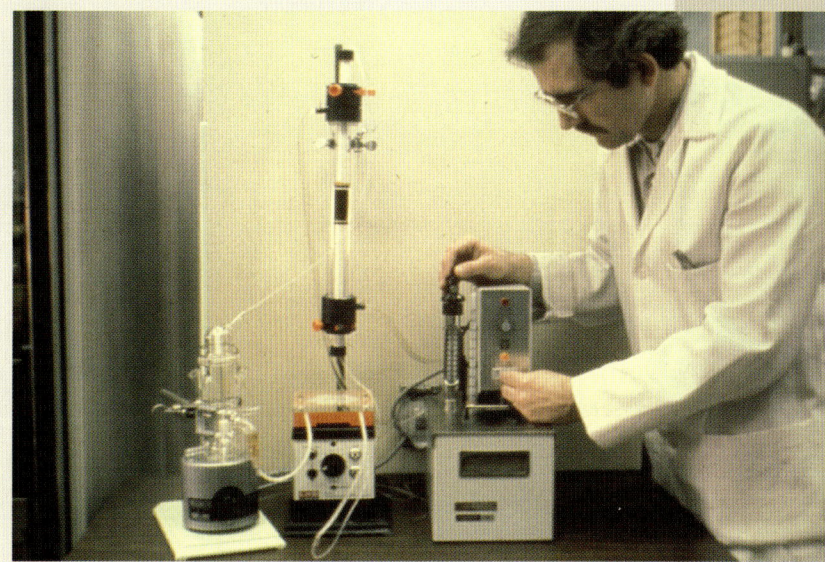

▶ Tom Jeffries demonstrating the fermentation of the five-carbon sugar xylose by the yeast *Candida tropicalis*. (1981) [16]

Recently Jeffries and collaborators from several laboratories sequenced the genome of a xylose-fermenting yeast, providing a leap in understanding and improving the potential for industrial application. [51]

Pichia stipitis Chromosomes

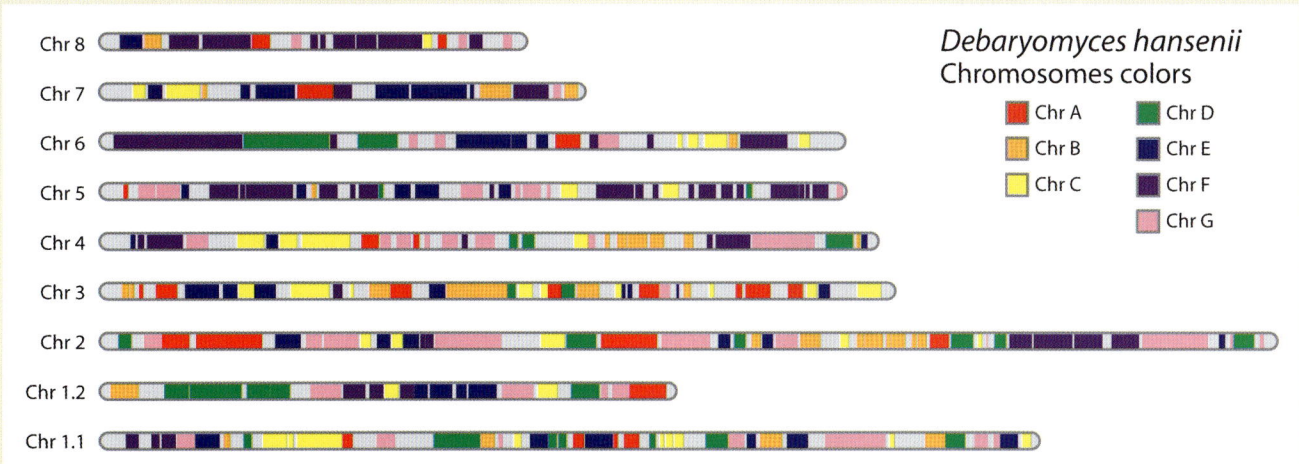

▲ This diagram shows the "synteny" between the chromosomes of *Pichia stipitis* and its closest completely sequenced relative, *Debaryomyces hansenii*. The term synteny is used to describe the extent to which related genomes preserve the order or sequence of genes in their chromosomes. In yeasts and other higher organisms, genes are undergoing constant rearrangement, and the orders of the genes on these two chromosomes are very different. *D. hansenii* is normally found in association with cheese; *P. stipitis* is found in association with wood-boring beetles. Both of these yeasts have diverged greatly from *Saccharomyces cerevisiae*, which is normally used for ethanol fermentations of glucose and sucrose. (2007) [44, 51]

New Trees from the Tissue from the Cambium Layer of the Tree

Early research by FPL's Karl Wolter on the culture of callus tissues from tree cells has contributed to genetic engineering of trees for a variety of applications. (1960–1970s) [3, 7, 10, 13, 14]

► Shoots (tiny trees) grown on aspen callus are examined by Adrian Richter. (1960s) {M 134 558 No.9}

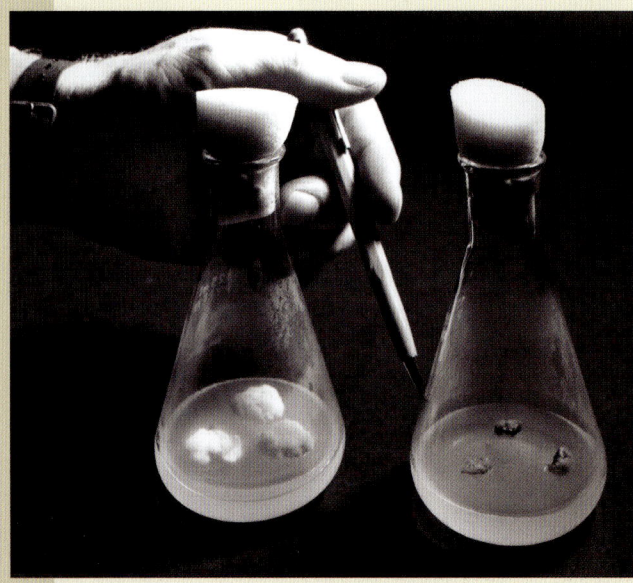

◄ Nutrients with (left) and without (right) myo-inositol show effect of this factor on cambium cell growth. (1960s) {M 128 313} [11]

1910

Little activity

2010

Advances in tree physiology were made, a better understanding of the structure of lignin has advanced, the chemistry and biochemistry of wood decays are being elucidated, and the DNA sequences of important wood decay fungi are being determined.

Further Information

[1] The manufacture of ethyl alcohol from wood waste / E.C. Sherrard / FPL No. 668 (1920)

[2] A technical review of developments in microbiological utilization of wood sugars / G.J. Hajny / Forest products journal Vol. 9(5) (May 1959)

[3] In vitro cultivation of ash, aspen, and pin oak callus tissue / Karl E. Wolter / Thesis (Ph. D.) University of Wisconsin–Madison (1963)

[4] Degradation of wood preservatives by fungi / Catherine G. Duncan and Flora J. Deverall / Applied microbiology Vol. 12(1) (1964)

[5] Food-Yeast Production from wood-processing byproducts / FPL–RN–065 (1964)

[6] Erythritol production by a yeast like fungus / G.J. Hajny, J.H. Smith, J.C. Garver / Applied microbiology Vol. 12(3) (May 1964)

[7] Nutritional requirements of *Fraxinus callus* cultures / K.E. Wolter / American journal of botany Vol. 53(3) (1966)

[8] Effect of light on the rate of decay of three wood-destroying fungi / Catherine G. Duncan / Phytopathology Vol. 57(10) (Oct. 1967)

[9] Factors contributing to heartwood-boundary stain in living oak / E.H. Bulgrin, J.C. Wood / Wood science Vol. 1(1) (July 1968)

[10] Root and shoot initiation in aspen callus cultures / K.E. Wolter / Nature Vol. 219(5153) (Aug 3, 1968)

[11] The fate of myo-inositol in *Fraxinus* tissue cultures / P. Jung, W. Tanner, K. Wolter / Phytochemsitry Vol. 11(5) (May 1972)

[12] Lignification in trees: indication of exclusive peroxidase participation / John M. Harkin, John R. Obst / Science Vol. 180(4083) (Apr. 20, 1973)

[13] Guaiacyl lignin associated with vessels in aspen callus cultures / Karl E. Wolter, John M. Harkin, T. Kent Kirk / Physiologia plantarum Vol. 31 (1974)

[14] Radioautography of myo-inositol in cultured *Fraxinus callus* / K.E. Wolter, L.L. Murmanis / New phytologist Vol. 78(1) (Jan 1977)

[15] Unique polysaccharide- and glycoside-degrading enzyme complex from the wood-decay fungus *Poria placenta* / Karl E. Wolter, Terry L. Highley, Faye J. Evans / Biochemical and biophysical research communications Vol. 97(4) (Dec. 31, 1980)

[16] Conversion of xylose to ethanol under aerobic conditions by *Candida tropicalis* / T.W. Jeffries / Biotechnology letters Vol. 3(5) (May 1981)

[17] Biological utilization of wood for production of chemicals and foodstuffs / George J. Hajny / FPL–RP–385 (1981)

[18] Mycor process for color removal from bleach plant effluent: bench scale studies / A.G. Campbell, E.D. Gerrard, T.W. Joyce, H.–M. Chang, T.K. Kirk / Tappi research and development division conference proceedings Atlanta: Tappi Press (1982)

[19] Lignin-degrading enzyme from the hymenomycete *Phanerochaete chrysosporium* Burds. / Ming Tien, T. Kent Kirk / Science, New Series, Vol. 221(4611) (Aug. 12, 1983)

[20] Lignin-degrading enzyme from *Phanerochaete chrysosporium*: purification, characterization, and catalytic properties of a unique H_2O_2-requiring oxygenase / T.K. Kirk / National Academy of Science, USA. Proceedings Vol. 81 (Apr. 1984)

[21] Discovery and promise of lignin-degrading enzymes / T.K. Kirk / New horizons for biotechnological utilization of the forest resource, Falun, Sweden: Marcus Wallenberg Foundation, Symposia proceedings 2 (1985)

[22] Oxidation of polycyclic aromatic hydrocarbons and dibenzo [rho]-dioxins by *Phanerchaete chrysosporium* ligninase / Kenneth E. Hammel, B. Kalyanaraman, T. Kent Kirk / Journal of biological chemistry Vol. 261(36) (1986)

[23] Micromorphology of degradation in western hemlock and sweetgum by the white-rot fungus *Coriolus versicolor* / T.L. Highley / Holzforschung Vol. 41(2)(1987)

[24] Enzymatic "combustion": The microbial degradation of lignin / T.K. Kirk, R.L. Farrell. / Ann. Rev. Microbiol. 41:465–505 (1987)

[25] Biotechnology in pulp and paper manufacture: applications and fundamental investigations /Proceedings of the fourth international conference on biotechnology in the pulp and paper industry / edited: T. Kent Kirk, Houmin Chang / Boston: Butterworth–Heinemann (1990)

[26] White rot fungi in the treatment of hazardous wastes / R.T. Lamar, J.A. Glaser, T.K. Kirk / In G F. Leatbam (ed.) Frontiers in industrial mycology: Proceedings of industrial mycology symposium, 1990 June 25–26; Madison, p. 127–143, EI. Routledge, Chapnan & Hall: New York (1992)

[27] The potential of white-rot fungi in bioremediation / T.K. Kirk, Richard T. Lamar, John A. Glaser / Biotechnology and environmental science: molecular approaches. Proceedings of an international conference on biotechnology and environmental science: molecular approaches; 1990 August 21–24; Bangkok, Thailand. New York: Plenum, (1992)

[28] Immuno-scanning electron microscopic localization of extracellular wood-degrading enzymes within the fibrillar sheath of the brown-rot fungus *Postia placenta* / Frederick Green III, C.A. Clausen, M.J. Larsen, T.L. Highley / Canadian journal of microbiology. Vol. 38(9) (Sept. 1992)

[29] Solid-phase treatment of a pentachlorophenol-contaminated soil using lignin-degrading fungi / Richard T. Lamar, James W. Evans, John A. Glaser / Environmental science & technology Vol. 27(12) (Nov. 1993)

[30] Field evaluation of the lignin-degrading fungus *Phanerochaete sordida* to treat creosote-contaminated soil / Mark W. Davis, J.A. Glaser, J.W. Evans, R.T. Lamar / Environmental science & technology. Vol. 27(12) (Nov. 1993)

[31] Biopulping: a glimpse of the future? / T. Kent Kirk Kirk, J.W. Koning, Jr., R.R. Burgess, M.V. Akhtar, R.A. Blanchette, D.C. Cameron, D. Cullen, P.J. Kersten, E.N. Lightfoot, G.C. Myers, I.B. Sachs, M.S. Sykes, M.B. Wall / FPL–RP–523 (1993)

[32] Strain selection, taxonomy, and genetics of xylose-fermenting yeasts / T.W. Jeffries, C.P. Kurtzman / Enzyme and microbial technology Vol. 16 (Nov. 1994)

[33] Xylitol formation and key enzyme activities in *Candida boidinii* under different oxygen transfer rates / Eleonora Vandeska, Slobodanka Kuzmanova, Thomas W. Jefferies / Journal of fermentation and bioengineering 80(5): 513–516 (1995)

[34] Development of improved xylose fermenting yeasts / Thomas W. Jeffries, K. Dahn, H.K. Sreenath, B.P. Davis / Proceedings of 2nd biomass conference of the Americas: energy, environment, agriculture, and industry, August 21–24, 1995, Portland, OR. / National Renewable Energy Laboratory, Golden, CO (1995)

[35] Roles for microbial enzymes in pulp and paper processing / T. Kent Kirk, Thomas W. Jeffries / In: Jeffries, Thomas W., Viikari, Liisa, eds. Enzymes for pulp and paper processing. ACS symposium series 655. Proceedings, 211th ACS national meeting; 1996 March 24–28; New Orleans, LA. American Chemical Society: Washington, DC: 2–14. Chap. 1 (1996)

[36] Recent advances on the molecular genetics of ligninolytic fungi / Daniel J. Cullen / Biotechnology 53: 273–289 (1997)

[37] Antibody-mediated immunochemistry and immunoassay in plant-related diseases / Carol A. Clausen, Frederick Green III / Methods in plant biochemistry and molecular biology. Boca Raton: CRC Press Chap. 6: 73–88. (1997)

[38] Overview of biomechanical and biochemical pulping research / M. Akhtar, G.M. Scott, R.E. Swaney, T.K. Kirk / In: Eriksson, Karl-Erik L.; Cavaco-Paulo, Artur, eds. Enzyme applications in fiber processing. Proceedings, ACS symposium; 1997 April; San Francisco, CA. Washington, DC; American Chemical Society: 15–27 (1997)

[39] Nondestructive elemental analysis of wood biodeterioration using electron paramagnetic resonance and synchrotron X-ray fluorescence / B.L. Illman, S. Bajt / International biodeterioration & biodegradation Vol. 39(2–3) (1997)

[40] Fungal degradation of lignin / K.E. Hammel / In: Cadisch, G., Giller, K.E., eds. Chap. 2. Driven by nature: plant litter quality and decomposition. United Kingdom: CAB International: 33–45. (1997)

[41] Enzymology and molecular genetics of wood degradation white-rot fungi / T.K. Kirk, D. Cullen / In R.A. Young and M. Akhtar (eds.) Environmentally friendly technologies for the pulp and paper industry, p. 273–307, John Wiley: New York (1997)

[42] Biokraft pulping of kenaf and its bleachability / Aziz Ahmed, Gary M. Scott, Masood Akhtar, Gary C. Myers / In: North American nonwood fiber symposium. Proceedings, 1998 Tappi proceedings; 1998 February 17–18; Atlanta, GA. Atlanta, GA: TAPPI PRESS: 231–238.(1998)

[43] Economic evaluation of biopulping / Gary M. Scott, Masood Akhtar, Ross E. Swaney / In: Proceedings, 7th international conference on biotechnology in the pulp and paper industry; 1998, June 16–19; Vancouver, BC, Canada. Montreal, Quebec, Canada: Canadian Pulp and Paper Assoc. Poster presentations Vol. B: B3–B6. (1998)

[44] Genetic engineering of *Pichia stipitis* for the improved fermentation of xylose / T.W. Jeffries, N.Q. Shi, J.Y. Cho, P. Lu, K. Dahn, J. Hendrick, H.K. Sreenath, B.P Davis / In: Proceedings, Bioenergy '98: expanding bioenergy partnerships; 1998 October 4–8; Madison, WI. Madison, WI: Great Lakes Regional Biomass Energy Program: 843–851 (1998)

[45] Biomechanical pulping: a mill-scale evaluation / Masood Akhtar, Gary M Scott, Ross E Swaney, Mike J. Lentz, Eric G Horn, Marguerite S. Sykes, Gary C. Myers / In: Proceedings, 1998 TAPPI international mechanical pulping conference; 1999 May 24–26; Houston, TX. Atlanta, GA: TAPPI Press: 1–10. (1999)

[46] Oxalic acid pretreatment for mechanical pulping greatly improves paper strength while maintaining scattering power and reducing shives and triglycerides / Ross Swaney, John H. Klungness / 2003 TAPPI Fall Technical Conference: Engineering, Pulping & PCE & I: Atlanta, GA: TAPPI Press, (2003)

[47] Differential stress-induced regulation of two quinone reductases in the brown rot basidiomycete *Gloeophyllum trabeum* / Roni Cohen, Melissa R. Suzuki, Kenneth E. Hammel / Applied environmental microbiology 70: 324–331. (2004)

[48] Enzymology and molecular biology of lignin degradation / D. Cullen, P.J. Kersten / Mycota: a comprehensive treatise on fungi as experimental systems for basic and applied research. III, Biochemistry and molecular biology. Berlin; London: Springer, 2004: p. 249–273 (2004)

[49] Genome sequence of the lignocellulose degrading fungus *Phanerochaete chrysosporium* strain RP78 / Diego Martinez, Luis F. Larrondo, Nik Putnam, Maarten D. Sollewijn Gelpke, Katherine Huang, Jarrod Chapman, Kevin G. Helfenbein, Preethi Ramaiya, J. Chris Detter, Frank Larimer, Pedro M. Coutinho, Bernard Henrissat, Randy Berka, Dan Cullen, Daniel Rokhsar / Nature biotechnology Vol. 22(6) (June 2004)

[50] Fungal hydroquinones contribute to brown rot of wood / Melissa R. Suzuki, Christopher G. Hunt, Carl J. Houtman, Zachary D. Dalebroux, Kenneth E. Hammel / Environmental Microbiology Vol. 8(12) (2006)

[51] Genome sequence of the lignocellulose-bioconverting and xylose-fermenting yeast *Pichia stipitis* / Thomas W. Jeffries, Igor V. Grigoriev, Jane Grimwood, Jose M. Laplaza, Andrea Aerts, Asaf Salamov, Jeremy Schmutz, Erika Lindquest, Paramvir Dehal, Harris Shapiro, Yong-Su Jin, Volkmar Passoth, Paul M.Richardson / Nature biotechnolgy Vol. 25(3) (March 2007)

[52] Chlorination of lignin by ubiquitous fungi has a likely role in global organochlorine production / Patricia Ortiz-Bermudez, Kolby C. Hirth, Ewald Srebotnik, Kenneth E. Hammel / PNAS Vol. 104(10) (March 6, 2007)

[53] Vapor-phase diethyl oxalate pretreatment of wood chips. Part 1, Energy savings and improved pulps / William Kenealy, Eric Horn, Carl Houtman / Holzforschung Vol. 61 (2007)

[54] Vapor-phase diethyl oxalate pretreatment of wood chips. Part 2, Release of hemicellulosic carbohydrates / William Kenealy, Mark Davis, Carl J. Houtman / Holzforschung Vol. 61 (2007)

[55] Xylose-fermenting recombiant yeast strains / Thomas W. Jeffries, Haiying Ni, Jose M. Laplaza / US Patent No. 7,285,403 (Oct. 10, 2007)

[56] Method for producing pulp / Masood Akhtar, Eric G. Horn, Carl J. Houtman / US Patent No. 7,306,698 (December 11, 2007)

Economics

With the growing U.S. population, now over 300,000,000, projections of public demand for forest products and trends in industry are very helpful in planning for future wood products and wood energy technology needs. By understanding these trends, we can also anticipate changes needed in forest management. For, example, in the United States, it is estimated that each person consumes the equivalent of 70 cubic feet of timber each year. Presently this wood is provided by what is harvested in the United States and imports of wood from around the world. Understanding the needs of the public and the options for meeting them can help in planning programs that enhance forest management and wood technologies to meet those needs. [10, 11, 13, 21, 22, 34, 44, 45]

Challenge: What are the effects of changing population demands, changing industry size and technology, and international competition on forestry? How do new or improved forest products play a role in changing forest stewardship and sustainability? The U.S. Forest Service is charged with managing 193 million acres of public forests and grasslands. The uses are many, including recreation, water, wood, wildlife, grazing, and wilderness. How best to manage these different and sometimes conflicting uses is complex.

FPL's Contributions: The FPL prepares economic studies on wood product consumption, trade, and technology changes. The results are available to the American Public. Studies also include economic evaluation of new products or processes that FPL develops.

Results: The public has an unbiased source of information, and planners have a solid source of basic information to help make better decisions in future planning, all of which greatly help to sustain the vast public—and privately owned—forests.

Selective Logging

Some of the earliest economic work (1930s) at FPL was to determine costs and returns for clear cutting timber versus selective logging at numerous commercial operations in both the north and the south. The results showed that it is unprofitable to harvest trees below a certain diameter limit. This information supplied an economic incentive for the adoption of the forestry practice of selective logging of larger trees in place of the then prevailing "cut out and get out" practice of the 1930s. [2, 4]

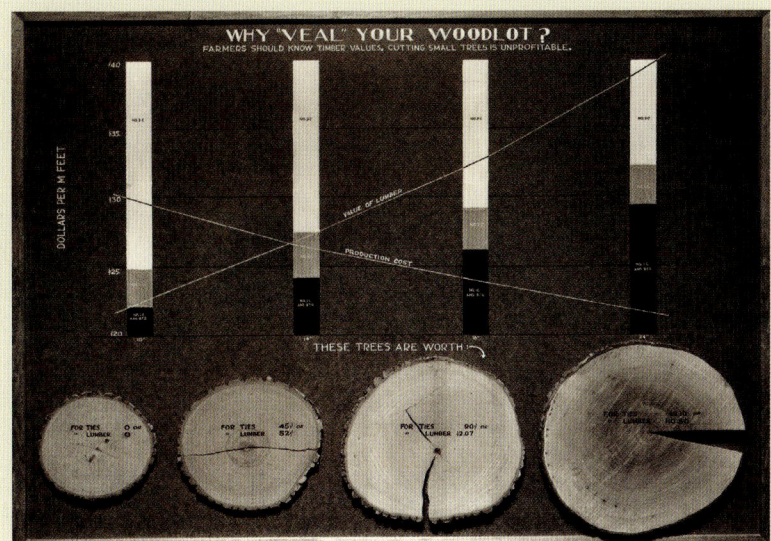

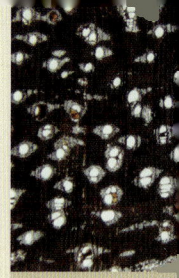

◀ Logging studies show a diameter limit for each forest type below which it is unprofitable to log. (1930s) {M-32144-F}

U.S. Wood-Based Industry

In 1984, Paul Ellefson, University of Minnesota, and Bob Stone wrote a book on the industrial organization and performance of the U.S. wood-based industry. Though many of the numbers are now out of date, the insight into what, when, why, where, and how of the industry is still useful for someone trying to understand this large, fragmented, and complex industry. [13]

U.S. Wood Products Consumption Compared with U.S. Production

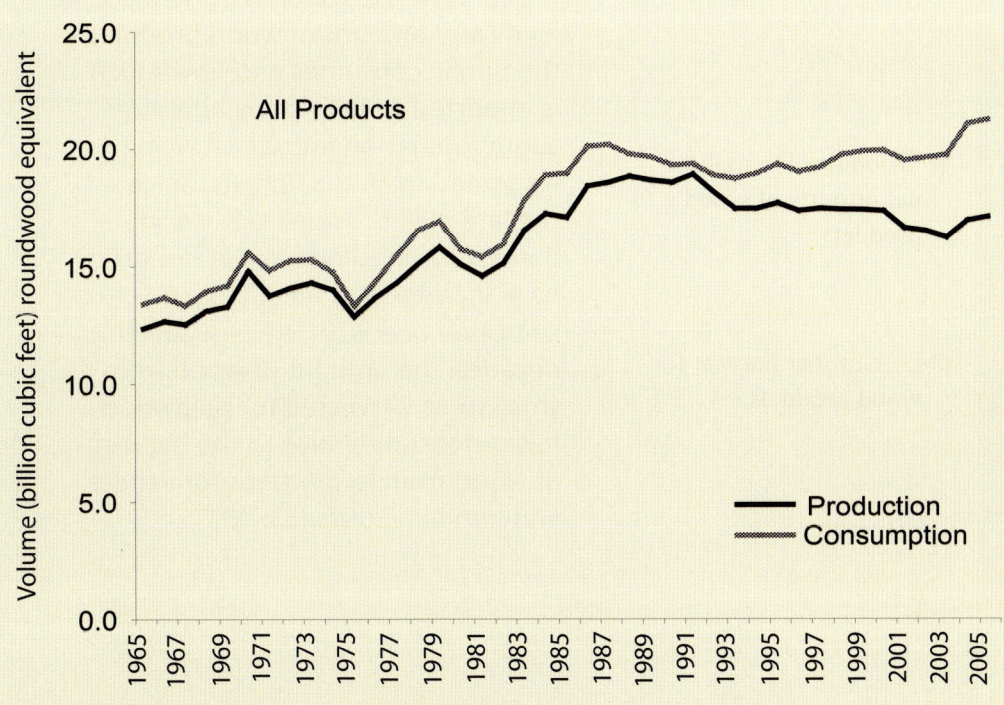

◀ Total roundwood production (U.S. harvest) increased to 17.2 billion cubic feet in 2005. Roundwood harvest ("Production" in the graph) has declined since the mid-1990s. The high point was 1991 when total roundwood production was 18.8 billion cubic feet. While U.S. harvest has declined, roundwood consumption has increased, with the difference between consumption and production made up by increasing imports. (2007) [45]

Tree Growth versus Removal

A key concern in monitoring the use of wood for products is how much timber remains in the forests. For more than half a century, the growth of timber in forests in the United States has exceeded removals with resultant increases in timber inventory.

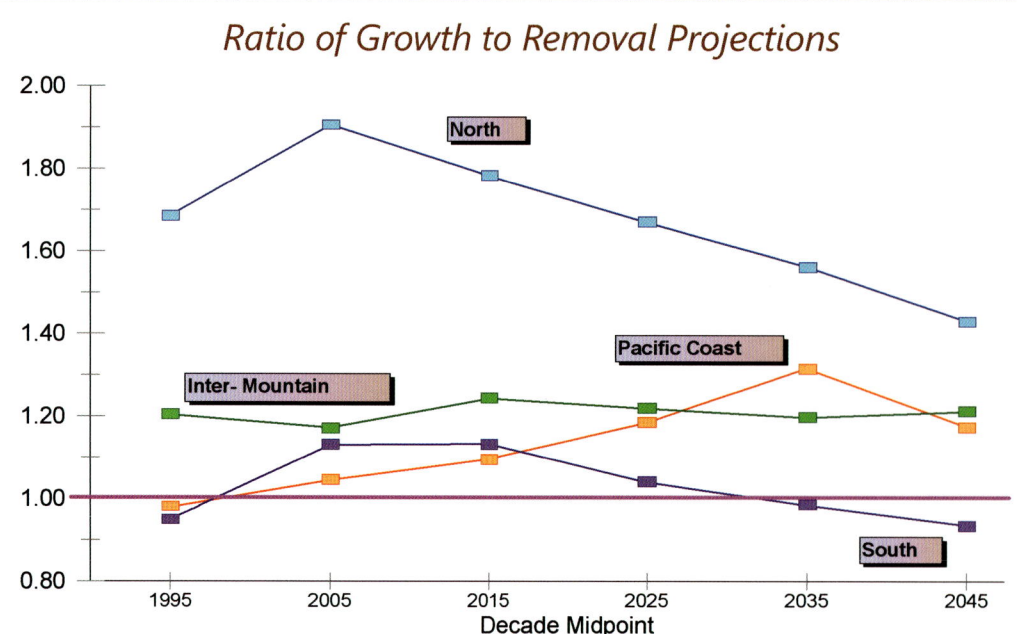

Ratio of Growth to Removal Projections

◀ One important measure of the impact of timber harvesting is whether growth of trees exceeds removal of trees. Example projections from an economic model suggest that the ratio of growth to removals will be above 1.0 for all regions through 2035 on U.S. private timber land. In 2045, heavy harvest of hardwoods may result in removals exceeding growth. This suggests we should closely watch hardwood harvest levels in the South. (2003) [36]

Survey of Wood Utilization in Major End Uses

FPL has collaborated for many years with industry groups to conduct surveys of wood use in housing, nonresidential construction, and manufacturing. The results show how the public is shifting consumption from lumber toward panel products and using more engineered wood products such as wood roof trusses, wood floor I-joists, oriented strandboard (OSB), engineered wood floors, laminated veneer lumber (LVL), and glulam beams. [32]

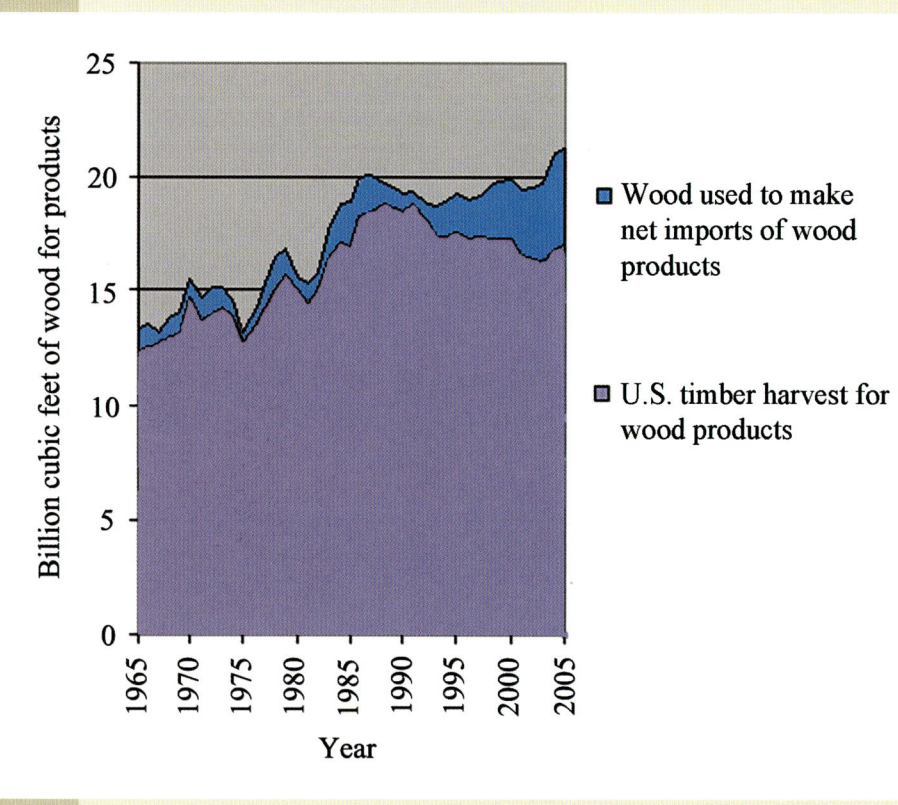

◀ Economics research reports the kinds and amount of wood products the public consumes and how much is imported to help show how consumption has an impact on wood harvested from U.S. forests. Since the late 1980s, overall U.S. wood harvest has declined as net imports have increased to meet U.S. consumption needs. In the past several decades, the amount of wood from small trees harvested for pulpwood has increased relative to the harvest of wood from larger trees for veneer and lumber. (2007) [33, 45]

Survey of Wood Used in Navy Buildings

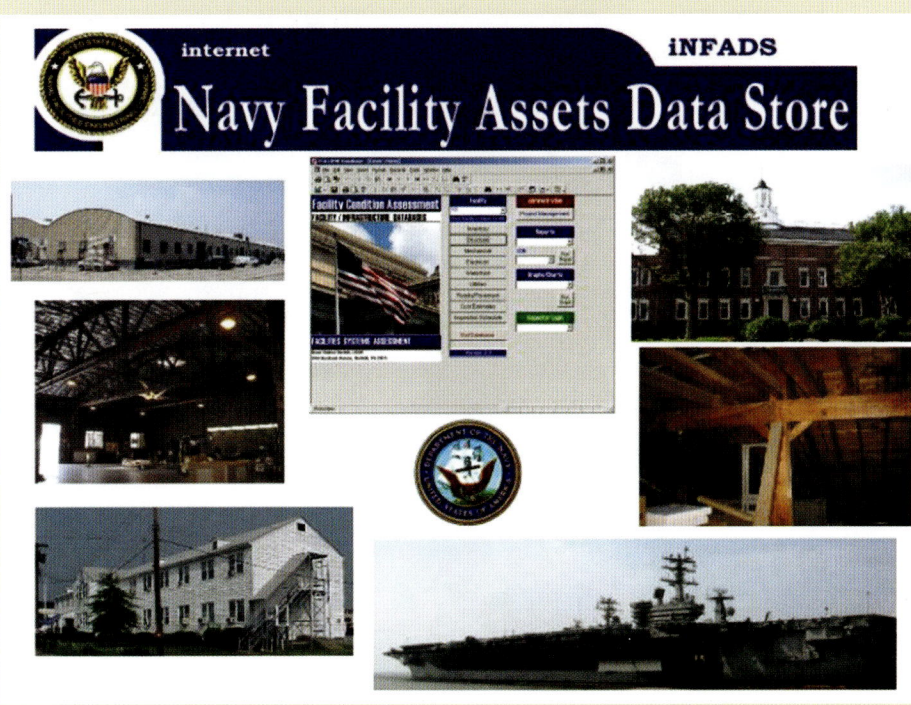

◄ Buildings at Naval Station (NAVSTA), Norfolk, Virginia, as part of the U.S. Navy's Advanced Wood Composites for Naval Facilities Project, sponsored by the Office of Naval Research. The principal objectives of this initiative were to identify areas and applications where newly developed wood–plastic composite materials could be used to extend the service life of buildings while reducing required maintenance. [37]

Potential of Forest Residues as an Energy Source

A study was conducted to evaluate the economic feasibility of using residues from harvesting operations, disease and insect mortality, and blow downs for energy production. This study helped various sectors of industry, manufacturers of combustion equipment and other research organizations in planning research and potential expansion. [19]

Residential Fuelwood Study

Working with the University of Wisconsin Survey Research Laboratory, the FPL conducted a nationwide telephone survey of about 5,500 households to learn about residential wood burning. The survey results showed that in 1980–1981, residential fuelwood use was 3.4 billion cubic feet, roughly one-fourth the amount used for all other wood products. [15]

Trends in Softwood Sawmill Capacity

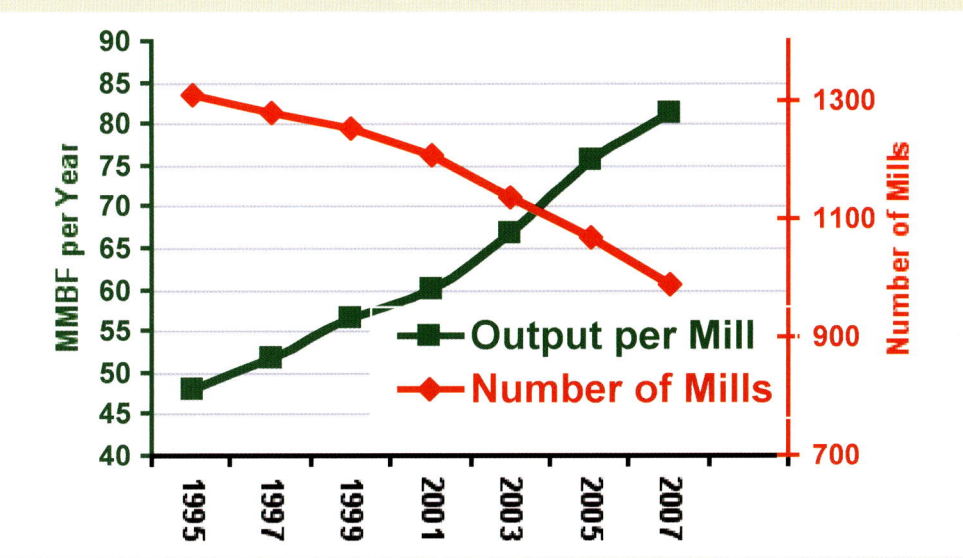

◄ Studies of individual industries like the softwood lumber industry show how the U.S. has responded to global competition by improving efficiency of production. Softwood sawmills have consolidated into fewer, larger, and more efficient mills. The number of permanent, significant operations has fallen below 1,000. Production has been maintained with more output per mill. (2007) [47]

Projecting Wood Demand in Housing

▲ Robert Stone, Tom Marcin, Cherilyn Hatfield, and George Harpole (left to right) discussing projections of wood products demand for housing in the United States. In the 1970s, Tom Marcin developed a computer model for projecting long-term demand for housing by type of unit and region. Projections were based on assumptions of population and economic growth. This research is continued by Dave McKeever. [5, 6, 25]

Evaluating Impact of Paper Recycling

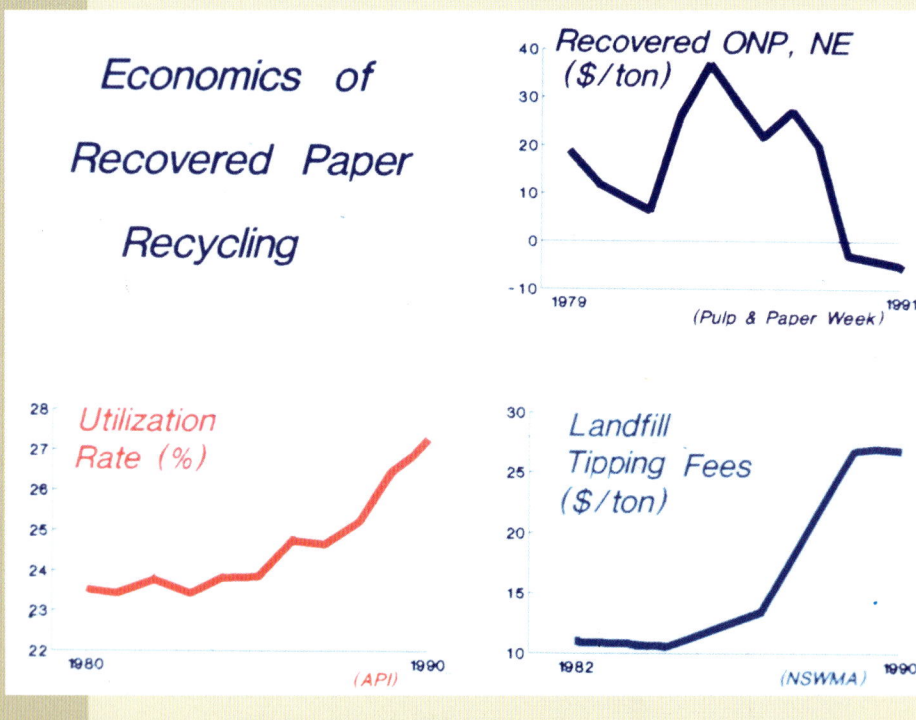

◀ Early examples of the type of information gathered and analyzed were trends associated with paper recycling. Identifying trends in industrial practices like the recycling of paper can affect forest management decisions such as species selection for replanting, length of time trees should grow before harvest, and utilization of existing species. (1980s) {12P MA} [23, 26, 27]

Evaluating Economic Potential for New Laminated Products

◀ Peter Ince discussing the economic potential of the FPL-developed Press-Lam process with Tom Ellis. (1970s) {27M} [7]

George Harpole prepared economic models for structural flakeboard production. [8]

North American Pulp and Paper Model (NAPAP)

A model was developed to make 50-year projections of markets for pulpwood, pulp, and paper. The model shows how changes in pulp and paper technology will increase use of hardwoods and recycled paper to meet increasing consumption needs. [45]

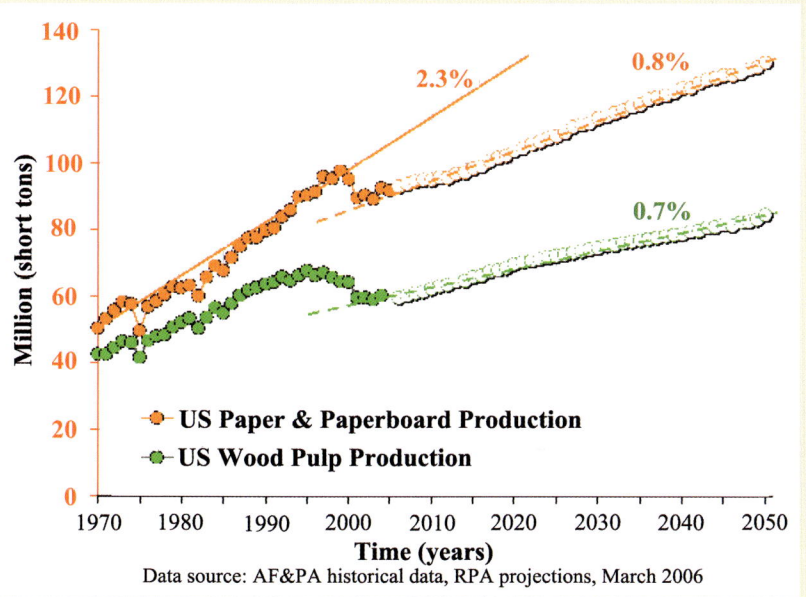

Data source: AF&PA historical data, RPA projections, March 2006

◀ U.S. paper, paperboard, and pulp production projected to 2050. Paper and paperboard = paper + paperboard + building paper. Wood pulp includes estimated dissolving pulp production. (2007) [45] (Data from AF&PA (historical data) and RPA projections, March 2006.)

Computer Model of an Unbleached Kraft Paperboard Plant

Peter Ince developed a detailed model of a Kraft papermill process and made it available to pulp and paper researchers. The model provides an opportunity to see what would occur in a real-life plant if certain changes were made at various stages in the Kraft paperboard manufacturing process. The following diagram shows the general steps in the model for Kraft papermaking. [12]

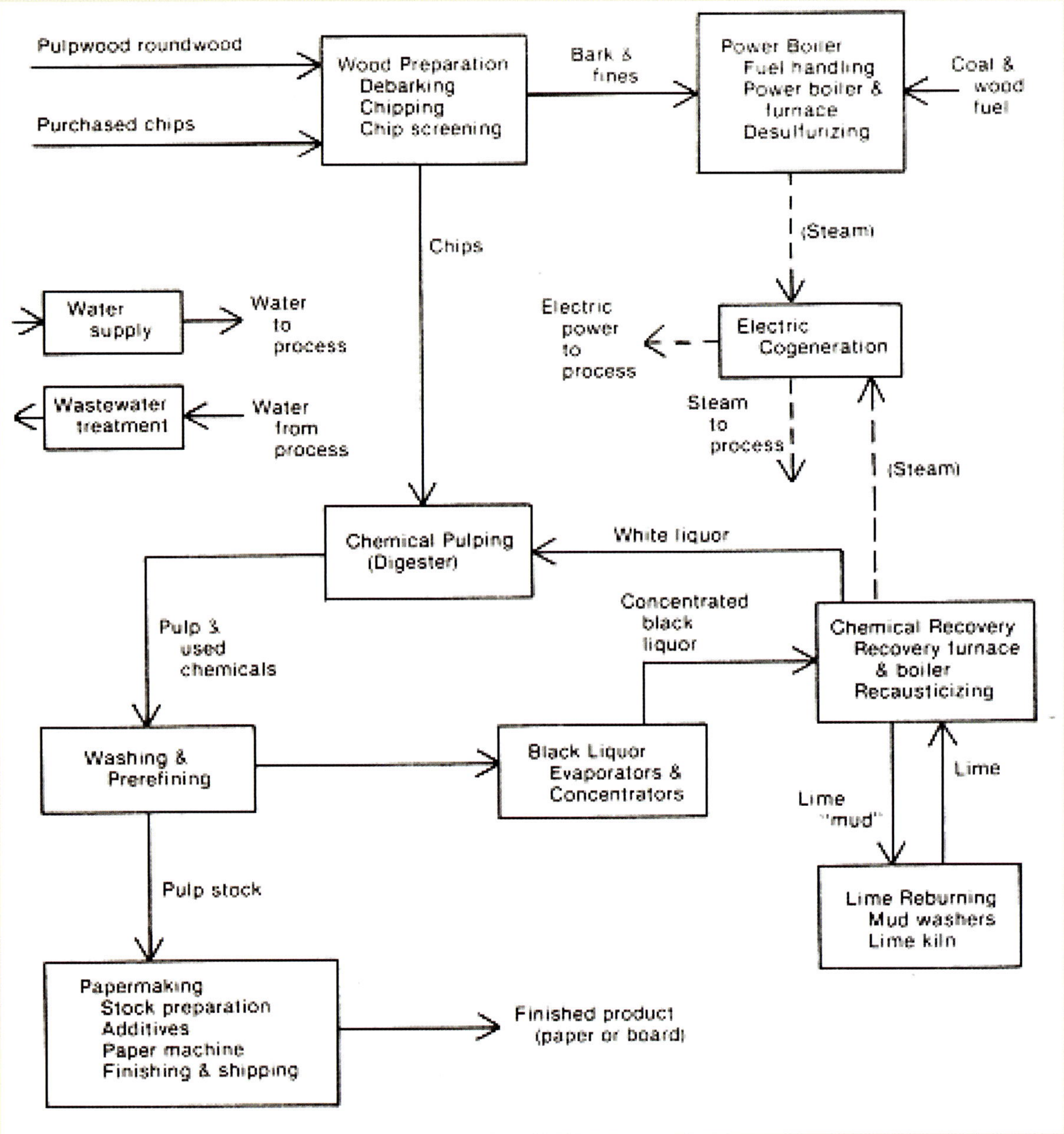

Diagram labels:

Pulpwood roundwood → Wood Preparation / Debarking / Chipping / Chip screening

Purchased chips →

Bark & fines → Power Boiler / Fuel handling / Power boiler & furnace / Desulfurizing ← Coal & wood fuel

Chips

Water supply → Water to process

Wastewater treatment ← Water from process

Electric power to process ← Electric Cogeneration

(Steam)

Steam to process

(Steam)

Chemical Pulping (Digester) ← White liquor

Pulp & used chemicals

Concentrated black liquor

Chemical Recovery / Recovery furnace & boiler / Recausticizing

Washing & Prerefining → Black Liquor Evaporators & Concentrators

Lime "mud"

Lime

Lime Reburning / Mud washers / Lime kiln

Pulp stock

Papermaking / Stock preparation / Additives / Paper machine / Finishing & shipping → Finished product (paper or board)

▶ To determine if a new technology can make money, economists work with engineers to build engineering process models and use them to estimate how costs and revenues change with improvements in the processes. This diagram shows the general steps in a model for Kraft papermaking. (1984) [12]

Evaluating New Wood Energy Technologies

Reducing Fuel Loading

An emerging issue is that at least 51 million acres of national forestland could benefit from some type of mechanical treatment to reduce fuel loading and wildfire risk. Biomass from these forest health treatments has the potential to be converted into 14 billion kilowatt-hours of electricity annually. One barrier to utilizing this fuel is the cost of transporting it. Electrical generators located on or near forest lands utilizing gasified wood would be one way to overcome the transport problem. [29]

Cost of Converting Wood to Electricity

To address the cost issues associated with converting wood to electricity, FPL economists conducted a study with support from the National Fire Plan. The study found that small wood gasifiers that generate electricity could be economic if placed at a facility that could use both the electricity and excess heat (such as for space heating). The report includes a spreadsheet tool that may be used to analyze any scale of project involving wood gasification and electrical generation. [39]

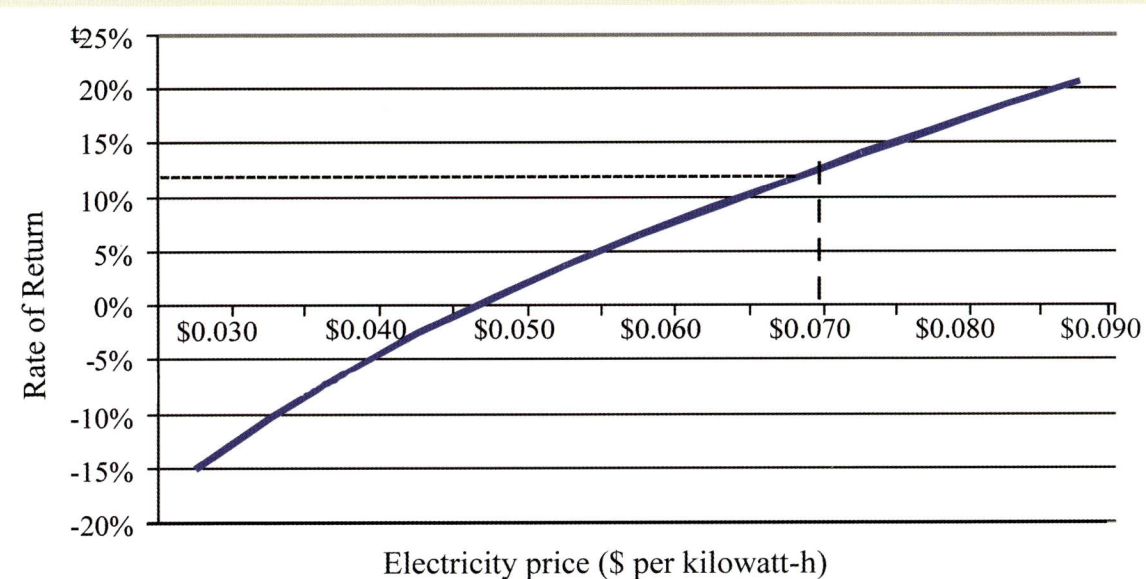

▲ A 2005 economic study suggested that mobile wood gasifiers that produce 100 kilo-watt-hours of electricity may be an economical way to use small trees from fire-hazard-reduction thinnings if the price of electricity in an area is high enough. This graph shows that investment in a 100-kW unit may provide a rate of return of 12% if electricity costs 7 cents per kilowatt-hour. [39]

Life-Cycle Inventory and Life-Cycle Assessment of Wood Products and Structures

Life-cycle inventory (LCI) provides an accounting of the energy and waste associated with the creation of a product through use and disposal. In this study, the gate-to-gate LCI tracks softwood lumber production from softwood logs stored in the log yard to planed dry lumber leaving the planing process. Life-cycle analysis (LCA) is a broader examination of the environmental and economic effects of a product at every stage of its existence, from harvesting to disposal and beyond. [49]

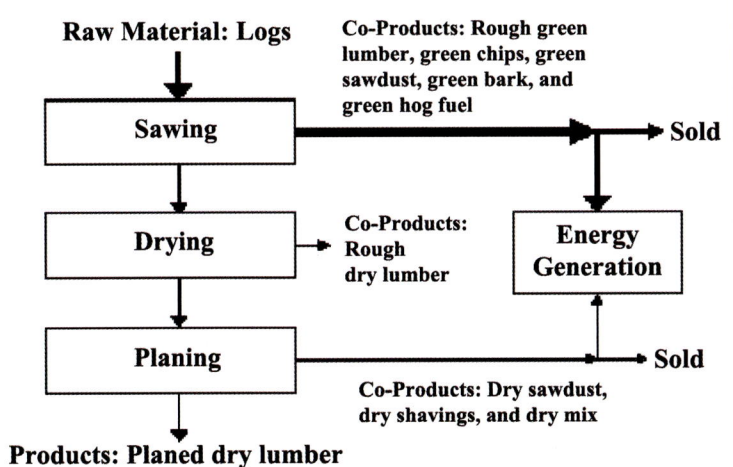

◀ Description of the three main processes for hardwood lumber manufacturing showing material flow. This LCI study showed that 44% of hardwood logs are converted to planed dry hardwood lumber. Energy consumption, per cubic meter of lumber, included 608 MJ of electricity and 5,800 MJ of thermal energy. Burning green wood residues on-site generated the most energy. Emission data, produced through modeling, estimated total biomass (biogenic) and fossil (anthropogenic) carbon dioxide production of 428 kg per m³ and 139 kg per m³, respectively. (2008) [49]

"Green building" is defined as the practice of improving energy efficiency for materials, construction, and operation while reducing the overall environmental impact of building. Two percent ($7.4 billion) of new residential starts in 2005 were classified as "green buildings," and the minimum market share is expected to increase to 5% ($19 billion) by 2010. Economic costs, energy consumption, and environmental impact of residential building products are playing an increasingly large part due to increased public awareness on environmental issues. In 2003, residential building used 21.7 billion board feet of softwood

lumber in the United States. Developing a sound policy for building practices, especially for green building, must be a priority if the United States is to decrease its environmental burden on the world's resources. However, more scientific evidence is needed to evaluate claims for green building materials. [42, 47–49]

Carbon Storage in Wood Products

In recent years all countries have been required to report their greenhouse gas emissions and additions of carbon to sinks such as forests. FPL scientist Ken Skog contributed to the work of the Intergovernmental Panel on Climate Change (IPCC) to provide methods every country can use to estimate annual additions of carbon to wood products in use and wood products in landfills. [46]

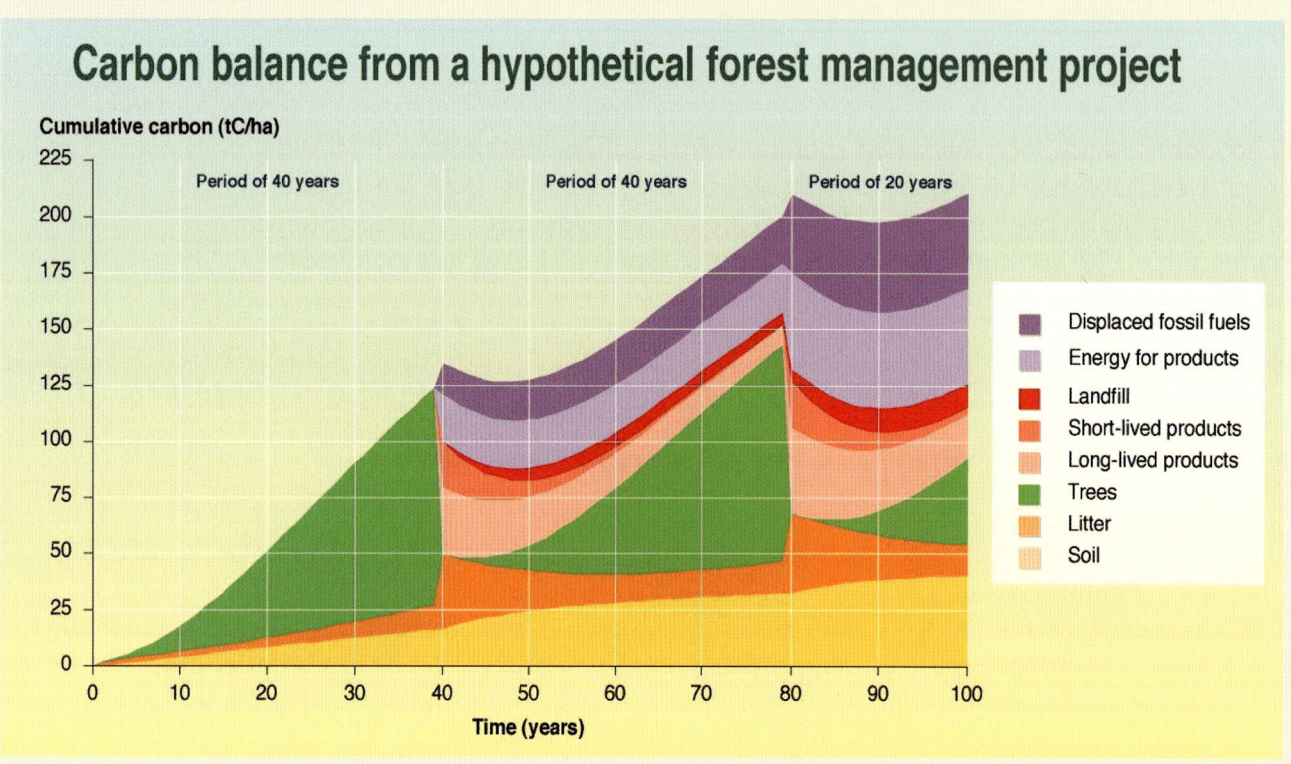

Carbon balance from a hypothetical forest management project

Cumulative carbon (tC/ha)

Period of 40 years   Period of 40 years   Period of 20 years

Displaced fossil fuels
Energy for products
Landfill
Short-lived products
Long-lived products
Trees
Litter
Soil

Time (years)

▲ Ken Skog participated on teams for two projects for the United Nations Intergovernmental Panel on Climate Change (IPCC), which was awarded the 2007 Noble Peace Prize along with former U.S. Vice President Al Gore. Skog has been involved with the IPCC since 1998, and his work contributed to the publication of two reports, *Good Practice Guidance for Land Use, Land Use Change and Forestry*, published in 2003, and the 2006 *IPCC Guidelines for National Greenhouse Gas Inventories*. [46]

◀ Ken Skog participating on a panel.

1910

Limited basic information on the utilization of wood.

2010

Detailed current and projected economic information on wood use and its impact on forest management

Further Information

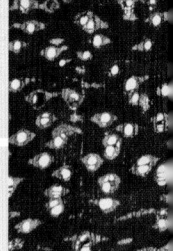

[1] Naval stores industry / A.W. Schorger, H.S. Betts / USDA Agriculture bulletin 229 (1915)

[2] Selective logging versus clear cutting in shortleaf pine / R.D. Garver / Southern lumberman Vol. 141(1789) (Oct. 15, 1930)

[3] Some accomplishments of the Forest Products Laboratory (February 1, 1937)

[4] Some Forest Products Laboratory developments of particular significance to the South / FPL report 0-8 W73Sf (April 15, 1940)

[5] Projections of demand for housing by type of unit and region / Thomas C. Marcin / USDA Agriculture Handbook No. 428 (1972)

[6] Outlook for housing by type of unit and region: 1978 to 2020 / Thomas C. Marcin / FPL Research paper 304 (1977)

[7] Economic feasibility of red oak press-lam for upholstered furniture framestock / W.L. Hoover, C.A. Eckelman, R.W. Jokerst, J.A. Youngquist / Forest products journal Vol. 29(11) (Nov. 1979)

[8] Economic models for structural flakeboard production / G.B. Harpole / Forest products journal Vol. 29(12) (Dec 1979)

[9] EVALUE: a computer program for evaluating investments in forest products industries / Peter J. Ince, Philip H. Steele / FPL–GTR–30 (January 1980)

[10] Trends and patterns in wood products consumption and production / C.D. Risbrudt / Timber demand: the future is now / R.N. Stone / Madison, WI: Forest Products Research Society (1981)

[11] A Competitive assessment of the U.S. solid wood products industry / U.S. Department of Commerce, International Trade Administration, Basic Industries Sector, Office of Forest Products and Domestic Construction. Washington, D.C. (1984)

[12] Computer model for economic study of unbleached Kraft paperboard production / Peter J. Ince / FPL–GTR–42 (1984)

[13] U.S. wood-based industry: organization of the nation's timber enterprises / Paul V. Ellefson, Robert N. Stone / New York Praeger {1984}

[14] Impacts of forest utilization research: an economic assessment. St. Paul: College of Forestry, University of Minnesota, 1983. Final report, project no. USDA–FP–81–0395. This report includes a supplementary analysis / K. Skog, R. Haynes, written and added in (1986)

[15] Residential fuelwood use in the United States: 1980–81 / Kenneth E. Skog, Irene A. Watterson / Forest Service resource bulletin WO 3 (1986)

[16] PAPYRUS: a model of the North American pulp and paper industry. Forest science. Vol. 33(1) suppl., monograph 28 (Mar. 1987)

[17] The stumpage market impact of timber utilization research / K.E. Skog / Forest products journal Vol. 37(6) (June 1987)

[18] Modeling technological change in wood products processing / Peter J. Ince / Forest sector and trade models: theory and application: proceedings of an international symposium, November 3–4, 1987. Seattle, Wash.: Center for International Trade in Forest Products, University of Washington, College of Resources (1988)

[19] Sources and uses of wood for energy / John I. Zerbe, Kenneth E. Skog / Proceedings from international symposium: energy options for the year 2000: contemporary concepts in technology and policy. Newark, Del. Center for Energy and Urban Policy Research, University of Delaware, (1988)

[20] National measures of forest productivity for timber / Peter J. Ince, H.E. Dickerhoof, H.F. Kaiser, J.F. Fedkiw / FPL–GTR–61 (1989)

[21] Using forest sector models to estimate the potential impact of current wood utilization research / K. Skog, I. Durbak, J.L. Howard, H.N. Spelter, D.P. Bradley, R.W. Haynes, D. Adams / Proceedings of the forest sector analysis symposium: August 13–18, 1989, Soderfors, Sweden. Uppsala, Sweden: Swedish University of Agricultural Sciences (1989)

[22] The Impact of technological change on projections of costs and recoveries in wood products processing / Henry Spelter / Healthy forests, healthy world: proceedings of the 1988 Society of American Foresters national convention, Rochester, New York, October 16–19.Bethesda, Md.: The Society (1989)

[23] Analysis of future high levels of wastepaper use in regional pulpwood markets in the United States / J.L. Howard, I. Durbak, P.J. Ince, W.J. Lange / Proceedings of the 1989 southern forest economics workshop: restructuring to cope with changing times: San Antonio, Texas, March 1–3, 1989. Huntsville, Tex.: Champion International (1989)

[24] Recent trends in the consumption of timber products / Robert B. Phelps, D.B. McKeever, H.N. Spelter, I. Durbak, K.E. Skog, D. Finkel / An analysis of the timber situation in the United States, 1989–2040. Fort Collins, Colo. USDA Forest Service, Rocky Mountain Forest and Range Experiment Station, General technical report RM 199 (1990)

[25] Wood used in new residential construction in the United States: a Forest Service end-use survey / David B. McKeever / Proceedings of the 1988 southern forest economics workshop: forest resource economics: past, present, and future: Orlando, Florida, May 4–6, 1988. Gainesville, Fla.: School of Forest Resources and Conservation, Institute of Food and Agricultural Sciences, University of Florida, (1988)

[26] North American pulp and paper model: market trends, technological changes and impacts of accelerated paper recycling / Peter J. Ince, Don G. Roberts, Romain Jacques / Forest sector, trade and environmental impact models: theory and applications: proceedings of an international symposium, April 30–May 1, 1992. Seattle, Wash.: Center for International Trade in Forest Products, University of Washington, College of Forest Resources (1992)

[27] Impacts of recycling technology on North American fiber supply and competitiveness / Peter J. Ince, Dali Zhang, Joseph Buongiorno / What is determining international competitiveness in the global pulp and paper industry? proceedings, third international symposium, September 13–14, 1994, Seattle, Washington. Seattle, Wash.: University of Washington, College of Forest Resources, Center for International Trade in Forest Products, SP; 17 (1994)

[28] Wood products technology trends: changing the face of forestry / K.E. Skog, P.J. Ince, D.J. S. Dietzman, C.D. Ingram / Journal of forestry. Vol. 93(12) (Dec. 1995)

[29] Economic feasibility of products from inland west small-diameter timber / Henry Spelter, Rong Wang, Peter J. Ince / FPL–GTR–92 (1996)

[30] An estimation of opportunity cost for sustainable ecosystems / Howard, James L. / In: Forests, biological diversity and the maintenance of the natural heritage: Protective and environmental functions of forests. Proc. 11th world forestry congress; October 13–22; Antalya, Turkey. Vol. 2: 41–47 (1997)

[31] Review of wood-based panel sector in United States and Canada / Henry Spelter, Dave McKeever, Irene Durbak / FPL–GTR–99 (1997)

[32] Engineered wood products: a response to the changing timber resource / David B. McKeever / Pacific rim wood market report 123, Gig Harbor, WA. (November 1997)

[33] U.S. timber production, trade, consumption, and price statistics 1965–2005 / James L. Howard / FPL–GTR–116 (1999) (2007 Update FPL–RP–637)

[34] Long-range outlook for U.S. paper and paperboard demand, technology, and fiber supply-demand equilibria / Peter J. Ince / Proceedings of the Society of American Foresters 1998 national convention, Traverse City, Michigan, September 19–23, 1998. Bethesda, MD: Society of American Foresters publication SAF 99–01 (1999)

[35] Urban tree and woody yard residues / David B. McKeever, Kenneth E. Skog / FPL–RN–0290 (2003)

[36] An analysis of the timber situation in the U.S. 1950:2050 / Richard W. Haynes / Forest Service, Pacific Northwest Experiment Station PNW–GTR–560 (2003)

[37] Structural wood products in onshore buildings at Naval Station Norfolk / David B. McKeever / FPL–GTR–140 (2003)

[38] Assessing the market potential of roundwood recreation buildings / Dorothy Paun, Randall Cantrell, Susan L. LeVan-Green / FPL–GTR–144 (2004)

[39] Fuel to burn: economics of converting forest thinnings to energy using BioMax in southern Oregon / E.M. (Ted) Bilek, Kenneth Skog / FPL–GTR–157 (2005)

[40] U.S. forest products annual market review and prospects 2001–2004 / James L. Howard / FPL–RN–0292 (2004)

[41] Status and trends; profile of structural panels in the United States and Canada / Henry N. Spelter, David B. McKeever, Matthew Alderman / FPL–RP–636 (2006)

[42] Green building smartmarket report / McGraw Hill Construction. New York. HD9715.U5 G69. 44 pp. (2006)

[43] Charge out! Determining machine and capital equipment charge-out rates using discounted cash-flow analysis / E.M. (Ted) Bilek / FPL–GTR–171 (2007)

[44] Globalization and structural change in the U.S. forest sector: an evolving context for sustainable forest management / Peter Ince, Albert Schuler, Henry N. Spelter, William Luppold / FPL–GTR–170 (2007)

[45] The 2005 RPA timber assessment update / R.W. Haynes, Darius Adams, R.J. Alig, P.J. Ince, R. John, Xiaoping Zhou / PNW–GTR–699 (2007)

[46] IPCC. 2006 Guidelines for national greenhouse gas inventories / K.E. Skog, K. Pingoud / Prepared by the national Greenhouse Gas inventories Programme, H.S. Eggleston, L. Buendia, K. Miwa, T. Ngara, K. Tanabe (eds)./ Published: IGES, Japan

[47] Profile 2007: softwood sawmills in the United States and Canada / Henry Spelter, David McKeever, Matthew Alderman / FPL–RP –644 (2007)

[48] McGraw–Hill construction outlook 2008 report / R. Murray / McGraw Hill Construction. New York, NY. 32 pp. (2008)

[49] Environmental impact of producing hardwood lumber determined by life-cycle inventory / Richard D. Bergman, Scott A. Bowe / Wood and Fiber Science 40(3):448–458 (2008)

Engineering

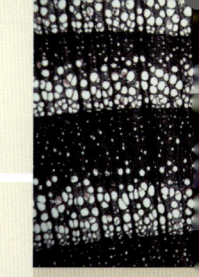

▲ A major result of FPL research on wood is contained in the Wood Handbook. A popular book that was first published in 1935, it has been updated six times, with the most recent copy published in 2010. It contains information on physical and mechanical properties of wood and how these properties are affected by variations in the wood itself and moisture. Information is also given on lumber, plywood, structural sandwich and laminated members, along with fasteners, structural analysis equations, gluing, wood drying, composites, biodeterioration, preservation, finishing, use of wood in buildings and bridges, fire safety, round timbers and ties, and specialty treatments. {30Q} [10, 33]

Prior to FPL, wood engineering research was conducted at several universities. With the establishment of FPL, much of this research was brought together to improve efficiency and help develop recognized standards for wood evaluation.

Challenge: How to efficiently use the various wood species in engineering applications.

FPL's Contributions: FPL helped demonstrate that wood can be used in engineering applications. Scientists determined the mechanical strength properties of U.S. wood species and many foreign ones. FPL determined physical properties (such as thermal, electrical, decay resistance, working qualities, weight, density, specific gravity, and shrinkage) of U.S. wood species. Scientists actively participated in development of internationally recognized standards for quantifying wood properties and assigning design properties for structural applications of wood. FPL provided guidelines to promote safe, efficient, long-term use of wood in a wide range of environments. FPL researchers performed basic research to promote the development of engineered wood composites such as glulam, parallel laminated lumber, and flake and strand composites. FPL developed prototype applications to demonstrate efficient use of sandwich panels, stress-skin panels, and truss-frame construction. FPL helped wood-using industries to improve product standards. FPL scientists provide unbiased evaluation of problems that helps consumers and industry better use wood products.

Results: Wood can now be treated as an engineering material and used in thousands of applications, from simple scaffolding to huge complex structures.

Wood Handbook Chapters

Wood Evaluation

Test Procedures

By 1937, FPL had evolved the procedures for determining mechanical properties of wood, procedures that have since been adopted as standard in many laboratories. [1, 17, 32]

◀ Wood engineering test laboratory for evaluating strength properties of small wood specimens. (1934) {M 23955 F}

▲ Tim Nelson and Director Chris Risbrudt explain the electronic setup for recording test data to U.S. Forest Service Associate Chief Sally Collins.

Wood Properties

FPL engineers determined the strength and related properties of 164 American woods, affording specific data for comparison and substitution of species. [3, 5, 7, 9, 14]

Evaluation of Clear Wood Specimens

◀ FPL display showing (top) many of the different tests used to determine engineering properties of clear wood specimens and (bottom) samples of wood species evaluated. These specific data provide direct comparison of wood properties and allow for species substitution. This research information allows the user of wood a wider choice of species, thus reducing costs. (1960s) [3]

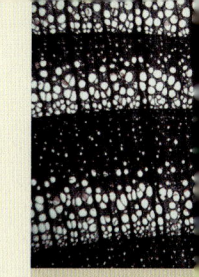

Strength Testing of Wood

FPL scientists established selection and testing procedures for determining strength properties of wood, which were adopted as standards by ASTM International (formerly the American Society for Testing and Materials, ASTM). These standards have in recent years had an important bearing on the development of comprehensive international standards sponsored by the Committee on Mechanical Wood Technology of the Food and Agricultural Organization of the United Nations. [16, 33]

◀ Test of a large wood cylindrical structure in the 1,000,000-pound capacity testing machine. This machine is also used to evaluate poles, piles, and large wood beams. (1950s) {M 115 904}

▲ Donald Doyle checking a specimen in a tester used to determine the effect of knots on tensile strength of lumber. (1960s) {M 132 102} [16]

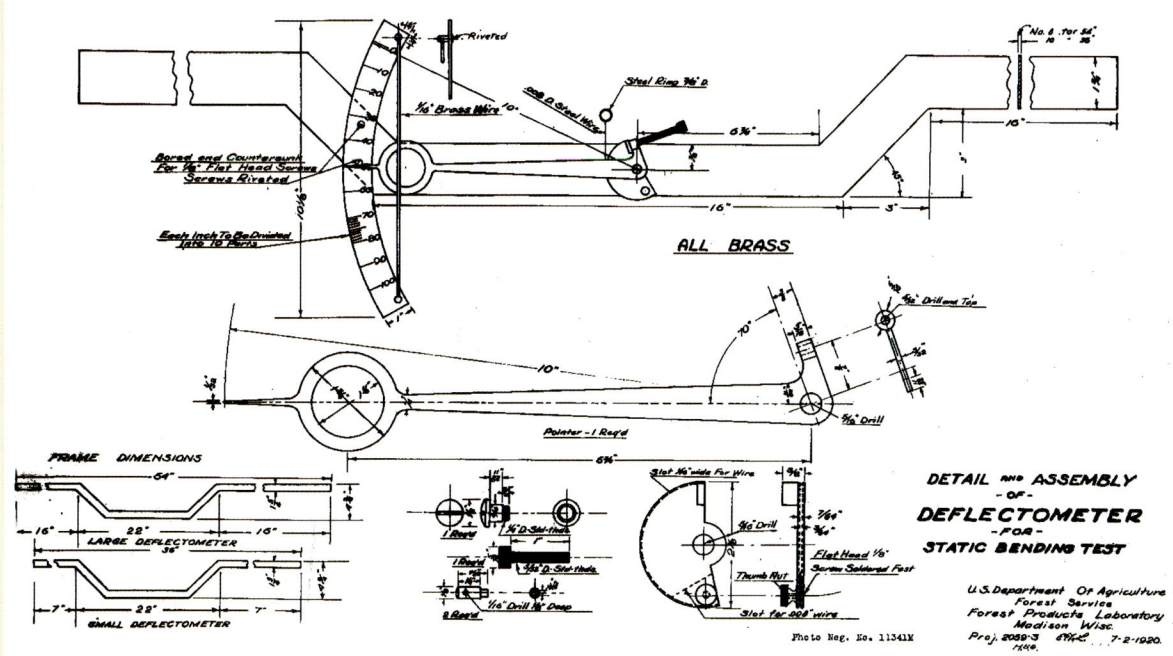

◀ Details of a bending test apparatus designed by FPL staff in 1920 to accurately determine the deflection of a beam during loading. By 1960, dial gauges accurate to 0.001 inch were used. {M 21430 F}

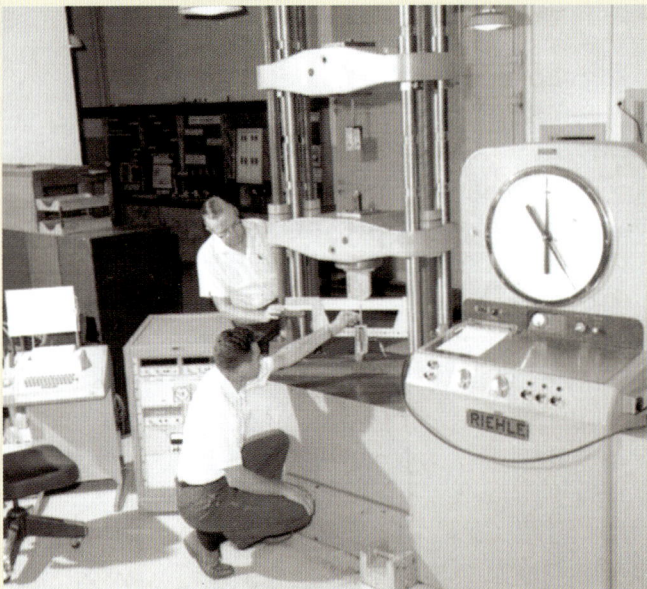

▲ Linear variable differential transducer (LVDT) used to determine the deflection of a beam during a bending test being conducted by Alan Bendtsen (kneeling) and Burt Munthe. (1970s)

▲ Karl Kanvik (left) and Russell Moody observing a large beam being tested in bending. James Vargo is filming the test. {27 D} (1970s)

Bending tests are one method of determining effects of knots, decay, and cross grain on strength properties of wood.

Toughness Testing

FPL developed a machine to test the ability of wood to absorb shock or impact loads. The toughness test procedure and machine have become standard both nationally and internationally. [33]

Strength Factors

The staff determined the effect on strength of knots, cross grain, preservative treatment, decay, moisture content, and other factors. This work has resulted in increased safety, marked improvement in efficiency, and generally increased satisfaction in wood use. [2, 6, 8, 11, 15, 18]

Low Temperature

FPL carried out research at temperatures as low as −300°F, which showed that—far from becoming weak and brittle at low temperatures—wood actually gets stronger. These data established wood's advantages for construction in frigid areas and have helped establish such new uses for wood as structural insulation in commercial barges that provide low-cost, world-wide transportation for liquid methane. [12]

Decayed Wood

FPL evaluated the properties of Douglas-fir lumber cut from timber infected with a fungus, called white pocket, to show how it could be used effectively. As a result, Douglas-fir sheathing and dimension grades are permitted to contain certain amounts of white pocket. Thus, over-mature timber previously left in the woods can be harvested for effective use. [33]

Long-Term Loading Effects

Most strength testing of wood reveals the reaction of wood to the application of loads over a very short time. However, most wood used in structures is expected to carry load for long periods of times. Thus, the FPL has carried out long-term loading experiments to develop data to support engineering design. [19–21, 33]

Instrumentation

▲ Richard Geier checking a test apparatus used to determine the long-term loading characteristics of lumber. {(1970s) {126 B}

▶ Al Motelet using a proofing ring to calibrate a testing machine. (1960s)

A critical component of any test is the instrumentation to record the property of interest. In early testing, most of the measuring instruments were mechanical in design. With the advent of electronic strain gages, Dunc Godshall developed a method for calibrating strain gages to help ensure accurate test measurements. [13]

Designing with Wood

One of the major contributions of FPL's research has been in the area of designing with wood. Following the intensive and extensive evaluation of wood strength properties, classification of lumber, and development of fasteners and connectors, design formulas were developed. Initially the emphasis was on structures using wood timbers and lumber, but then it was extended to products such as laminated arches, plywood, sandwich panels, and other composite products. Lately the emphasis has been on round timbers to provide outlets for trees removed in thinning operations being conducted to improve the health of forests and decrease the intensity of wildfires. [23, 33, 35]

Plywood Design: FPL developed design criteria for plywood that provide a basis for its effective use as an engineered material in box beams, girders, and other highly stressed applications, as well as in component parts of airplanes. [33]

Columns: FPL established design criteria for various types of wood columns that make more effective use of wood in compression members of timber trusses for airplane hangars, warehouses, and factories. [4, 26]

Wood Decks

In the 1980s and 1990s, many material and structural failures in decks (with associated injuries and fatalities) throughout the United States indicated the need for a comprehensive guide for properly designing, constructing, and maintaining wood decks. Prior to FPL studies, the only information on designing wood decks was from an architectural standpoint. The engineering information then available was either so oversimplified or so inconsistent in its approach that it was of no real value. Through a cooperative effort initiated with the wood products industry,

FPL generated the first nationally approved engineering data and design recommendations to safely design, detail, and construct these structures. This work has been incorporated into the comprehensive manual *Wood Decks: Material, Design, and Finishing* [29]. The results have been accepted for use in existing and new building code standards, demonstrating its originality and practicality. In 2000, all of Chapter 3, "Structural Design and Construction of Decks," was adopted (verbatim) into the Georgia Residential Building Code. More recently, and more broadly, the engineering design data and construction design recommendations are being incorporated into the 2007 International Building Code and the 2007 International Residential Code. These codes hold jurisdiction over the entire United States. [25, 27–30]

◀ Failure of a wood deck. (Photo courtesy of Kalamazoo Press Gazette.)

Safe Working Stresses and Building Codes

FPL developed basic strength data to establish safe working stresses for commercial timbers, thus removing chaos in this highly controversial field and promoting efficiency in the use of timber. FPL research helped lay the foundation for establishing sound building codes for wood throughout the Nation. These codes help protect the consumer from unsafe homes and buildings. [33]

Nondestructive Evaluations

Historically it has been necessary to destroy wood specimens to determine their properties. After years of research by the FPL, universities, and industry, new techniques have been developed to make these evaluations by nondestructive means. These methods include determining electrical resistance, wave propagation, acoustic emissions, x-ray, and proof loading.

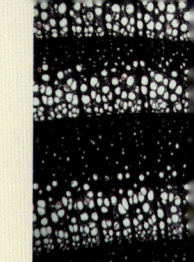

Better Defect Detection through Nondestructive Techniques Means More Efficient Use of Hardwood Lumber

The earlier in the manufacturing process defects in wood are identified, the easier it is to decide how to process the lumber correctly, minimize costs of manufacture, and reduce waste. The FPL, Minnesota's Natural Resources Research Institute, and industry are advancing ultrasound-based defect detection technology for hardwood lumber. Ultrasound can be used efficiently to detect internal flaws such as cracks, splits, ring shake, and honeycomb in wood. This technology will help wood products manufacturers make better decisions about their lumber resource, resulting in improved yield while minimizing reject material further downstream in the manufacturing process. [40]

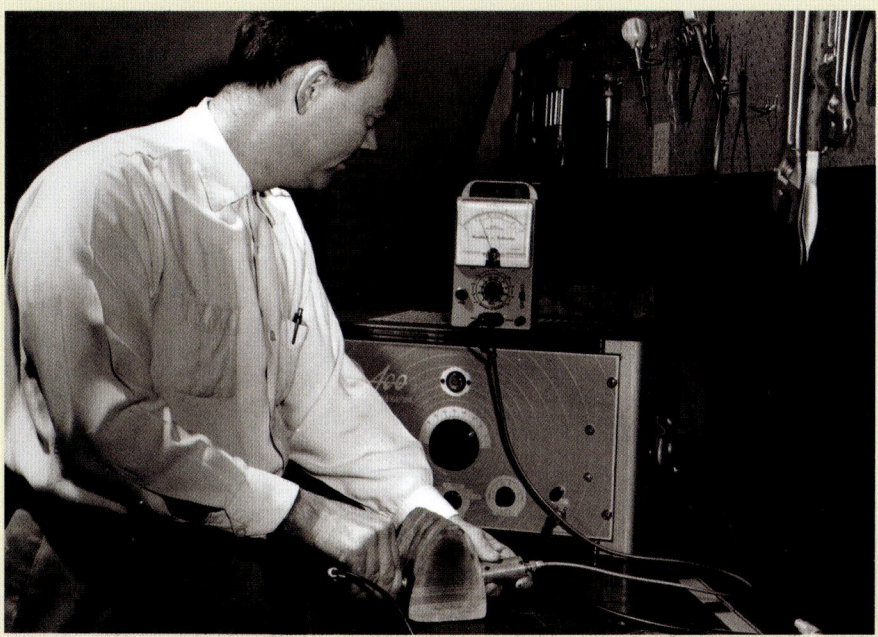

◀ William James passes ultrasonic vibrations through a seasoned hard maple shoe-last blank as a nondestructive means of finding drying defects before costly carving exposes them. (1960s) {M 101 213} [24]

Machine-Stress-Rating Studies Increase Product Values and Let Western Sawmills Make Economic Use of Small-Diameter Western Softwoods

Substantial quantities of small-diameter, mixed species stands exist in the Intermountain West. These stands are at high risk from wildfires. Approximately 51,000,000 U.S. acres are in need of fuels reduction. Research at the FPL is being focused on developing economic uses for this material in forest products. This research would significantly increase carbon sequestration of the estimated 10 tons of carbon emitted per acre from wildfire. In addition, rural communities can benefit from economic development opportunities provided by an increased array of product options, and land managers have a broader range of economical management options for achieving forest ecosystem objectives. [35]

Some lumber from these small-diameter stands may have desirable mechanical properties because of its slow growth, which would increase its value and make it more economical to harvest. FPL research showed that high-quality round timbers and lumber can be produced from small-diameter trees, with the help of the machine-stress-rating (MSR) grading process. The MSR grading process nondestructively measures product stiffness, resulting in a more accurate assessment of mechanical properties than the traditional visual grading procedures. The improved accuracy permits recognition of pieces with the best mechanical properties. [37]

Deterioration in Wood Products

Wood in use can sometimes develop decay, like cavities found in teeth. Locating these decayed areas, so they can be treated, is now possible through the use of nondestructive methods. These evaluations and proper treatments can lead to greatly extending the life of a structure and ensuring that it is safe. [25, 31, 37, 38]

Sound Douglas Fir

200–300 µs/ft

Incipient Decay

400–600 µs/ft

Severe Decay

700–1000 µs/ft

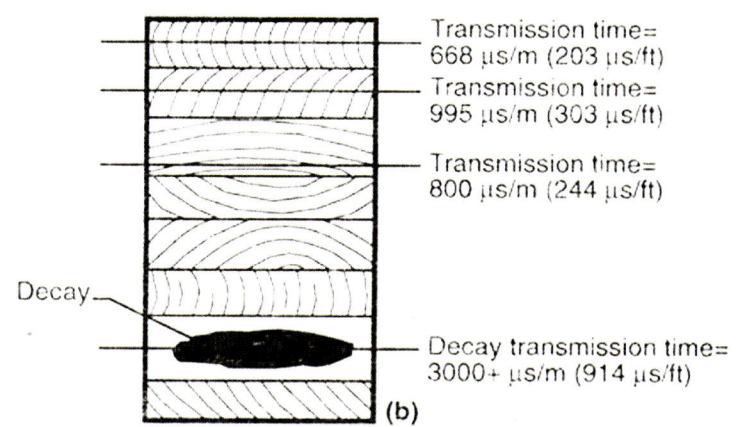

Transmission time= 668 µs/m (203 µs/ft)
Transmission time= 995 µs/m (303 µs/ft)
Transmission time= 800 µs/m (244 µs/ft)

Decay

Decay transmission time= 3000+ µs/m (914 µs/ft)

(b)

◀ Use of ultrasonics to determine the decay in a wood beam. (1990s)

Standing Tree Evaluation

Through use of recently developed nondestructive methods, it is now possible to assess the wood in a standing tree or logs at a sawmill. These assessments help make the decision as to what to harvest and how to process the logs that are at the mill, saving both time and money. [39–41]

▲ Xiping Wang evaluating the quality of a standing tree using a nondestructive testing device. FPL researchers developed the foundation for tools that use acoustic, laser, ultrasound, and wireless technologies to evaluate the strength and soundness of the wood in standing trees. This technology allows forest managers to determine the quality of wood without cutting down the tree. (2004) [42, 43]

RECEIVER PROBE

Acoustic wave-front through tree

VISUAL LASER
Used to align probes

ULTRASOUND
Measures the distance between the pins

INFRARED LED
Activates timing switch

TRANSMITTER PROBE

12,000 ft/sec

LCD
Displays acoustic velocity through timber between probes

RUGGEDISED PDA
Cordless connection to probe
Displays and stores test information

▲ Example of one commercial nondestructive testing device for evaluating the quality of a standing tree. The time it takes for the impact wave to transit between the two probes is related to the quality of the wood.

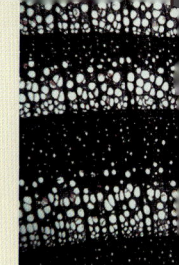

1910

Limited wood property data and design information was available.

2010

Wood Handbook providing engineering properties and design formulas of all commercials U.S. species and many foreign woods.

Further Information

[1] Deductions from strength tests of woods / J.A. Newlin / American lumberman (Jan 16, 1915)

[2] Factors affecting the strength of wooden members / Arthur Koehler / FPL No. 363 (1919)

[3] Comparative strength properties of woods grown in the United States / L.J. Markwardt / USDA, Forest Service technical bulletin No.158 (1930)

[4] Tests of large timber columns and presentation of the Forest Products Laboratory column formula / J.A. Newlin, J.M. Gahagan / USDA, Forest Service technical bulletin No. 167 (1930)

[5] The distribution and the mechanical properties of Alaska woods./ L.J. Markwardt / USDA, Forest Service technical bulletin No. 226 (1931)

[6] Strength-moisture relations for wood / T.R.C. Wilson / USDA, Forest Service technical bulletin No. 282 (1932)

[7] The strength and related properties of redwood / R.F. Luxford L.J. Markwardt / USDA, Forest Service technical bulletin No. 305 (1932)

[8] Causes of brashness in wood / Arthur Koehler / USDA, Forest Service technical bulletin No. 342 (1933)

[9] Specific gravity and related properties of softwood lumber / Edward C. Peck / USDA, Forest Service technical bulletin No. 343 (1933)

[10] Wood handbook: basic information on wood as a material of construction with data for its use in design and specifications / Forest Products Laboratory (1935)

[11] Structure, occurrence, and properties of compression wood / M.Y. Pillow, R.F. Luxford / USDA, Forest Service technical bulletin, No. 546 (1937)

[12] Wood at low temperatures / Kenneth H. Boller / Modern packaging Vol. 28(1) (Sept. 1954)

[13] The FPL linear deadweight accelerometer calibrator / W.D. Godshall / FPL No. 2239 (1962)

[14] Strength and related properties of white fir / C.C. Gerhards / FPL–RP–14 (1964)

[15] Some observations on perpendicular-to-grain rheology of Northern red oak (Quercus rubra L.) / R.L. Ethington, R.L. Youngs / Holz als Roh-und Werkstoff Vol. 23(5) (May 1965)

[16] A horizontal machine for testing structural wood elements in tension / K.H. Boller / FPL–RN–0150 (1967)

[17] Method for evaluating shear properties of wood / B.P. Munthe / FPL–RN–0195 (1968)

[18] Is wood hardness affected by preservative treatment? / Robert L. Ethington / Forest products journal Vol. 22(5) (May 1972)

[19] Modeling the creep of wood in a changing moisture environment / E.L. Schaffer / Wood and fiber Vol. 3(4) (Winter 1972)

[20] Accelerated aging: residual weight and flexural properties of wood heated in air at 115° to 175°C. / M.A. Millett, C.C. Gerhards / Wood science Vol. 4(4) (Apr. 1972)

[21] Time-related effects of loading on wood strength: a linear cumulative damage theory / C.C. Gerhards / Wood science Vol. 11(3) (Jan. 1979)

[22] Wood utilization priorities as set by energy and materials shortages / Jerome F. Saeman, Robert L. Youngs / North America's forests: gateway to opportunity: proceedings of 1978 Joint Convention of the Society of American Foresters and the Canadian Institute of Forestry. Washington: SAF (1979)

[23] Study of lumber used for bracing trenches in the United States / W.L. Galligan / National Bureau of Standards, Building science series: 122 (1980)

[24] Feasibility of using speed of sound in wood to monitor its drying progress in a kiln / William L. James, R. Sidney Boone, William L. Galligan / FPL (1981)

[25] Methods for inspecting wood structures in-place, serviceability and durability of construction materials / R.H. Falk, R.J. Ross / Proceedings of the 1st Materials Engineering Congress, August 13–15, Denver, Colorado, p.324–330. (1990)

[26] New column design formula / John J. Zahn / Wood design focus Vol. 2(2) (Summer 1991)

[27] Fasteners for outdoor wood structures / R.H. Falk, A. Baker / Wood design focus, Vol.4(3): 14–17 (1994)

[28] Controlling moisture in wood decks / R.H. Falk, K. McDonald, J. Winandy / Fine homebuilding, Taunton Press, September, No. 97 (1995)

[29] Wood decks: materials, construction, and finishing / K. McDonald, R.H. Falk, S. Williams, J. Winandy / Forest Products Society technical manual No. 7298 (1996)

[30] Details for a lasting deck / R.H. Falk, S. Williams / Fine homebuilding, Taunton Press, No. 102, April/May (1996)

[31] Nondestructive elemental analysis of wood biodeterioration using electron paramagnetic resonance and synchrotrons X-ray fluorescence / B.L. Illman, S. Bajt / International biodeterioration and biodegradation Vol. 39(2–3) (1997)

[32] Standards for structural wood products and their use in the United States / David W. Green, Roland Hernandez / Wood design focus Vol. 9(3): 3–12. (1998)

[33] Wood Handbook: wood as an engineering material / FPL–GTR–113 (1999)

[34] Wood products utilization—A call for reflection and innovation / John A. Youngquist, Thomas E. Hamilton / Forest products journal Vol. 49(11/12): 18–27 (1999)

[35] Research challenges for structural use of small-diameter round timbers / R. Wolfe / Forest products journal Vol. 50(2): 21–29 (2000)

[36] Industrial wood productivity in the United States, 1900–1998 / Peter J. Ince / FPL–RN–0272 (2000)

[37] Nondestructive evaluation of wood / R.F. Pellerin, R.J. Ross / Forest Products Society publication No. 7250 (2002)

[38] Wood and timber condition assessment manual / Robert J. Ross, Brian K. Brashaw, Xiping Wang, Robert H. White, Roy F. Pellerin / Forest Products Society (2004)

[39] Assessment of decay in standing timber using stress wave timing nondestructive evaluation tools / Xiping Wang, Ferenc Divos, Crystal Pilon, Brian K. Brashaw, Robert J. Ross, Roy F. Pellerin / FPL–GTR–147 (2004)

[40] Nondestructive evaluation of incipient decay in hardwood logs / Xiping Wang, Jan Wiedenbeck, Robert J. Ross, John W. Forsman, John R. Erickson, Crystal Pilon, Brian K. Brashaw / FPL–GTR–162 (2005)

[41] System for and method of performing nondestructive evaluation techniques on a log or round timber / Xiping Wang, Robert Ross, James Mattson, John Erickson, John Forsman, Earl Geske, Michael Wehr / US Patent No. 7,043,990 (May 16, 2006)

[42] Acoustic assessment of wood quality of raw forest materials: a path to increased profitability / Xiping Wang, Robert J. Ross / Forest products journal Vol. 57(5) (May 2007)

[43] Method and apparatus for the evaluation of standing timber / Xiping Wang, Robert Ross, Peter Carter, Nigel Sharplin / US Patent No. 7,418,866 B2 (September 2, 2008)

Nondestructive Evaluation of Wood

Nondestructive evaluation is the science of identifying the physical and mechanical properties of materials, products, and structures without altering their end-use capabilities. Such evaluations rely upon a variety of testing technologies to provide accurate information pertaining to the properties, performance, or condition of the product in question.

The Forest Products Laboratory, with industry and university cooperators, has been actively involved in this area of research and development since its founding. For example, as part of its extensive effort in support of the Department of Defense during the early part of the 20th century, studies at FPL led to improved airplane strut designs. Researchers uncovered information relating to exact strength requirements of struts, compared the suitability of different wood species, identified the defects in rejected struts, and determined to what extent existing methods of inspection needed revision. Most importantly, they revealed that a strictly visual system of inspection was unreliable. Accepted struts did not always prove strong enough, and rejected struts often had adequate strength. FPL developed a nondestructive method of testing struts that was subsequently incorporated into specifications.

FPL researchers were, and are, actively involved developing improved visual grading criteria for products including logs, timbers, and lumber. Its basic research on wood physics, particularly vibration and sound propagation properties of various wood species, serves as the foundation for many automated assessment and grading technologies that are widely used today.

FPL was a leader in the development and use of ultrasound technologies to locate defects in hardwood and softwood lumber. It has also been a leader in the development of mechanical grading procedures and technologies for softwood and hardwood structural lumber.

Recently, FPL and cooperators worked to develop advanced technologies for assessing the quality of standing timber and trees. Their efforts resulted in a series of portable devices for field use that are currently used throughout the world.

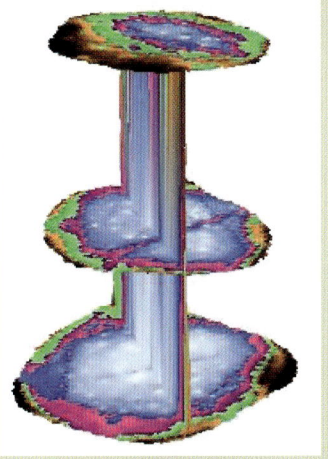

◀ Three-dimensional tomographic image obtained from one of the park trees inspected at the Capitol Park, showing significant internal decay that originated from the root system and progressed up to the trunk. Further information is found in "Acoustic tomography for decay detection in red oak trees" / Xiping Wang, R. Bruce Allison, Lihai Wang, Robert J. Ross / FPL–RP–642 (2007)

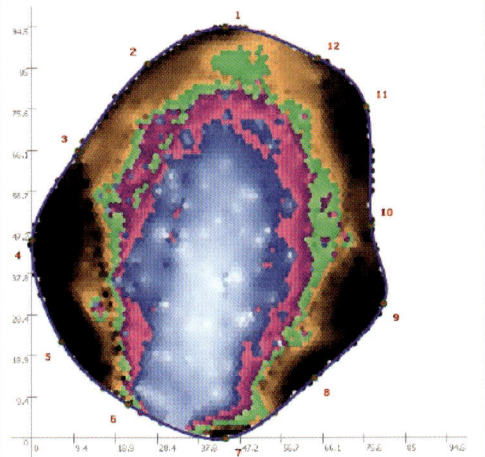

◀ Tomographic cross-sectional image of a tree.

▶ Xiping Wang (left), research scientist from the University of Minnesota Duluth stationed at FPL, and Bruce Allison, arborist, were conducting acoustic tomography testing on century-old red oak trees at the Capitol Park, Madison, Wisconsin, to detect internal structural defects of the park trees. (2000s)

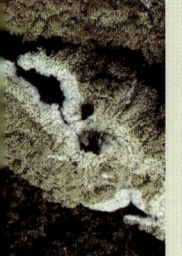

FPL has also worked to develop nondestructive evaluation techniques for assessing hazard trees. Trees in urban areas, parks and campsites require frequent inspection to ensure public safety. Coupling sound propagation properties with computer-aided tomography, field testing units are capable of determining the extent of decay in standing trees.

The use of nondestructive evaluation for assessing the condition of structural members while still in-service has been a significant contribution from FPL. Technologies developed, including those that utilize vibration, sound propagation, and instrumented probes, are used to determine the extent of deterioration of wood in homes, public buildings, and historic structures.

Over the years, much progress has been made in the way the properties of wood are determined, starting with visual analysis, progressing to destructive physcial testing of the specimen, to the present effort to use nondestructive means. The accomplishments have been significant, and research on new approaches is continuing.

One reason for the success of this effort has been FPL's work with industry and university scientists and engineers to improve the approaches and then sharing the information through the Nondestructive Testing Symposium Series.

Another method of sharing information has been through the *Wood and Timber Condition Assessment Manual,* a practical, comprehensive, useful manual on inspection of wood in service.

▲ *Nondestructive Evaluation of Wood* (2002) and *Wood and Timber Condition Assessment Manual* (2004) were published by Forest Products Society.

Statistics

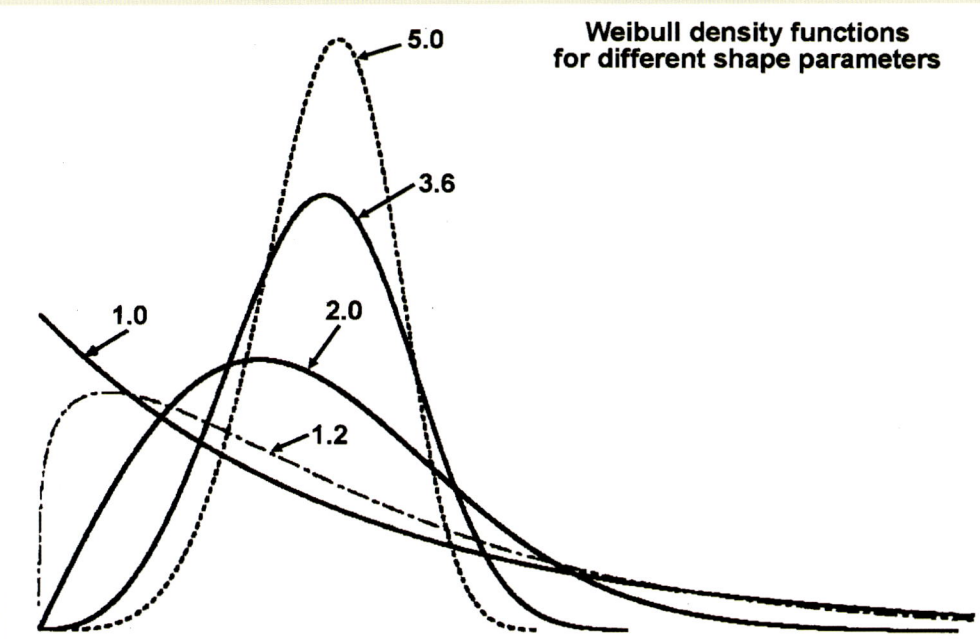

◀ Different shapes for distributions that may characterize research data. Knowing the distribution of the data greatly facilitates the quality of the analysis of the data. (2000s) [39]

Use of the Weibull Distribution in Wood Research

Sometimes new statistical techniques must be developed or competing statistical techniques must be evaluated in order to advance research efforts. One example is modeling distributions of wood property data. Originally, when modeling the distribution of wood properties, all data were assumed to be normally distributed. Current data shows that many variables might be better fitted by non-normal distributions, such as a Weibull distribution. This led to many questions. Can we test the goodness-of-fit of data to a Weibull distribution? How do we calculate common information used in normal distributions (such as tolerance limits used to represent a lower strength property) to report values to engineers designing with wood? How do we model differences in strength of various sizes of boards from the same species? Can we use a non-normal joint distribution of two wood properties to help more effectively model wood performance in structures? [24, 39]

Statistics is commonly viewed as the collection, collation, and presentation of numerical data. However, the Forest Products Laboratory has long recognized that the field of statistics is critical for appropriately testing research hypotheses and making inferences to untested populations. The field has provided extensive and powerful tools for designing studies, analyzing data, summarizing or modeling data, and interpreting results for many research studies at the FPL.

Challenge: Because of the vast number of wood species and growing conditions, complexity of wood structure, and variety of environmental exposures, how should research experiments be designed to address research objectives and hypotheses most efficiently to maximize the value content of the resulting data?

FPL's Contributions: Developed carefully crafted experimental designs and modeling methods, new statistical techniques, new theory and comparisons associated with both old and new statistical techniques, and computer software that make statistics more available to scientists. FPL statisticians continue to work with scientists to design efficient and effective research studies.

Results: Use of statistics has allowed for more accurate and precise estimations of effects of factors in completing meaningful research experiments. This has resulted in more efficient and cost-effective research programs and more reliable results.

Evolution of FPL Statistic Section

◀ The first section was commonly known as computing. Their job was to summarize test information generated by engineers and scientists. Because much of this was done by hand and simple adding machines, it required a number of people. (1912)

◀ Early computing laboratory. Note mechanical calculators on the table. In 1926, the FPL installed tabulating equipment, consisting of one electric punch, one sorter, and one tabulator. All machines were for 45-column cards.

In 1958 the section was named "Statistics and Computing." Prior to that, statistical design was handled in the research projects. The staff consisted of nine women. In 1959, Dave Yandle, statistician, transferred to the new section.

◀ Later computing laboratory. In 1930, 80-column equipment, including four electric punch machines, a hand verifier, a sorter, and a tabulator were acquired. Note mechanical card tabulator in background. In 1943, 23 women and one man worked in the section.

▼ Diana Smith with Fisk University sophomore Ethel Hunter using a 1960s IBM computer.

◀ Vicki Herian and Patty Lebow use an advanced UNIX work station for analyzing data. (2000s).

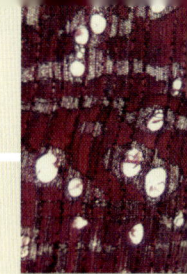

Statistical Analysis at FPL

Through the 1960s and 1970s, the section was known by a number of names, such as "Administrative & Technical Services," "Mathematical and Computing Services," "Biometrics Section," "Statistics and Numerical Analysis," "Statistical Research," and "Statistics and Automated Data Processing." The statistics and computing functions diverged into separate groups in 1980. The computer group (led over the years by James Haskell, David Lewis, and Jane Evans), known as "Systems and Automated Data Processing" and later as "Management Systems," took on the responsibility of implementing and supporting the national computer program, focusing on administrative processes with the Data General computer system and later IBM servers. The group ceased to exist in 2005 when computer support was consolidated at the Washington Office.

The "Statistics Group" began in 1980 with one statistician (James Evans). By 2001, the Statistics Group consisted of three mathematical statisticians and two statisticians (James Evans, Steve Verrill, Patricia Lebow, Cherilyn Hatfield, and Vicki Herian) and had been made a research work unit (RWU), "Statistical Methods in Wood and Fiber Research," at FPL to better reflect the importance of statistics to FPL research. In 2007, the Statistics and Economics RWUs were merged into a combined RWU.

Statistical Program

The present statistical program includes (1) providing statistical help to engineers and scientists on designing experiments and interpreting research data, (2) working as partners on major studies requiring extensive designs with multiple research groups, and (3) exploring new statistical methods for improving the accuracy, efficiency, and usefulness of research experiments. Following are a few examples of research studies in which statistical design played a major role.

Tree Growth and Properties

◄ Anemometer mounted on 60-foot tower to record wind velocity at crown height of red pine at Crandall Woods Plantation, Wisconsin Dells, Wisconsin. Research results were used to help understand the effect of wind on tree growth and wood properties. (1970s) [10]

Wood Density Study

The density of a tree significantly affects the properties of wood products. Density is a result of wood species and growth conditions, such as type of soil, available water, sunlight, and competition from other trees and plants.

◀ Southern pine logs; the one on the left was slow grown (high density), and the one on the right was fast grown (low density). (1960s) (M 123 457). [9]

▶ Increment borer wood sample. Note the annual growth rings. By counting the number of rings, the age of the tree can be determined; by measuring the space between rings, the rate of growth can be estimated. The wood density can also be estimated by weighing the sample and knowing its volume. (M 123 102) [2]

◀ Richard L. Nielsen, Pacific Northwest Forest and Range Experiment Station, using increment borer to sample the wood in the tree to determine its age and density. (1960s) (M 123 101) [5]

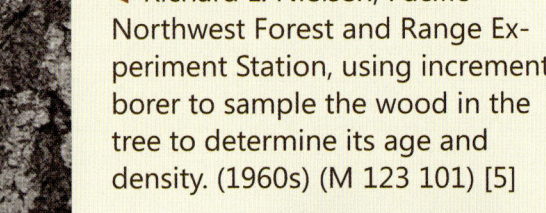

▶ Extracting the wood core from the increment borer.

◀ A sample of Engelmann spruce being packed out of the forest as part of the western wood density survey conducted by the FPL, industry, and universities. Some timber was commercially inaccessible. (1960s) (M 135 387-6) [5]

▲ Dimitri Pronin and Arnie Okkonen weigh disks of wood to determine specific gravity of the wood. (1960s) (M 133 457-10) [5, 9]

In-Grade Program

This major study, referred to as the "In-grade testing of structural lumber program," was one of the largest cooperative research programs ever undertaken by the North American wood engineering community. It included universities, the Southern Pine Inspection Bureau, the West Coast Lumber Inspection Bureau, the Western Wood Products Association, a number of companies, and the FPL. [18, 20–23]

Physical and mechanical properties information was obtained on 33 species or species groups of visually graded structural lumber. Over an eight-year period, nearly 70,000 pieces of lumber—approximately 1,000,000 board feet—were tested to destruction in bending, tension, or compression. The information provided the basis for more accurately estimating mechanical properties of lumber and revising allowable design properties. To accomplish this and provide meaningful data required statistical design for selecting the wood samples, testing them, and analyzing the results so that the results were useful and applicable to structural lumber in general. The data are now part of the National Design Standard.

► Failure of a piece of structural lumber that was subjected to bending forces to help determine allowable design properties for structural lumber. (1980s)

Preservative Testing Program

FPL researchers are evaluating wood preservative treatments for both efficiency and environmental impacts. FPL statisticians are involved in many aspects of this research, including experimental design, analysis of results, and development of predictive models. In many cases, wood preservative effectiveness is evaluated by using preservative-treated stakes in plots located in decay-prone environments. Further evaluating potential environmental impacts of treated wood in sensitive exposures can require very complex experimental designs. An example is the determination of leaching and environmental accumulations of preservatives in soils (see Preservation section). [33, 38]

◄ Example of treated boardwalk and stair structure with soil contact in an outdoor environment where both the efficiency and environmental impact of a preservative may be important. (2000s)

Paperboard

Paperboard constitutes about half of all the paper manufactured in the United States. It is the major component of corrugated fiberboard. How it is made has a direct effect on how the containers made from it will perform. FPL researchers, using a statistically designed experiment, were able to evaluate six different variables, including wood species, pulp yield, refining, freeness, press pressure, and machine wire, at two levels, and their interactions on the properties of paperboard handsheets (see Paper section). (1970s) [12]

◀ Rolls of paperboard ready for converting into corrugated fiberboard containers.

Understanding Wood Structure

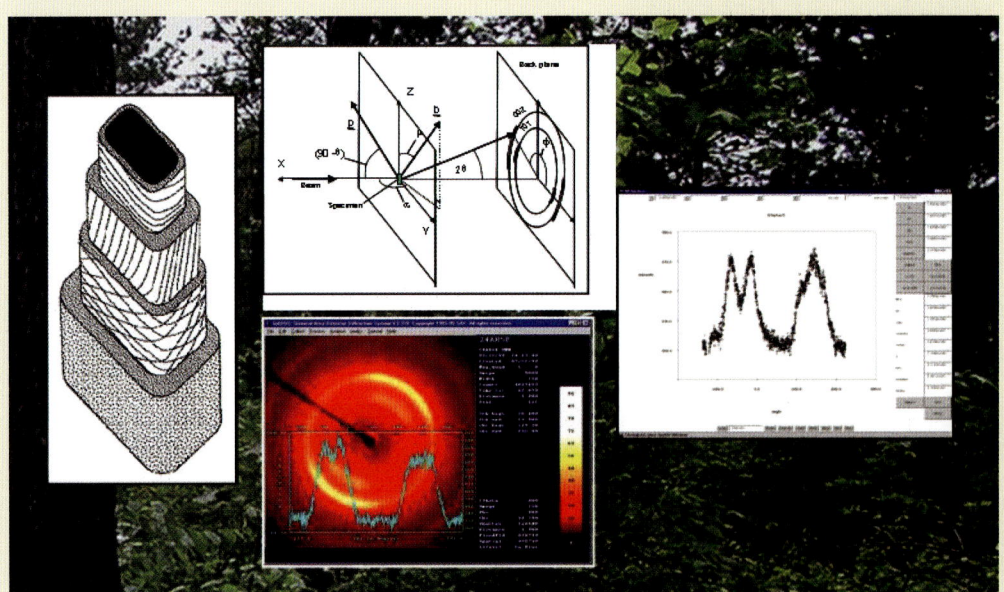

◀ Using statistical modeling to represent physical properties of wood cells to predict wood properties.

Microfibril Angle Research

Microfibril angle between cells of various layers of wood is an important predictor of wood properties. X-ray diffraction measurements can theoretically permit scientists to obtain accurate microfibril angle estimates in minutes instead of the hours required by previous procedures. However, estimating this value is very complex, and initial procedures required several simplifying assumptions, including only using part of the data. Statisticians developed new methods that incorporated all the available data and still could be completed in minutes. (2000s) [37]

Recycling Fiber

One of the continuing challenges in recycling paper is the separation of contaminants from the recovered paper products. FPL scientists, working with the University of Wisconsin, developed a disk separation process that allows for separation of usable fiber from mixed waste. Statistically designed methods of evaluating the process were used to determine the effectiveness of the process (see Recycle section). (1980s) [17]

Use of Pine Bark Beetle Killed Trees

◄ The rattlesnake creek bridge, near Missoula, Montana, built using 6-inch-diameter lodgepole pine from a beetle-killed forest near Elk City, Idaho. (2000s)

This unique bridge was designed and built to demonstrate the potential use of the large volume of lodgepole pine in the Pacific Northwest that has been killed by the pine bark beetle. Presently these standing dead trees are going to waste. FPL, working with the University of Idaho (UID) and Timber Products Inspection (TPI) of Vancouver, Washington (a major U.S. grading agency), developed the technical basis for the mechanical grading procedure for round timbers. The FPL provided the statistical experimental design for comparing the visual grading with mechanical grading of lodgepole pine logs for use in two cable suspension bridges. Nondestructive testing and property estimation were completed by the FPL–UID team, with TPI providing visual grading of the logs for comparison with mechanical grading. The results of the experiment indicated that mechanical grading provides a more efficient property assignment than does visual grading. [34–36]

The statistics group provides invaluable service to all scientists at the laboratory in planning their research. A review of the report subjects ("Further Information") at the end of this section gives some indication of the breadth of the help provided.

1910

Research studies completed with minimal statistical design.

2010

Research studies incorporate statistical design, thus making the experiments more efficient, useful, and reliable.

Further Information

[1] Statistical evaluation of the effect of age on specific gravity in loblolly pine / David O. Yandle / FPL No. 2049 (1956)

[2] Estimating tree specific gravity from a single increment core / Harold E. Wahlgren. FPL No. 2146 (1959)

[3] Elementary statistical methods for foresters / Frank Freese / USDA Agriculture handbook 317 (1964)

[4] Elementary Forest Sampling / F. Freese / Agricultural handbook 232 (1962)

[5] Western wood density survey: report number 1 / Forest Service / Research paper FPL 27 (1965)

[6] A forest sampling method for wood strength / B.A. Bendtsen, Frank Freese, R.L. Ethington / Forest products journal Vol. 20(11) (Nov. 1970)

[7] Taper of wood poles / Billy Bohannan, Hermann Habermann, Joan E. Lengel / General technical report FPL 2 (1974)

[8] Press-drying green, flat sliced walnut veneer to reduce buckling and end waviness / J.F. Lutz, H. Habermann, H.R. Panzer / Forest products journal Vol. 24(5) (May 1974)

[9] Properties of major southern pines: Part I—Wood density survey / H.E. Wahlgren, D.R. Schumann / (Part I (FPL–RP–176); Part II—Structural properties and specific gravity / B.A. Bendtsen, R.L. Ethington, W.L. Galligan. (FPL–RP–177) (1975)

[10] Effect of mechanical stress on growth and anatomical structure of red pine: stem vibration / J.T. Quirk, F. Freese / Canadian journal of forest research Vol. 6 (1976)

[11] Structural flakeboards using ring flakes from fingerling chips / Bruce G. Heebink, E.L. Schaffer, J. Chern, J.H. Haskell / Research paper FPL 296 (1977)

[12] Papermaking factors that influence the strength of linerboard weight handsheets / John W. Koning, Jr., James H. Haskell / Research paper FPL 323 (1979)

[13] Evaluation of lumber properties in the United States and their application to structural research / W.L. Galligan, D.S. Gromala, D.W. Green, J.H. Haskell, / Forest products journal Vol. 30(10) (Oct. 1980)

[14] Estimating correlation between strength properties / David W. Green, James W. Evans / Recent advances in engineering mechanics and their impact on civil engineering practice: proceedings of the fourth engineering mechanics division specialty conference, May 23–25, 1983, West Lafayette, Indiana, Vol. II. New York, NY : American Society of Civil Engineers (1983)

[15] Investigation of the procedure for estimating concomitance of lumber strength properties / D.W. Green / Wood and fiber science Vol. 16(3) (1984)

[16] Estimating the correlation between variables under destructive testing, or how to break the same board twice / James W. Evans, Richard A. Johnson, David W. Green / Technometrics Vol. 26(3) (Aug. 1984)

[17] Disc separation: the effect of disc geometry / John H. Klungness, James W. Evans / Tappi journal Vol. 70(8) (Aug. 1987)

[18] Mechanical properties of visually graded lumber / David W. Green, James W. Evans / Report on cooperative studies between the USDA Forest Service, Forest Products Laboratory ... [et al.]. (September 1987)

[19] Statistical considerations in duration of load research / Carol L. Link / FPL–RP–48 (1988)

[20] Tensile strength of laminating grades of lumber / Catherine M. Marx, James W. Evans / Forest products journal Vol. 38(7/8) (July/Aug. 1988)

[21] Evaluating lumber properties: practical concerns and theoretical constraints / David W. Green, James W. Evans / 1988 conference on timber engineering held September 19–22, 1988 in Seattle, Washington (editor Rafik Y. Itani) (1988)

[22] In-Grading testing of structural lumber / David W. Green, Bradley E. Shelley, Harvey P. Vokey / Forest Products Research Society (1989)

[23] Temperature adjustments for the North American in-grade testing program / J. David Barrett, David W. Green, James W. Evans / In-grade testing of structural lumber: Proceedings. Madison, Wis. Forest Products Research Society (1989)

[24] Two- and three-parameter Weibull goodness-of-fit tests / James W. Evans, Richard A. Johnson, David W. Green / FPL–RP–493 (1989)

[25] Procedures for deriving allowable properties for species groupings / James W. Evans, David W. Green / In-grade testing of structural lumber: Proceedings. Madison, Wis. Forest Products Research Society, (1989)

[26] Computer programs for adjusting the mechanical properties of 2-inch dimension lumber for changes in moisture content / James W. Evans, Jane K. Evans, David W. Green / FPL–GTR–63 (1990)

[27] Moisture content adjustment procedures for engineering standards / D.W. Green, J.W. Evans / Proceedings, 25th meeting of international council for building structures; 1992 August 24–27; Ahus, Sweden. Ahus, Sweden: CIB–W18, (1992)

[28] Solid-phase treatment of a pentachlorophenol-contaminated soil using lignin-degrading fungi / Richard T. Lamar, James W. Evans, John A. Glaser / Environmental science & technology Vol. 27(12) (Nov. 1993)

[29] Field evaluation of the lignin-degrading fungus *Phanerochaete sordida* to treat creosote-contaminated soil / Mark W. Davis, J.A. Glaser, J.W. Evans, R.T. Lamar, / Environmental science & technology Vol. 27(12) (Nov. 1993)

[30] Compressive creep behavior of corrugating components affected by humid environment / John M. Considine, D.L. Stoker, T.L. Laufenberg, J.W. Evans, / Tappi journal. Vol. 77(1) (Jan. 1994)

[31] Mechanical properties of fire-retardant-treated plywood after cyclic temperature exposure / Susan L. LeVan, J.M. Kim, R.J. Nagel, J.W. Evans / Forest products journal. Vol. 46(5) (May 1996)

[32] Patterns of long-term performance: how well are they predicted from accelerated tests and should evaluations consider parameters other than averages? / Rodney C. De Groot, James W. Evans / Paper prepared for 29th annual meeting, international research group on wood preservation, Maastricht, The Netherlands, 14–19 June (1998)

[33] Comparison of wood preservatives in stake tests: 2000 progress report / compiled by D.M. Crawford, B.M. Woodward, C.A. Hatfield / FPL–RN–02 (2002)

[34] Improved grading system for structural logs for log homes / David W. Green, James W. Evans, Joseph F. Murphy, Thomas M. Gorman / Forest products journal 54(9):53–62 (2004)

[35] Mechanical grading of 6-Inch-diameter Lodgepole pine logs for the travelers' rest and Rattlesnake Creek Bridges / David W. Green, James W. Evans, Joseph F. Murphy, Cherilyn A. Hatfield, Thomas M. Gorman / FPL–RN–0297 (2005)

[36] Mechanical grading of round timber beams / David W. Green, James W. Evans, Joseph F. Murphy / Journal of materials in civil engineering (Jan./Feb. 2006)

[37] JMFA 2 – A graphically interactive Java program that fits microfibril angle X-ray diffraction data / Steve P. Verrill, David E. Kretschmann, Victoria L. Herian / Research Paper FPL 635 (2006)

[38] Statistical analysis of the influence of soil properties and wood source on arsenic and copper depletion in laboratory soil-contact leaching of CCA-C treated wood / P. Lebow, R. Ziobro, L. Sites, T. Schultz, D. Pettry, D. Nicholas, S. Lebow, P. Kamden, R. Fox, D. Crawford. Wood and fiber science. 38(3):439–449. (2006)

[39] Applications of the Weibull distribution in wood engineering / James W. Evans, Richard A. Johnson, David W. Green, Steve P. Verrill / FPL report in press.

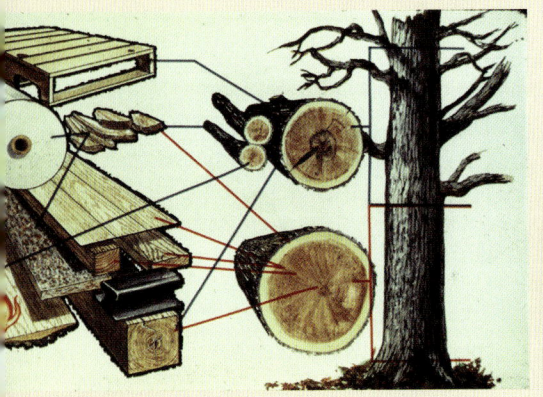  

Processes and Products

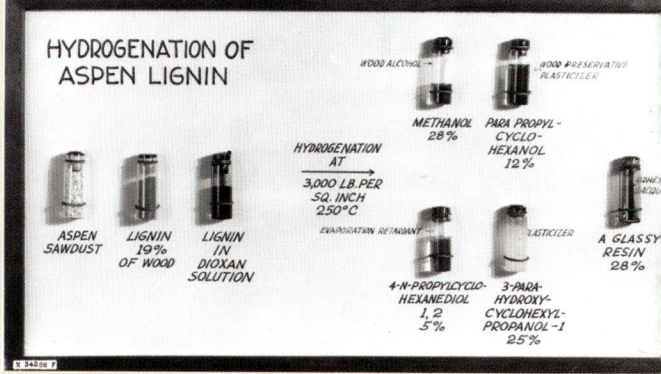

Processes and Products

This part of the book is organized, in general, by the wood particle size.

Logs   **1-20 m**

Lumber   **1-20 cm**

Veneer

Long Flakes

Chips   **1-20 mm**

Flakes

Excelsior

Particles

Fiber Bundles

Paper Fiber   **1-20 μm**

Wood Flour

Cellulose   **< 1 nm**

▲ Basic wood elements from largest to smallest. Wood comes in many shapes and sizes. This depiction of wood shows wood as a solid material and the ways it can be modified in decreasing particle size all the way to its chemical constituents. The Forest Products Laboratory conducts research on all these various sizes of wood particles. The idea of looking at wood use as a function of particle size was described by George Marra as the Non-Periodic Table of Wood (in "The Future of Wood as an Engineered Material," Forest Products Journal 22(9):43, 1972).

Wood

◄ Major products from wood: poles, ties, lumber, composite products, paper, and chemicals. {20 L SA}

Wood is so common in our lives that we take it for granted. Approximately 350,000,000 tons of wood are harvested annually in the United States. The value of shipments for the timber-based manufacturing industry in 2006 was $319,042,000,000, and the companies that make these products provided 1,290,000 jobs. (2006 Annual Survey of Manufacturers)

Challenge: Wood is a very complex material. Efficiently using wood to its fullest potential is a major challenge.

FPL's Contributions: The following sections will give examples of what FPL research has contributed to processes and products over the past 100 years.

Wood is complex, and its properties can vary depending on many things, such as the way that it is cut, wood grain, species, growth conditions, age, and processing.

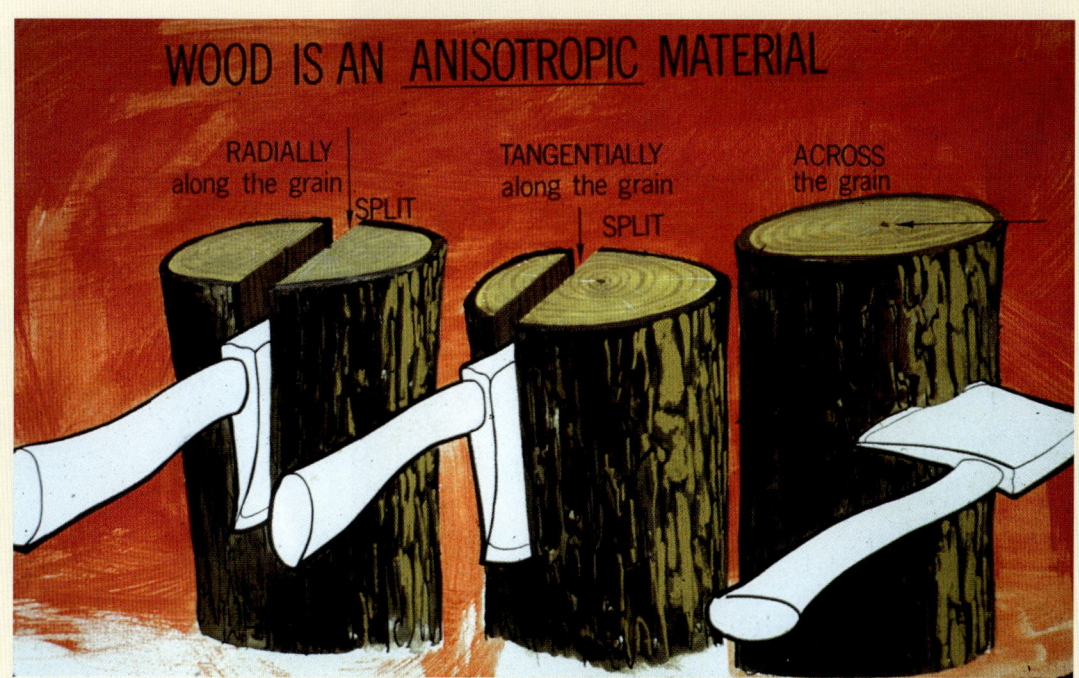

◄ The three major ways of cutting wood. Each results in unique appearance and properties. {111 F}

The following photos show how grain pattern is affected by method of cutting: (left to right) tangential, radial, and cross.

PONDEROSA PINE

◄ Ponderosa pine. {73 E}

QUAKING ASPEN

◄ Quaking aspen. {73 B}

BLACK WALNUT

◄ Black walnut. {73}

HONEYLOCUST

◄ Honey locust. {73 C}

SUGAR MAPLE

◄ Sugar maple. {73 A}

The grain of the wood can have an effect on wood properties.

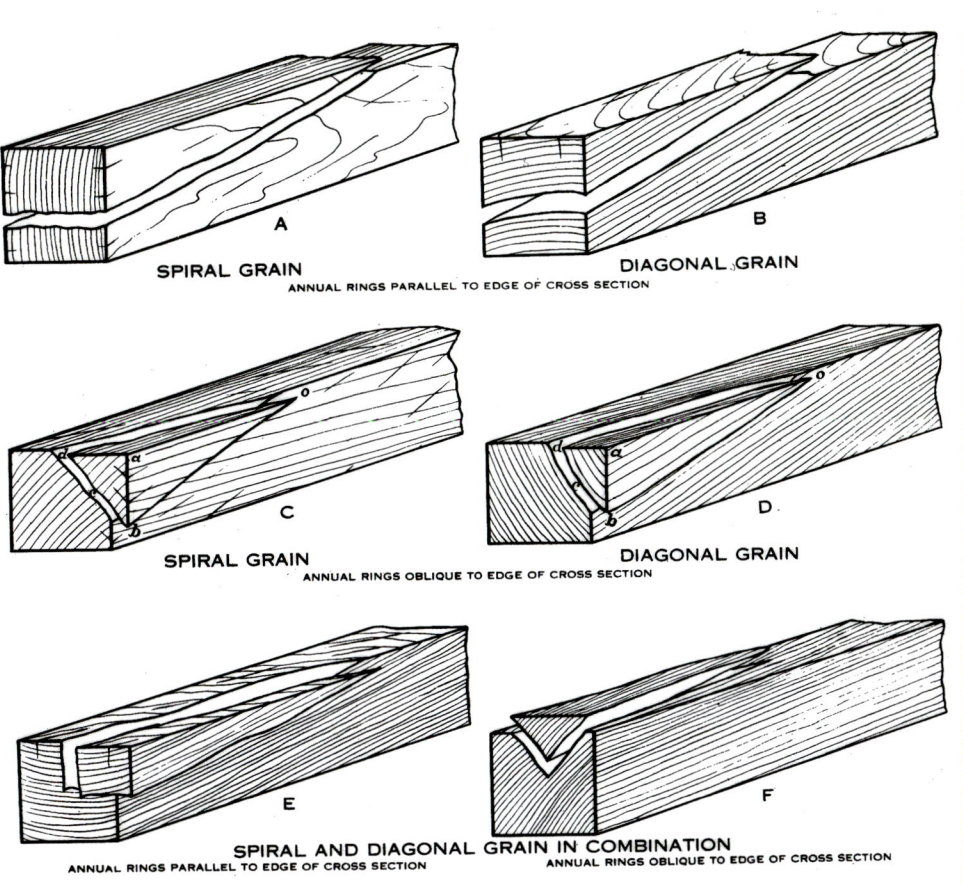

◄ The effect of spiral and diagonal grain on the splitting of wood. {ZM 6758 F}

▼ Cross section of a hardwood, left, and a softwood, right. {24 O}

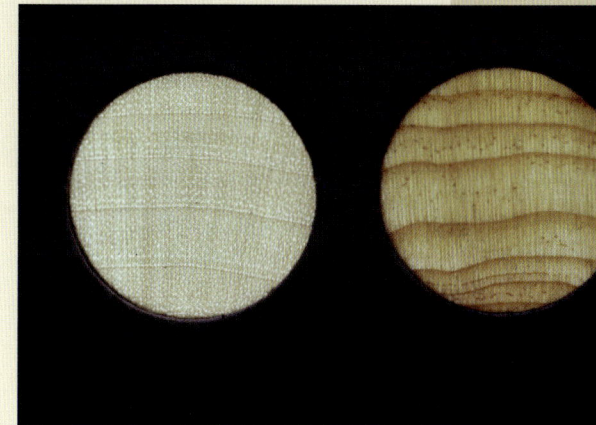

Another factor is the type of wood—hardwood or softwood—and the species.

The cellular structure of each wood species influences its properties. Growing conditions, including soil type, available moisture, light, altitude, length of growing season, temperature, and fertilizer, also significantly affect wood properties.

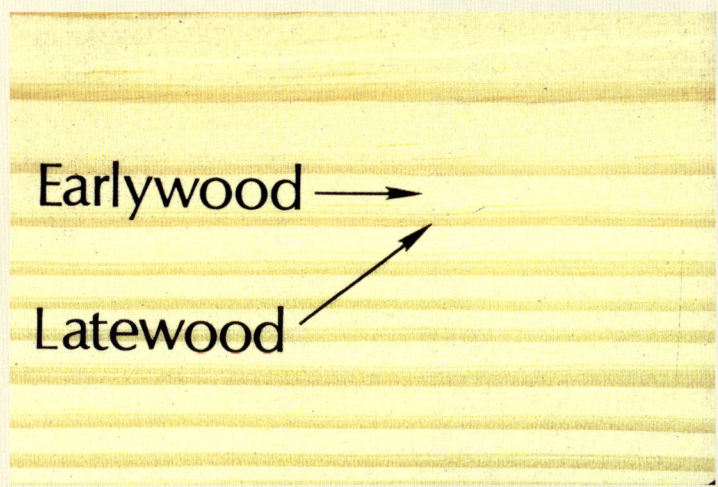

◄ Slice of wood showing the difference between earlywood and latewood fibers produced by this softwood tree each year; all tree species are different, as are their conditions for growth. {124 M}

The age of the tree can be an important factor in the strength properties of the wood.

Young trees have a higher percentage of juvenile wood. Juvenile wood is the wood a tree lays down in its first 5 to 20 years of growth. All trees have juvenile wood, but it had little significance when the timber supply was primarily old-growth trees grown in natural forest conditions. In such trees, the juvenile wood core was small because early growth was suppressed by competition from surrounding trees. In addition, the percentage of juvenile wood in the total volume was small because trees grew to very large diameters before they were harvested.

Now, improved trees grown on intensively managed plantations reach sawtimber size and are harvested at a young age. Because diameter growth is generally fastest during the years juvenile wood is produced, the juvenile wood core may be a very significant part of the harvested tree.

With efforts to grow trees faster and harvest them younger, such as in plantations, it is important to understand how products from this wood are affected. FPL researchers have studied the effect on strength properties and have found that dimension lumber cut from the juvenile wood core may have only 50% to 70% of the strength and stiffness of lumber cut from mature wood, depending upon grade and species. These results are helping lumber standards groups modify design stresses to account for the changing timber resource. [15–17, 29, 31, 32]

◄ Tracing an interesting grain pattern on the finished cross section on display in the lobby of the FPL are, left, George Englerth (FPL) and Fred Ihlenfeld, Jefferson, Wisconsin, who cut the section from a large boxelder burl and treated and dried it in accordance with instructions from the FPL staff. (1960s) {M 126 475}

◄ When Canada's Birch Girl, Miss Linda Barks of Ottawa, visited the FPL, Director Edward Locke showed her a cross section of a California redwood 864 years old when felled. Also shown, on the left, are Clark Leith, Assistant Consul, and Donald Cheney, Canadian Consul. (1960s) {M 128 935}

1910	2010
A few preferred wood species were over-used, leading to potential shortages and excessive waste.	Based on research, most wood species are presently used, greatly enhancing forest management and extending the timber supply.

The following sections give some insight to what we can do with the marvelous material WOOD.

Further Information

[1] Deductions from strength tests of woods / J.A. Newlin / American lumberman (Jan 16, 1915)

[2] Factors affecting the strength of wooden members / Arthur Koehler / FPL No. 363 (1919)

[3] Wood Handbook: basic information on wood as a material of construction with data for its use in design and specifications / prepared by Forest Products Laboratory. Washington, D.C.: United States, Dept. of Agriculture (1935)

[4] Wood at low temperatures / Kenneth H. Boller / Modern packaging Vol. 28(1) (Sept. 1954)

[5] Inside wood: a short trip into the interior for the layman / Research note FPL 014 (1963)

[6] Properties and behavior of wood / Harold Tarkow / Journal of forestry Vol. 68(7) (July 1970)

[7] Relationships of specific gravity to tree height in commercially important species / E.A. Okkonen / Forest products journal Vol. 22(7) (Jul 1972)

[8] Wood / E.H. Bulgrin / McGraw-Hill yearbook of science and technology New York: McGraw-Hill (1974}

[9] Density of tropical timbers as influenced by climatic life zones / Martin Chudnoff / Commonwealth forestry review Vol. 55(3) (1976)

[10] Time-related effects of loading on wood strength: a linear cumulative damage theory / C.C. Gerhards / Wood science Vol. 11(3) (Jan. 1979)

[11] Wood utilization priorities as set by energy and materials shortages / Jerome F. Saeman, Robert L. Youngs. / North America's forests: gateway to opportunity: proceedings of 1978 joint convention of the Society of American Foresters and the Canadian Institute of Forestry. Washington: SAF (1979)

[12] Structure of wood / Research note FPL 04 (1980)

[13] Wood: colors and kinds / D.A. Zischke, B.F. Kukachka / USDA Agriculture handbook No. 101 (1981)

[14] Biological utilization of wood for production of chemicals and foodstuffs / George J. Hajny / FPL–RP–385 (1981)

[15] Quality impacts of the changing timber resource on solid wood products / B. Alan Bendtsen / Managing and Marketing the Changing Timber Resource (1986: Ft. Worth, Tex.). Proceedings, Forest Products Research Society, Madison, Wis.: (1987)

[16] The Influence of juvenile wood on the mechanical properties of 2 X 4s cut from Douglas-Fir plantations / B.A. Bendtsen, P.L. Plantinga, T.A. Snellgrove / Proceedings of the 1988 International Conference on Timber Engineering: Westin Hotel, Seattle, Washington, U.S.A., September 19–22, 1988. Forest Products Research Society, Madison, Wis. (1988)

[17] Effect of juvenile wood on grading of fast grown North American species / David E. Kretschmann / XIX World Congress, 5–11 August 1990: science in forestry, IUFRO's second century: report. Hull, Quebec: Canadian IUFRO World Congress Organizing Committee (1990)

[18] Hardwoods of North America / Harry A. Alden / FPL–GTR–83 (1995)

[19] Wood decks: materials, construction, and finishing / Kent A. McDonald, Robert H. Falk, R. Sam Williams, Jerrold E. Winandy / Madison, WI: Forest Products Society (1996)

[20] Finishes for exterior wood: selection, application, and maintenance / R. Sam Williams, Mark T. Knaebe, William C. Feist / Forest Products Society (1996)

[21] Softwoods of North America / Harry A. Alden / FPL–GTR–102 (1997)

[22] Outdoor furniture, artwork, fences, and play equipment / FPL / The finish line, A FPL finishing fact sheet (January 1998)

[23] Standards for structural wood products and their use in the United States / David W. Green, Roland Hernandez / Wood design focus 9(3): 3–12. (1998)

[24] Wood Handbook: wood as an engineering material FPL–GTR–113 (1999)

[25] Wood products utilization--a call for reflection and innovation / John A. Youngquist, Thomas E Hamilton / Forest products journal 49(11/12): 18–27 (1999)

[26] Research challenges for structural use of small-diameter round timbers / Ronald Wolfe / Forest products journal 50(2): 21–29 (2000)

[27] Industrial wood productivity in the United States, 1900–1998 / Peter J. Ince / FPL–RN–0272 (2000)

[28] Machine grading of lumber: practical concerns for lumber producers / William L. Galligan, Kent A. McDonald / FPL–GTR–7 (2000)

[29] The impact of juvenile wood content on shear parallel and compression and tension perpendicular-to-grain strength for loblolly pine / David E. Kretschmann / Forest Products Society 58th Annual Meeting: June 27–30, 2004, Grand Rapids, Michigan, USA : biographies & abstracts. Forest Products Society, Madison, WI (2004)

[30] Handbook of wood chemistry and wood composites / R.M. Rowell, ed. / CRC Press, New York, NY (2005)

[31] Effect of press-drying on static bending properties of plantation-grown loblolly pine lumber / Denise L. Stoker, David E. Kretschmann, William T. Simpson, / Forest products journal. Vol. 57(11) (Nov. 2007)

[32] Influence of juvenile wood content on shear parallel, compression, and tension transverse to grain strength and mode I fracture toughness for loblolly pine / David E. Kretschmann / FPL–RP–647 (2008)

Poles, Piling, Posts, & Railroad Ties

Solid wood is generally used in poles, piles, posts, and railroad ties. One common requirement of all these wood products is their ability to withstand exposure to the outdoor environment. Much FPL research has been directed to evaluating natural resistance to decay and enhancing that resistance through chemical preservative treatment. Two areas of FPL research have been determining the durability of these treatments and their effect on the strength of the wood. [5, 15] (See also the Preservation section.)

Challenge: How to help ensure that power poles are strong enough to support wires and humans during installation and maintenance, and how to evaluate pilings, posts, and railroad ties for strength and long life?

FPL's Contributions: Evaluated the strength properties of poles made from different species and subjected to different preservative treatments and provided information on pole specifications. Designed low-cost structures using poles. Evaluated posts made from different species and treated with various preservatives. Evaluated railroad ties for strength and preservative treatment effects. Active participation and leadership roles in American National Standards Institute (ANSI) 05.1 and ASTM International (ASTM). Compiled initial database for wood pole tests that was later expanded to provide a basis for derivation of design properties.

Results: Wood poles continue to be a major component of our electric energy distribution system, wood ties support our trains, posts support our fences and signs, and piles support our buildings and piers. All these applications of wood significantly improve our lives.

Early Wood Pole Research

As early as 1914, the FPL tested Rocky Mountain woods for telephone poles and in 1915 tested treated and untreated poles.

According to L.J. Markwardt in the 1950s, "One of the largest and most important cooperative programs undertaken by the laboratory was a so-called ASTM Wood Pole Research Program under written at a cost of about $300,000. The magnitude of the program can be better comprehended when it is noted that some 70 companies, organizations, and associations contributed funds and services to the program. The importance of wood poles to the forest utilization program is indicated by the fact that some 6 million treated wood poles are produced annually, and that the distribution systems of the Nation represent a multibillion dollar investment." (The number of poles produced today is 2 to 3 million, because more wires are placed underground and other materials are used for poles.)

"Methods for two types of wood pole tests were developed and approved by ASTM as standards (Crib Test Method and Machine Test Method). Pole tests were started in 1951. The final report was published and distributed by ASTM in 1960. In all, the program included tests of some 600 full-sized poles and about 14,000 tests of small clear specimens from the pole material. The results of the study were invaluable in the revision of pole specifications and allowable stresses

▲ Wood power poles delivering electrical energy. This new bicycle trail adjacent to FPL has solar-powered lighting (on the right). (2008)

for different species of wood poles." (Based on Lorraine J. Markwardt's (1975) report, "A Rewarding Career at the FPL.")

L.J. Markwardt played a major role in the conception, implementation, and supervision of this program. At that time, he was Assistant Director of the Laboratory, Chairman of Committee D–7 on Wood, and a member of the Board of Directors of the American Society for Testing and Materials (now ASTM International). These associations put him in an ideal position to carry out this research project. [1, 2]

Much of the FPL's research is conducted this way—cooperating with a number of interested parties, designing the research, carrying it out, and reporting the results. Having the comprehensive facilities and background knowledge of wood pays off repeatedly as the Laboratory strives to use wood in the most efficient, effective, and safe manner. [18–21]

Pole Preservation Standards

FPL scientists conducted research on commercial preservative treatment of poles and helped develop specification requirements that have been adopted by industry. These are (1) minimum sapwood and penetration requirements for lodgepole pine, (2) limitations on machine peeling of certain western pole species, and (3) minimum penetration requirements for Douglas-fir poles. Piles—ASTM D 7381; poles—ASTM D 1036. (See Preservation section.)

More Recent Pole Research

Most of the recent research done on poles—and for that matter, piling and posts—has been in the area of obtaining data to enhance better design information and hence standards. This work is critical to the safety of people that service electric power lines and helps provide more reliable service to consumers. (ASTM D 1036–99 (2005) Standard Test Methods of Static Tests of Wood Poles) [24–26]

◀ Power pole under load in testing machine. (84 I)

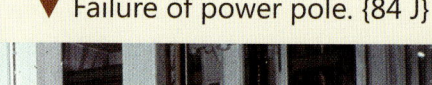
▼ Failure of power pole. {84 J}

◀ Pole being tested in bending. The pole was so long that part of it extended into the courtyard at FPL. {84 H}

Pole Drying

FPL researchers established the effectiveness of a device for measuring moisture content of air-dried lodgepole pine and southern pine poles to determine when the poles are dry enough for preservative treatment. An open-piling method was developed for poles that decreases drying time, wood stain, and mold. [7, 8]

◀ Based on engineering evaluation of poles, strength data are available to build safe pole barns. {6 A}

Laminated Poles

Laminating wood together is another method of using wood for poles. [17]

▲ Gazebo made of small-diameter timber and preservative-treated poles at FPL, Madison, Wisconsin (2000s)

◀ Laminated power pole being prepared for testing. [24, 25]

Wood Pole Standards

FPL representatives have served on national committees that develop standards for pole use and design. For example, for many years FPL representatives chaired American National Standard Institute Committee 05 on Wood Pole Use and Design. (ASTM D 3200–74 (2005) Standard Specification and Test Method for Establishing Recommended Design Stresses for Round Timber Construction Poles) [18, 20, 21]

Pilings

FPL scientists helped develop tests and standard guidelines for the design and installation of pile foundations. (ASTM D 7381–07 Standard Practice for Establishing Allowable Stresses for Round Timbers for Piles from Tests of Full-Size Material) [3–5, 16, 19]

◀ Example of the use of wood pilings to support a railroad bridge.

Posts

In an effort to find outlets for unused wood, various species of Appalachian woods were evaluated for use as highway posts. [11]

More recently studies have indicated the potential for expanding the use of small-diameter timber for wood posts in highway applications. [23, 27]

▲ Michael Wiemann and David Kretschmann using an increment borer to evaluate the use of small-diameter softwood as guardrail posts. (2001) [27]

◀ Joe Murphy conducting a nondestructive test to determine the strength of a post. (2000s) [23]

◀ Moving boggie (ram) impacting a post to determine its resistance to dynamic forces such as would be experience when impacted by an out-of-control car or truck. (2000s)

Railroad Ties

Research on railroad ties is confined mainly to finding preservative treatments that are more effective and have a minimum negative impact on the environment. [5] (See Preservation section.)

▲ Typical rail line using treated wood ties.

▲ Close-up of railroad ties. Note the large crack in the second tie. This is an excellent opening for decay organisms to invade the tie.

Durable Railroad Ties

"One of the longest—if not the longest—service records emphasizing the value of preservative treatment for wood under severe decay-hazard conditions was concluded in 1962—exactly 51 years after it had begun. In 1911, the Milwaukee railroad entered into a cooperative agreement with FPL to install and keep service records on creosoted railroad ties of several species in a section of track just south of Madison. The final results demonstrated that well treated ties can have an average life of more than 50 years compared with a 12-year average for untreated white oak and 9 years for untreated red oak under Wisconsin conditions." (1962 FPL Annual Report, p. 28)

1910

Millions of additional trees were used because untreated wood poles, piles, posts, and railroad ties had to be replaced in just a few years because of decay.

2010

Effective treatment of wood with preservatives greatly extends the life of wood in use and decreases the drain on the forests.

Further Information

[1] Tests of rocky mountain woods for telephone poles / Norman deW. Betts, A.L. Heim / USDA Bulletin 67 (1914)

[2] Service tests of treated and untreated telephone poles / Clyde H. Teesdale / In cooperation with the American Telephone and Telegraph Company (1915)

[3] Results of experiments in the treatment of piling specimens against attack by marine borers / J.D. MacLean / Engineering and contracting (Mar. 1924)

[4] The selection of marine piling / T.R.C. Wilson / The constructor Vol. 12(7) (July 1930)

[5] Poles, piling, and ties / FPL / FPL No. 1903–6 (1951)

[6] Effects of treating variables on absorption and distribution of chemicals in pine posts treated by double diffusion / R.H. Baechler / Forest Products Research Society journal Vol. 3(5) (Dec. 1953)

[7] Comparison of three methods of determining whether southern pine poles are well air seasoned / J.S. Mathewson, Paul J. Berger / Forest Products Research Society journal Vol. 7(5) (May 1957)

[8] Air seasoning of lodgepole pine poles / J.S. Mathewson / FPL No. 1922 (1958)

[9] Strength and related properties of wood poles / Lyman W. Wood, E.C.O. Erickson, A.W. Dohr (with a foreword by L.J. Markwardt) / American Society for Testing Materials (1960)

[10] Groundline treatments and their evaluation / J. Oscar Blew, Jr. / For presentation at the South Atlantic wood pole conference, North Carolina State College, Raleigh, N.C. (April 1961)

[11] Evaluating Appalachian woods for highway posts / D.V. Doyle / FPL–RP–111 (1969)

[12] Pole groundline preservative treatments evaluated / J.O. Blew / Transmission and distribution (Aug. 1970)

[13] Utility pole decay. Part II, Basidiomycetes associated with decay in poles / Wallace E. Eslyn / Wood science and technology Vol. 4(2) (1970)

[14] Survey of totem poles in southeast Alaska and recommendations for their preservation or recovery / Joe W. Clark / Preprinted for the International Institute for Conservation of Historic and Artistic Works, June 1970, New York, N.Y. Apparently never published. Manuscript only (1970)

[15] Wood poles and piles / Billy Bohannan / Wood structures: a design guide and commentary / compiled by task committee on status of the art: wood, committee on wood, American Society of Civil Engineers, Structural Division. New York: ASCE, (1975)

[16] Appraising deterioration in submerged piling / Wallace E. Eslyn, Joe W. Clark / Material und organismen. Beiheft No. 3 (1976)

[17] Bending properties of reinforced and unreinforced spliced nail-laminated posts / David R. Bohnhoff, Russell C. Moody, Steven P. Verrill / FPL–RP–503 (1991)

[18] ANSI pole standards: development and maintenance / Ronald W. Wolfe, Russell C. Moody / Proceedings of the first southeastern pole conference. Madison, WI: Forest Products Society, Proceedings; 7314 (1994)

[19] Standard guidelines for the design and installation of pile foundations / American Society of Civil Engineers. ASCE 20–96 (1997)

[20] Standard specifications for wood poles / Ronald Wolfe, Russell Moody / Utility pole structures conference and trade show: Nov. 6–7, Reno/Sparks, NV. (1997)

[21] Design stress derivation for ANSI poles / Ronald W. Wolfe / In: Proceedings, international conference on utility line structures; March 20–22; Fort Collins, CO: 189–203 (2000)

[22] Nondestructive methods of evaluating quality of wood in preservative-treated piles / Xiping Wang, Robert J. Ross, John R. Erickson, Rodney C. De Groot / Research note FPL 0274 (2000)

[23] Potential for expanding small-diameter timber market: assessing use of wood posts in highway applications / Dorothy Paun, Gerry Jackson / FPL–GTR–120 (2000)

[24] Nondestructive evaluation of Young's moduli of full-size wood laminated composite poles / Cheng Piao, Chung-Yun Hse / 7th pacific rim bio-based composites symposium, Nanjing, China, October 31, November 2, 2004. Nanjing, China: Science & Technique Literature Press, c2004: pages 291– 298 (2004)

[25] Finite element analyses of wood laminated composite poles / Cheng Piao, Chung-Yun Hse / Wood and fiber science Vol. 37(3) (2005)

[26] Designated fiber stress for wood poles / Ronald W. Wolfe, Robert O. Kluge / FPL–GTR–158 (2005)

[27] Investigating the use of small-diameter softwood as guardrail posts: static test results / David E. Kretschmann / Dick Shilts, Tim Nelson / FPL–RP–640 (2007)

Preservation

When natural woods are exposed to the outdoor environment, they are subjected to degradation by a number of natural causes. Though some wood species have some natural resistance to degradation, such as decay, the supply of these naturally durable species is limited. Common wood species such as southern pine, ponderosa pine, and Douglas-fir have little resistance to decay and need some form of protection to extend their useful life. [1, 29, 30, 46, 53, 56, 66]

Challenge: How to keep decay fungi, insects, and other organisms from destroying wood structures.

FPL's Contributions: From the beginning, the FPL has had a research program on how best to preserve wood. This research has led to improved treatments and processes, and decreased cost.

Results: Today the wood used in railroad ties, poles, pilings, posts, and some other wood products is chemically treated to extend the useful life of the various products. These treatments have significantly decreased the drain on the timber resource.

Extending the Life of Railroad Ties

In the early days of the railroads, the railroad ties would last only a few years and had to be replaced because of decay. This was costly and used vast amounts of wood. Thus, in the early 1910s, FPL scientists focused on developing methods of increasing the durability of railroad ties. They worked with industry partners to put their findings into practice. Through these efforts, the useful life of railroad ties was increased by a factor of five. The volume of wood treated with preservatives increased from 43 million cubic feet in 1909 to 189 million cubic feet in 1925. [2, 3]

▲ FPL research has helped to make wood a durable, economical, and environmentally friendly building material. (2000s)

▲ Experimental treatment of railroad ties at FPL's original location on the University of Wisconsin campus. (1910s) [5]

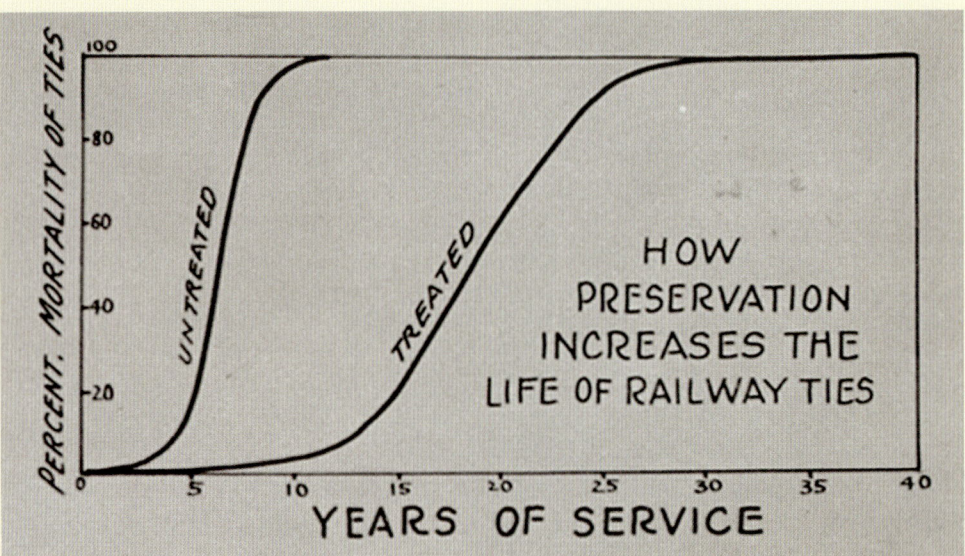

Preservatives for Marine Piles

Two-Stage Pressure Treatment: FPL researchers demonstrated that two-stage (dual) pressure treatments with certain types and retentions of waterborne preservatives followed by certain types and retentions of creosotes can prevent marine borer damage even in subtropical waters where creosote or waterborne preservatives, such as chromated copper arsenate (CCA), alone cannot. [27]

Research also showed how simple field treatments could be used to prevent fungal decay in the above-water portion of piles. Brush-on surface treatments for the top surface of cut piles, application of caps to protect pile tops, and internal treatments with liquids and fumigants were shown to extend the useful life of piles by at least 10 years. [36, 37, 45, 48]

◄ Damage to creosote-treated piling caused by marine borers. {91 F}

◄ In the 1960s, Roy Baechler treated a small section of a post by nonpressure double diffusion, followed by pressure treatment with creosote, and installed it in a Key West, Florida, harbor. Bruce Johnson inspected this specimen in 2007 and found no marine borer damage after 40 years in the harbor. Untreated, it would have been destroyed in one to two years. [20, 27]

Improving Treatment Quality

FPL researchers determined the treating properties of important commercial wood species and preservatives, comparing the effectiveness of more than 100 preservatives and 10 treating methods under actual service conditions. Materials and methods for treating wood with preservatives were improved and refined with efficiency and economy, contributing greatly to the continued use of durable treated wood by the railroads and thus extending the wood supply. [11, 22, 24, 67, 78]

▼ Difference in penetration of wood preservative because of wood species, chemical, and process. {116 E} [46]

▲ Experimental pole treatments in FPL's "new" treatment facility. (1930s)

◀ Harley Davidson removing a load of incised lumber after treatment. The incising of the lumber increases the penetration of the preservative into the wood and extends the wood's service life.

Low-Cost Treatment Methods

FPL scientists researched simple, low-cost treatment methods that could be used by farmers and small businesses or in isolated areas. Simple methods, such as steeping or soaking fence posts in barrels, were improved when researchers demonstrated the practicality and effectiveness of hot-and-cold bath treatments with creosote and other oil-type preservatives.

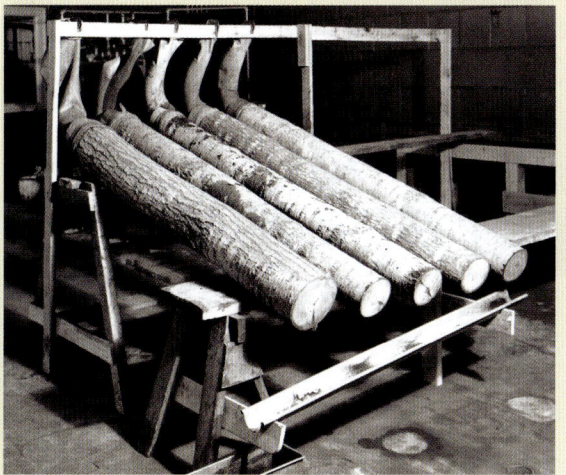

◀ Tire-tube treatment for the preservation of fence posts developed by the FPL. Sections of tight-fitting inner tubes were placed over the butt ends of freshly cut posts. Preservative poured into the tubes displaced the sap within the posts. (1930s) {M 32857 F} [6]

▶ Roy Baechler using low-cost double diffusion method of treating wood fence posts. (1950s) {M 94018 F}

Double Diffusion Treatment Process

Roy Baechler pioneered the use of double diffusion for preservative treatment of wood. The double diffusion treatment process utilizes a chemical reaction between two salts to form a leach-resistant precipitate within the wood. In some wood species, the double diffusion treatments yielded more uniform penetration than pressure treatment. Long-term tests were performed to document the durability of wood treated with these methods, and some of these technologies are still in use today. This approach was further developed by Lee Gjovik, Bruce Johnson, and other staff at FPL [12, 25]

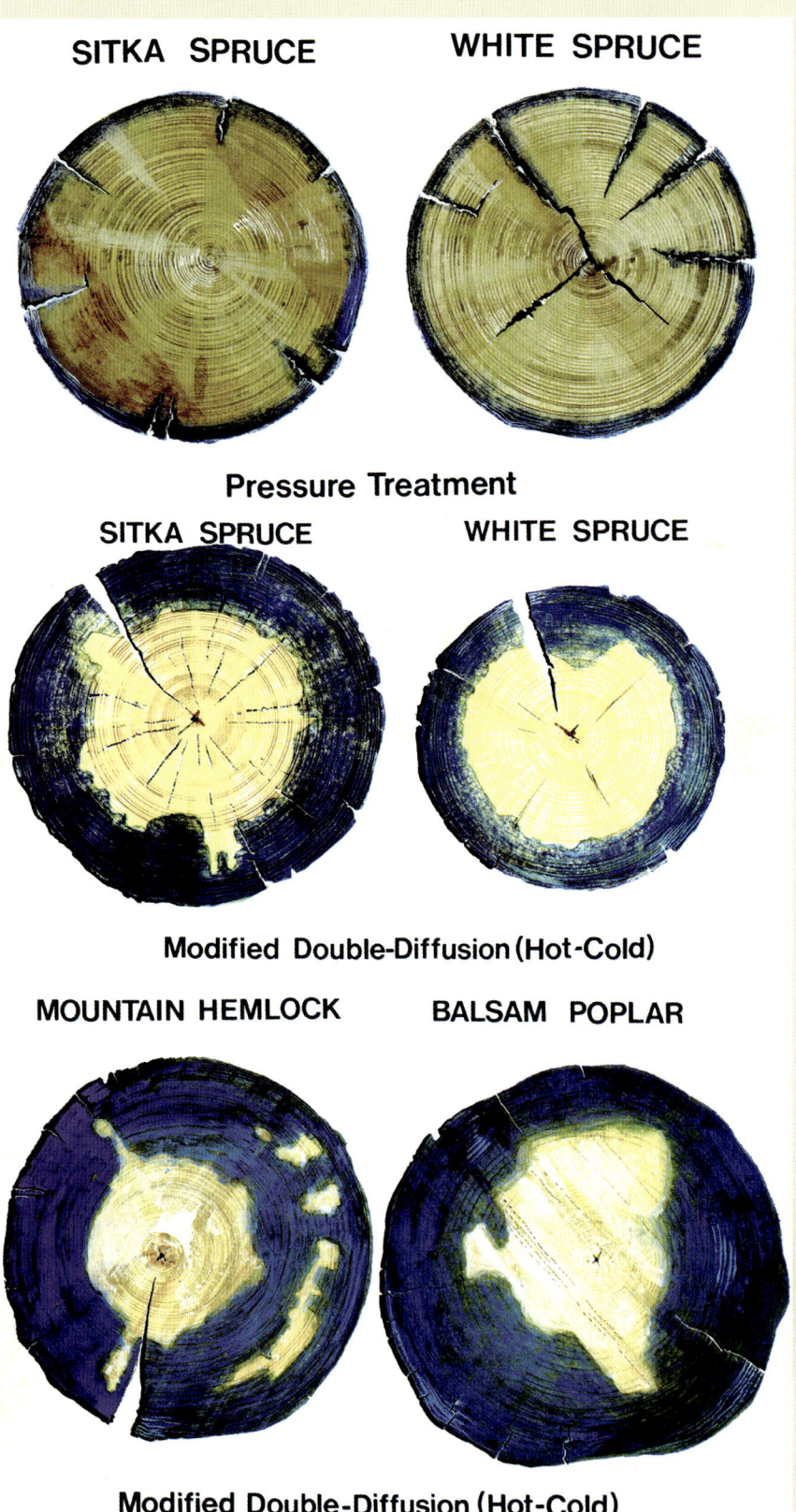

SITKA SPRUCE **WHITE SPRUCE**

Pressure Treatment

SITKA SPRUCE WHITE SPRUCE

Modified Double-Diffusion (Hot-Cold)

MOUNTAIN HEMLOCK **BALSAM POPLAR**

Modified Double-Diffusion (Hot-Cold)

◀ Penetration with double diffusion treatments sometimes exceeds that obtained with conventional pressure treatments. (1950s)

Wood Preservation Approaches

Toxic Chemical Treatments

Over the past century, FPL researchers have been involved in many aspects of the development and evaluation of new wood preservatives. FPL scientists discovered relationships between the chemical constituents of benzene derivatives and their toxicity to wood-destroying fungi and suggested to manufacturers around the 1930s that the more highly chlorinated phenols would be of value for wood preservation. To this day, pentachlorophenol remains one of the most widely used preservatives for treatment of utility poles. However, its handling and treatment are tightly controlled to protect human health. [4, 7]

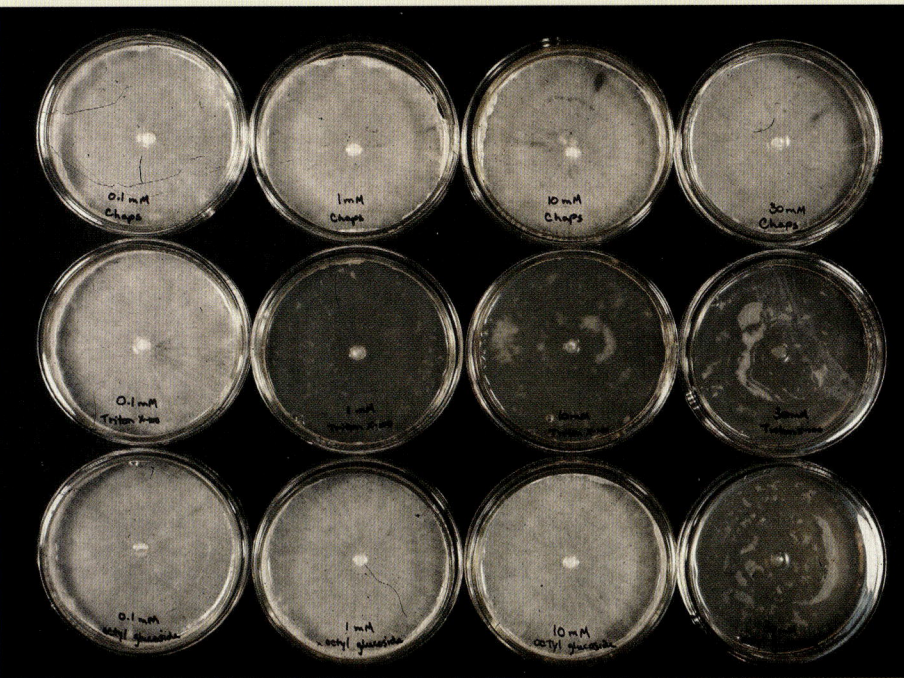

◄ FPL scientists determined the toxic thresholds of many chemicals against common decay fungi using the agar plate method. (2000s)

Chitin Inhibition

FPL researchers also studied a new approach to wood preservatives based on chitin inhibitors. Chitin inhibitors prevent formation of chitin, an essential building block in the cell walls of most fungi, in the exoskeleton of insects, and in some marine borers. The chemical does not poison the organism, as conventional preservatives do, but prevents the organism from growing. Recently this approach has gained attention from international researchers seeking to find preservatives with decreased toxicity to non-target organisms. [35, 38, 39]

◄ Bruce Johnson explaining chitin inhibitors to Rebecca Schumann and George Chen. (1980s) {M 150 677-8}

Acetylation

As an alternative approach for protecting wood, FPL scientists developed methods such as acetylation and modification with alkylene oxides to make the wood resistant to fungal attack and less likely to swell, shrink, or warp when moisture levels change. [40]

Borates

The use of borates is another method of preventing decay, though leaching of the borate from the wood is a challenge. Rodney De Groot studied the distribution of borates around point source injections in dry wood members. [61]

Chelating Compound (NHA)

Another approach discovered by FPL researchers was the use of a chelating compound (NHA) to deprive wood-attacking organisms of calcium needed for growth. This approach was shown to be effective against both decay fungi and termites in field tests and was subsequently licensed as a component of technology used to prevent damage from termites. [75]

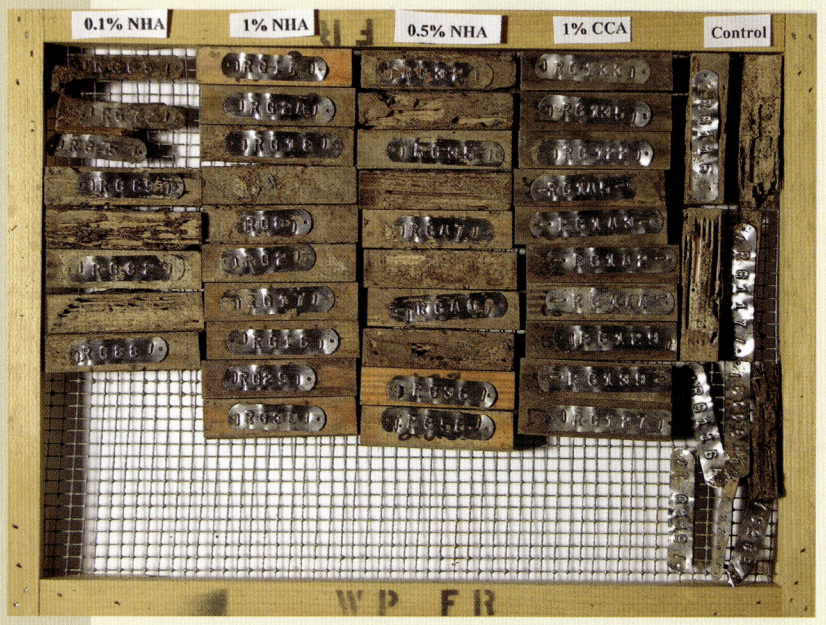

▲ Test specimens showing the effect of different preservative treatments, including NHA, against termite damage. (2000s)

▲ Frederick Green installs test decking treated with an experimental preservative based on calcium chelation.

Inspection and Remedial Treatments

Wooden structures, even those treated with wood preservatives, can develop decay during service. Inspection and remedial treatment are important in maximizing service life. FPL researchers developed techniques to make these processes more efficient and effective. One innovative approach to detecting incipient decay was the development of an immunoassay method that produced a color change if enzymes associated with decay fungi were present in the wood. This method, which was subsequently patented, allows detection of the presence of decay before significant structural damage occurs. [57, 58, 76]

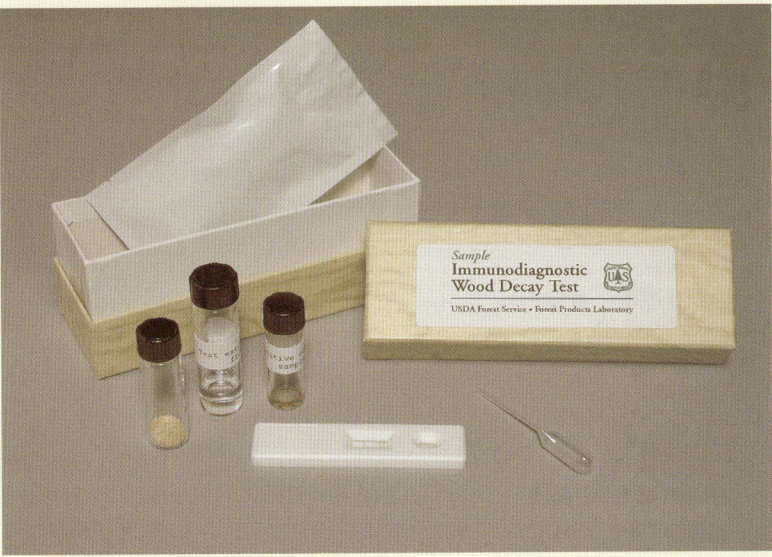

▼ Color change indicates the presence of decay fungi. (2003)

▲ Test kit used for the immunoassay indicator checks for the presence of decay fungi. (2003) [76]

Historic Preservation

◄ B. Francis Kukachka inspecting the timber supporting the Liberty Bell for decay. (1950s) {M 92064} [49]

◄ Terry Highley inspecting a portion of the U.S.S. Constitution for decay.

▲ The historic U.S.S. Constitution, also known as "Old Ironsides," is moored in the Charlestown Navy Yard in Boston Harbor. FPL provided information on live oak used to construct the ship, advice on inspection and repair methods, and methods to manufacture live oak glued-laminated replacement timbers. {15 K} [8, 42, 54, 59]

Remedial Treatment

Remedial treatments are supplemental preservative treatments that are applied to wood while it is in service. They include surface treatments that are brushed or sprayed on the wood; diffusible chemicals in the form of liquids, pastes, or solid; or fumigants. Supplemental in-place treatments extend the useful life of wood products and wood structures. [17, 18, 28, 50, 52, 63, 68, 71, 73]

◄ Wally Eslyn installs a test for the effectiveness of fumigant treatments in pine timbers (1980s) [44]

Developing Methods for Evaluating Durability

The FPL has long been a leader in developing methods to better evaluate the durability of wood products. The FPL maintains one of the world's most extensive and longest running field durability research programs. Since the early 1900s, post and stake test specimens have been placed in test sites in Mississippi, Wisconsin, and other locations to evaluate treated wood exposed to soil and to the aboveground environment. Some existing specimens were placed over 70 years ago. Data from these tests have proven invaluable to researchers considering potential new wood preservatives. [26, 62, 74]

▲ Wallace Hegge (left) and Oscar Blew rate stakes at one of FPL's plots in Harrison Experimental Forest, Mississippi. (1940s) [9]

◄ Test specimens before and after exposure to the elements. Results are used to determine the durability of various wood species and preservative treatments. (1990s) [13, 41, 60, 64, 65]

The world's most widely used climate index for estimating the decay potential of wood used aboveground in the United States was developed by FPL. [23]

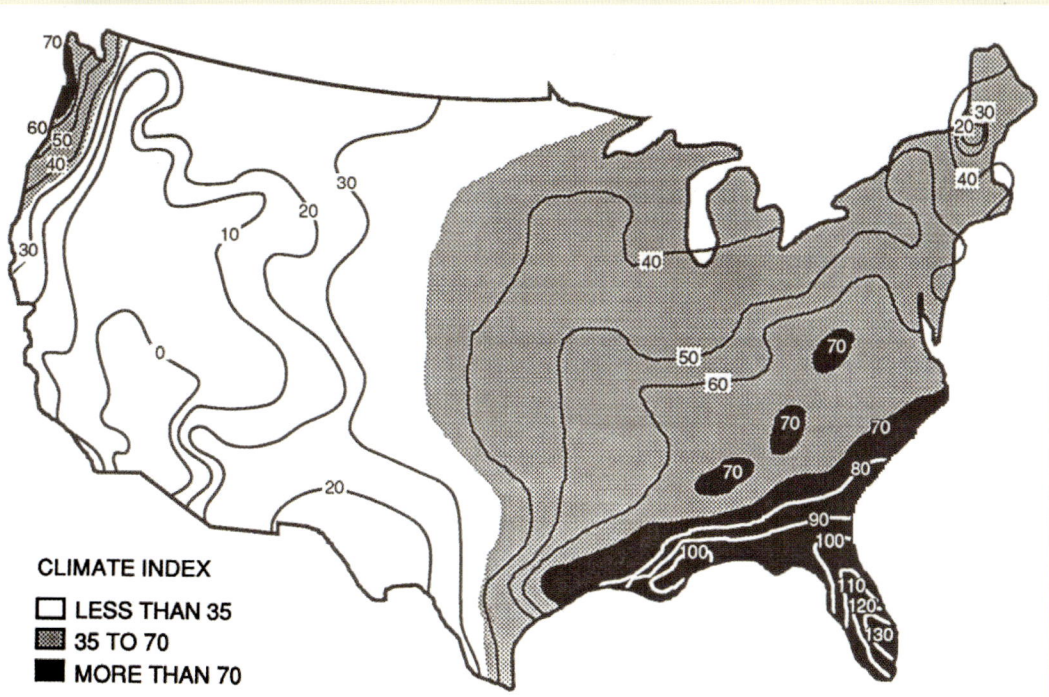

◄ Climate index–decay hazard map developed by FPL scientist Theodore Schaffer. (1970s)

CLIMATE INDEX
☐ LESS THAN 35
▨ 35 TO 70
■ MORE THAN 70

Soil-Block Decay Test

FPL researchers played a major role in development of the soil-block decay test, which remains the mostly widely used laboratory test to assess the ability of treated wood to resist fungal degradation. [10]

◄ Ralph Lindgren and Catherine Duncan inspecting the growth of a fungus that is part of the fungi collection. (1950s) {M 99 655}

▲ Soil-block tests showing the effects of fungi on a specimen: (A) fungus growing on untreated wood specimen; (B) fungus fruiting on untreated wood specimen; (C) preservative-treated wood specimen. (1950s)

The FPL houses the largest wood fungus collection in the world, with over 5,000 species of fungi. Many of these fungi are used in research to determine the effectiveness of wood preservatives.

▲ Oscar Blew inspecting posts at the Harrison Experimental Forest near Saucier, Mississippi. (1940s)

Evaluations of Natural Durability

FPL scientists generated and compiled data that are used by the forest products industry as a basis for selecting wood species for use under conditions favorable to decay. The data cover the decay resistance of most native and more than 60 foreign tree species. Use of decay-resistant species decreases the need for chemical preservatives. [16, 21]

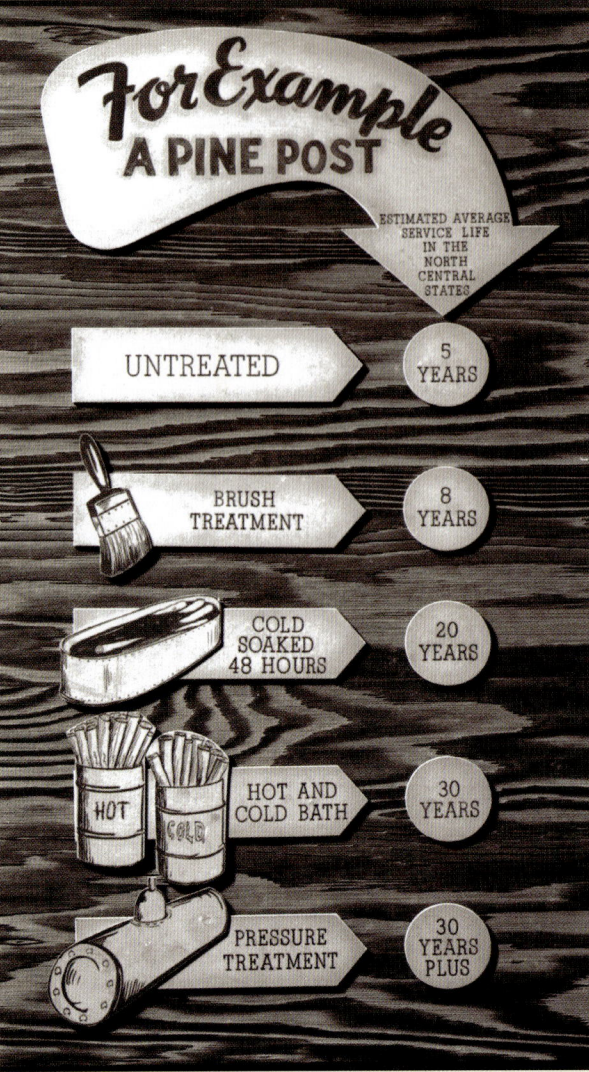

◄ Effectiveness of various treatment methods based on field tests of posts. (1950s) {M 107 538} [46]

▲ Terry Highley generated widely used data on the durability of native wood species exposed aboveground. (1990s) [55]

▲ Bessie Woodward and Douglas Crawford install a stake plot to evaluate the durability of Alaskan yellow cedar. (1990s)

Assessing and Minimizing Environmental Impacts

In the past three decades, addressing environmental concerns associated with wood preservatives has become an increasing emphasis for FPL researchers. FPL staff, in cooperation with an assessment team of more than 20 specialists from other government agencies and universities, published a 700 page report, *Biologic and Economic Assessment of Penta, Inorganic Arsenicals and Creosote*. This report was requested by the U.S. Environmental Protection Agency, who used it as a basis for decisions regarding the registration and allowable uses of these preservatives. [34]

► Lee Gjovik and David Gutzmer discuss research results. The display in background shows the effectiveness of treating various species using a dip tank process. (1980s) {135 K}

Use of Bacteria to Protect Landfills from Preservative-Treated Wood

FPL researchers are using bacteria to protect landfills from preservative-treated wood. Since the 1970s, an estimated 60 billion board feet of chromium-, copper-, and arsenic- (CCA-) treated wood has been placed in service. Now, almost 40 years later, this preservative-treated waste wood is accumulating in landfills. FPL scientists are helping to solve this problem by using preservative-tolerant bacteria to extract the metals so that the remaining wood fiber can be utilized for other purposes. [69, 80]

Environmental Impacts of Treated Wood

Stan Lebow evaluated the potential for environmental impacts from preservative-treated wood used in sensitive aquatic environments. In a cooperative Forest Service and Bureau of Land Management project, several types of treated wood were used in the construction of a large wetland boardwalk in western Oregon. Leaching and environmental accumulations of preservative were monitored, as were the effects on populations of aquatic invertebrates. The study showed that although some leaching of preservative into the wetland did occur, it was not enough to affect aquatic invertebrates. [70]

▲ Carol Clausen conducts tests during bioremediation of waste treated wood. After bacterial and acid extraction of the preservative, cleaned flakes were formed into flakeboard. (2000s)

◀ This test section of a wetland boardwalk was treated with chromated copper arsenate (CCA) and used to evaluate leaching and environmental impacts from treated wood. (2000s) [51, 72, 77, 79]

Wood used in honeybee hives is susceptible to decay. FPL researchers evaluated various treatments that would protect the wood without harming the bees or the quality of the honey produced. [43]

◀ Decay of a honeybee hive constructed with untreated wood (left), compared with a sound hive treated with preservative (right). (1980s) {113 C}

◀ Field treatments to preserve historic covered bridges. [32, 44, 49]

For over 100 years of FPL research, durable wood products have evolved from limited use in railroad ties to thousands of applications that benefit consumers, industry, and government.

1910

Large losses of wood because of decay. Ineffective preservatives or processes were used. No standard tests for effectiveness were available.

2010

New test methods to evaluate treatments were developed that resulted in improved preservatives and processes, and decreased environmental impacts.

Further Information

[1] Records on the life of treated timber in the United States / Howard F. Weiss, Clyde H. Teesdale / American Wood Preservers' Association, Proceedings of the 11th annual meeting. Baltimore, MD: AWPA (1915)

[2] Tests of wood preservatives / Howard F. Weiss / USDA Forest Service Bulletin No. 145 (1915)

[3] Selecting the wood for treatment / G.M. Hunt / Forest Products Laboratory (1920?)

[4] The effect of concentration on the toxicity of chemicals to living organisms / Ernest Bateman / USDA, Forest Service technical bulletin No. 346 (1933)

[5] Manual on preservative treatment of wood by pressure / J.D. MacLean / USDA Miscellaneous publication No. 224 (1935)

[6] Old inner tubes used in low-cost fence post treatment / G.M. Hunt, R.M. Wirka / FPL No. 1158 (1937)

[7] Some toxicity data and their practical significance / Ernest Bateman, R.H. Baechler / FPL No. 1222 (1937)

[8] Readiness and care of vessels in inactive status / Washington, D.C., U.S. Navy Dept., Bureau of Ships Manual (1945)

[9] Comparison of wood preservatives in stake tests / J. Oscar Blew / American Wood-Preservers' Association. Proceedings of the 44th annual meeting, Washington, DC: AWPA (1948)

[10] Evaluating wood preservatives by soil-block tests / Catherine G. Duncan, G. Audrey Richards / Proceedings of the American Wood-Preservers' Association No. 46:131–145 (1950) and 47:264–292 (1951)

[11] Preservative treatment of wood by pressure methods / J.D. MacLean / USDA Agriculture handbook No. 40) (1952)

[12] How to treat fence posts by double diffusion / R.H. Baechler / FPL No. 1955 (1953)

[13] Effectiveness for different wood species and climates of on-the-job preservative treatment of exterior millwork / Theodore C. Scheffer, A.F. Verrall, George Harvey / Forest Products Laboratory (1962)

[14] A half century of service testing crossties / J. Oscar Blew, Jr. / American Wood-Preservers' Association. Proceedings of the 59th annual meeting Washington, DC (1963)

[15] Effect of gamma radiation on the decay resistance of wood / Theodore C. Scheffer / Forest products journal Vol. 13(5) (May 1963)

[16] Natural resistance of wood to microbial deterioration / Theodore C. Scheffer, Ellis B. Cowling / Annual review of phytopathology Vol. 4 (1966)

[17] Application and use of fungicides in wood preservation / R.H. Baechler / Fungicides / edited by DeWayne C. Torgeson / Academic Press (1967)

[18] On-site preservative treatment for exterior wood of buildings / T.C. Scheffer, J.W. Clark / Forest products journal Vol. 17(12) (Dec. 1967)

[19] PEG is the sweetheart of the wood craftsman / Harold L. Mitchell / Science for better living, USDA Yearbook of Agriculture (1968)

[20] Studies of several methods for determining suitability of creosote for marine use / R.H. Baechler, L.R. Gjovik, H.G. Roth / American Wood-Preservers' Association. Proceedings of the 65th annual meeting Washington, DC: AWPA (1969)

[21] Natural decay resistance of fifteen exotic woods imported for exterior use / J.W. Clark / FPL–RP–103 (1969)

[22] Vacuum treatment of lumber / J.O. Blew, E. Panek, H.G. Roth / Forest products journal Vol. 20(2) (Feb. 1970)

[23] A climate index for estimating potential for decay in wood structures above ground / Theodore C. Scheffer / Forest products journal Vol. 21(10) (Oct. 1971)

[24] Analyzing creosote by gas chromatography: relationship to creosote specifications / L.F. Lorenz, L.R. Gjovik / American Wood-Preservers' Association. Proceedings of the 68th annual meeting, Washington, DC: AWPA 68: 32–34, (1972)

[25] Treatment of Alaskan species by double-diffusion and modified double-diffusion methods / L.R. Gjovik, H.G. Roth, H.L. Davidson / Research paper FPL 182 (1972)

[26] Comparison of wood preservatives in Mississippi post study: 1973 progress report / L.R. Gjovik, H.L. Davidson / FPL–RN–01 (1973)

[27] Single- and dual-treated panels in a semi-tropical harbor: preservative and retention variables and performance / B.R. Johnson, L.R. Gjovik, H. Roth / American Wood-Preservers' Association. Proceedings of the 69th annual meeting Washington, DC: AWPA (1973)

[28] Effectiveness of groundline treatments of creosoted pine poles under tropical exposure / M. Chudnoff, L.R. Gjovik, R. Wawriw. Forest products journal Vol. 23(9) (Sept. 1973)

[29] Nontoxic wood preservation treatments / Roger M. Rowell / Wood & wood products Vol. 83(2) (Feb. 1978)

[30] Ammoniacal copper borate: a new treatment for wood preservation / B.R. Johnson, D.I. Gutzmer / Forest products journal Vol. 28(2) (Feb 1978)

[31] Overseas protection of logs and lumber: part I / J.M. McMillen / Import/export wood purchasing news Vol. 4(5) (Apr/May 1978)

[32] Principles for protecting wood buildings from decay / T.C. Scheffer / FPL–RP–190 (1979)

[33] Effects of wood preservatives on electric moisture-meter readings / William L. James / FPL–RN–0106 (1965, Revised 1980)

[34] The biologic and economic assessment of pentachlorophenol, inorganic arsenicals, creosote / USDA Technical bulletin No. 1658 (1980)

[35] Responses of wood decay fungi to polyoxin D, an inhibitor of chitin synthesis / Bruce R. Johnson / Material und organismen Vol. 15(1) (1980)

[36] In-place treatments for control of decay in waterfront structures / Terry L. Highley / Forest products journal Vol. 30(9) (Sept. 1980)

[37] Using fumigants to control interior decay in waterfront timbers / Terry L. Highley, Wallace Eslyn / Forest products journal Vol. 32(2) (Feb 1982)

[38] Effects of polyoxin inhibitors of fungal chitin synthesis on the decay of wood / B.R. Johnson / International biodeterioration bulletin Vol. 18(2) (Summer 1982)

[39] Occurrence and inhibition of chitin in cell walls of wood-decay fungi / B.R. Johnson, G.C. Chen / Holzforschung Vol. 37(5) (1983)

[40] Chemical modification of wood / Roger M. Rowell / Forest products abstracts Vol. 6(12) (Dec. 1983)

[41] The role of water repellents and chemicals in controlling mildew on wood exposed outdoors / William C. Feist / FPL–RN–0247 (1984)

[42] Evaluation of live oak submerged underwater for 50 years and proposed for use in rebuilding the U.S.S. Constitution / Jerry Winandy / Forest products journal Vol. 34(5) (May 1984)

[43] Effect of wood preservative treatment of beehives on honey bees and hive products / Martins A. Kalnins and Benjamin F. Detroy / Journal of agricultural and food chemistry Vol. 32(5) (Sept.–Oct. 1984)

[44] Efficacy of fumigants in the eradication of decay fungi implanted in southern pine timbers / T.L. Highley, W.E. Eslyn / International research group on wood preservation annual meeting Stockholm, Sweden: IRG Secretariat, Document No. IRG/WP/3365 (1986)

[45] Fifteen-year test of in-place treatments for control of decay in waterfront structures / Terry L. Highley, Theodore Scheffer / Material und organismen Bd. 21, heft 3 (1986)

[46] Looking back at 75 years of research in wood preservation at the U.S. Forest Products Laboratory / R.H. Baechler, L.R. Gjovik / American Wood-Preservers' Association. Proceedings of the 82nd annual meeting Stevensville, MD: AWPA (1986)

[47] *Sphaeroma terebrans* Bate: a note on distribution and preservative tolerance in Florida coastal waters / B.R. Johnson / Eighteenth meeting of the international research group on wood preservation (Working group IV: marine wood preservation) Honey Harbour, Ont., Can. (May 17–22, 1987)

[48] Controlling decay in waterfront structures: evaluation, prevention, and remedial treatments / Terry L. Highley, Theodore Scheffer / FPL–RP–494 (1989)

[49] Archaeological wood: properties, chemistry, and preservation / R.M. Rowell, J. Barbour, eds. / American Chemical Society Advances in chemistry series 225, Washington, DC (1990)

[50] Evaluation of bacteria for biological control of wood decay / Riana Benko, Terry L. Highley / Stockholm, Sweden: IRG Secretariat, (1990)

[51] Chromium-containing waterborne wood preservatives: fixation and environmental issues / J.E. Winandy, H.M. Barnes / Madison, WI: Forest Products Society (1993)

[52] Prevention of enzyme stain of hardwoods by log fumigation / Elmer L. Schmidt, Terry L. Amburgey / Forest products journal Vol. 44(5) (May 1994)

[53] Biochemical approaches to wood preservation / Philip J. Kersten / In: Proceedings, Wood preservation in the '90s and beyond; 1994; September 26–28; Savannah, GA. Proceedings 7308. Madison, WI: Forest Products Society: 153–157 (1994)

[54] Inspection and evaluation of structural members and connections in the USS Constitution. Lawrence A. Soltis / Restructuring: America and beyond: proceedings of structures congress XIII, Boston, Massachusetts, April

2–5, 1995. New York, N.Y.: American Society of Civil Engineers (1995)

[55] Comparative durability of untreated wood in use above ground / T.L. Highley / International biodeterioration & biodegradation Vol. 47 (1995)

[56] Evaluation of new creosote formulations / Douglas M. Crawford, Rodney C. De Groot / In: Ritter, Michael A.; Duwadi, Shella Rimal; Lee, Paula D. Hilbrich, eds. Proceedings, national conference on wood transportation structures—new wood treatments; 1996 October 23–25 / FPL–GTR–94 (1996)

[57] Method and apparatus for immunological diagnosis of fungal decay in wood / Carol A. Clausen, Fredrick Green / U.S. Patent No. 5,563,040 (Oct. 8, 1996)

[58] Immunological detection of wood decay fungi: an overview of techniques developed from 1986 to the present / Carol A. Clausen / International biodeterioration & biodegradation Vol. 39(2–3) (1997)

[59] Assessing wood members in the USS Constitution using non-destructive evaluation methods / Robert J. Ross, Lawrence A. Soltis, Patrick Otton / APT Bull.–Journal of Preservative Technology 29(2): 21–25. (1998)

[60] Field durability of CCA- and ACA-treated plywood composed of hardwood and softwood veneers / Rodney C. De Groot, Lee R. Gjovik, Douglas M. Crawford, Bessie Woodward / Forest products journal 48(2) (1998)

[61] Distribution of borates around point source injections in dry wood members / Rodney C. De Groot, Colin C. Felton / Holzforschung 52(1) (1998)

[62] Ten-year performance of treated northeastern softwoods in above ground and ground-contact exposures / Douglas M. Crawford, Rodney C. DeGroot, Lee R. Gjovik, / FPL–RP–578 (1999)

[63] Mildew and mildew control for wood surfaces / Steve Bussjaeger, George Daisey, R. Simmons, Saul Spindel, R. Sam Williams / Journal of coatings technology 71(890) (1999)

[64] Water repellents and water-repellent preservatives for wood / R. Sam Williams, William C. Feist / FPL–GTR–109 (1999)

[65] Effect of water repellents on long-term durability of millwork treated with water-repellent preservatives / R. Sam Williams / Forest products journal 49(2) (1999)

[66] Wood Handbook: wood as an engineering material / FPL–GTR–113 (1999)

[67] Treatability of U.S. wood species with pigment-emulsified creosote / Douglas M. Crawford, Rodney C. De Groot, John B. Watkins, Harry Greaves, Karl J. Schmalzl, T.L. Syers / Forest Prod. J. 50(1) (2000)

[68] Restoration of severely weathered wood / R. Sam Williams, Mark Knaebe / J. Coatings Technol. 72(902): 43–51 (2000)

[69] CCA removal from treated wood using a dual remediation process / C.A. Clausen / Waste Management and Research 18 (5):485–488 (2000)

[70] Environmental impact of preservative-treated wood in a wetland boardwalk / Stan T. Lebow, Patricia K. Lebow, Daniel O. Foster, Kenneth M. Brooks / FPL–RP–582 (2000)

[71] Wood preservation based on in situ polymerization of bioactive monomers–Part 1. Synthesis of bioactive monomers, wood treatments and microscopic analysis / Rebecca, E. Ibach, Roger M. Rowell / Holzforschung 55(4) (2001)

[72] Guide for minimizing the effect of preservative-treated wood on sensitive environments / Stan T. Lebow, Michael Tippie / FPL–GTR–122 (2001)

[73] Protecting wood decks from biodegradation and weathering: evaluation of deck finish systems / J.J. Morrell, P.F. Schneider, R. Sam Williams / Forest products journal Vol. 51(11/12) (2001)

[74] Comparison of wood preservatives in stake tests: 2000 progress report / compiled by D.M. Crawford, B.M. Woodward, Cherilyn A. Hatfield / FPL–RN–02 (2002)

[75] New environmentally-benign concepts in wood protection: the combination of organic biocides and non-biocidal additives / Frederick Green III, Tor P. Schultz / Wood deterioration and preservation: advances in our changing world. Washington, DC : American Chemical Society, ACS symposium series; 845 (2003)

[76] Evaluating wood-based composites for incipient fungal decay with the immunodiagnostic wood decay test / C.A. Clausen, L. Haughton, C. Murphy / FPL–GTR–142 (2003)

[77] Polycyclic aromatic hydrocarbon migration from creosote-treated railway ties into ballast and adjacent wetlands / Kenneth M. Brooks / FPL–RP–617 (2004)

[78] Alternatives to chromated copper arsenate for residential construction / Stan Lebow / FPL–RP–618 (2004)

[79] Rate of CCA leaching from commercially treated decking / Stan Lebow, Daniel Foster, Patricia Lebow / Forest products journal 54(2) (2004)

[80] Bioremediation of treated wood with bacteria / Carol A. Clausen / Environmental impacts of treated wood. Boca Raton, FL: CRC p. 401–411. Taylor & Francis, (2006)

Lumber Manufacturing

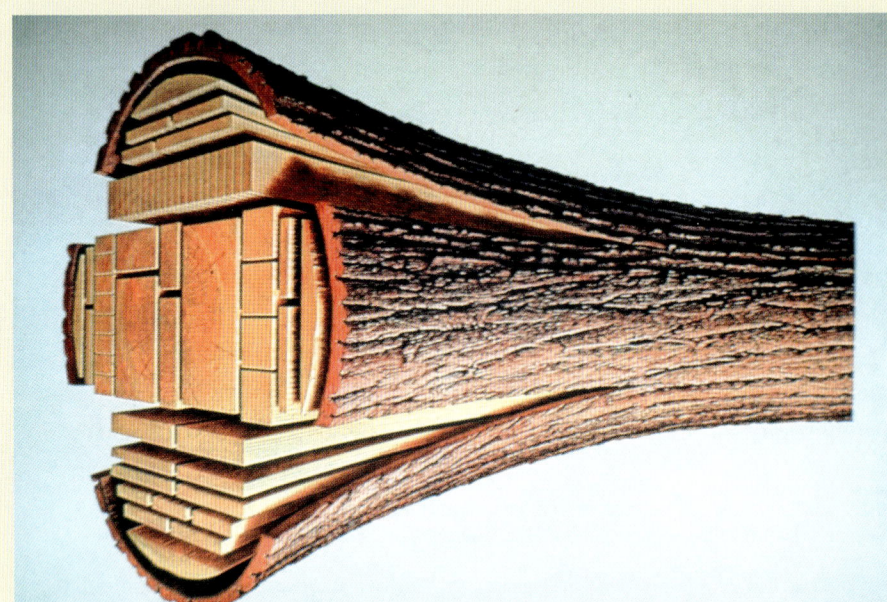

◄ Schematic of lumber being recovered from a log. The combination of sizes and shapes of the lumber that can be manufactured is large. Optimizing the yield from a given log can mean the difference between profit and loss, wise use and the creation of waste. Research at FPL has been aimed at maximizing the volume and quality of material that can be recovered from a log along with reducing waste.

Lumber manufacturing starts with cutting the tree down (felling), removing the branches and limbs (swamping), and cutting (bucking) the tree to proper log length. The logs are transported from the forest to the sawmill, where they are sawn into timbers and rough lumber. Then using secondary saws, the rough lumber is further cut to improve the quality of the lumber. In the period around 1910, the primary criterion for sawing lumber was maximizing volume. From about 1920 to 1960, sawing concentrated on visually graded lumber. Since the 1960s, sawing processes have been computerized to maximize both volume and quality.

Challenge: How to most efficiently and effectively produce timbers and lumber from a tree. Sawing logs into lumber had been practiced long before the FPL was established, so why was any research on sawmilling necessary? One reason was that early saws were thick and generated substantial amounts of sawdust instead of boards—a real waste. Another was that the processes were more of an art form, relying heavily on the skill and experience of the head sawyer. Those not skilled produced excessive waste of our wood resource. And unfortunately, even a skilled sawyer wasted wood.

FPL's Contributions: In the 1950s, FPL prepared a small sawmill operating manual and pocket guide. Later, FPL scientists developed the taper tension saw to decrease the amount of sawdust, solved the screaming saw problem, and conducted research on the use of water jets and lasers to cut wood. A major development was computer programs to improve lumber recovery, called Best Opening Face (BOF) sawing. Also, FPL developed the Saw, Dry, and Rip (SDR) and Edge, Glue, and Rip (EGAR) concepts.

State and Private Forestry Contributions (S&PF): Through the Sawmill Improvement Program (SIP), S&PF staff and State Utilization Foresters worked closely with lumber mills to improve their sawmilling practices based on the FPL's research and the S&PF and State Utilization Foresters experience. (See Technology Marketing Unit section.)

Results: With efficient sawmilling, more and higher quality lumber is produced from a given log, thus extending the timber supply and providing consumers with higher quality lumber and engineers with stress-graded lumber for use in construction.

Primary Sawmilling

If the first step in successful sawmilling—removal of the tree from the forest—is not done properly, large quantities of wood will be wasted. Just cutting the log too short eliminates the possibility of cutting a board of the desired length. Not making the right choices as to how best to cut the tree into logs can lead to excessive waste, requiring more trees to be cut than are necessary. Not getting the logs to the mill in a timely manner can lead to blue stain in the wood, which can result in significant losses in the value of the wood.

▶ Lumber manufacturing really begins in the woods. After the tree is felled, it must be de-limbed (swamped) and then cut (bucked) into log lengths. Not taking defects into account and errors in cutting the proper log lengths have significant impacts in the sawmill. The stick laying on top the tree was used to determine the proper length of the log. The stick has been replaced with a tape measure for increased accuracy.

Log Grades and Hardwood Yields

It might be assumed that just one method is used to estimate the net yield for logs of a given diameter and length. But in 1973, Frank Freese reviewed 95 recognized rules bearing about 135 names. [24]

One major contribution of the FPL was determining the yield of standard lumber from hardwood logs. Working with 28 sawmills in the northern, central, and southern hardwood regions, more than 11,000 logs were sawn and yields measured to develop the relationship between log characteristics and end product yield. Results of the study provided information for foresters, timber sellers, and buyers to separate logs suitable for manufacture into standard factory-grade lumber from other hardwood logs. Furthermore, logs could be ranked into categories of high, medium, and low value. This information is essential for obtaining the best use of timber. [18]

These values are the basis of appraisal and acquisition work of the Forest Service and of hardwood log grades applied in the inventory phase of the Forest Survey. [14, 16, 18]

▶ One of the early diagrams to educate loggers in the proper way of cutting a felled tree to maximize quantity and quality of lumber. Major losses of timber value can occur if the tree is improperly cut into logs. {M 15475 F}

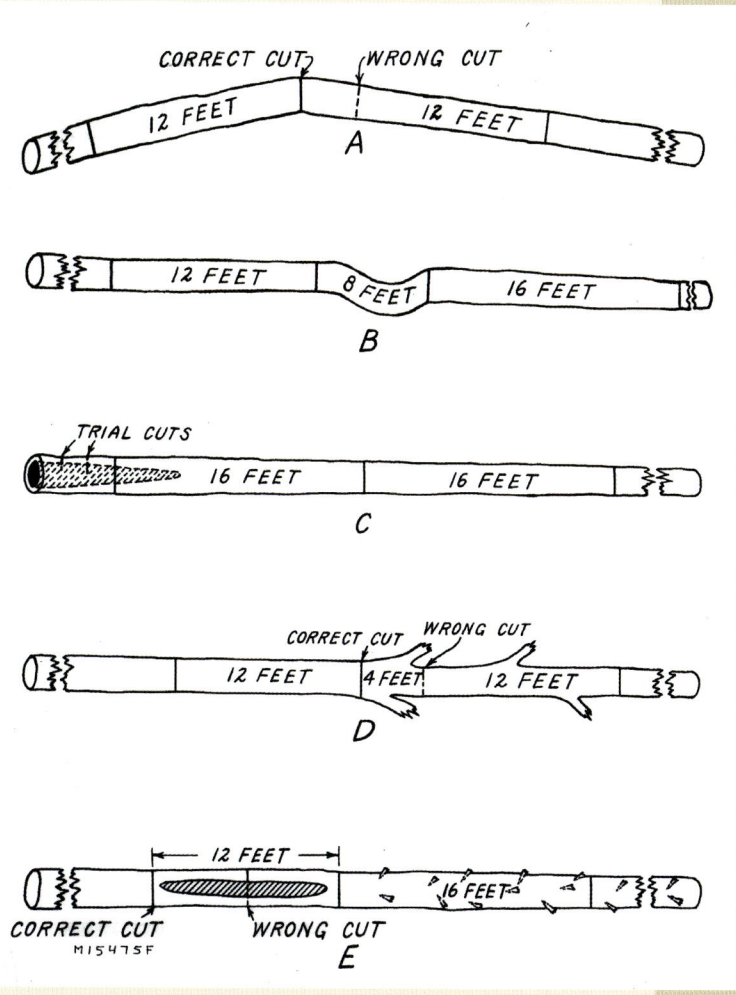

▲ The log sort yards provide many services in marketing wood and fiber. It provides a means of ensuring the wood is utilized for its highest value. This helps return the greatest value to the landowner, logger, and mill. (2000s) [43]

▲ Fred Malcomb operating the 1952 FPL experimental circular sawmill and Angelo Merli off-bearing the lumber. The sawmill was donated to the FPL by the Corely Mfg. Co. It had a variable-speed 250 HP headsaw motor. {M 133 202} [6–8]

The experimental sawmill was used to study improved cutting methods, saw blade designs, and operating processes that led to increased volume and quality of lumber obtained from a log. FPL staff helped solve the screaming saw problem. [22]

◀ Hiram Hallock checking a large circular saw to improve its cutting accuracy. (1960s) [11, 31, 41]

Thin Kerf, Taper-Tension Sawing

FPL research found that using thin kerf, taper-tension sawing can reduce sawdust up to 30%, resulting in increased lumber recovery. [19]

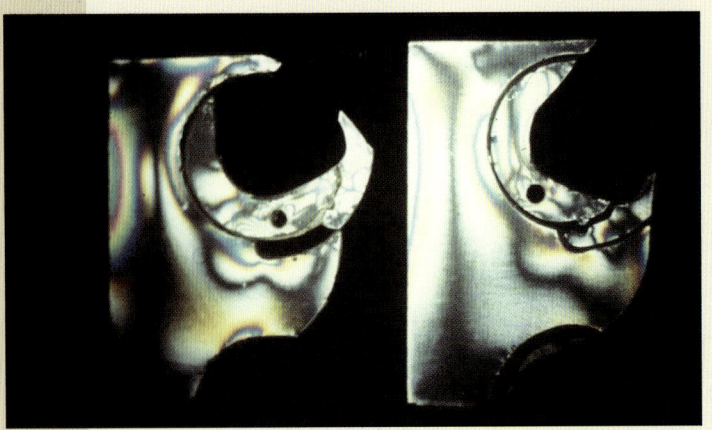

◀ Saw blade showing the installation of an inserted tooth, and the related stresses in the blade {36 I} [12, 20]

▲ High-speed photograph of sawdust exploding from the gullet of a circular saw tooth. {35 K} [8]

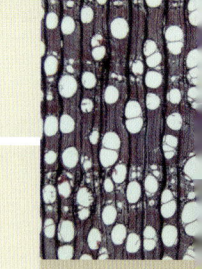

Intermittent Tooth Saws

Studies on intermittent tooth saws for the production of large sawdust particles satisfactory for pulping established criteria for rates of speed and types of saws that convert this normally wasted material to a usable product. Where saws must be used under present technology for the reduction in size of round wood or lumber, it is advantageous to produce such pulpable material. [9]

▼ Stan Lunstrum operating a commercial band saw at FPL. The band saw blade is much thinner than the circular saw blade. This results in significantly less sawdust and more lumber. {36 H} (1990s) [42]

▲ Portable band-saw mill designed at the FPL to aid small operators. (1940) [2–4]

Improved Sawmill Operations

FPL scientists developed new knowledge and technologies to improve the efficiency of lumber recovery in the sawmilling process. To transfer this research, State and Private Forestry and State Utilization Foresters have analyzed more than 2,000 sawmills. Implementation of lumber size quality control programs with little capital investment can increase yields by an estimated 10%. Adoption of new manufacturing and machinery systems could increase efficiency by another 15%. Cooperative action by the FPL staff, State and Private Forestry, State Utilization Specialists, and industry is the major reason many of these sawmill improvement plans have been successful. [36, 41]

Building Studs

A continuing problem in the production of 2 by 4 studs, commonly used in building, is the warp encountered because of grain orientation developed in the sawing process. A sawing procedure was developed to increase the number of straight studs cut from standard log sizes, which increases the value of the lumber obtained from each log. [17]

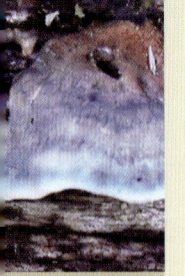

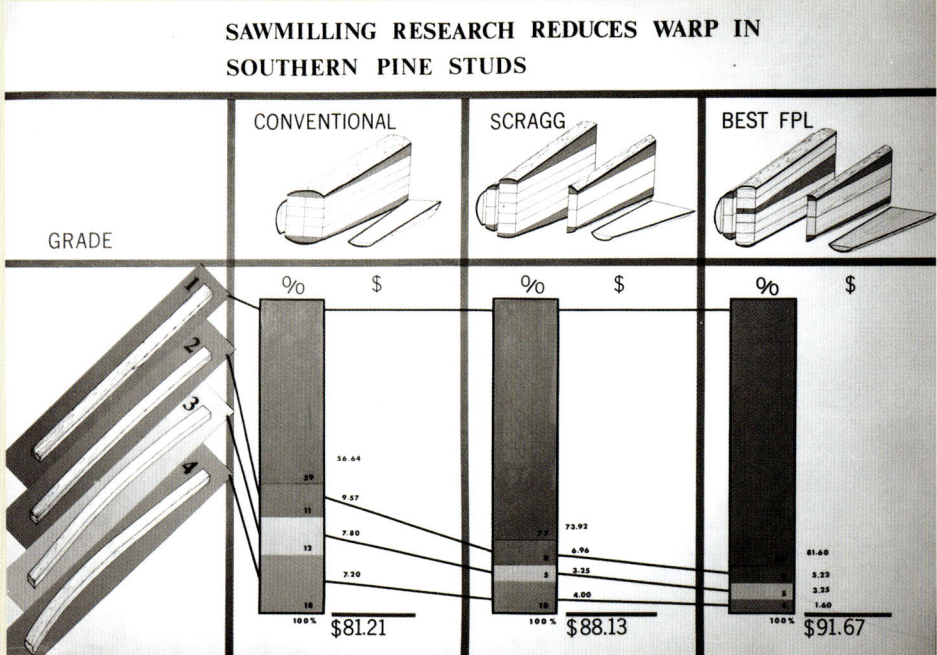

SAWMILLING RESEARCH REDUCES WARP IN SOUTHERN PINE STUDS

GRADE

CONVENTIONAL

SCRAGG

BEST FPL

$81.21

$88.13

$91.67

◀ Display showing three methods of cutting a southern pine log and the economic impact of each. (1960s) {M 128 085}

▼ Hiram Hallock carefully marking the ends of experimental logs to indicate the specific sawing method to be used. This research was part of a study on the effect of sawing method on the warping tendency of southern yellow pine studs. {M 133 238-3} [17]

Best Opening Face

Of the hundreds of different ways a log can be sawn into lumber, each results in a given amount and size of lumber. Best Opening Face maximizes the volume that can be cut from a given log. [38, 39]

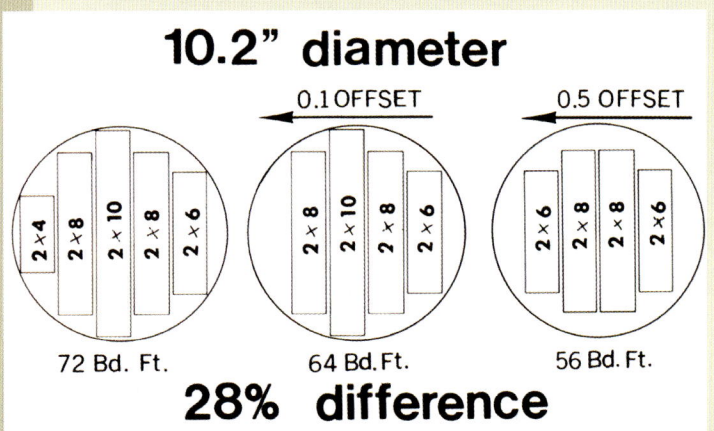

10.2" diameter

0.1 OFFSET

0.5 OFFSET

2×4 2×8 2×10 2×8 2×6

2×8 2×10 2×8 2×6

2×6 2×8 2×8 2×6

72 Bd. Ft.

64 Bd. Ft.

56 Bd. Ft.

28% difference

◀ FPL research on the effect of cutting patterns on yield resulted in the concept of Best Opening Face (BOF). In essence, where the log is opened up on the first cut (that is, the opening face) is critical to the yield of lumber that can be expected. Best Opening Face is a computerized sawing system that controls the sawing of the log and significantly decreases waste. [21, 27, 34]

▶ Display of the best opening face concept. (1970s) {75 F}

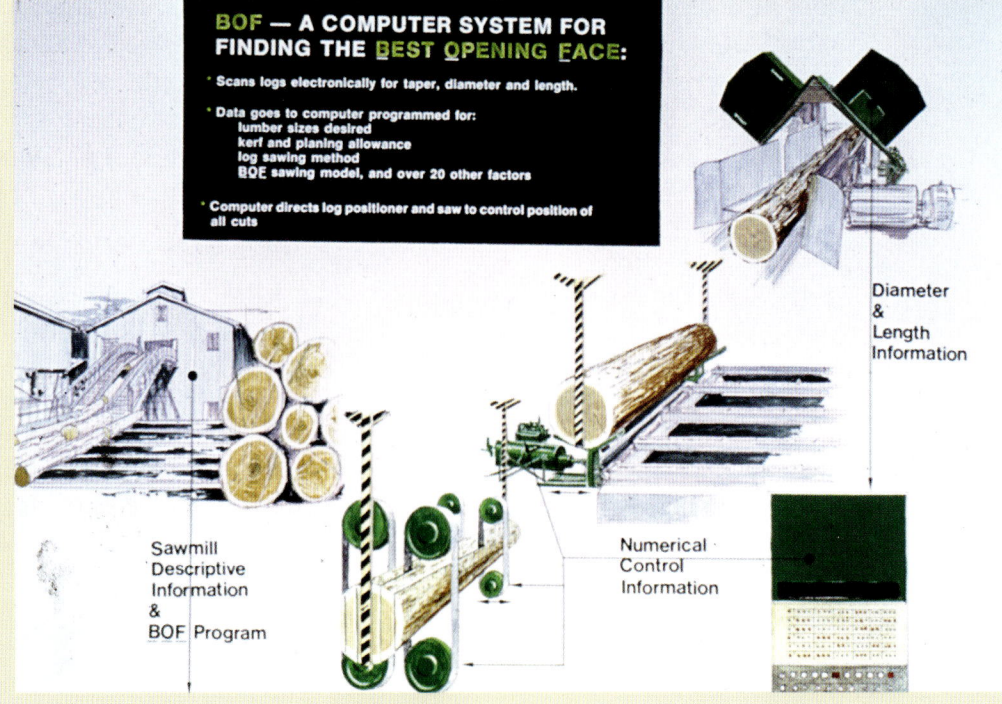

BOF — A COMPUTER SYSTEM FOR FINDING THE BEST OPENING FACE:

* Scans logs electronically for taper, diameter and length.

* Data goes to computer programmed for:
  lumber sizes desired
  kerf and planing allowance
  log sawing method
  BOF sawing model, and over 20 other factors

* Computer directs log positioner and saw to control position of all cuts

Diameter & Length Information

Sawmill Descriptive Information & BOF Program

Numerical Control Information

Remanufacturing

Following primary sawing of the log into rough lumber, the boards are many times resawn to improve the quality of the lumber or to cut it into shapes for specific uses, such as millwork, furniture, moulding, cabinets, doors, window frames, and a host of other wood products. [1, 10]

Edge, Glue, and Rip (EGAR)

LIVE SAW BY BOF

EDGE FULL WIDTH

EDGE GLUE PANEL

RIP FOR HIGHEST QUALITY

◄ Another process for improving quality is to saw using Best-Opening-Face, edge the boards full width, dry, and then edge-glue the boards creating a panel. Then the panel is rip-sawn to maximize volume and grade. This process is called EGAR for Edge, Glue, and Rip. (1970s) [28]

► George Englerth and Claudia Wodzinski study data on hard maple cutting yields using a computer. (1970s) {M 131 312}

Dimension Stock Cutting Yields

Compiled utilization information is used to predict expected cutting yields from factory grades of hardwood lumber, which is important to furniture parts manufacturers. [23]

Predicting Millwork and Moulding Product Yields

High-quality lumber, suitable for millwork and moulding products, is always in short supply. FPL, and the Northern, and Pacific Northwest Research Stations developed computer programs that predict the yield of high-value products from mixed and lower grades of shop lumber. These programs permit the successful use of lower grade lumber for high-value millwork products such as cabinets, doors, and window frames. [30, 33, 35]

Reducing Warp by using Saw, Dry, and Rip (SDR)

▲ Example of extreme warp in 2 by 4 studs because of species and poor sawmilling practices. {110 B}

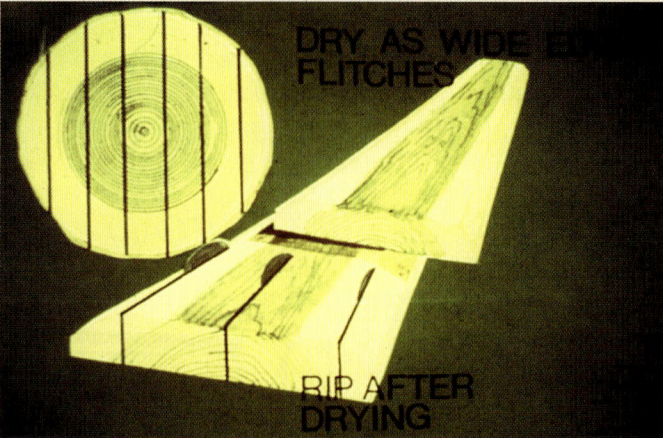

▲ FPL developed the process of Saw, Dry, and Rip (SDR) to decrease warp.

Saw, Dry, and Rip

The SDR method begins by live-sawing green logs into 1-3/4-inch, unedged planks called flitches. These are dried to an average moisture content of 12%, then are rip-sawn into studs. High-temperature drying ensures minimum stress within the flitches. Results are straight studs, free of warp and twist. SDR can increase yields over conventional methods. Suitable species include aspen, sycamore, cottonwood, basswood, soft maple, red alder, and several others. [37, 40]

◄ Comparison of yellow poplar studs. Those on the left were conventionally sawn and those on the right were sawn using the saw–dry–rip process. (1980s) {47 E}

Grading Lumber by Computer

Use of computers for accurately grading hardwood lumber has been successfully accomplished and is far faster than visual grading. The standard hardwood grading rules were modeled mathematically and programmed for the computer, as were descriptions of board characteristics. When perfected, scanning devices will permit reading board defects directly into a computer that will control the saw in edging, ripping, and trimming operations. This will allow for maximizing the quality of lumber cut from a board. This early FPL research provided the basis for today's commercial applications of visual computer grading of lumber at the sawmill. [35]

▲ Using electronic sensors to locate and map defects in lumber. {29 B}

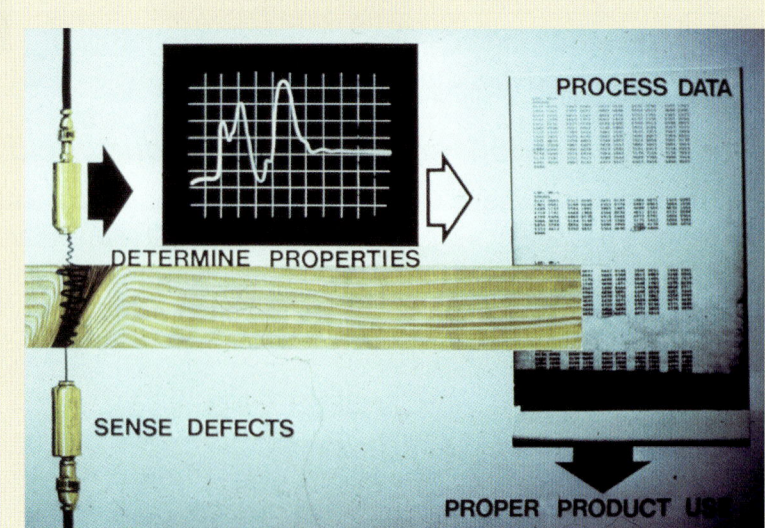

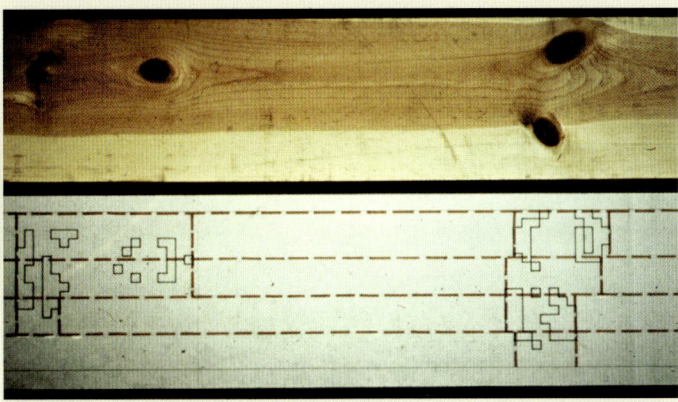

▲ Piece of lumber with defects that have been sensed and plotted by a computer. Information on defects can be used by the computer to determine the best way to resaw the board to maximize quality.{110 B}

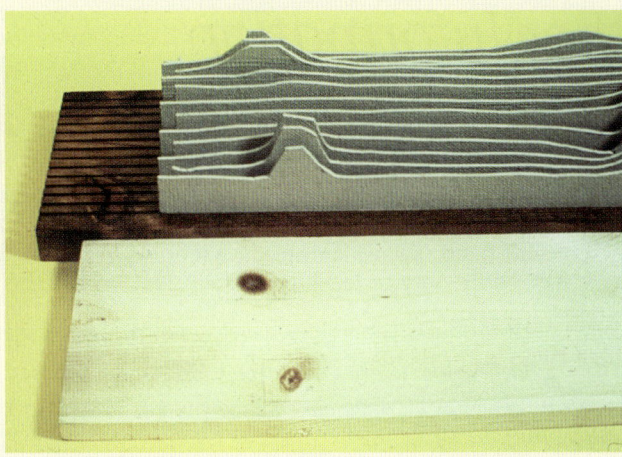

▲ Three-dimensional model of defects located in a board. (1980s) {29 E}

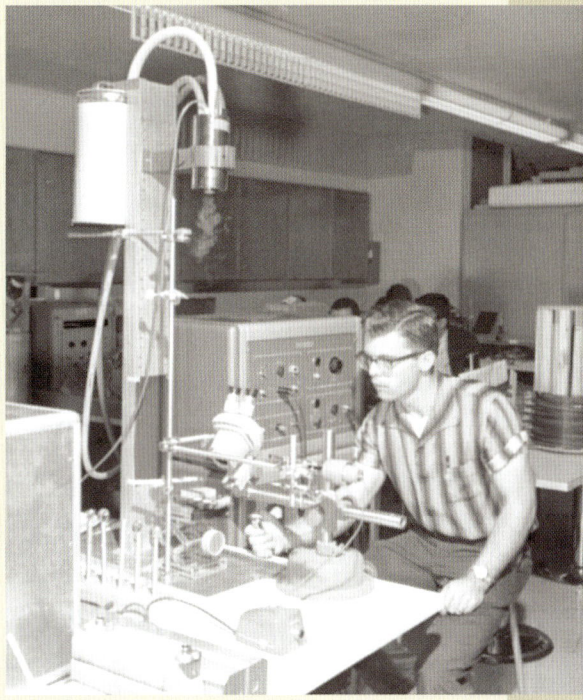

▲ In 1963, Eugene Bryan, University of Michigan graduate student, conducted experiments in the use of a high-pressure jet of water (left) and a ruby laser (right) to cut wood. Experiments were conducted under a FPL-funded grant to University of Michigan, Department of Wood Technology. Both techniques are commercially used today. {M 125 255} {M 122 127} [13, 26]

1910

Yield and quality of lumber cut was an art and primarily depended on the ability of the head sawyer. Waste was substantial because of poor equipment and thick saw blades. There was little standardization of lumber sizes or grades.

2010

Developed Best Opening Face process, which changed sawmilling from an art to a science. Played a major role in establishing industry standards for lumber sizes and grading rules.

Companies employed 170,293 people and shipped products with a total value of $32,162,583,000. (2006 Annual Survey of Manufacturers)

Further Information

[1] Small dimension stock: its production from lumber at the factory / Arthur T. Upson / Southern lumberman Vol. 107(1398) (July 8, 1922)

[2] Portable band sawmills / R.D. Garver. / Journal of forestry Vol. 28(8) (Dec. 1930)

[3] The portable band sawmill and selective logging in the loblolly pine forests of North Carolina / R.D. Garver, J.B. Cuno / USDA, Forest Service Technical bulletin, No. 337 (1933)

[4] Some Forest Products Laboratory developments of particular significance to the South / FPL report 0–8 W73Sf (April 15, 1940)

[5] Saw teeth in action / L.H. Reineke / Excerpted as "The Duo-Kerf rip saw" in Southern lumber journal 54(8): 20, 24, 80–81 (Aug. 1950)

[6] Small sawmill operator's manual / C.J. Telford / USDA Agriculture handbook No. 27 (1952)

[7] Small sawmills: a pocket guide / C.J. Telford / USDA Agriculture handbook No. 70 (1954)

[8] Sawing rates, sawdust chambering, and spillage / L.H. Reineke / Forest products journal Vol. 6(9) (Sept 1956)

[9] Kerf chip sawing—a mixed blessing? / Hiram Hallock / Southern lumberman Vol. 203(2528) (Aug. 1, 1961)

[10] Machining and related characteristics of United States hardwoods / E.M. Davis / USDA Technical bulletin No.1267 (1962)

[11] A mathematical analysis of the effect of kerf width on lumber yield from small logs / Hiram Hallock / FPL No. 2254 (1962)

[12] Improvement in the design and operation of inserted-tooth saws / F.B. Malcolm / Presented at the meeting of Section 41, Forest Products, of the International Union of Forestry Research Organizations, September 11–13,1963, Forest Products Laboratory, Madison, Wis. (1963)

[13] High energy liquid jets as a new concept for wood machining / Eugene Lee Bryan / Thesis (Ph.D.)—University of Michigan 1963

[14] A guide for the selection and control of study sawmills / E.H. Bulgrin, H. Hallock / Prepared for use in the Forest Service log and tree grade program (Nov. 1963)

[15] Some thoughts on marginal sawmill logs / Hiram Hallock / Forest products journal Vol. 14(11) (Nov. 1964)

[16] A simplified procedure for developing grade lumber from hardwood logs / F.B. Malcolm / FPL–RN–098 (1965)

[17] Effect of sawing methods and other factors on warp of studs from small loblolly pine logs / Hiram Hallock / For presentation at second colloquium of wood technology at Braunschweig, West Germany in May (1966)

[18] Hardwood log grades for standard lumber / C.L. Vaughan, A.C. Wollin, K.A. McDonald, E.H. Bulgrin / Research paper FPL 63 (1966)

[19] "Taper-tension" saw: a new reduced kerf saw / H. Hallock / FPL–RN–0185 (1968)

[20] Locating maximum stresses in tooth assemblies of inserted-tooth saws / F.B. Malcolm, A.L. Koster / Forest products journal Vol. 20(10) (Oct. 1970)

[21] Increasing softwood dimension yield from small logs: best opening face / H. Hallock / FPL–RP–166 (1971)

[22] Symmetrically placed expansion slots solve the problem of screaming saws / F. B. Malcolm / Forest industries Vol. 98(6) (June 1971)

[23] Dimension stock yields from lumber of three hardwood species / D.R. Schumann / Forest products journal Vol. 23(3) (Mar 1973)

[24] A collection of log rules / Frank Freese / General technical report FPL 1 (1973)

[25] The sawmill improvement program / S.J. Lunstrum / Southern lumberman Vol. 229(2848) (Dec. 15, 1974)

[26] Cutting wood materials by laser / C.C. Peters, H.L. Marshall / FPL–RP–250 (1975)

[27] Individual log yields by eight sawing systems / H. Hallock / FPL–RP–280 (1976)

[28] Yield and strength of softwood dimension lumber produced by EGAR system / Kenneth C. Compton, H. Hallock, C.C. Gerhards, R. Jokerst / Research paper FPL 293 (1977)

[29] Individual log yields by four centered sawing systems / H. Hallock / Research paper FPL 321 (Supplement) (1978)

[30] Computer optimization of cutting yield from multiple-ripped boards / A.R. Stern, K.A. McDonald / Research paper FPL 318 (1978)

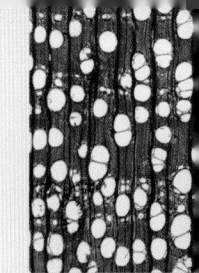

[31] Improving sawing accuracy does help / Abigail R. Stern, Hiram Hallock, David W. Lewis / FPL–RP–320 (1979)

[32] Does gang ripping hold the potential for higher clear cutting yields? / Hiram Hallock, Pamela Giese / FPL Research paper FPL 369 (1980)

[33] 5/4 ponderosa pine shop grade cutting yields / Kent A. McDonald, Pamela J. Giese and Richard O. Woodfin / Research paper FPL 394 (1981)

[34] What have we learned from the sawmill improvement program after nine years / S.J. Lunstrum / Southern lumberman Vol. 243(3028) (Dec 1982)

[35] OPTYLD: a multiple rip-first computer program to maximize cutting yields / Pamela J. Giese, Kent A. McDonald / FPL–RP–412 (1982)

[36] Payback as an investment criterion for sawmill improvement projects / George B. Harpole / FPL–GTR–34 (May 1983)

[37] Manufacture of quality yellow-poplar studs using the saw–dry–rip (SDR) concept / R.R. Maeglin, R.S. Boone / Forest products journal 33(3): 10–18 (1983)

[38] Sawmill simulation and the best opening face system: a user's guide / David W. Lewis / FPL–GTR–48 (December (1985)

[39] Best opening face system for sweepy, eccentric logs: a user's guide / David W. Lewis / FPL–GTR–49 (1985)

[40] Saw–Dry–Rip improves quality of random-length yellow-poplar 2 by 4's / Robert R. Maeglin, R. Sidney Boone / FPL–RP–490 (1988)

[41] Circular sawmills and their efficient operation / J. Stanford Lunstrum / U.S. State and Private Forestry, 1981 (reprinted 1993)

[42] Analyzing investments in thin-kerf saws / Philip H. Steele, Philip A. Araman / Putting research to work for the hardwood industry: new technology available today: proceedings of the twenty-fourth annual Hardwood Symposium. (Memphis, TN: National Hardwood Lumber Association) (1996)

[43] Review of log sort yards / John Rusty Dramm, Gerry L. Jackson, Jenny Wong / General technical report FPL 132 (2002)

Best Opening Face (BOF) and Sawmill Improvement Program (SIP)

To address increased demands for improved conversion efficiency, the Best Opening Face (BOF) system was developed at the Forest Products Laboratory to conserve and extend the Nation's timber supply. The FPL developed BOF algorithms (a mathematical sawing model) to maximize lumber volume recovery from small logs. In the 1970s and 1980s, this FPL-developed technology aided the retooling and automation of the sawmill industry to successfully adapt to the shift from large old-growth to small-diameter second-growth timber supplies.

Since the mid-1950s, the U.S. Forest Service State and Private Forestry (S&PF) Unit has maintained liaison with researchers at the FPL. In 1971, Stan Lunstrum joined the unit as the National Sawmill Specialist. Stan was charged with developing a plan for a national program to improve sawmill conversion efficiency. Stan, along with Hiram Hallock and Dave Lewis of the FPL, developed the Sawmill Improvement Program (SIP).

In 1973, Hiram Hallock, Dave Lewis, and Stan Lunstrum discussed with industry the possibility of a sawmill improvement program based on the BOF sawing system as a yardstick to a mill's conversion efficiency.

The concept of the program was simple—initial studies would determine a mill's current lumber recovery factor (LRF) and identify areas in the milling process where changes in manufacturing conditions or practices would result in increased lumber recovery. Follow-up studies would analyze the improvements realized by a mill as a result of making changes identified by the initial study.

Maximum recovery of highest value products depends on making correct log breakdown decisions. With literally thousands of judgments possible, costly mistakes are inevitable in regard to maximizing lumber grade, volume yield, and sawmill efficiency. Sawyer, edger, and trimmer operators cannot consistently guess the best sawing solution and apply those solutions hour after hour, day after day. SIP provided a method to measure a mill's current conversion efficiency and theoretical maximum efficiency as measured by lumber recover factor (LRF). Follow-up studies confirmed improvements made in processing and resulting LRFs.

Success of BOF's adoption by industry was made possible by a successful national technology transfer effort through SIP. The SIP program, led by the S&PF Unit at FPL worked closely with FPL researchers and the National Wood Products Extension Program (NWPEP) managed by Ted Peterson.

SIP studies were conducted by Forest Products Utilization, S&PF, and State Utilization and Marketing Specialists (such as Terry Mace in Wisconsin); data were analyzed by the S&PF Unit at FPL. Confi-

◄ John (Rusty) Dramm selecting a sample of sawn wood as part of one of more than 2,000 Sawmill Improvement Program (SIP) studies.(1970s) [2]

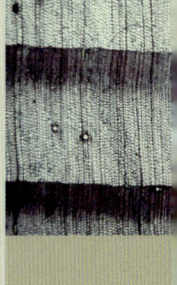

dential recommendations for specific mill improvements were reported to each of the participating mill owners. NWPEP also provided education programs—such as the "Tightening Up!" slide series—to aid sawmill owners in making these mill improvements. Improvements were made by mill owners at their own expense—no federal funding was used for making mill improvements and retooling sawmill equipment.

In June 1981, SIP study results were published in "Softwood Sawmill Improvement Program: Selected Study Results (1973–1979)." The SIP program conducted more than 2,000 lumber recovery studies. SIP reached over half of the U.S. softwood lumber sawmill production capacity. Nationally, SIP was responsible for an improvement in lumber recovery of about 4% of total U.S. softwood production—over 1,000,000,000 board feet annually, or enough lumber to build all the houses in Madison, Wisconsin, each year. [1]

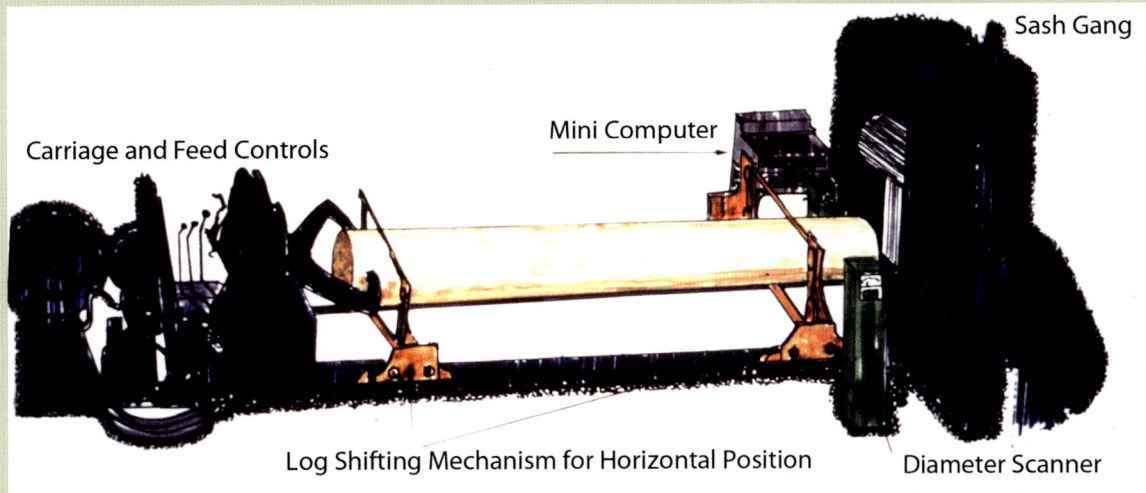

▲ Positioning the log to produce the highest yield using best opening face sawing. This concept of log positioning is the basis for small-log single-pass sawing (breakdown) systems and curve sawing. (1970s) [3]

Further Information

[1] Best Opening Face (BOF) & Sawmill Improvement Program (SIP) / John (Rusty) Dramm / USDA Forest Service State and Private Forestry (April 6, 2006)

[2] The sawmill improvement program / S.J. Lunstrum / Southern lumberman Vol. 229(2848) (Dec. 15, 1974)

[3] Individual log yields by four centered sawing systems / H. Hallock / Research paper FPL 321 (Supplement) (1978)

Machining

◄ Finished wooden bowls and gavel. {30 P} [6, 8]

The ability to shape wood into countless forms using simple tools goes back thousands of years. With the advent of power tools, the options increased—from complex furniture forms to fancy millwork to unusual art work. Each wood species exhibits its own particular properties when exposed to cutting and shaping tools.

Challenge: When the FPL was started, there was little information on the factors that affect the machining of wood. Much of the information was based on artisan experience and folklore, and much wood was wasted.

FPL's Contributions: Evaluated the machining characteristics of wood species, as related to cutting tool shapes, angles, and travel speed of wood with various density, moisture content, and growth rates.

Results: Much of the millwork equipment has been automated by industry, and many wood species that were previously wasted are now used.

► Use of wood in carvings. This wood was treated with polyethylene glycol (PEG). FPL experimented with the use of PEG as a treatment to stabilize wood. (1970s) {30 M} [10]

◄ Examples of carvings in various stages of finishing.

Machining Characteristics of Wood

◄ Kent McDonald examines annual rings of hardwood specimen being stored under controlled temperature and relative humidity before machining to evaluate effects of soil fertilization nearly 30 years previous. (1960s) {M 123 126} [7]

► George Englerth evaluating the machining characteristics of wood specimens.(1960s) {M 127 253} [4, 7]

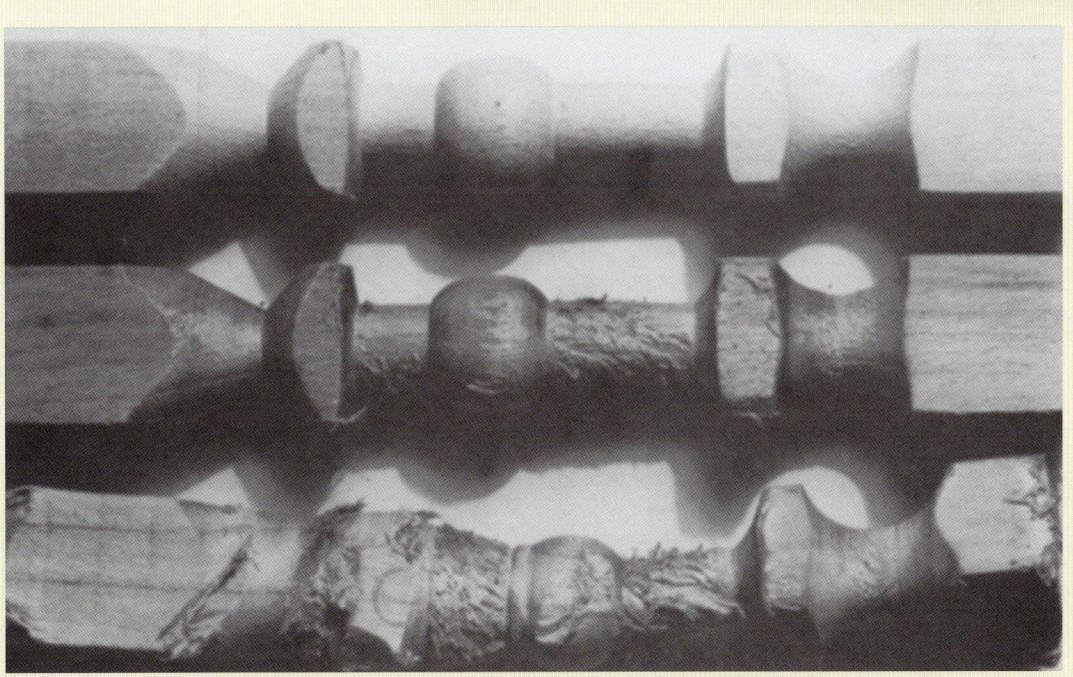

► Results of turning wood specimens (6%, 12%, and 20% moisture content, top to bottom) on a lathe. (1960s) {ZM 91693 F} [4, 7]

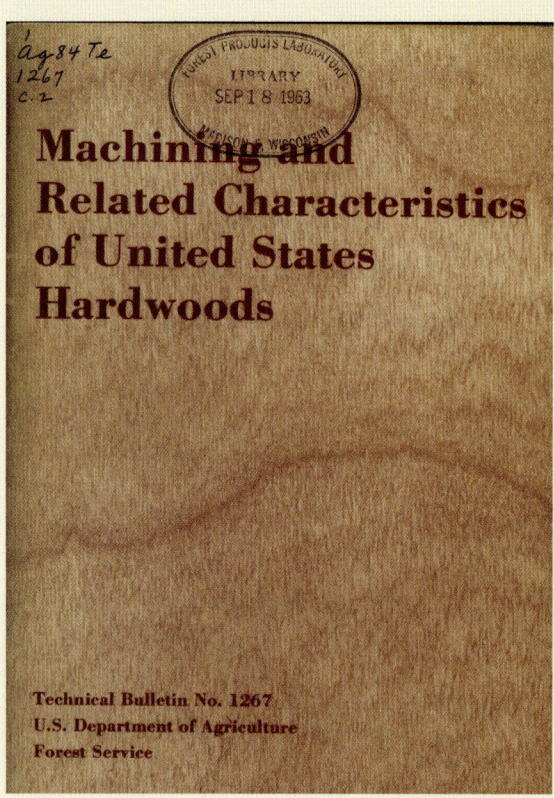

Machining and Related Characteristics of United States Hardwoods

Technical Bulletin No. 1267
U.S. Department of Agriculture
Forest Service

◄ Cover of the report summarizing the results of the FPL's research on machining characteristics of U.S. hardwoods. (1962) [1–4]

Shop Grade Lumber Yields

In an effort to maximize yield—in terms of both volume and quality—FPL research has studied methods of secondary wood manufacturing. [11, 12]

Furniture and Cabinetry

Efforts have also been made to provide diagnostic guidance to the furniture and cabinetry industry. This technology transfer activity is carried out primarily by the Forest Service, State and Private Forestry Technology Marketing Unit. [5, 9, 13]

1910

Little scientific information on the machining characteristics of various wood species, leading to large losses in wood.

2010

Wood species classified in terms of machining characteristics. Companies employed 433,410 people and shipped products with a total value of $66,713,433,000. (2006 Annual Survey of Manufacturers)

Further Information

[1] Machining and related characteristics of southern hardwoods / E.M. Davis / USDA, Forest Service Technical bulletin No. 824 (1942)

[2] Advances in sawing from forest to shop / L.H. Reineke / FPL No. 2100 (1958)

[3] Feed-speed relationships for cutting tools / L.H. Reineke / FPL No. 2184 (1960)

[4] Machining and related characteristics of United States hardwoods / E.M. Davis / USDA Technical bulletin No.1267 (Supersedes USDA Technical bulletin No. 824) (1962)

[5] Portable apparatus for surface evaluation of furniture panels / B.G. Heebink / Research note FPL 015 (1963)

[6] New horizons in bowl turning / George H. Englerth, Harold L. Mitchell / Forest products journal Vol. 13(2) (Feb. 1963)

[7] Machining and other properties of fast-versus slow-grown black walnut / George H. Englerth / Black walnut culture: papers presented at the walnut workshop, Carbondale, IL, (Aug 2 & 3, 1966)

[8] Bowls from scrapwood / George H. Englerth, Harold L. Mitchell / Workbench (July–Aug. 1967)

[9] Measuring wood surface smoothness: a proposed method / C. Peters, A. Mergen / Forest products journal Vol. 21(7) (July 1971)

[10] How PEG helps the hobbyist who works with wood / H.L. Mitchell / FPL Unnumbered report / (1972)

[11] 5/4 ponderosa pine shop grade cutting yields / Kent A. McDonald, Pamela J. Giese, Richard O. Woodfin / Research paper FPL 394 (1981)

[12] Western hardwoods: value-added research and demonstration program / edited by David Green, William Von Segen, Susan Willits / Forest Products Laboratory (1995)

[13] Diagnostic guide for evaluating surface distortions in veneered furniture and cabinetry / Alfred W. Christiansen, Mark Knaebe / FPL–GTR–143 (2004)

Drying

◀ Cross section of a log showing how a piece of wood removed from a specific area in the tree changes its shape when dried. {109 C SA}

Wood, while growing and when first cut, contains a large amount of water. This water greatly influences the properties of wood products, thus it is necessary to remove a substantial amount of it before satisfactory use is possible. [11, 14, 16, 23, 25, 39, 49]

▲ The buckets hold the same amount of water as the green lumber shown on the table.

Wood Shrinkage

▼ Typical shrinkage resulting from drying a board. (1940s) {M 50734 F}

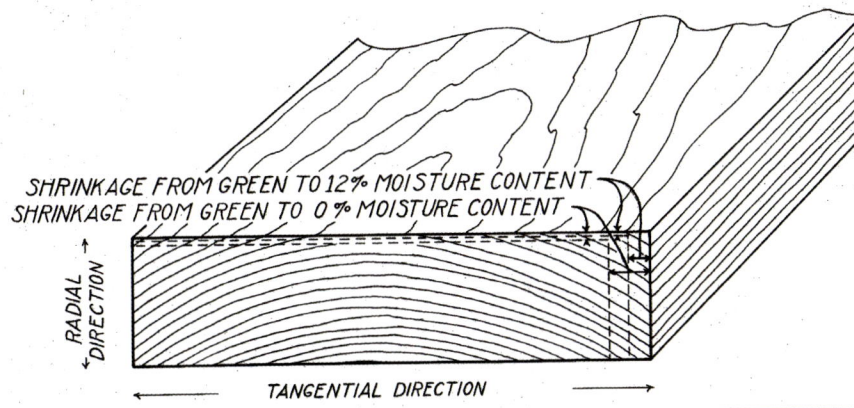

Wood also shrinks when it dries, resulting in dimensional change; but it does not shrink the same in all directions. Research has shown that this shrinkage difference results from the orientation of the cellular structure of wood and the structure of the cells themselves. [16, 28]

Challenge: To achieve the best performance of wood, it is important to use it at the proper moisture content. The challenge is how to remove moisture from green wood with minimal drying defects and stresses.

FPL's Contributions: Wood-drying research has explained moisture movement during the drying process, developed effective methods of air and forced-air drying that minimize the defects and stresses and that greatly decreased lumber degrade, and effectively transferred the information to industry through wood-drying classes and the Dry Kiln Operator's Manual. [53]

Results: Proper drying results in significantly decreased manufacturing waste and higher quality of wood in service.

Water should be removed from wood (by drying or seasoning) for several reasons:

- Minimizes insect damage
- Improves dimensional stability
- Improves preservative treatments
- Controls warping, splitting, checking
- Improves machinability
- Makes wood lighter—easier to handle and cheaper to ship
- Makes wood (less than 20% moisture content) less susceptible to decay, mold, and stains
- Improves strength
- Improves joinery, such as nail-holding ability and glueability [56]

Wood Drying

The underlying mechanism of wood drying is not simple; when the FPL opened, loss of wood due to improper drying was huge.

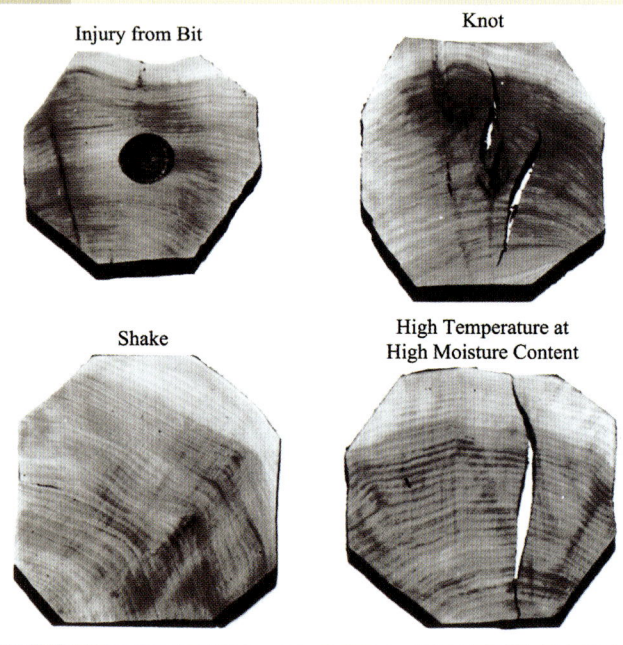

▲ Factors causing splits and honeycombing in artificial-limb blanks manufactured from Ohio red willow {Z M 71496 F} [17]

▲ Honeycombing can occur in any wood but is more prevalent in "wetwood"—a type of wood in standing trees that has been internally infused with water mostly caused by bacterial infections. Wetwood is particularly difficult to dry without causing defects such as honeycombing and collapse. (1980s) [38, 48]

◄ Total collapse of cell structure is a severe drying defect. {130 D SE}

▲ Other forms of defects that can occur if wood is not dried properly. Defects lower the wood's value and may, in serious cases, produce only waste wood. (1950s) [20, 21]

Wood Moisture Relationships

▲ One of the pioneers in the field of wood drying and wood–moisture relationships was Harry D. Tiemann. (1920s)

Harry D. Tiemann

Before coming to the FPL, Harry Tiemann conducted extensive research at Yale University on the effect of wood moisture content on physical and mechanical properties of clear wood. He coined the phrase "fiber saturation point of wood" as the point at which the free water found in a cell lumen of wood disappears and bound water in the cell walls remains. As the cell walls begin to dry and shrink the strength of the wood increases. [8] He also helped develop the water-spray dry kiln, which was successfully commercialized and led to substantial reductions in lumber degrade. [4, 5]

Stain

▶ Spruce log with blue stain. Blue stain can greatly decrease the value of wood. Rapid harvesting and processing and proper drying can eliminate blue stain. Application of antiseptic sprays and coatings to the ends of hardwood and softwood logs can provide some protection. Spraying logs with water may prevent some blue stain. [51]

Air Drying

One of the earliest methods of water removal was by air drying. [1, 10, 24, 25]

◀ Improper placement of spacing stickers in a stack of lumber results in poor quality lumber. (1920s) {2305 F} [7]

▶ Proper way to stack lumber for air drying. Note the sloping base so that water will run off. Also, the ends of the lumber are coated to minimize end checking.

▲ Air drying lumber. Note the flat roofs placed on top of the piles to decrease the effects of rain and direct sunlight and how clean the area is of plants and debris. {55 G} [6]

Air Drying of Lumber Report

The report entitled *Air Drying of Lumber* describes how lumber can be air dried most effectively under outdoor conditions and illustrates the principles and procedures of air-drying lumber that were developed through field investigations and observations of industrial practices. Particular emphasis is placed on the yarding of lumber in unit packages. (1990s) [54]

Estimates of Lumber Air-Drying Times

Published data on estimated air-drying times of lumber are of limited usefulness because they are restricted to a specific location or time of year when the lumber is stacked for drying. At best, these estimates give a wide range of possible times over a broad range of possible locations and stacking dates. A study done at FPL developed an analysis method for estimating air drying times for specific locations by optimizing a drying simulation using existing experimental air drying times for northern red oak, sugar maple, American beech, yellow poplar, ponderosa pine, and Douglas-fir. The results of the analysis are parameters for a computer simulation used to estimate air-drying times of these species regardless of when they are stacked, for any location where average temperature and relative humidity are known, and for lumber of any thickness dried to any final moisture content. (2000s) [58, 61]

Chemical Seasoning

Early research at FPL found that the use of chemicals, such as common salt, could be used to assist the drying of refractory (difficult) species (such as as southern swamp oak) and large timbers (such as 12- by 12-inch Douglas-fir columns and 12-inch-diameter redcedar poles). [12, 22]

Kiln Drying Lumber

In an effort to speed up the drying of lumber and control drying defects, FPL scientists conducted extensive research on kiln-drying techniques.

This early internal-fan model was the precursor of a host of forced-circulation commercial dry kilns employing similar principles. This breakthrough resulted in significantly faster drying than with air drying and the ability to dry wood to lower moisture contents with minimum defects. This made it possible to use many underutilized species of wood and significantly decreased defective lumber resulting from improper drying. More than 7,500 commercial dry kilns in the United States employ the internal-fan system to dry lumber rapidly and safely for most species. Drying times were decreased from years to weeks—and in some cases days—and defects decreased by 10% to 80%, resulting in a greater quantity of higher quality and lower priced lumber for the consumer. [2, 3, 5, 13] FPL also helped developed the water-spray dry kiln. [4]

◀ One of FPL's major contributions to the effective drying of wood was the development of the internal-fan dry kiln by Rolf Thelen and Harry Tiemann. (1920s) {M 12297 F} [5]

Kiln-Drying Schedules

Lumber kiln-drying schedules were developed by FPL scientists for the principal commercial species grown and used in this country and for some of the more important foreign woods. Each wood species requires its own drying conditions because of its different wood cell structure, so the FPL developed specific drying schedules for each species. This amounts to about 300 recommended kiln schedules that cover domestic hardwoods, domestic softwoods, and tropical woods in various thicknesses. New schedules have substantially decreased kiln-drying time and costs, as well as losses because of warp and other drying degrade. Estimated kiln schedules were also developed using wood specific gravity and are available from the FPL web site. [2, 9, 31, 43, 52, 53, 55]

▲ Three early FPL brick dry kilns used to establish drying schedules for various species of wood. (1930s) {M 23963 F}

Dry Kiln Inspection Program

During WWII, the FPL was extensively involved in drying wood. To help ensure that good drying techniques were being used in industry, the FPL conducted a dry kiln inspection program. This helped save thousands of board feet of lumber. (See National Defense section.)

Automated Kiln Drying System

Two-thirds to three-fourths of the harvested wood resource is dried before it is put into use. Research aimed at reducing drying costs has led to the development of a unit that enables automatic operation of lumber dry kilns based on the changing moisture content of the wood during the drying process. This early improved kiln system cut labor costs and gave faster drying. This unit was designed so that almost any company, large or small, could install it on its kilns. [30, 32]

Drying Hardwood Lumber

The FPL report *Drying Hardwood Lumber* focuses on common methods for drying hardwood lumber of different thicknesses, with minimal drying defects, for high-quality applications. The manual also includes predrying treatments that, when part of an overall quality-oriented drying system, decrease defects and improve drying quality, especially of oak lumber. Special attention is given to drying white wood, such as hard maple and ash, without sticker shadow or other discoloration. [55]

Drying Stresses in Lumber

Research by John McMillen and Bob Youngs focused on drying stresses in lumber. This knowledge added to a better understanding of how to dry lumber and the effects of drying on mechanical properties. [19, 21, 23]

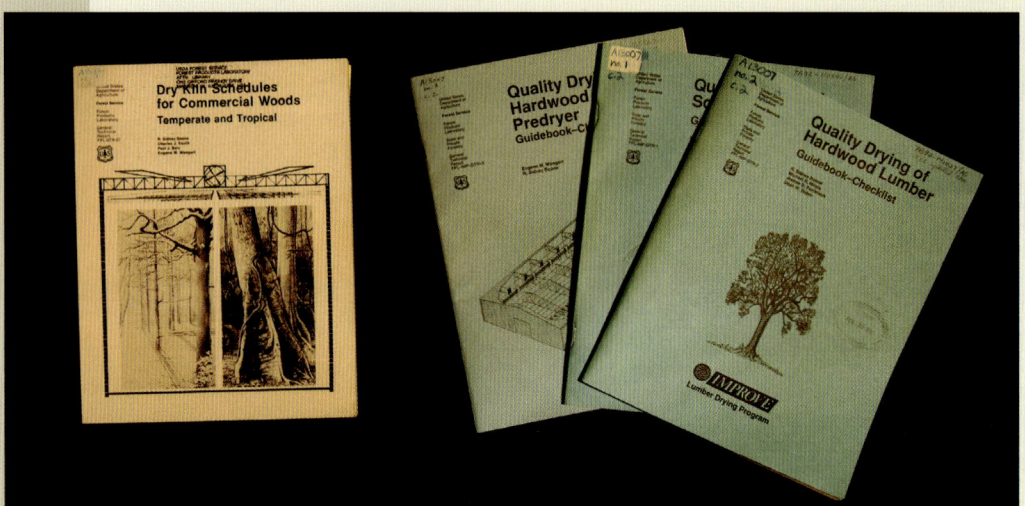

◄ Early publications on drying tropical woods and guidebook and checklist for drying hardwoods. [43–46, 50] FPL researchers also determined and published the principles governing accelerated drying of hardwoods. [15, 31]

Moisture Content Monitoring in Kiln Drying

Accurate measurement of moisture content in lumber is important to good kiln schedule control. Additional desirable qualities of a measurement system are remote and non-destructive monitoring, the ability to measure moisture content above 30 percent, flexibility to place sensors anywhere across the width of a lumber stack, and the ability to direct a kiln control system. Research was conducted at FPL on the use of speed of sound to estimate moisture content of red oak, hard maple, and ponderosa pine during drying. Two useful findings were (1) the speed of sound gives good estimates of moisture content all the way from green moisture contents well above 30 percent to dry, with a linear relationship between the transit time of sound waves and moisture content; and (2) that the sensors have the ability to be easily mounted on the ends of full-length boards in the center of a lumber stack in a dry kiln. This latter ability allows one to track the drying progress of the slower-drying center boards rather than be limited to the faster-drying edge boards. [26, 37, 57]

Energy Saving Drying Methods

Existing drying methods for hardwood lumber were not energy efficient because the methods were designed to prevent visual defects in furniture-grade hardwoods. Scientists at FPL have determined that conservative kiln schedules were not necessary for drying of hardwood structural lumber. Research confirmed that even the most severe (that is, faster) drying test schedules caused no more grade loss than the milder, traditional schedule. More energy-efficient drying helps make structural applications of low-grade maple lumber more economic. In addition, finding uses for low-grade hardwoods supports good ecosystem management by making underutilized tree removals economically feasible. [31]

Drying Lumber and Rounds from Small-Diameter Softwood Trees

Dense stands of small-diameter softwood trees in the western United States are creating a fire and forest health hazard. Removing these trees is expensive, so there is an interest in finding value-added uses. One problem in using lumber from these trees is that it is notorious for warping during drying. FPL has conducted two studies to characterize and control warp in nominal 2 by 4 inch dimension lumber sawn from small-diameter ponderosa pine trees. One study was conducted at a commercial sawmill with trees harvested in central Arizona. The other study was conducted in experimental kilns at FPL using lumber harvested in central Idaho. The three main variables in the studies were top loading, presteaming, and a high-temperature kiln schedule. A limited study of hot press drying was also included. The high-temperature kiln schedule in the experimental kilns decreased drying time to about half that of the conventional temperature schedule. Press drying time was slightly more than three hours. Crook and bow cause most of the grade loss from warp. There was no evidence that presteaming affected warp or grade loss from warp. Top loading had a modest effect in reducing warp and grade loss from warp. High-temperature drying did not affect measured warp immediately after drying compared with the conventional temperature schedule. Lumber dried using a high-temperature kiln schedule, and moisture equalization in storage, had a reduction of about 50% in grade loss from warp compared with lumber dried using a conventional temperature schedule. [27, 59, 62–64]

Dry Kiln Operator's Manual

The *Dry Kiln Operator's Manual* describes both basic and practical aspects of kiln drying lumber. The manual is intended for several types of audiences. First and foremost, it is a practical guide for the kiln operator—a reference manual to turn to when questions arise. It is also intended for mill managers so they can see the importance and complexity of lumber drying and thus be able to offer kiln operators the support they need to do their job well. Finally, the manual is intended as a classroom text, either for a short course on lumber drying or for the wood technology curriculum in universities or technical colleges. [53]

◀ The results of years of research are found in the *Dry Kiln Operator's Manual*. (1990s) [53]

Technology Transfer

One of the significant aspects of drying research at the FPL was its technology transfer program. Losses because of drying defects and long drying times using air-drying were so great that the industry sought alternative drying methods, such as kiln drying. To facilitate this transfer of important information, the Laboratory in cooperation with Forest Service State and Private Forestry (S&PF) held two-week-long dry kiln courses to provide hands-on training to wood products industry personnel. These courses were held annually for many years.

The FPL and S&PF also developed an "IMPROVE" program to help assist companies in improving their manufacturing process including drying. (1984) [44] (See also the Technology Marketing Unit section.)

◀ Ninety-ninth class in kiln drying of lumber, March 25–29, 1974. The class was conducted by State and Private Forestry (S&PF) and the Forest Products Laboratory.

Solar Drying

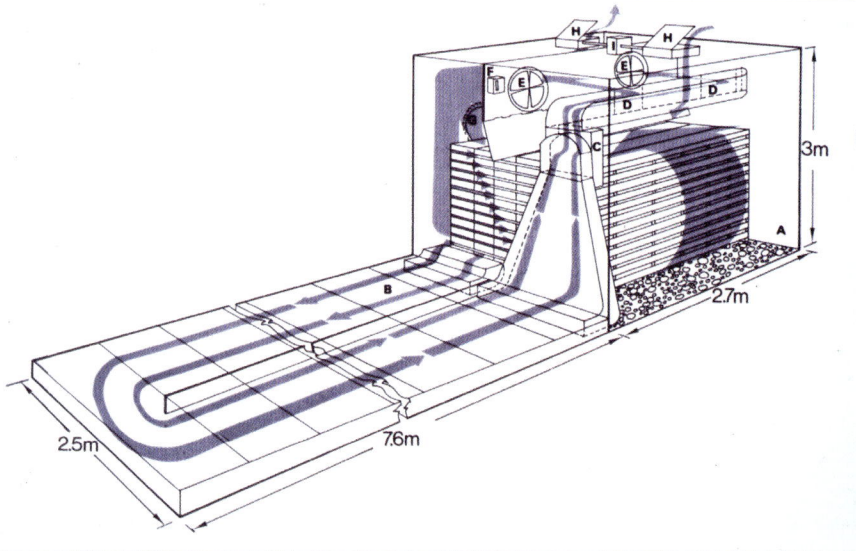

▲ Small solar dryer designed and built by Curt Johnson. (1970s) {87 P} [36]

▲ Solar dryer developed by FPL and built in Sri Lanka (1980s) {120 L}

Bill Simpson designed a low-cost solar collector, built into the ground, that supplies heat to a kiln; a novel design that lowers the cost. [40, 41]

A computer analysis for estimating solar energy needed for wood drying using direct or indirect collection with supplemental heating was prepared by John Tschernitz. [42]

◀ A small-scale dehumidification dry kiln. This dry kiln was developed to meet the needs of people who are looking for an inexpensive and simple kiln design that can cost-effectively dry as little as a few hundred board feet of lumber or as much as several thousand board feet of lumber each year. (2000s) [65]

Heat Sterilization of Wood

The increased importation of invasive insects through infested wood used in pallets and packaging is a critical issue. One prevention method is heat sterilization of the wood. FPL conducted research to determine how long it takes for different species of wood to reach lethal temperatures for these insects. By killing these pests through heat sterilization, markets are kept open for wood products, such as pallets and packaging, while preventing infestations of other areas of the United States and other countries. (2002) [60]

1910

Huge losses of wood because of improper drying or the lack of drying.

2010

Improved methods of air drying lumber.

Effective kilns developed and drying schedules determined, thus greatly reducing waste and improving the quality of lumber for consumers.

Further Information

[1] The relation of the moisture content of wood and rate of drying to the humidity of the surrounding atmosphere / Arthur Koehler / FPL No. 509 (1919)

[2] Drying schedules and other essential information of use to the dry kiln operator in the drying of hardwood stock for furniture and similar uses / prepared by Section of Timber Physics; D.R. Brewster, L.K. Pomeroy, L.V. Teesdale, N.B. Eckbo, H.D. Tiemann / FPL Library has manuscript and reprints of part 1, to part 5 at B3089 (1919)

[3] The kiln drying of lumber: a practical and theoretical treatise / Harry Donald Tiemann, 3d ed. Philadelphia, J.B. Lippincott Company (1920)

[4] Manual of design and installation of Forest Service water spray dry kiln / L.V. Teesdale / USDA Bulletin 894) (1920)

[5] The Forest Service internal fan kiln / Rolf Thelen, H.D. Tiemann / US Patent No. 1,451,747 (April 17, 1923)

[6] Circulation and piling of lumber / FPL No. 551 (1925)

[7] Ample stickering keeps lumber straight / Forest products research in pictures No. 65 (1925)

[8] The movement of moisture in wood. Part 1, Analysis of various theories of drying wood / Harry D. Tiemann / Forest Products Laboratory (1926)

[9] The kiln drying of southern yellow pine lumber / L.V. Teesdale / USDA, Forest Service Technical bulletin No. 165 (1930)

[10] The air seasoning of wood / J.S. Mathewson / USDA, Forest Service Technical bulletin No. 174 (1930)

[11] Wood–liquid relations / L.F. Hawley / USDA, Forest Service Technical bulletin No. 248 (1931)

[12] Chemical seasoning of wood / W. Karl Loughborough / American lumberman 3099:66–67 (May 8,1937)

[13] The reversible-circulation internal-fan kiln / FPL Technical note No. 208 (1940)

[14] Passage of liquids, vapors, and dissolved materials through softwoods / Alfred J. Stamm / USDA, Forest Service Technical bulletin No. 929 (1946)

[15] Accelerating the kiln drying of hardwoods / Raymond C. Rietz / Southern lumberman Vol. 181(2262) (July 1,1950)

[16] Control of moisture content and shrinkage of wood / FPL No. 1903–7 (1953)

[17] Seasoning of Ohio "red" willow artificial-limb blanks / O.W. Torgeson / FPL No. 1636 (1956)

[18] Comparison of three methods of determining whether southern pine poles are well air seasoned / J.S. Mathewson, Paul J. Berger / Forest Products Research Society Journal Vol. 7(5) (May 1957)

[19] Mechanical properties of red oak related to drying / Robert L. Youngs / Forest products journal Vol. 7(10) (Oct. 1957)

[20] Detection and relief of casehardening and final moisture content in kiln–dried lumber / Forest Products Laboratory / FPL Technical note No. 213 (1958)

[21] A method of calculating internal stresses in drying wood / R.L. Youngs, C.B. Norris / FPL No. 2133 (1958)

[22] Special methods of seasoning wood: chemical seasoning / J.M. McMillen / FPL No. 1665–6 (1960)

[23] Stresses in drying lumber / John M. McMillen, Robert L. Youngs / Southern lumberman Vol. 201(2513) (Dec. 15, 1960)

[24] Lumber seasoning: equipment, techniques and financial requirements: an operation manual for cooperative program use / (prepared for the International Cooperation Administration by Edward C. Peck and John M. McMillen). Washington, D.C.: Technical Aids Branch, Office of Industrial Resources, International Cooperation Administration (1961)

[25] Comparative strength of air-dried and kiln-dried wood / FPL–RN–055 (1964)

[26] News and views of this kiln-drying business: operation and maintenance of lumber dry kilns / FPL–RN–0118 (1966)

[27] Press drying nine species of wood / M.E. Hittmeier, G.L. Comstock, R.A. Hann / Forest products journal Vol. 18(9) (Sept 1968)

[28] Longitudinal shrinkage in seven species of wood / R.A. Hann / Research note FPL 0203 (1969)

[28] Gray–brown chemical stain in southern hardwoods / P.J. Bois / Forest products utilization technical report No. 1 (1970)

[30] Computer cuts kiln time 50% at Forest Lab / FPL / Furniture design & manufacturing Vol. 42(4) (Apr. 1970)

[31] Accelerating the kiln drying of hardwoods / R.C. Rietz / Southern lumberman. Vol. 221(2741) (July 1, 1970)

[32] Dry kiln automation / E.M. Wengert / Southern lumberman Vol. 223(2776) (Dec. 15, 1971)

[33] Effect of steaming on the drying rate of several species of wood / W.T. Simpson / Wood science Vol. 7(3) (Jan 1975)

[34] Control of reddish-brown coloration in drying maple sapwood / John M. McMillen/ Research note FPL 0231 (1976)

[35] Kiln-drying lumber from American elm trees killed by Dutch elm disease / R.S. Boone / Forest products journal Vol. 27(5) (May 1977)

[36] Constructing and operating a small solar-heated lumber dryer / Paul J. Bois / Forest products utilization technical report No. 7 (1977)

[37] Using speed of sound in wood to monitor drying in a kiln / W.L. James / Forest products journal Vol. 32(9) (Sep. 1982)

[38] Bacterial oak: drying problems / J.C. Ward / Forest products journal Vol. 33(10) (Oct 1983)

[39] Drying wood: a review / W.T. Simpson / Drying technology Vol. 2(2 and 3) (1983-1984)

[40] Solar dry kiln for tropical latitudes / William T. Simpson, John L. Tschernitz / Forest products journal Vol. 34(5) (May 1984)

[41] FPL design for lumber dry kiln using solar/wood energy in tropical latitudes / John L. Tschernitz, William T. Simpson / FPL–GTR–44 (April 1985)

[42] Solar energy for wood drying using direct or indirect collection with supplemental heating: a computer analysis / J.L. Tschernitz / FPL–RP–477 (1986)

[43] Dry kiln schedules for commercial woods, temperate and tropical / R. Sidney Boone, C.J. Kozlik, P.J. Bois, E.M. Wengert / FPL–GTR–57 (1988)

[44] IMPROVE lumber drying program / Jeanne D. Danielson / Proceedings, Western Dry Kiln Association joint meeting, May 3–5, 1989, Corvallis, Oregon. Western Dry Kiln Association (1989)

[45] Quality drying of softwood lumber: guidebook–checklist / Michael R. Milota, J.D. Danielson, R.S. Boone, D.W. Huber / FPL; IMP–GTR–1 (1991)

[46] Quality drying of hardwood lumber: guidebook–checklist / R. Sidney Boone, M.R. Milota, J.D. Danielson, D.W. Huber / FPL; IMP–GTR–2 (1992)

[47] Grouping tropical wood species and thicknesses by similar estimated kiln drying time using mathematical models / William T. Simpson / Understanding the wood drying process: a synthesis of theory and practice: 3rd IUFRO Conference on Wood Drying, August 18–21, 1992, Vienna, Austria. Wien, Austria: IUFRO (1992)

[48] Identifying bacterially infected oak by stress wave nondestructive evaluation / Robert J. Ross, James C. Ward, Anton TenWolde / FPL–RP–512 (1992)

[49] Specific gravity, moisture content, and density relationship for wood / William T. Simpson / FPL–GTR–76 (July 1993)

[50] Quality drying in a hardwood lumber predryer: guidebook–checklist / Eugene M. Wengert, R. Sidney Boone / FPL; IMP–GTR–3 (1993)

[51] Response of Lutz, Sitka, and white spruce to attack by *Dendroctonus rufipennis* (Coleoptera: Scolytidae) and blue stain fungi / Richard A. Werner and Barbara L. Illman / Environmental entomology Vol. 23(2) (Apr. 1994)

[52] Estimating kiln schedules for tropical and temperate hardwoods using specific gravity / William T Simpson, Steve P. Verrill / Forest products journal 47(7/8): 64–68 (1997)

[53] Dry Kiln Operator's Manual / Edmund F. Rasmussen / USDA Agriculture handbook No. 188 (1961) / Revision edited by William T. Simpson / USDA Agriculture handbook No. 188 (1991) Revision edited by William T. Simpson / USDA Agriculture handbook No. 188 (1999)

[54] Air drying of lumber: a guide to industry practices / Raymond C. Rietz, Rufus H. Page / USDA Agriculture handbook No. 402 (1970) Reviewed and updated as: Air drying of lumber / Edward C. Peck / FPL–GTR–117 (1999)

[55] Drying hardwood lumber / Joseph Denig, Eugene M. Wengert, William T. Simpson / FPL–GTR–118 (2000)

[56] Proceedings: linking healthy forests and communities through Alaska value-added forest products / T.L Laufenberg, B.K. Brady / PNW–GTR–500 (2000)

[57] Relationship between longitudinal stress wave transit time and moisture content of lumber during kiln-drying / William T. Simpson, Xiping Wang / Forest products journal Vol. 51(10): 51–54 (Oct. 2001)

[58] Method for estimating air-drying times of lumber / William T. Simpson, C. Arthur Hart / Forest products journal Vol. 51(11/12) (Nov./Dec. 2001)

[59] Effect of drying methods on warp and grade of 2 by 4's from small-diameter ponderosa pine / William T. Simpson, David W. Green / FPL–RP–601 (2001)

[60] Lumber drying and heat sterilization research at the U.S. Forest Products Laboratory / William T. Simpson / 30th Hardwood Symposium Proceedings (May 30, 2002)

[61] Estimating air drying times of lumber with multiple regression / William T. Simpson / FPL–RN–0293 (2004)

[62] Effect of drying temperature on warp and downgrade of 2 by 4's from small-diameter ponderosa pine / William T. Simpson / FPL–RP 624 (2004)

[63] Method for interconverting drying and heating times between round and square cross sections of ponderosa pine / William T. Simpson / Forest products journal Vol. 55(6) (June 2005)

[64] Using acoustic analysis to presort warp-prone ponderosa pine 2 by 4s before kiln-drying / Xiping Wang. William T. Simpson / Wood and fiber science Vol. 38(2) (2006)

[65] Dehumidification drying for small woodworking firms and hobbyists / Scott Bowe, Patrick Molzahn, Brian Bond, Richard Bergman, Terry Mace, Steve Hubbard / University of Wisconsin Extension Pub–FR–396 (2007)

Lumber Grading

◀ Ronald Knipsel and Vilma Canto visually grading a board. Of concern are the number and location of knots and presence or absence of other defects, such as rot, stain, bark, and splits. {4 I}

Challenge: In the early 1900s, there were multiple grading rules for softwood based on regional standards. Hardwoods were also graded using various methods. Thus, it was very difficult for a consumer to acquire the right size, quality, and species of wood to maximize its utility at a minimum cost.

FPL's Contributions: Working with industry and associations, FPL developed common sizes of lumber and grades so the lumber could be classified. FPL conducted research to help establish fair and uniform grade standards.

Results: FPL helped establish size and grade standards for the public and industry. See related article *ASTM Committee D–7: Wood*, by Dave Green, Robert Ethington, Edward King, Bradley Shelley, and David Gromala.

Grading and Properties of Lumber

Beginning in 1910, FPL developed test data on strength of most species of wood grown in the United States. This information was first published in 1919 and was revised as new data were added. The effort was led by John Newlin and T.R.C. Wilson. They then proposed a grading system that related wood characteristics such as knots and slope-of-grain to lumber test data and developed equations for calculation of allowable properties. By 1934, these calculation procedures had been adopted in the United States and many foreign countries.

▶ "Mechanical Properties of Woods Grown in the United States," by J.A. Newlin and T.R.C. Wilson, 1923. [1]

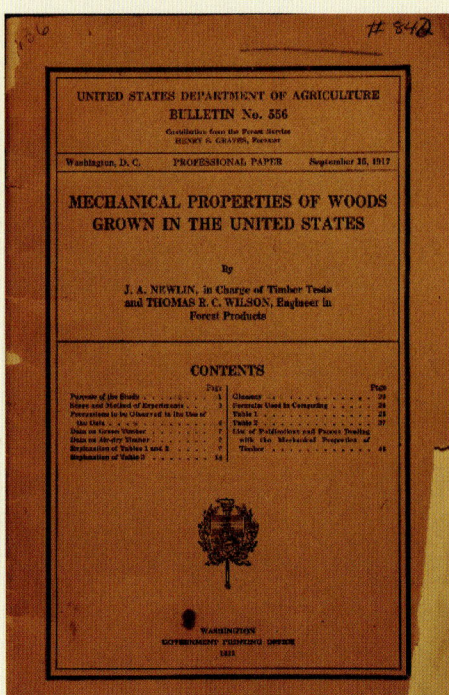

◀ "Guide to the Grading of Structural Timbers and the Determination of Working Stresses," by T.R.C. Wilson, 1934 [4]

Softwoods

Working with industry and associations, FPL scientists designed a single set of national grading standards for softwood building lumber, which replaced the many regional standards formerly used. [3]

Hardwoods

Grading rules evolved that were based on the cutting sizes required in the principal wood fabrication industries. These rules formed the basis for commercial adoption in 1928 of a radically improved hardwood grading system. Estimated annual savings in 1937 was $3,000,000. [2]

Moisture Content

FPL research was used to establish the basis for moisture content specifications for building lumber and brought about their adoption by two important regional associations. This provided consumers a degree of confidence in knowing what they were buying.

Grading

Hardwood and softwood lumber are graded differently. In very general terms, hardwood lumber is graded on its appearance and softwood lumber is graded on its strength. The reason for the difference is that hardwood is usually used in nonstructural applications (such as furniture and cabinets), whereas softwood is usually used in construction where strength is important. However, as more hardwood is being used in situations where strength is a requirement, it can be graded according to softwood rules. [20]

Hardwood Grading

The grade of hardwood lumber is determined by the clear areas of the board—the higher the percentage of clear areas with no knots or other defects, the higher the grade. The grader also looks at the length and width of the clear areas in determining the grade. [12]

▲ Visual grading of green lumber during manufacture. Early in the history of FPL, buying lumber was difficult because there were no grading standards. FPL, working with industry and associations, established lumber standards; throughout its history, FPL has continued to assist in modifying grade standards and is the unbiased source of data and information for various organizations. {30 G}

▲ The many steps involved in a computer program for grading hardwood lumber are shown in this flow chart being checked by Lynn Galiger and Hiram Hallock. (1970s) {M 136 822-7} [8]

Softwood Grading

For softwood lumber, the species, slope of grain, and knot size and location are determined and a visual grade assigned.

Ninety years ago, the FPL developed procedures for grading structural lumber. Working with the U.S. Department of Commerce and regional grading associations, agreement on standard grades of structural lumber were developed. This was the first system that was based on observations from lumber strength tests. The system showed that a 20% increase in design values was justified for major species and identified many other species that were suitable for structural use. [2]

◄ Stamp found on a board that has been visually graded. It indicates that the structural lumber was produced by Mill No. 12, was graded No. 1 & Better, the supervising agency was WWP (Western Wood Products Association), the moisture content at time of surfacing was S-GRN (unseasoned), and wood species combination (grouping) was Doug. Fir-L (Douglas Fir–Western Larch). (2000s)

Using machines that test the lumber without destroying it can lead to a machine grade. [6, 7, 10, 16]

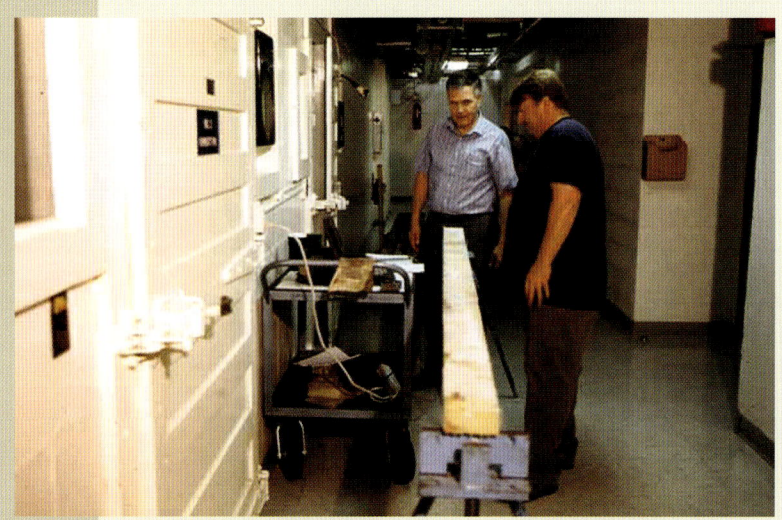

▲ Commercial machine used for determining the rating of lumber. (2000s)

▲ David Green (left) and William Nelson measuring the stiffness of a board using transverse vibration. (2000s) [18]

► Stamp found on a board that has been graded by machine, rather than visually. It indicates that the softwood lumber was machine rated, the supervising agency was WWP (Western Wood Products Association), it was produced by Mill No. 12, the moisture content at time of surfacing was S-dry (19% maximum moisture content), the species or species combination was Hem–Fir (Hemlock–Fir), and the strength rating was 1650f 1.5E (design value for fiber (f) stress in bending was 1650 pounds per square inch (lb/in²) and stiffness (E) was 1,500,000 lb/in²). {100 J}

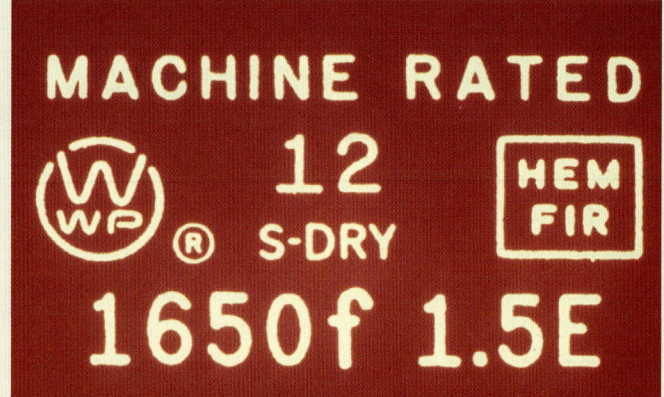

◀ Grading rules—the results of FPL, grading organizations, and industry working together. As more species are included and new techniques developed, the grading rules are revised and updated. (2000s)

▶ FPL has a long history of working with Committee D-7: Wood, of ASTM International (formerly the American Society for Testing and Materials) to develop test methods and procedures to evaluate lumber. This basic research in test methods and analysis has contributed significantly to building codes and standards that help protect the public. (2000s)

In-Grade Testing Program

In the 1980s, a changing resource base and demands for more precise establishment of design values resulted in another comprehensive study of lumber properties. Working with industry and associations, this FPL-led "in-grade" program changed the historic basis for property assignments. As a result, design values assigned to 2 by 4's generally increased significantly, which is of particular importance for the utilization of lumber from small-diameter trees in engineered wood products. (See also the Statistics section.) [11]

Grading Recycled Lumber

Current grading research at FPL is bringing the use of recycled lumber and timbers closer to commercial reality by establishing grading rules for recycled lumber and timbers. [13]

Improving Construction Safety

FPL scientists helped establish grading systems for new and used trenching lumber. Continued upgrading of lumber reuse practices in trenching and excavating provides guidance to minimize construction accidents and encourages reuse of lumber. [9]

1910

Lack of or confusing lumber grades and standards.

2010

Standard hardwood and softwood lumber grades established and adopted by industry. This standardization has allowed for increased commerce and helps the consumer know what they are buying.

Further Information

[1] Mechanical properties of woods grown in the United States / J.A. Newlin, T.R.C. Wilson / USDA Bulletin No. 556 (1923)

[2] The standardization of lumber grades and sizes / L.J. Markwardt, R.P.A. Johnson / To be presented before the Engineering Society of Wisconsin (February 21, 1925)

[3] Progress in standardization of softwood factory lumber grades / Edward M. Davis / West coast lumberman Vol. 48 (Sept. 15, 1925)

[4] Guide to the grading of structural timbers and the determination of working stresses / T.R.C. Wilson / USDA Miscellaneous publication No. 185 (1934)

[5] Grade defects in hardwood timbers and logs / C.R. Lockard, et al. / USDA Forest Service Agriculture handbook No. 244 (1963)

[6] Machine grading: theory and practice / H.C. Hilbrand, D.G. Miller / Forest products journal Vol. 16(11) (Nov 1966)

[7] Some stress-grading criteria and methods of grade selection for dimension lumber / R.L. Ethington / Research paper FPL 148 (1970)

[8] Grading hardwood lumber by computer / H. Hallock, L. Galiger / Research paper FPL 157 (1971)

[9] Potential structural quality of trenching lumber / W.L. Galligan, B.A. Bendtsen, L.I. Knab, F.Y. Yokel, J.F. Senft / Forest products journal Vol. 31(5) (May 1981)

[10] Evaluating lumber properties: practical concerns and theoretical restraints / David W. Green, James W. Evans / Proceedings of the 1988 international conference on timber engineering, Westin Hotel, Seattle, Washington, U.S.A., September 19–22, 1988, Volume 1. Madison, Wis.: Forest Products Research Society (1988)

[11] In-grade testing of structural lumber. Madison, WI: Forest Products Research Society, 1989 Sponsored by the in-grade testing program technical committee" -- p. [1]. Host Institution: Forest Products Laboratory (1989)

[12] An introduction to grading hardwood lumber / The National Hardwood Lumber Association, Memphis, Tenn. (1994)

[13] Stress grading of recycled lumber and timber / Robert H. Falk, David W. Green / 1999 structures congress: structural engineering in the 21st century: Apr. 18–21, 1999, New Orleans, LA. Reston, VA : American Society of Civil En

[14] Mechanical grading of oak timbers / David E. Kretschmann, David W. Green / Journal of materials in civil engineering Vol. 11(2) (May 1999)

[15] Monitoring of visually graded structural lumber / David E. Kretschmann, James W. Evans, Linda Brown / Research paper FPL 576 (1999)

[16] Machine grading of lumber—practical concerns for lumber producers / William L Galligan, Kent A. McDonald / FPL–GTR–7 (Rev.). (2000)

[17] Grading options for western hemlock "Pulpwood" logs from Southeastern Alaska / David W. Green, Kent A. McDonald, John Dramm, Kenneth Kilborn / FPL–RP–583 (2000)

[18] Nondestructive evaluation of wood / Roy F. Pellerin, Robert J. Ross / Forest Products Society publication No. 7250 (2002)

[19] Lumber / David E. Kretschmann / McGraw–Hill encyclopedia of science & technology. 9th ed. New York: McGraw–Hill (2002)

[20] ASTM Committee D-7: Wood, Promoting safety and standardization for 100 years / D.W. Green, R.L. Ethington, E.G. King, B.E. Shelley, D.S. Gromala / Forest products journal, Vol. 54(9) (Sept. 2004)

[21] Grading timber and glued structural members / David E. Kretschmann, Roland Hernandez / Primary wood processing: principles and practice. Dordrecht: Springer (2006)

Pallets

◀ Richard Geier assembling a wood pallet used to increase the efficiency in transporting finished products. (1970s)

In 1910, the closest thing to a pallet was a wood skid that was generally used to move heavy equipment. Around the 1930s, with the evolution of forklift trucks, two-faced skids (pallets) were developed. During WWII, with the great need to move large quantities of goods, the use of pallets was greatly expanded.

Challenge: Pallets are generally reused, and it is important to determine how to extend a pallet's useful life for moving goods that are produced. [11]

FPL's Contributions: Through research in pallet design and fasteners, and the development of an impact panel for forklift trucks, the useful life of a pallet can be significantly extended.

Results: Millions of wooden pallets are manufactured annually, utilizing a significant amount of lumber, particularly low-quality hardwood. This use enhances sustainable forestry by providing an economic market for lumber that in the past was wasted.

Properly designed, manufactured, and handled, pallets can be reused many times, thus extending the timber supply. [4–6]

Pallet Evaluation

Bin Pallets

▲ Milo Schimming observes how a wood pallet reacts to the effects of rough handling by subjecting the pallet to a series of drops provided by a 14-foot-diameter revolving drum tester developed at the FPL. The drum test accelerates the handling of the pallet so that scientists can understand the weak points in its construction and improve its performance through better materials and design. (1960s) {M 123 435}

▶ Tom Heebink observes the performance of a bin pallet designed for vegetables. Of particular interest is designing the pallet for minimal bulge so it can be effectively fitted into transportation and storage facilities. (1950s) {M 101 145} [1, 3]

▶ Example of a bin pallet that is used in the harvesting and transportation of field crops. By adding the diagonal crosspiece, the rigidity of the pallet is significantly improved and useful life extended. (1950s) [1, 3]

Load-Carrying Capacity of Pallet Deck Boards

Early research resulted in a method for determining the load-carrying capacity of the deck boards of a pallet. However, this assumed the pallet is on the floor. If the pallet is placed in racks that support just the edges of the pallet, the method is not applicable. [2]

Computer-Designed Pallets

FPL worked with the National Wooden Pallet and Container Association (NWPCA), Northeastern Forest Experiment Station, and Virginia Polytechnic and State University to develop a design procedure for pallets. The objective was to develop a sound design that could be used by both small and large pallet manufacturers. This design procedure is now a "user-friendly" computer program that many manufacturers are using. [8–10]

Impact Panel

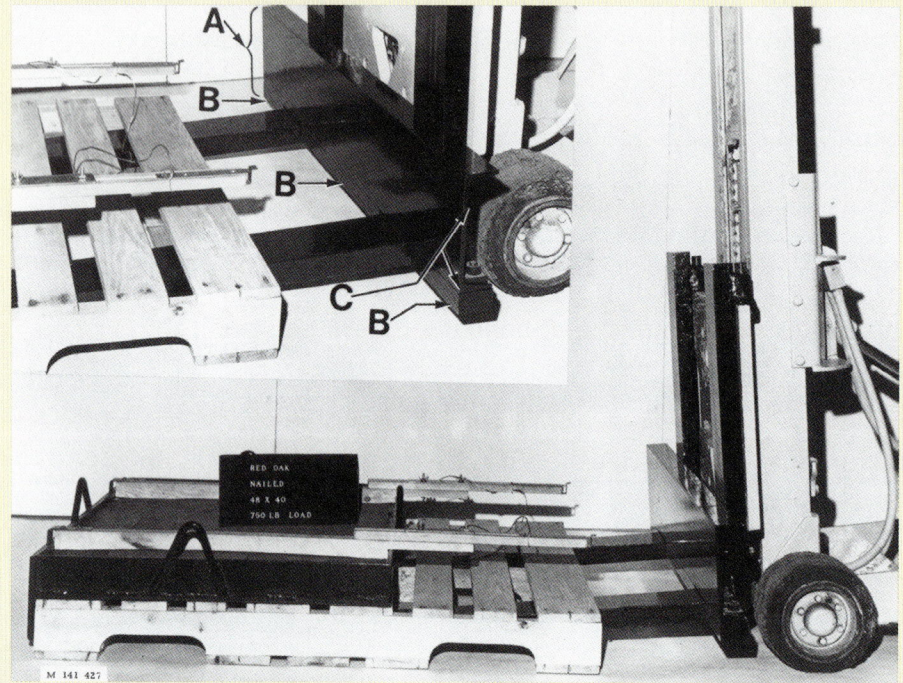

◄ Based on tests and actual use information, it was found that the leading edge deckboard of the pallet was prone to fail first. Robert Stern developed this impact panel that fits on a fork lift truck. The panel distributes the impact over the entire width of the leading edge deckboard and the stringers, significantly extending the life of the pallet. A, the impact panel; B, extension of the panel so contact is made with the stringers; C, 8- by 2-1/4-inch, 13.75-lb/ft steel channel. (1970s) {M 141 427} [7]

Used Pallets

In the past, used pallets generally ended up in landfills. Today, many pallets are disassembled, broken parts replaced, and reassembled. Many of those that are beyond refurbishing are ground up and converted into landscaping mulch or used for fuel.

1910

Virtually nonexistent.

2010

Developed methods to extend the life of pallets, thus saving lumber and reducing repair costs.

Use of pallets helps decrease shipping costs and saves consumers money.

Companies employed 51,642 people and shipped pallets with a total value of $6,541,546,000. (2006 Annual Survey of Manufacturers)

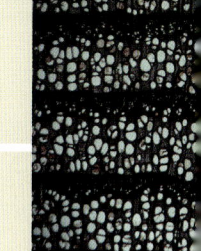

Further Information

[1] Bin pallets for agricultural products / T.B. Heebink / Woodworker 77(8): 6–7, 16–24, (Oct 1958

[2] Load-carrying capacity of deck boards for general-purpose pallets / T.B. Heebink / FPL No. 2153 (1959)

[3] An evaluation of eleven bin-pallet designs / T.B. Heebink / FPL No. 2216 (1961)

[4] Performance comparison of slender and standard spirally grooved pallet nails / T.B. Heebink / FPL No. 2238 (1962)

[5] Wood pallet manufacturing / FPL–RN–0213 (1971)

[6] Evaluation of methods of assembling pallets / R.S. Kurtenacker / FPL–RP–213 (1973)

[7] Increasing serviceability of wood pallets / R.K. Stern / FPL–RP–258) (1975)

[8] Strength and stiffness analysis of notched, green oak pallet stringers / Terry D. Gerhardt / FPL–RP–452 (Study was part of a cooperative research program involving the National Wooden Pallet and Container Association, Virginia Polytechnic and State University and the Forest Service) (1984)

[9] Deckboard bending theory for three-stinger wood pallets in drive-in racks / Thomas J. Urbanik / Journal of testing and evaluation Vol. 13(1) (Jan. 1985)

[10] A method for determining the effect of fasteners on the stiffness and strength of wood drive-in-rack pallets / T.J. Urbanik / Journal of testing and evaluation Vol. 13(5) (Sep 1985)

[11] Unitizing goods on pallets and slipsheets /James F. Laundrie / FPL–GTR–52 (1986)

Bending Wood

Wood usually comes in flat, straight boards, if properly manufactured. However, it is sometimes desirable to have a bend in the wood. In some instances, bent wood may end up as unusual art, but more likely as furniture or ship hulls.

Challenge: How to bend wood without splintering or breaking it.

FPL's Contributions: FPL scientists evaluated methods and the effects of different wood species on the bending properties of wood.

Results: The ability to form wood into unique shapes and designs has led to new products and art forms.

▲ Karl Loughborough showing a piece of bent wood that was pretreated with urea. (1940s) {M 39429 F}

◄ Karl Loughborough demonstrating a wood spring made of urea-plasticized oak. (1940s) [5]

Bending Process

Early FPL Research resulted in improvements in working with steaming and bending processes for the furniture industry. [1, 2]

Ship Hulls

Processes and equipment were also developed to bend large wood members that could be used in structures and in ships. [6]

◄ Clarence Jordan and Edward C. Peck watch the piece of wood as it is bent by pulling the wood from both ends around the wooden form until the desired position is reached. (1940s) {M 66379 F}

▲ Technician operating the winding mechanism used to apply force to each end of the wood section. (1940s) {M 66374 F}

▲ Later stage of bending the wood section. {M 66378 F}

Scientists improved the method for bending wood to curved forms, which greatly decreased waste of material in the bending operation. (1940s) [6]

▲ Clarence Jordan with a bent boat member prepared for drying and fixing, showing strap, tie rods, and wood stays. (1940s) {M 66373 F}

▲ Bent wood members used in a boat hull.

▲ Samples of bent wood.

▶ Samples of plasticized oak. Wood soaked in a chemical solution and heated can be easily bent and twisted to very sharp curvatures, which are retained on cooling. Further, the hot material can be quickly pressed into dense molded or flat shapes. The figure-eight-shaped sample was bent with the grain; the open loop to the right was bent across the grain; the large disk was made by pressing together small pieces; the small disk was made from blackjack oak sawdust. (1940s) {M 37222 F} [3, 4]

1910

Large losses during processing were common.

2010

Developed new methods to enhance the bending of wood and decrease waste.

Further Information

[1] Preparation of wood for bending /Technical note / FPL No. B 9 (1919)

[2] Bending southern oak chair posts in a hot-plate bender / W.K. Loughborough / Canadian woodworker and furniture manufacturer Vol. 22(6) (June 1922)

[3] Some Forest Products Laboratory developments of particular significance to the South / FPL 0–8 W73Sf (April 15, 1940)

[4] Molding wood to man's will: new plasticizing treatment may open way to use of low-quality timber / F.J. Champion / Wood products Vol. 46(9) (Sept. 1941)

[5] Forest Products Laboratory urea-plasticized wood (uralloy) / FPL No. 1277 / Also, Further information on the plasticizing of wood as developed at the Forest Products Laboratory / Wood products 46(11) (Nov 1941)

[6] Bending solid wood to form / Edward C. Peck / USDA Agriculture handbook No. 125 (This is a revision of FPL Mimeo report 1764) (1957)

[7] Wood handbook: wood as an engineering material / FPL–GTR–113 (1999)

Finishes

◀ Modern house finished with the Forest Products Laboratory Natural Finish, a semitransparent penetrating oil-based stain formulation (1970s) {M 147 331-5} [8]

Challenge: How to protect wood products from the effects of moisture, sun, wind, and other elements. One method is to select species of wood that have unique properties that make them naturally resistant. Another is to coat the surface with paint or treat the wood with a penetrating oil or stain. The total cost to U.S. homeowners to keep their property painted is estimated at over $375,000,000 per year.

FPL's Contributions: Over the years, research at FPL has concentrated on understanding the interaction of the wood surface, finish characteristics, and effects of weather and time. Much of this research involves the interaction of wood species and various treatments and pretreatments using test fences to expose various combinations to weather and time.

Results: Developed FPL stain, that has good appearance, requires minimum upkeep, and is easily reapplied with a minimum of surface preparation. An understanding of the causes of paint failure and remedies decreases house maintenance and extends service life of structures.

◀ Paint flaking off house siding, caused by moisture problems. {18 A}

▶ John Black inspecting paint failure on wood structure. (1980s) {M 88 9000}

Paint Failure

Early paint maintenance research discovered that paints had a critical film thickness, which when exceeded, causes them to fail by "cross-grain" cracking. This is a type of paint failure found on older homes and buildings. (Old paint should be checked for lead so that adequate safety measures are taken prior to preparing the surface for repainting.) (1933) [2]

FPL staff demonstrated that rainwater and dew are responsible for many instances of paint failure and established the effectiveness of pretreatment of wood siding with a water-repellent preservative in preventing such paint failure. [25]

Wood Discoloration

◀ Discoloration of wood can be caused by moisture and uncoated metal fasteners. The discoloration shown here (iron tannate) will show right through a coating of paint. This problem can happen especially when improper fasteners are used in woods that contain tannin, such as oak, cedar, or redwood. (1990s) {99 B} [15, 18]

Early Research on Finishes

◀ Frederick Browne, head of the wood finishing research project from 1922 to 1962. (1950s) {M 88 440} [1]

▶ Early laboratory for preparing various coating formulations to be evaluated when applied to various wood species. (1930s) {M 23953 F} [15, 18]

Blister Box

◀ Blister box designed in 1954 by Donald Laughnan and Frederick Browne. Paint panels on front were exposed to extremes of cool dry air on the outside and warm moist air on the inside to induce paint blistering and early failure. (1950s) {M 96651} [15, 18]

Exposure Evaluation

◀ Edward Mraz and Peter Sotos inspecting paint test panels at the wood products exposure site in Madison, Wisconsin. (1978) A similar site is located on the Harrison Experimental Forest maintained by the Southern Forest and Range Experiment Station in Gulf Port, Mississippi. (1970s) {M 146 491-8}

▶ An example of the FPL test fence at the Harrison Experimental Forest, used to evaluate various finishes and pretreatments. (1985) {M 85 0008-16}

Following the Sun

◀ Paul Evans inspects wood specimens mounted on the panel of "sun follower" designed by Arthur Koster. The device tracked the sun to increase exposure to daily solar radiation, thus accelerating tests of finishes on the specimens. (1960s) {18 F}

Effect of Preweathering Wood Prior to Painting [16, 17, 19–22, 35]

▶ Painted western redcedar after 17 years. Boards were preweathered (exposed to direct sunlight) for 1, 2, 4, 8, or 16 weeks prior to being painted. Boards were painted with one coat of alkyd-oil or acrylic-latex primer and one coat of acrylic-latex top coat. (2000s) [14]

▲ Control, no exposure prior to painting. Alkyd-oil primer (one coat), acrylic-latex top coat (one coat)

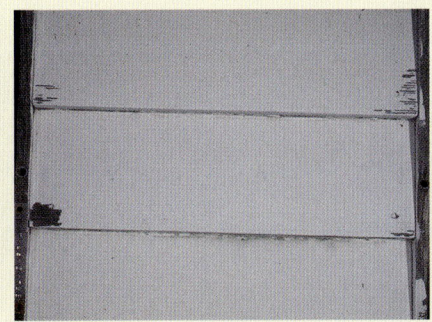

▲ One week of exposure prior to painting. Alkyd-oil primer (one coat), acrylic-latex top coat (one coat)

▲ Sixteen weeks of exposure prior to painting. Alkyd-oil primer (one coat), acrylic-latex top coat (one coat)

Instrumented Exposure House

◄ In 1964, a house of many parts was built to serve as an exposure-site instrumentation structure. One purpose of the structure was to study the effects of climate and interior conditions on exterior paints and other finishes. Walls, floors, and roof contain many different experimental materials and building components. Cassandra Steinkopf and Robert Hann are reading a thermometer registering maximum and minimum temperatures. Results of experiments such as these were used to prepare various manuals published by FPL on housing. (1960s) {M 124 960} [15]

Accelerate the Weathering of Paints and Coatings

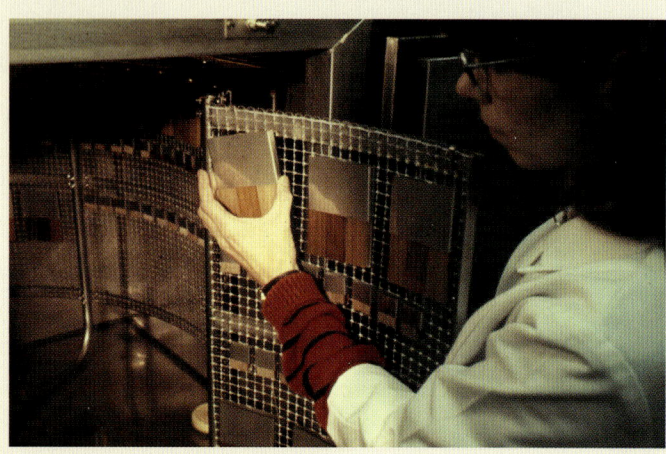

◄ Jill Wennesheimer placing a test specimen in a weatherometer, which will expose the coating to ultraviolet light for several months. The test is used to accelerate the weathering of paints and stains. {111 Q} [7]

Water-Repellent Treatment

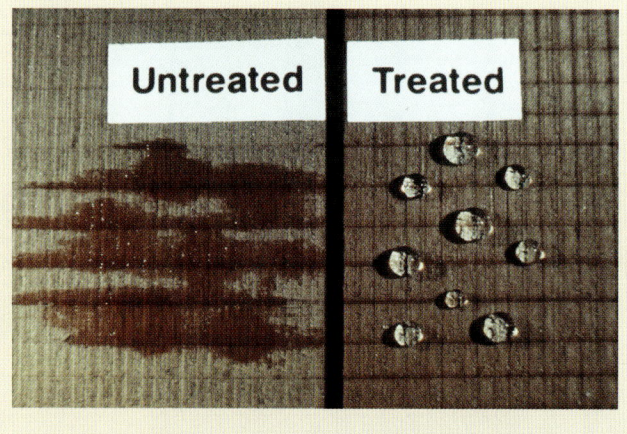

Untreated   Treated

◄ Droplets of water are absorbed in untreated wood (left) but repelled by wood that is treated with a water-repellent preservative (right). (1960s) {67 N} [6]

Working with industry in the 1950s, FPL scientists developed a pretreatment system consiting of preservative, paraffin wax, resin, and solvent, which when applied to millwork prior to painting protects painted surfaces from decay for up to 20 years of exposure in Madison, Wisconsin. [6]

▲ Bill Feist examining water-repellent preservative treatments of window frame corners. This style of test sample is used extensively by the FPL, industry, and other researchers, for outdoor exposure studies. (1990s) [31, 32]

FPL Natural Finish

The FPL Report 2096 originally issued in 1957 outlined the formulation guidelines for making wood stains for use on exterior wood. The finish, originally called the "Madison formula," was a whole new concept for finishing wood because the finish penetrated the wood's surface rather than forming a film. The report was revised and republished in 1961, 1964, 1970, and 1972. Using the concept and the formulation guidelines in these publications, the paint industry began marketing penetrating stains for use on wood. The breakthrough made by FPL scientists in developing this finish was instrumental in the development of a whole semitransparent stain industry in the United States. Almost all paint manufacturers in the United States that formulate finishes for wood include semitransparent stains for wood in their product lines; some companies manufacture only wood stains. [8]

◄ Donald Laughnan (right), developer of FPL natural finish, and Joe Terry evaluating the performance of natural finishes placed on the FPL's outdoor test fence. (1950s) {M 100 469} [8]

Paper Overlaid Lumber

► One successful method to improve the paint characteristics of wood, particularly knotty wood, is through laminating a resin-impregnated paper to the wood. (1960s) {M 136 048} Today this approach is used on fiberboard and oriented strandboard (OSB). [3]

Decks

◄ Greg Meeker applying another deck finish to determine the service life of different types of deck stains on several wood species. (2000s) [5, 23–25, 27, 33, 34]

Technology Transfer

One of the most important contributions of the FPL exterior wood finishing research program has been disseminating results of basic and applied research in a wide variety of scientific journals. The research has practical application for painters and homeowners. The FPL has also disseminated research results in a variety of nontechnical publications, informational guides, trade journals, popular magazines, and newspapers. As a result, many thousands of architects, builders, and homeowners have benefited from FPL research. Improved exterior finishes have resulted in substantial savings in maintenance time and costs.

Shortly after its creation in 1922, the FPL wood finishing research group made strides in characterizing wood surfaces, clarifying the functions of primers and paints, and understanding the interaction of wood and moisture in paint failure. These principles proved to be the group's greatest contribution to the Armed Services efforts in WWII, saving money in the maintenance of thousands of buildings used to house troops, equipment, and materials. This fundamental research work continues today.

As far back as 1932, farmers were provided with painting recommendations in articles published in farm and other practical journals and through the USDA Extension Service in cooperation with universities. In addition, the FPL has issued thousands of copies of General Technical Reports and Research Notes free to the public, addressing topics such as exterior finishes for homes and log cabins, proper application procedures, causes and cures for most finish discolorations and failures, and refinishing recommendations. These reports serve as valuable references for professional builders, architects, and wood finishers. [4, 9–13, 15, 18, 26, 28–30, 36]

1910

Little understanding of wood–finish interactions.

2010

Increased understanding of wood–finish interactions, leading to improved performance of wood finishes.

Development of FPL stain, semi-transparent stains, and water-repellent treatments to help prevent weathering of wood.

Further Information

[1] Properties of wood that determine the service given by exterior paint coatings / F.L. Browne / American paint journal (Apr. 7, 1930)

[2] Some causes of blistering and peeling of paint on house siding / F.L. Browne / FPL Report [Mimeo] No. 6. (Superseded by 1933 edition, same title) (1933)

[3] Paper overlaid lumber / Bruce G. Heebink / Forest products journal Vol. 11(4) (Apr. 1961)

[4] Finishing exterior plywood / Research note FPL 0133 (1966)

[5] Microscopic observations of ultraviolet irradiated and weathered softwood surfaces and clear coatings / V.P. Miniutti / Research paper FPL 74 (1967)

[6] Water repellents and water-repellent preservatives / William C. Feist, Edward A. Mraz / Research note FPL 0124 (1968)

[7] Comparison of outdoor and accelerated weathering of unprotected softwoods / W.C. Feist, E.A. Mraz / Forest products journal Vol. 28(3) (Mar. 1978)

[8] Forest Products Laboratory natural finish / John M. Black, Don F. Laughnan, Edward A. Mraz. / (Rev. ed Research note FPL 046) (Supersedes earlier editions of FPL report No. 2096, dating back to the 1950's) (1979)

[9] Outdoor wood finishes / William C. Feist / Fine woodworking No. 42 (Sept./Oct. 1983)

[10] Finishing wood exteriors: selection, application, and maintenance / Daniel L. Cassens, William C. Feist / USDA Agriculture handbook No. 647 (1986)

[11] Effects of acid rain on painted wood surfaces: the importance of the substrate / R.S. Williams / In: Baboian, R., ed., Materials degradation caused by acid rain; ACS symposium series 318; 1985 June 17–19, Arlington, VA, Washington, DC; p. 310–331 (1986)

[12] Impact of acid precipitation on wood and painted wood. Part I—Effect of dilute acid on photochemical degradation of wood / R.S. Williams, J.W. Spence / Proceeding of the NAPAP review of materials research, April 7–11 1985, Arlington, VA (1986)

[13] Acid effects on accelerated wood weathering / R.S. Williams / Forest products journal Vol. 37(2) (Feb 1987)

[14] Adhesion of paint to weathered wood / R. Sam Williams, Jerrold E. Winandy, William C. Feist / Forest products journal Vol. 37(11/12) (Nov./Dec. 1988)

[15] Chronicle of 65 years of wood finishing research at the Forest Products Laboratory / Thomas M. Gorman, William C. Feist / FPL–GTR–60 (1989)

[16] Adhesion of paint to weathered wood / R.S. Williams, J.E. Winandy, W.C. Feist / American paint and coatings journal 73(34):36–41 (1989)

[17] Photodegradation of wood affects paint adhesion / R.S. Williams, P.L. Plantinga, W.C. Feist / Forest products journal 40(1):45–49 (1990)

[18] 80 years of wood research: Forest Products Laboratory searches for answers to weathering challenge / W.C. Feist, R.S. Williams / American paint and coatings journal. Convention daily 75(18):30–33 (1990)

[19] Effect of weathering of new wood on the subsequent performance of semitransparent stains / M. Arnold, W.C. Feist, R.S. Williams / Forest products journal 42(3):10–14 (1992)

[20] Durability of paint or solid-color stain applied to preweathered wood / R.S. Williams, W.C. Feist / Forest products journal 43(1):8–14 (1992)

[21] Determining paint adhesion to wood using uniform double-cantilever beam specimens / M. Knaebe, R.S. Williams / ASTM Journal of testing and evaluation. 21(4) (1993)

[22] Effect of weathering of wood prior to finishing on paint bond strength and durability / R.S. Williams, W.C. Feist / Proceedings of the Polymeric Materials Science and Engineering Division, American Chemical Society spring meeting, Denver CO. March 29 – April 2 (1993)

[23] Wood decks: materials, construction, and finishing / K.A. McDonald, R. Falk, R.S. Williams, J.E. Winandy / Forest Products Society, Madison WI. 93p. (1996)

[24] Details for a lasting deck: Government scientists study outdoor structures and report on which details, fasteners, and finishes hold up best / R. Falk, S. Williams / Fine homebuilding 102:78–81 (April/May, 1996)

[25] Finishes for exterior wood: selection, application, and maintenance: a comprehensive guide to the painting/staining and maintenance of homes, decks, log structures, and more / R. Sam Williams, Mark T. Knaebe, William C. Feist / Madison, WI: Forest Products Society (1996)

[26] The Finish Line: practical facts on wood finishing from FPL / R.S. Williams, M.T. Knaebe / American paint and coatings journal. Convention daily (Nov. 3, 1997)

[27] Cleaners & restorers for wood decks / Alan Ross, George Daisey, Charles Jourdain, Sam Williams / The paint dealer. 7(4): 30–33. (1998)

[28] Selection and application of exterior stains for wood / R. Sam Williams, William C. Feist / FPL–GTR–106 (1999)

[29] Changing nature of wood products—what does it mean for coatings and finish performance? / Charles Jourdain, Jack Dwyler, Keith Kersell, Douglas Mall, Ken McClelland, Robert Springate, Sam Williams / Journal of coatings technology 71(890) (March, 1999)

[30] Mildew and mildew control for wood surfaces / S. Bussjaeger, G. Daisey, R. Simmons, S. Spindel, S. Williams / Journal of coatings technology 71(890):67–69 (1999)

[31] Effect of water repellents on the long-term durability of millwork treated with water-repellent preservatives / R.S. Williams / Forest products journal 49(2):52–58 (1999)

[32] Water repellent and water-repellent preservative finishes for wood / R.S. Williams, W.C. Feist / FPL–GTR–109 (1999)

[33] Restoration of severely weathered wood / R.S. Williams, M. Knaebe / Journal of coatings technology 72(902): 43–51 (2000)

[34] Protecting wood decks from boidegradation and weathering: evaluation of deck finish systems / J.J. Morrell, P.F. Schnieder, R.S. Williams / Forest products journal 51(11/12):27–32 (2001)

[35] Correlation of adhesive strength with service life of paint applied to weathered wood / R.S. Williams, J.E. Winandy, W.C. Feist / Paper 161. In: Proceedings 9th durability of building materials and components conference 17–20 March, Brisbane, Australia (2002)

[36] Finishes checklist—A guide to achieving optimum coating performance on exterior wood surfaces / Tony Bonura, Steve Bussjeager, Lynne Christensen, George Daisey, Tom Daniels, Mark Hirsch, Charles J. Jourdain, Douglas, Mall Sr., Bob Springate, Louis E. Wagner, Harry Warren, R. Sam Williams / Journal of coatings technology 1(3): 36–49 (2004)

Fasteners and Connectors

The joint is usually the weakest link in a timber structure. Small wood members are usually nailed, whereas large pieces are bolted, but with any mechanical fastener it is difficult to utilize the full strength of the wood.

Metal connectors, such as shear plates or split rings, greatly improve the strength of a joint and permit the design of large timber structures.

Challenge: How to build efficient structures of all sizes using wood. Wood is an amazing structural material, but to maximize its use, pieces of wood must be fastened together. Those fasteners can be as simple as nails, pins, or bolts or as complex as designed split rings, plate connectors, or dowel nuts. The goal is the same: to provide a safe, efficient, effective structure.

FPL's Contributions: Through research on various mechanical fasteners such as nails, bolts, rings, and plates, the strengths of various joints were determined and fastener design criteria were improved. This research also allowed the use of a greater number of species in construction. Much of the FPL research is the basis for fastener design criteria used in the U.S. building codes.

Results: Applying this information, coupled with the basic research on strength of various species of wood, it is possible to build large structures of wood.

▲ Communications tower made from wood members joined with "modern" connectors. (Radio station WRVA, Richmond, Virginia) (1930s) {M 22059 F}

Nails [12, 13, 16, 21, 43, 44]

◀ One of the most common methods of fastening wood is through the use of nails. Early research at FPL studied the nail-holding ability of various species and splitting tendencies. (1920s) [1, 2, 5, 9]

Joint Design

Tests on joints and fastenings have resulted in the establishment of design equations for all kinds of joints. With specialized connectors, entirely new designs of timber construction are possible. Design equations center on determining the withdrawal or lateral connection capacity.

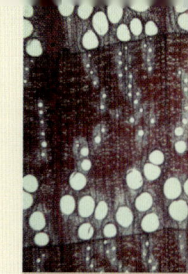

◄ The basic design data included in U.S. Department of Agriculture Technical bulletin No. 865, "Timber-Connector Joints, Their Strength and Design," were established through FPL research. These data are the basis for design criteria used in all timber-connected structures and have brought wood into use in structures that previously could be built only with steel. (1936) {M 28811 F} [8]

Fastener and Connector Evaluation

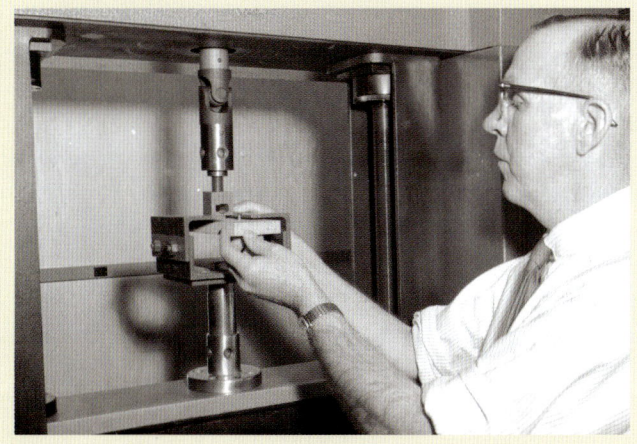

▲ Bert Munthe setting up a test to determine the screw-holding ability of 3/4-inch-thick, high-density particleboard (resistance to withdrawal from the face of the board). (1960s) {M 124 989}

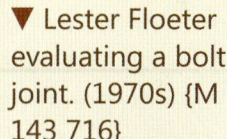

◄ Nail test to determine the fastener's resistance to bending.

▼ Lester Floeter evaluating a bolt joint. (1970s) {M 143 716}

◄ Roy Pien evaluating the strength and deformation of a metal butt joint splice plate connector. (1960s) {M 124 779}

Withdrawal Load Capacity

Nail withdrawal force is related to wood specific gravity, nail diameter, wood moisture content, and amount of nail penetration into the wood.

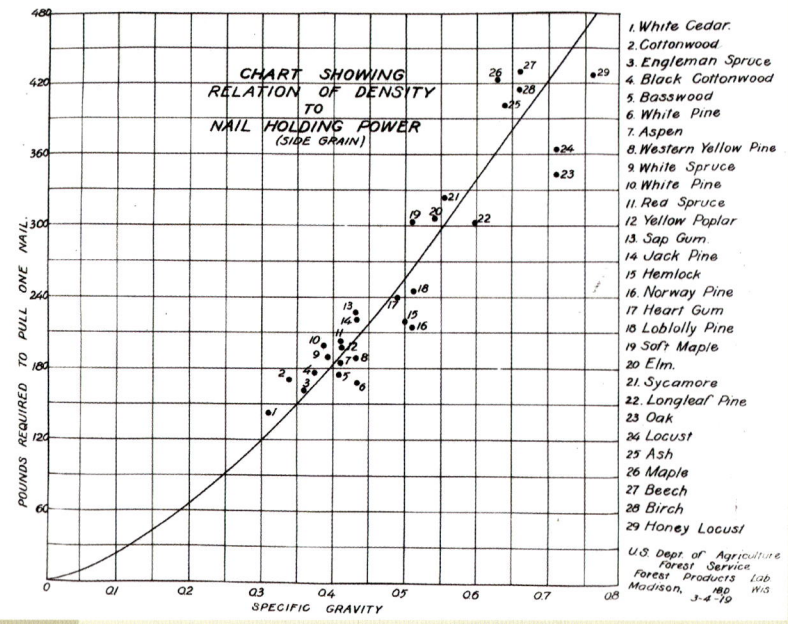

This work has extended to other fastener types such as wood screws, lag screws, and other fasteners with threaded characteristics. [12, 16]

Evaluation of nail shank types, including deformed and coated shanks, for box making showed the superior withdrawal resistance of spirally and annularly threaded nails, especially after the containers had been repeatedly wetted and dried. These results were reflected in military specifications for pallets, where the improved nails were specified. The pallet industry, recognizing the improved performance of these nails, has adopted these specifications and has described the nails in bulletins and standards. Currently, the post-frame construction industry is advocating the use of annularly threaded nails in their market. [15, 17, 21, 42, 43, 45, 60]

▲ The curve shows that the force required to remove nails from wood increases as the specific gravity (an expression of density) of the wood increases. The chance of wood being split by nails also increases with the density of the wood, making it necessary to use smaller nails or pre-drill holes in the denser species. It was also found that nails can be more easily driven into and withdrawn from green wood than dry wood. Nails driven into green wood tend to become loose as the wood dries out. (1919) [1]

Lateral Load Capacity

Pre-1991

The empirical design approach used prior to 1991 was based on a tabular value for a single bolt in a wood-to-wood, three-member connection where the side members are each a minimum of one-half the thickness of the main member. The single-bolt value must then be modified for any variation from these reference conditions. [14, 18, 25–27, 29–31]

◄ Single bolted connection.

Post-1991

After 1991, a theoretical design approach was adopted that is more general and is not limited to reference test conditions. This approach is based on work done in Europe and referred to as the European Yield Model (EYM). The EYM uses the joint geometry and material properties to determine a number of possible yield modes that can occur in the dowel-type connection. [32, 33]

► Examples of possible yield modes that can occur in a dowel-type connection.

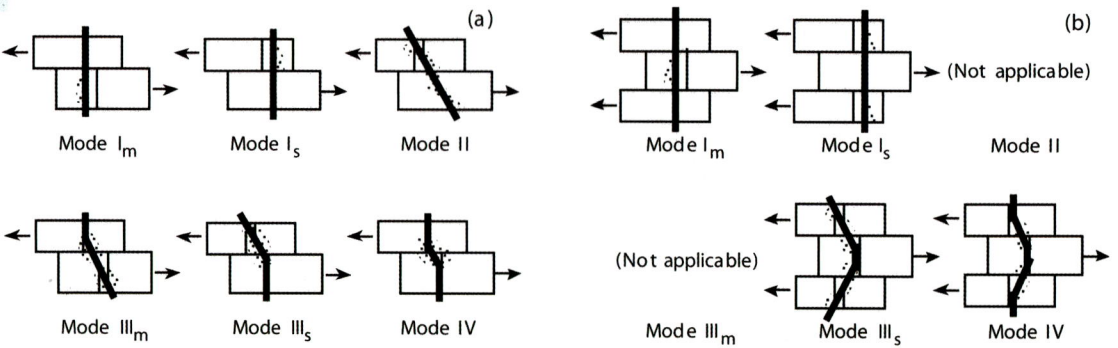

The yield strength of these different modes is determined from a static analysis that assumes the wood and the bolt have yielded. The yield mode that results in the lowest yield load for a given geometry is the theoretical connection yield load. [32, 33, 36, 37, 40, 41, 43, 46, 55]

The initial EYM approach focused on understanding of single fastener performance, but rarely do joints consist of a single fastener. The influence of material properties and the bolt geometry work to decrease capacity of multiple fastener joints. FPL research is concentrating on the ultimate fastener failure mechanism of multiple bolted connections and the influence of the wood material and bolt ductility on ultimate joint performance.

◀ Setup for evaluating a multiple bolted connection test specimen. (2000s)

▶ Douglas Rammer with results of the test showing failure of the multiple bolted specimen. (2000s)

Wood Fasteners and Connectors

In addition to dowel-type fasteners, such as nails and bolts, FPL has evaluated the performance of various other connector types, including split rings, shear plates, traditional truss plates, and glued-nail systems. FPL developed design procedures for split rings and shear plates that allowed the construction of large wood structures. [4, 6–8, 11–13, 15, 16, 19, 20, 23, 48, 49, 54]

▶ Some typical methods of connecting wood members. {137 A}

Specialized:
A. Dog plate
B. Claw plate
C. Split ring
D. Toothed

Common:
E. Bolt
F. Plate
G. Screw
H. Nail
I. Pin

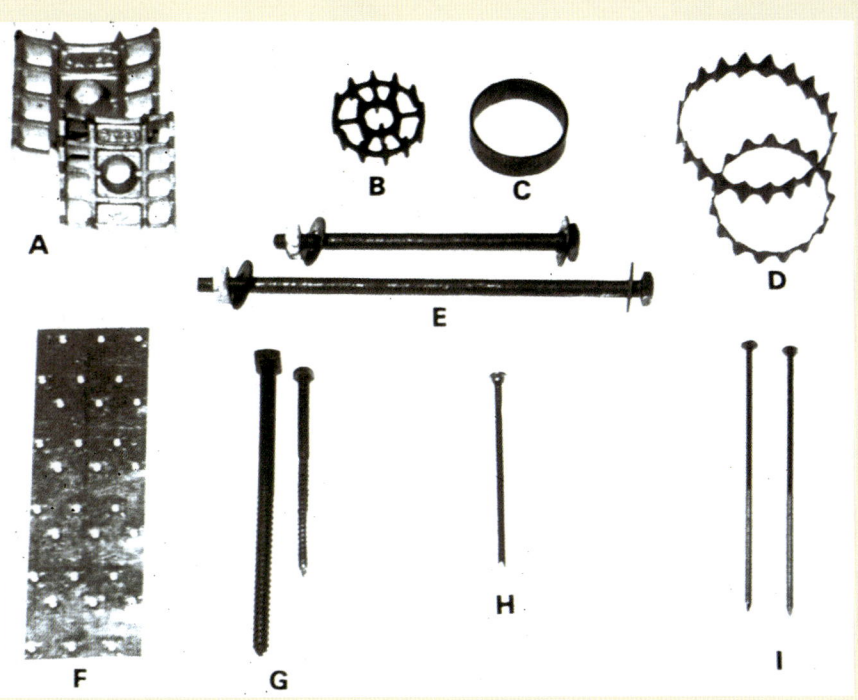

Details on many of these are found in the FPL *Wood Handbook*. [43]

► Wood fasteners allow for the design and construction of large wood structures such as this composite trestle-type highway bridge at Port Angeles, Washington (length, 755 feet; maximum height, 100 feet). (1930s) {M 29155 F}

Use of Some Types of Connectors

◄ Use of metal plate connectors in construction. {17 M} [34, 35]

◄ Close-up of a metal plate connector. {17 H}

▲ Truss plate connectors. This truss is made completely of 2 by 4 lumber. {17 O}

▼ Connection of a pole and beam using plywood gusset plates and bolts. {71 J}

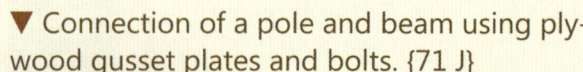

▲ Metal joist hangers to connect the composite joists to the beam. {37 D}

Corrosion of Fasteners

In many applications, fasteners used in wood are exposed to the natural environment or are in contact with chemical treatments used to preserve the wood. This exposure can cause the metal fasteners to corrode.

Roy Baechler, in 1949, reported on the corrosion of metal fasteners in zinc-chloride-treated wood after 20 years of exposure. This early research pointed out the importance of understanding the potential negative effects of chemical wood preservatives or fire retardants on metal connectors used in wood construction. Andrew Baker continued the research including metal coated connectors.

With the increased use of metal fasteners and new chemical wood treatments it is even more important to understand the interaction between the wood, environment, and type of fastener. If conditions are right, the effectiveness of the metal fasteners used in wood construction can be significantly decreased because of corrosion. This is a complex problem, and many tests are used to evaluate these interactions. [10, 22, 24, 28, 39, 50, 53]

As a result of the voluntary withdrawal of chromated copper arsenate (CCA) for residential applications, many designers are now using alternatives to CCA such as alkaline copper quaternary (ACQ) or alkaline copper azole (CuAz). Because of the lack of chromium and the increased copper in their formulations, these new preservatives are potentially more corrosive than CCA. The current standard method for measuring corrosion of metals in contact with treated wood involves placing wood in contact with metals at high temperatures and humidity. However, it has been shown that this method has poor correlation to in-service performance. Corrosivity of treated wood needs to be quantified for proper design and operation of residential and nonresidential wood structures to prevent failures. [47, 51, 52, 56–59]

◀ Joist hanger in boardwalk one month after installation. (2000s)

▲ Corrosion of galvanized bridge hardware in contact with amine-copper-treated lumber after three years of service. (2000s)

Zinc Plated

◀ Comparative corrosion of metal fasteners in wood treated with a copper-based preservative in service for one year. (2000s)

Hot Dipped

Stainless

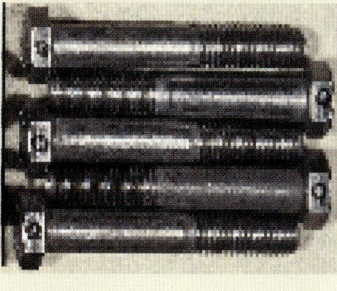

◀ Laboratory test set-up for determining the reaction of metal fasteners to exposure to various chemicals. (2000s)

1910

Limited design information was available.

2010

Connectors developed and test methods developed to evaluate new wood–connector designs. Information found in the *Wood Handbook*.

Further Information

[1] Nails and nailing / J.B. Frear / FPL Report B 34672, (1919)

[2] Tests and suggestions for the nailing of boxes / J.A. Newlin / FPL No. 488 (1925)

[3] The bearing strength of wood under bolts / George W. Trayer / USDA, Forest Service technical bulletin No. 332 (1932)

[4] Modern connectors for timber construction / Report prepared jointly by the National Committee on Wood Utilization, U.S. Dept. of Commerce and FPL / (Twenty-fourth report of committee's series) Washington: U.S. Government Printing Office (1933)

[5] Good practice in nailing wood / L.J. Markwardt, J.M. Gahagan / National safety news / FPL No. 1160 (March 1935)

[6] Lag-screw joints: their behavior and design / J.A. Newlin, J.M. Gahagan / USDA, Forest Service technical bulletin, No. 597 (1938)

[7] Some FPL developments of particular significance to the South / FPL 0-8 W73Sf, (February 28, 1940)

[8] Timber-connector joints, their strength and design / John A. Scholten / USDA, Forest Service technical bulletin, No. 865 (1944)

[9] How to nail a house / Otto C. Heyer, T.R.C. Wilson / In collaboration with the technical staff of the U.S. Housing and Home Finance Agency. Washington, DC: (Also ref. to as: Technique of house nailing) (1947)

[10] Corrosion of metal fastenings in zinc chloride-treated-wood after 20 years / R.H. Baechler / American Wood-Preservers' Association. Proceedings of the 45th annual meeting, Washington, DC: AWPA, (1949)

[11] The effects of fire on selected structural timber joints / Erwin L. Schaffer / Thesis (M.S.) University of Wisconsin, (1961)

[12] Design with nails, staples, and screws / John A. Scholten / Forest Products Laboratory (1963)

[13] Performance of nail-glued joints of plywood to solid wood / D.V. Doyle, / Research note FPL 042 (1964)

[14] Performance of joints with eight bolts in laminated Douglas-fir / D.V. Doyles / Research paper FPL 10 (1964)

[15] Performance of container fasteners subjected to static and dynamic withdrawal / R.S. Kurtenacker / Research paper FPL 29 (1965)

[16] Strength of wood joints made with nails, staples, or screws / J.A. Scholten / Research note FPL 0100 (1965)

[17] Nail-withdrawal resistance of American woods / FPL Research note FPL 093 (1965)

[18] Load distribution in multiple-bolt tension joints / Calvin O. Cramer / Journal of the Structural Division, ASCE Vol.94, No.ST5, Proc. Paper 5939 (May 1968)

[19] Use of mechanical fasteners to provide interaction in building constructions / E.W. Kuenzi. Proceedings of the Symposium on Research to Improve Design of Light-Frame Structures, February 25–26, 1970, U.S. Forest Products Laboratory, Madison, Wisconsin (1970)

[20] Bearing strength of wood under embedment loading of fasteners / T.L. Wilkinson / Research paper FPL 163 (1971)

[21] Effect of deformed shanks, prebored lead holes, and grain orientation on the elastic bearing constant for laterally loaded nail joints / T.L. Wilkinson / Research paper FPL 192 (1972)

[22] Degradation of wood by products of metal corrosion / A.J. Baker / Research paper FPL 229 (1974)

[23] Fasteners for timber construction in high wind areas / G.E. Sherwood, T.L. Wilkinson / Building to resist the effect of wind: Vol. 3, a guide for improved masonry and timber connections in buildings / S. George Fattal, G.E. Sherwood, T.L. Wilkinson. Washington, DC: National Bureau of Standards NBS building science series 100; Vol. 3 (1977)

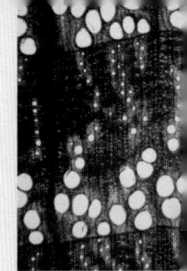

[24] Corrosion of metal in wood products / A.J. Baker / Durability of building materials and components / edited by P.J. Sereda, G.G. Litvan. Philadelphia: American Society for Testing and Materials, 1980 ASTM Special technical publication 691 (1980)

[25] Assessment of modification factors for a row of bolts or timber connectors / Thomas Lee Wilkinson / Research paper FPL 376 (1980)

[26] Load distribution among bolts parallel-to-load / Journal of structural engineering Vol. 112(4) (Apr 1986)

[27] Bolted-connection design / Lawrence A. Soltis, Thomas Lee Wilkinson / FPL–GTR–54 (1987)

[28] Corrosion of metals in preservative-treated wood / Andrew J. Baker / Wood protection techniques and the use of treated wood in construction. Madison, Wis.: Forest Products Research Society (1988)

[29] Duration-of-load effect on toothed metal-plate connections / Lawrence A. Soltis, Rodney M. Shea / Proceedings of the 1988 International Conference on Timber Engineering, Westin Hotel, Seattle, Washington, U.S.A., September 19-22, 1988, Volume 2. Madison, Wis. Forest Products Research Society (1988)

[30] Metal-plate connections loaded in combined bending and tension / Ronald W. Wolfe / Forest products journal Vol. 40(9) (Sept. 1990)

[31] Design equation for multiple-fastener wood connections / John J. Zahn / Journal of structural engineering Vol. 117(11) (1991)

[32] United States adaptation of European Yield Model to large-diameter dowel fastener specification / Lawrence A. Soltis, Thomas L. Wilkinson / Proceedings of the 1991 International timber engineering conference, September 2–5, London, Vol. 3 London: TRADA, (1991)

[33] European Yield Model for wood connections / Lawrence A. Soltis / Structures Congress '91 compact papers: ninth structures congress. New York: American Society of Civil Engineers (1991)

[34] Test apparatus for simulating interactive loads on metal plate wood connections / Ronald W. Wolfe, Michael Hall, DeAndrea Lyles / Journal of testing and evaluation. Vol. 19(6) (Nov. 1991)

[35] Stiffness and strength properties of shear transfer plate connections / Ronald Wolfe, David Bohnhoff, Robert Nagel / FPL–RP–517 (1993)

[36] Bolted connection design values based on European Yield Model / T.L. Wilkinson / Journal of structural engineering Vol. 119(7) (July 1993)

[37] Bolted connection strength and bolt hole size / T.L. Wilkinson / FPL–RP–524 (1993)

[38] Fasteners for exposed wood structures / Robert H. Falk, Andrew J. Baker / Wood design focus Vol. 4(3) (Fall 1993)

[39] Corrosion of metal fasteners in wood / Andrew J. Baker / Technical aspects of maintaining, repairing, and preserving historically significant ships: proceedings of a conference, September 12–14, 1994, Charlestown, MA. Washington, DC : Naval Historical Center, (1994)

[40] Bolts, drift bolts, and pins / Lawrence A. Soltis, T.L. Wilkinson / In: Mechanical connections in wood structures. Chap. 4. ASCE manuals and reports on engineering practice No. 84. New York, NY: American Society of Civil Engineers (1997)

[41] Multiple-bolted joints in wood members--A Literature Review / Peter James Moss / FPL–GTR–97 (1997)

[42] Withdrawal and lateral strength of threaded nails / Douglas R. Rammer, Donald A. Bender, David G. Pollock / 5th world conference on timber engineering, August 17–20, 1998, Montreaux, Switzerland: proceedings, Volume 2. Lausanne: Presses Polytechniques et Universitaires Romandes, c1998. 2: 238–245 (1998)

[43] Wood handbook: wood as an engineering material / FPL–GTR–113 (1999)

[44] Effect of moisture content on nail bearing strength / Douglas R. Rammer / FPL–RP–591 (2001)

[45] Withdrawal strength of threaded nails / Douglas R. Rammer, Steve G. Winistorfer, Donald A. Bender / Journal of structural engineering Vol. 127(4):442–449 (2001)

[46] Wood: mechanical fasteners / D.R. Rammer / Encyclopedia of materials: science and technology. Amsterdam; New York: Elsevier (2001)

[47] Review of test methods used to determine the corrosion rate of metals in contact with treated wood / Samuel L. Zelinka, Douglas R. Rammer / FPL–GTR–156 (2004)

[48] Timber rivet connections in U.S. domestic species / Marshall Begel, Ronald W. Wolfe, Douglas Stahl / FPL–RP–619 (2004)

[49] Timber rivets in structural composite lumber / Ronald W. Wolfe, Marshall Begel, Bruce Craig / FPL–GTR–153 (2004)

[50] The use of electrochemical impedance spectroscopy (EIS) to measure the corrosion of metals in contact with wood / Samuel L. Zelinka, Douglas R. Rammer / TMS letters Vol. 2(1) (2005)

[51] Review of test methods used to determine the corrosion rate of metals in contact with treated wood / Samuel L. Zelinka, Douglas R. Rammer / FPL–GTR–156 (2005)

[52] Corrosion avoidance with new wood preservatives / Samuel L. Zelinka, Douglas R. Rammer / Wood design focus Vol.16(2) (summer 2006)

[53] Fastener corrosion: testing, research, and design considerations / Douglas R. Rammer, Samuel L. Zelinka, Philip Line / WCTE 2006: 9th World Conference on Timber Engineering, August 6–10, 2006, Portland, OR, USA: program and abstracts. Portland, OR: World Conference on Timber Engineering (2006)

[54] Wood construction codes issues in the United States / Douglas R. Rammer / Biohousing Symposium. Session 1, Eco-material & structural engineering (2006)

[55] Development of failure mechanisms for fasteners in the United States / Douglas R. Rammer, Philip Line / WCTE 2006 9th world conference on timber engineering, August 6–10, 2006, Portland, OR, USA: program and abstracts. Portland, OR: World conference on timber engineering, (2006)

[56] Corrosion avoidance with new wood preservatives / Samuel L. Zelinka, Douglas R. Rammer / Wood design focus Vol. 16(2) (summer 2006)

[57] Exposure test of fasteners in preservative-treated wood / Samuel L. Zelinka, Douglas R. Rammer / WCTE 2006: 9th world conference on timber engineering, August 6–10, 2006, Portland, OR, USA: program and abstracts. Portland, OR: World conference on timber engineering (2006)

[58] Direct current testing to measure corrosiveness of wood preservatives / Samuel L. Zelinka, Douglas R. Rammer, James T. Gilbertson / Corrosion science. Vol. 49 (2007)

[59] Uncertainties in corrosion rate measurements of fasteners exposed to treated wood at 100% relative humidity / Samuel L. Zelinka / Journal of testing and evaluation. Vol. 35(1) (2007)

[60] Withdrawal strength of bright and galvanized annulary threaded nails / Douglas R. Rammer, Alina M. Mendez / Frame building news (Apr. 2008)

Trusses

◀ Wood-trussed rafters undergoing 15-year loading test to determine the effect of aging. Note the trussed rafters used in the building. (1960s) {M 134 435-11} [11]

Early wood houses were constructed using various sizes of lumber one stick at a time. If one desired large open rooms, larger and larger pieces of lumber or timbers were needed to span the desired space. Large roofs also required large pieces of lumber. Thus, buildings were limited in size and open spaces. With a knowledge of wood properties, connectors, and engineering design, industry assembled smaller boards into trusses and significantly increased the options for building structures from wood. [7, 12, 13, 15]

This knowledge also allows for the use of previously unused wood species. [18]

Challenge: How to effectively use small dimension lumber to support roofs or floors over long spans. In 1910, spanning a long distance required cutting a large tree and sawing out a solid wood beam to span that distance. Many barns were built in this manner. As the need for these structures began to exceed the availability of large trees, it became necessary to design wood trusses to replace the large solid wood beams.

FPL's Contributions: Working with industry, FPL designed, built, and evaluated different trusses to determine their strength and stiffness, with particular emphasis on long-term loading effects. This research assisted in establishing standards for truss construction.

Results: It is now possible to build a complete house using trusses that are made from lumber no larger than a 2 by 4.

Nailed and Glued Trusses

FPL researchers established the adequacy of nailed and glued roof trusses for house construction. Such trusses are widely used because they permit roof construction independent of bearing partitions and provide for simpler and more economical installation of floors and interior finish. [1]

▶ Construction of a truss for evaluation. (1960s) {M 129 334}

Double Chord Truss

FPL researchers developed a unique roof truss employing double chords and single struts and gussets. [2, 3]

► Thomas Wilkinson checking a truss that is undergoing evaluation. (1960s) {M 129 567}

Factors Affecting Lumber Use by Truss Fabricators

FPL revealed a number of obstacles to increased use of trusses—builder and architect prejudices, raw material procurement practices, warp and variation with lumber grades, and builders' lack of knowledge of trusses. FPL also showed that substantial improvements can be made in matching species and grades of lumber with truss applications, which have been used by several large lumber manufacturers. [4–6, 8, 9]

Computerized Wood Engineering System

Dr. Stanley K. Suddarth, Purdue University, developed the first computer model for the analysis of wood structures. Known as the Purdue Plane Structures Analyzer (PPSA), it used recognized engineering analysis methods to determine stresses and deflections of wood trusses and frames. Dr. Suddarth cooperated with Ron Wolfe and other FPL engineers to modify the program to be consistent with recommendations in the 1982 *National Design Specifications for Wood Construction*. The PPSA program laid the foundation for development of computerized design of structural wood assemblies. Its influence has added to the cost competitiveness and engineering efficiency of structural wood components. (10)

Assessment of Truss Plate Performance

One major factor in truss performance is the connectors used to join the wood members together. Because there are many designs of wood connectors, an assessment of performance was made to provide information for use in truss design models. [14, 16]

Small-Diameter Timber Trusses

With the need to find new ways to utilize overstocked, small-diameter trees, the FPL developed ways to successfully use them in trusses.

◄ Example of using small-diameter timbers, connectors, and engineering design to make trusses. (2000s) [17]

1910

Limited use in the United States.

2010

Through design, advanced lumber grading technology, and efficient connectors, the use of trusses has increased significantly.

Further Information

[1] Glued and nailed roof trusses for house construction / R.F. Luxford, Otto C. Heyer / FPL No. 1992 (1954)

[2] Developing an improved system of wood-frame house construction (Nu-frame) / L.O. Anderson / Presented at Illinois small homes council short course in residential construction, January 28, 1966, Urbana, Ill (1966)

[3] Construction of Nu-frame research house: utilizing new wood-frame system / L.O. Anderson / FPL–RP–88 (1968)

[4] Lumber truss design: views of designers and regulatory agencies / William L. Galligan, Edwin Kallio, Isaac Sheppard / Forest products journal Vol. 27(12) (Dec. 1977)

[5] A study of truss fabrication, truss materials / William L. Galligan, Edwin Kallio / Professional builder Vol. 42(9) (Aug. 1977)

[6] Factors affecting the use of lumber by truss fabricators in the United States / E. Kallio / Forest products journal Vol. 28(3) (1978)

[7] The truss-framed system for residential and light commercial buildings / Roger L. Tuomi / Southern lumberman. Vol. 241(2991) (Dec. 15, 1980)

[8] Characterizing lumber properties for truss research / Robert J. Hoyle, Jr., William L. Galligan, James H. Haskell / Metal plate wood truss conference. Madison, WI: Forest Products Research Society proceedings No. P–79–28 (1981)

[9] Species, grades, and mechanical properties of lumber sampled from truss fabricators / Charles C. Gerhards, Donald H. Percival / Metal plate wood truss conference. Madison, WI: Forest Products Research Society proceedings No. P–79–28 (1981)

[10] Purdue Plane Structures Analyzer II: a computerized wood engineering system / Stanley K. Suddarth and Ronald W. Wolfe / FPL–GTR–40 (1984)

[11] Longtime performance of trussed rafters with different connection systems / Thomas Lee Wilkinson / FPL–RP–444 (1984)

[12] Influence of changes in allowable stresses on wood truss design / W.L. Galligan, P.W. McClellan, F.E. Woeste / Forest products journal Vol. 35(5) (May 1985)

[13] Strength and stiffness of light-frame sloped trusses / Ronald W. Wolfe, Donald H. Percival, Russell C. Moody / FPL–RP–471 (1986)

[14] Assessment of truss plate performance model applied to southern pine truss joints / M. McCarthy, R.W. Wolfe / FPL–RP–483 (1987)

[15] Wood-frame house construction / Gerald E. Sherwood, Robert C. Stroh / USDA Agriculture handbook No. 73 (1989)

[16] Tensile loading characteristics of truss plate joints after weathering and accelerated aging / Robert H. McAlister / Forest products journal Vol. 40(2) (Feb. 1990)

[17] Trussed assemblies from small-diameter round timbers / Ronald W. Wolfe, Roland Hernandez / Forest research bulletin No. 212. (1999)

[18] Structural use of red maple / Forest Products Laboratory / Techline: properties and use of wood, composites, and fiber products VI–16 (2001)

Adhesives

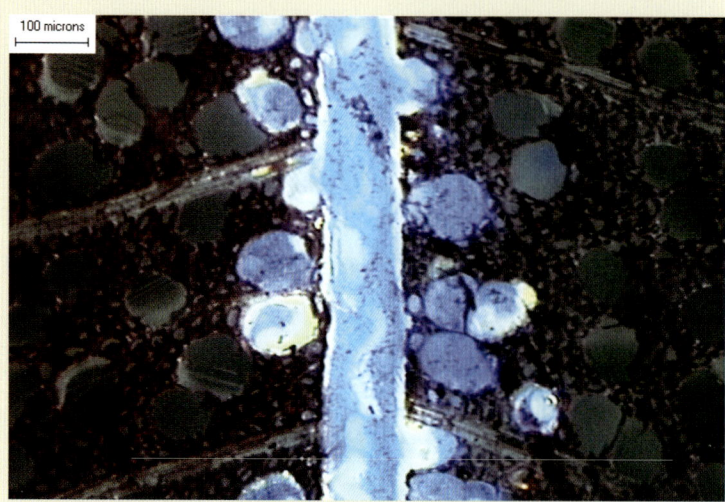

◄ Photomicrograph of an adhesive bond of two pieces of wood. The blue areas show the adhesive penetration into the wood structure. (2000s)

Wood adhesive bonding goes back to early recorded history and has become of increasing importance to the wood products industry. Ancient Egyptians bonded wood veneer. The use of adhesives to bond wood into furniture and other useful products continues to the present day. In the past 100 years, improvements in adhesives, and particularly the development of synthetic resin adhesives, have immensely expanded the types of glued wood products that can be produced and the range of service environments in which those products are durable.

Thomas Truax, an early pioneer in adhesives research, wrote in 1940 that "The use of glue in the fabrication of wood products brings about more complete utilization of timber through the use of lower grades, inferior species, and small sizes of material; it conserves supplies of clear material and of the scarcer and more valuable wood; and it makes possible a saving of material in the production of articles of unusual form, dimensions, and properties." [27]

These changes have resulted in great reductions in the amount of wood wasted from stump to consumer, and led to new uses for wood to serve our needs while conserving forest resources.

Until the mid-20th century these adhesives were biobased, being produced from plants and animals. After World War II, synthetic adhesives were pioneered using products derived from the large petroleum industry built to sustain the war effort. Synthetic adhesives have led to the production of a large variety of strong, durable wood products, including panel products (such as plywood, particleboard, and oriented strandboard) and structural wood products (such as glued-laminated beams, strand lumber, and wooden I-joists) important to the housing industry. None of these modern building materials could exist without suitable adhesives, which make up only about 1% to 6% of their weight.

FPL's Contributions: The FPL has played an integral role in developing the technical understanding of adhesives and the setting of product and performance standards by organizations such as the ASTM International (formerly American Society for Testing and Materials), American Institute of Timber Construction (AITC), APA–The Engineered Wood Association (APA), and the American Forest and Paper Association (AF&PA). Suitability of adhesives for housing, other buildings, timber bridges, and other structures has always been important. During World Wars I and II, there was emphasis on wooden airplanes and boats (mine sweepers and PT boats). The knowledge of bonding led FPL researchers to develop a method referred to as "FPL etch" for forming a strong aluminum-to-aluminum bond used for aircraft production. This "FPL etch" was essential to the construction of aircraft after World War II and is still used for some commercial aircraft construction and for most academic studies. This research was an outgrowth of epoxy bonding of wood. The FPL has often been a leader in testing new adhesive systems.

Results: Many of the wood adhesive standard test methods are an outgrowth of FPL's fundamental research program. The FPL has long been a strong contributor to the setting of performance standards because it is an unbiased source of information. (See list of ASTM Standards.) The FPL continues to work with industry to develop new adhesive systems, from early-on development of more water-resistant casein glues, to petroleum-based adhesives (phenolics, epoxies, and polyvinyl acetates), to the resurgence of biomass-based adhesives, including soybean adhesives.

The Forest Products Laboratory has played, and continues to play, a major role in these developments. The FPL's contributions to extending knowledge fall into five main categories: (1) adhesive properties, (2) adhesive techniques, (3) adhesive–wood product development, (4) test method and standards development, and (5) technology transfer.

Adhesive Properties

Natural Adhesives

The adhesive choices for gluing wood in 1910 were animal glue, vegetable starch, vegetable protein, casein, blood albumin, and liquid glues, all derived from natural resources. None are waterproof. The impetus for adhesive research at FPL was World War I and the need for water-resistant adhesives for aircraft wood structural frameworks (wings and fuselage), plywood skins, and aircraft propellers. Under request from the military to assist with aircraft production, the FPL surveyed adhesive manufacturers for information on water-resistant adhesives. The respondents' usual candidates were either blood albumin or casein. Two manufacturers did mention phenolic resin as a waterproof adhesive but stated that it was too expensive for commercial use. [1]

Nevertheless, FPL researchers bonded birch joints with phenol-formaldehyde resin in simulated propeller blanks, stored the blanks in various environments for some time, and then in 1919 cut and tested specimens for average shear strength and wood failure. These were the only recorded tests of synthetic resin glues until the 1930s. [3]

The first glue development research at the FPL in 1917 was to improve the water resistance of the best glues available—animal glue, casein, and blood albumin—for the manufacture of WWI aircraft components. [7, 8, 14, 16]

The improved adhesives generally used paraformaldehdye to react with the protein in the adhesives to improve resistance to redissolving in water and to provide some measure of mold resistance. These improved adhesives were patented for public use in the 1920s (U.S. Patent Nos. 1,456,842; 1,459,541; and 1,712,077) [7, 8, 17] and formed the basis for the glued wood products industry during the 1920s and 1930s. [5, 10, 16, 19, 23, 24]

An example of the type of information about glue joints discovered during WWI is the relationship between shear strength and the density of the wood.

▶ Early graph of relationship between shear strength and specific gravity (an expression of density) for 31 species of wood. (1930) {M 15547 F} [18]

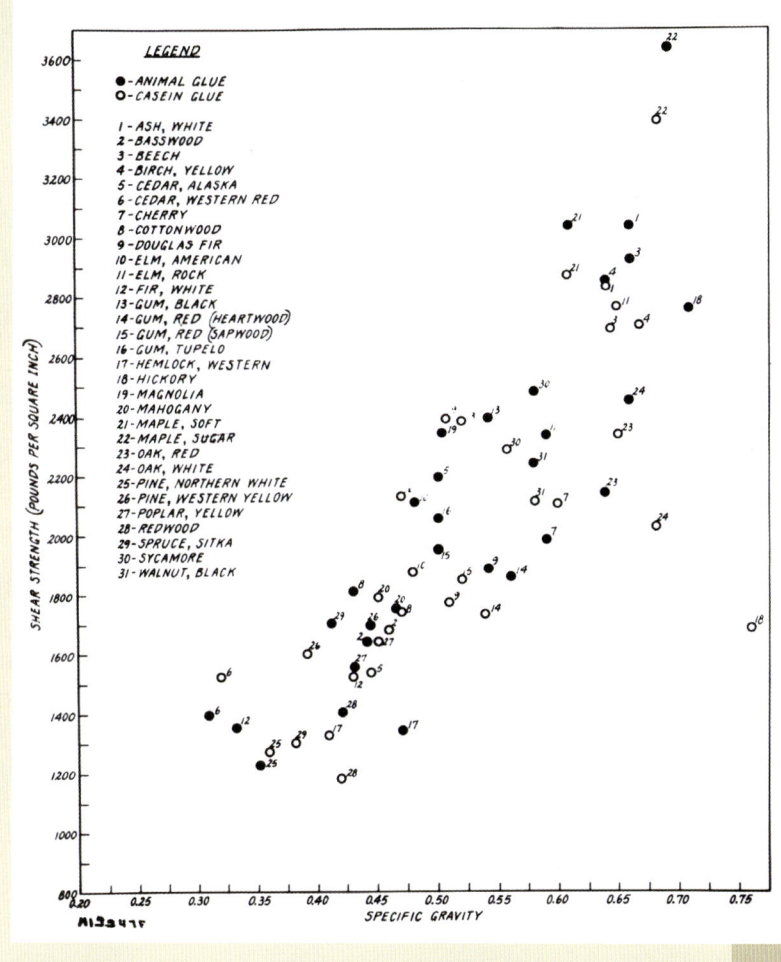

Synthetic Resin Adhesives

Commercial synthetic resin glues began to arrive in the early 1930s. All these synthetic resin adhesives could provide bonds more resistant to water than any of the natural adhesives, and most are waterproof. Phenol- and melamine-formaldehyde both require hot pressing. Urea-formaldehyde (UF) could be cured at room temperature or by hot pressing. Researchers found that bonds of melamine- and phenol-formaldehyde could withstand extended soaking, soaking and drying, or even boiling in water. On the other hand, urea-formaldehyde adhesive degrades rapidly in hot moist conditions. [36]

Some adhesives, when properly cured, are as durable as the wood. [68]

A fourth member of this class of thermosetting synthetic resin glues is resorcinol-formaldehyde. This is a much more reactive combination capable of curing at room temperature. Its commercialization had to wait until procedures were developed to form a stable resin that could be handled in a commercial gluing operation. The Laboratory tested and evaluated dozens of commercial synthetic resin adhesives and provided unbiased information about their capabilities. [56, 77]

Epoxy Adhesives

FPL research on epoxy adhesives for bonding wood to metal [48] for the military revealed that these adhesives had attractive properties for gluing wood, such as room-temperature cure and gap-filling ability. Research was undertaken to formulate epoxy adhesives specifically for wood. One of the formulations, FPL 16 [66], became a favorite with home builders of wooden aircraft and is still in use today (Hughes Glue). The Laboratory also tested many commercial adhesives for bonding metal and undertook the development of improved adhesives for metal-to-metal bonding for use in high performance military aircraft for the Navy. [52] A critical property for bonded metal joints in the maritime environment is resistance to saltwater corrosion.

◄ Bonded test specimens in a salt-spray test chamber ready to test for resistance of the adhesive bond to corrosion

Polyvinyl Resin and Elastomer-Based Adhesives

FPL began to evaluate polyvinyl resin adhesives during WWII as a substitute for animal glue, which was in short supply. This new type of adhesive had working properties much like animal glue and was found to be well suited to furniture manufacture and wood moulding. [41]

Elastomer-based adhesives were developed in the 1970s as adhesives with gap-filling ability that could be used to bond sheathing to lumber framing members under a wide range of outdoor conditions for building construction. Scientists at FPL evaluated the durability and mechanical properties of these adhesives to advise builders and government agencies. [86, 95, 102, 108]

◄ Applying adhesive to floor trusses to provide a rigid system. (1970s) {121 D}

Improved Urea-Formaldehyde Adhesive

Urea resin adhesives are waterproof in the short term but break down over time, especially in moist, warm exposures. They also emit formaldehyde, a health hazard, with the amount and rate of emission depending on the way the adhesives are made and cured as well as the warm, moist environment. Urea-formaldehyde (UF) adhesives are also brittle, which reduces their resistance to swelling and shrinking stresses. FPL researchers and colleagues developed improved UF adhesives, incorporating flexible polyamines in the chemical structure of the resin that increased hydrolysis resistance and improved swelling and shrinking resistance. The improvements greatly decreased formaldehyde emission in aging tests and increased durability in swelling and shrinking tests. [126, 134, 135]

Adhesives Derived from Renewable Resources

Synthetic resin adhesives, such as phenol-formaldehyde, dominated the glued wood products industry from about 1940 onward. However, in the 1970s, oil shortages with consequent resin shortages and rising adhesives costs meant alternatives to petroleum-based adhesives were needed. FPL researchers and their colleagues have been leaders in the effort to develop viable adhesives from renewable resources that have durability equal to the synthetic resins, while reducing dependence on petroleum and natural gas. The goal has been to make adhesives from forest or agricultural biomass (including carbohydrates, kraft lignin, tannin, furfuryl and soy protein) and by substitution of plant phenolics for some, or all, of the phenol requirement in an adhesive. [109, 113, 116, 117, 119, 125, 131, 139, 140, 144, 155, 156, 158, 164, 165]

◀ The white powder on the left is soy flour used to make soy-based adhesive (in the beaker), developed in cooperation with industry as a lower cost wood adhesive that is water resistant. Soy flour was once used in interior plywood but was displaced due to performance, cost, and ease-of-use issues. Although not completely eliminating the use of petroleum-based phenol-formaldehyde, the technology uses soybean flour as the main component in the adhesive. (2000s) [164–167]

Adhesive Preservatives

FPL scientists determined the effectiveness of preservatives for soybean and casein glues in increasing the resistance of plywood and other products to delamination by molds. The results of this work served as a basis for the preservative requirement for protein glues. [33]

Adhesive Techniques

Most adhesives must exist in two states of matter. First, they need to be liquid and have the right physical properties for forming a molecular-level contact with the substrate. Once the adhesive has flowed over and into the surface, the second step requires a change of state from liquid to solid that may be due to cooling, drying, or chemical cross-linking. The resulting bond needs to be strong enough to resist forces that are applied to the bond. Of course, the adhesive also needs to be economical and withstand most use conditions, which can include heat and moisture exposure.

▲ James Wescott mixing a new variation of soy-flour-based wood adhesive with no added formaldehyde, as part of FPL and industry cooperative research. (2000s)

Natural Adhesives

In 1910, gluing was an art, not a science. The first task for FPL researchers in 1917 was to scientifically determine the best methods for preparing and applying the best glues they could find to form the strongest and most durable bonds for use in aircraft. The candidate adhesives included animal, casein, and blood albumin glues. Variables that affect adhesive performance include glue preparation, allowable glue working life, method of glue application, allowable assembly time, glue spread rate, wood surface preparation, wood moisture content, allowable open and closed assembly times, pressure, temperature of cure, clamp time, cure time, and conditioning after bonding. [2, 9, 11, 20, 21, 37, 53, 55, 57, 60]

The research findings were collated in two publications: "The gluing of wood," Technical Bulletin No. 1500 [27], and "The gluing of wood in aircraft," Technical Bulletin No. 205. [18]

◀ Early hand screw press used to bond test specimens for measuring effects of pressure on bond quality. (1940s)

FPL researchers were principals in the early debate on mechanisms of adhesion between wood and glue, stating "that specific (chemical/physical) adhesion is essential for satisfactory gluing, though mechanical adhesion is also present." [13] Understanding adhesion in general has broadened immensely into many mechanisms and many fields of interest, but the subject is still an important area of research at the FPL today after more than 80 years. [133, 159, 163, 168, 169]

Synthetic Resin Adhesives

In the early 1930s, the Laboratory began to develop a gluing technique for the new synthetic resin adhesives (urea-, melamine-, and phenol-formaldehyde), primarily in hot-press bonding of plywood. [25]

When resorcinolic adhesives became available in the early 1940s, the FPL began to evaluate the adhesive and develop a gluing technique for laminating lumber. [44]

With the advent of the World War II, the FPL was again called upon to support the military (particularly the U.S. Navy) to develop a gluing technique for glued-laminated timbers for ship construction and other marine and outdoor applications. [47, 73, 76, 79]

Wood-to-Metal and Metal Bonding

During WWII and the Korean War, FPL was requested by the military to develop a glue technique for bonding metal to itself and other materials for military vehicles and other uses. Primarily, this involved developing methods to prepare the metal surface to be receptive to the adhesive. One development is of particular note—the FPL Etch. [46]

The FPL Etch

Two major requirements for strong durable joints are a chemically clean (easily wetted) and mechanically sound surface. With wood, the best (strongest and most durable) joints are generally obtained by freshly knife-planing the wood just before gluing.

With metal surfaces, which quickly oxidize after machining and are likely to be coated with machining oil, it is necessary to solvent-wash and chemically clean the metal surface to obtain the best bonds. During work at the FPL with metal bonding for aircraft applications, a superior method was developed for preparing aluminum for bonding. It became widely used and acquired the title the "FPL Etch." [46]

The method worked so well it became standardized (ASTM Test Method 2674). It was the standard the aerospace industry grew up on. The inspiration for the FPL Etch was the chemist's formula for cleaning lab glassware. [120] (See also the National Defense section.)

Wood Surface Quality for Gluing

The surface characteristics of wood pieces to be bonded are very important. At the turn of the century, roughening the surface of a piece of wood to be glued by scratching, tooth planing, or some other means was thought to improve the bond by exposing more surface to the adhesive. However, FPL research showed the amount of wood surface exposed by roughening was small compared to the amount accessible by penetration of the cell cavities on a cleanly planed surface. [6, 13, 27]

Too much roughening actually can block the adhesive from reaching sound wood. On the other hand, light sanding does actually improve adhesive "wetting" of wood surfaces of veneer that have been dried or hot-pressed at high temperature, or simply stored in the open air for a long time. [72]

Al Christiansen conducted an extensive, critical literature review of the research on surface inactivation from high temperatures, grouping the causes into physical and chemical effects. He found surface inactivation does not arise from a single cause and varies by species. [118, 121]

► A water drop test shows differences in wettability of a yellow birch veneer surface: left, aged and unsanded surface; middle, surface renewed by two passes of 320-grit sandpaper; right, surface renewed by four passes of sandpaper. (1984) {M84-0270-15}

Abrasive planing machinery, developed in the 1970s (a process, different from sanding by planing with a hard roller backing a coarse abrasive), was found by FPL researchers to cause significant surface and subsurface cell damage that may reduce joint strength, and definitely reduces durability of the joint in environments inducing cyclic swelling and shrinking stresses. The crushed wood cells caused by abrasive grit in abrasive planing lead to poor bond durability in swelling and shrinking environments. [87, 91, 106]

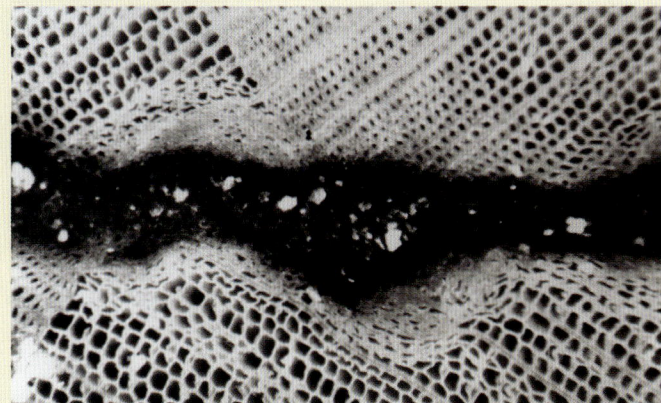

◄ Douglas-fir abrasively planed with 36-grit abrasive showing crushing of the cells and consequent poor durability of the adhesive bond in swelling and shrinking environments. (1980s) {M 151111}

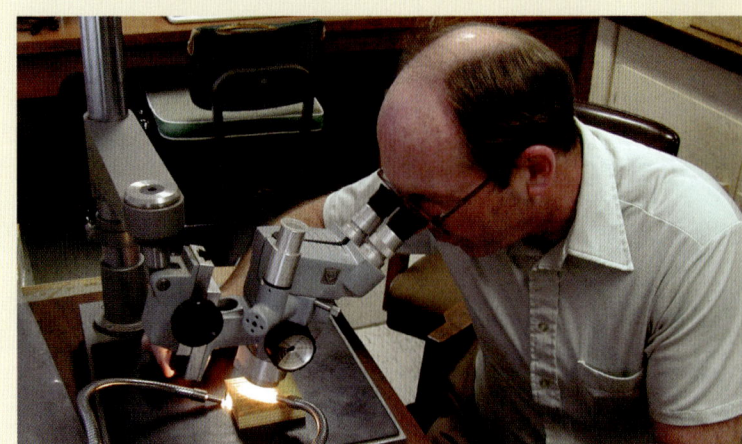

◀ Chuck Frihart studying surface characteristics of wood to better understand the interactions between the wood and adhesive interface. (2000s)

Hydroxymethylated Resorcinol (HMR) Coupling Agent

FPL experimented with various modifications of epoxies to improve performance ,which resulted in a particular formulation called FPL 16. [66]

In spite of FPL 16's superior performance with dry wood and its gap-filling capabilities, epoxy bonds in general, including FPL 16, do not provide durability on a par with resorcinol-formaldehyde in wet service environments. Failure in exposure tests generally occur in what appears to be a weak layer, or interphase, especially between the adhesive and the dense latewood.

This prompted FPL researchers to look for a primer that might improve the properties of the interphase. They developed a surface primer called hydroxymethylated resorcinol (HMR). [138]

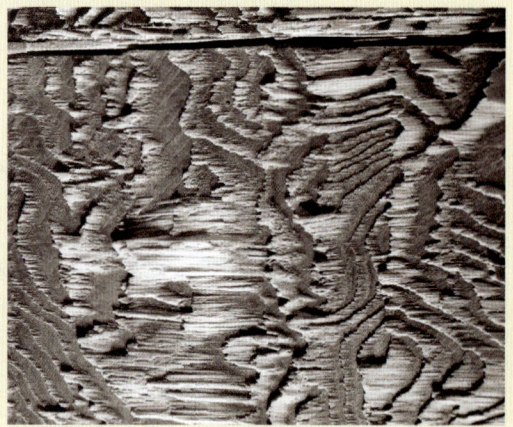

Subsequent research has shown that this surface treatment improves the wet strength and durability of not only epoxy-bonded joints, but also one-part moisture-curing polyurethane bonded joints. HMR also improves the bond of difficult-to-glue woods, such as very dense, high extractive-content, and treated woods bonded with conventional synthetic resin adhesives. [143, 145, 149, 153]

▲ Pattern of wood failure in the earlywood and adhesive failure in the latewood areas caused by ineffective bonding of epoxy resin adhesive to the latewood. (1980s) {M 150265}

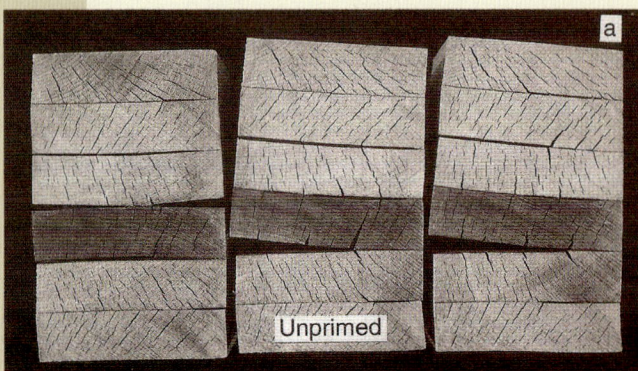

Unprimed

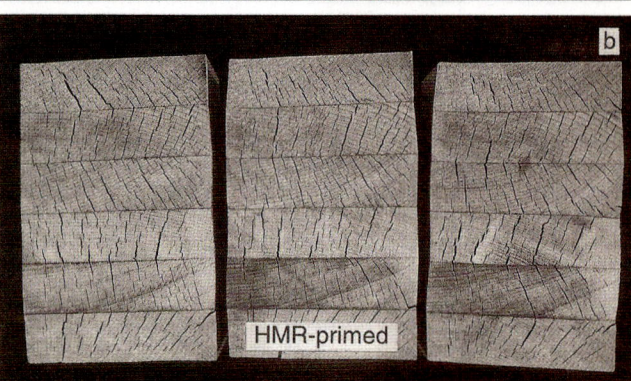

HMR-primed

◀ Upper photo (a) shows laminated beams that were bonded with an epoxy adhesive and then subjected to severe exposure conditions (ASTM D 2559 durability testing). Lower picture (b) shows laminated beams that were made with lumber pretreated (primed) with hydroxymethylated resorcinol (HMR) prior to bonding with epoxy adhesive. The use of HMR resulted in excellent bonding and little or no delamination in severe exposure conditions. (1990s)

▲ Linda Lorenz showing accelerated durability test specimen treated with hydroxymethylated resorcinol (HMR)

Steam Injection Pressing

Robert Geimer developed the process of injecting steam into a mat of particles during pressing to increase the rate of heat transfer and thus speed curing of phenolic adhesives, especially within the core of thick panels (U.S. Patent No. 4,393,019, July 12, 1983). The process had potential to improve the economics of manufacturing thick flakeboard panels. [104]

In conventional adhesive-bonded particle products, such as oriented strandboard, resin droplets form point bonds (filets) between the wood particles. A significant problem posed by steam injection was migration of adhesive droplets away from the flake surface, resulting in reduced bond strength. Adhesive scientists at FPL developed an understanding of the complex interactions between adhesive properties and heat and moisture regimes during pressing to help overcome this problem and help steam injection pressing become a commercial process. [123, 132, 142]

Wood Fiber–Plastic Composites

Waste wood fiber is the major constituent of municipal solid waste. FPL researchers had the idea that the combination of wood fiber (such as newspaper) and thermoplastic polymer (such as milk bottles) could be turned into useful thermoformable products such as decking and automobile door and head moldings. High-volume products like these could remove significant amounts of wood fiber and plastic from the waste stream. Achieving the strength necessary for useful proucts requires that the fiber and the plastic must form an adhesive bond between them. [124]

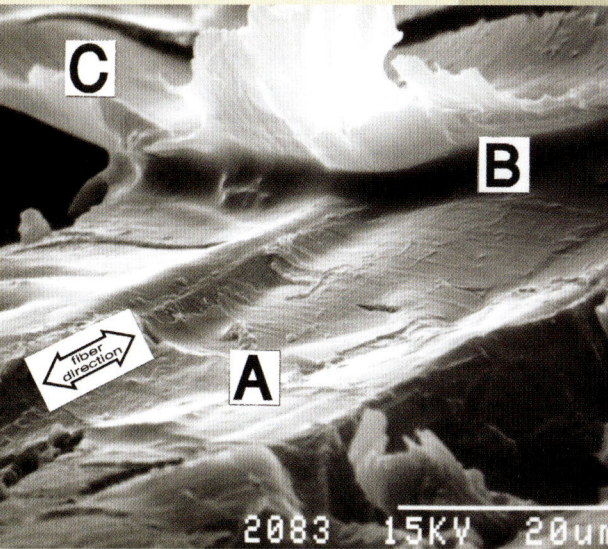

▶ Scanning electron micrograph of a filet of adhesive (B) between two particles or flakes of wood (A and C) in a flakeboard. (1993) {M 930031}

Unfortunately, the two materials are not adhesively compatible. FPL adhesive scientists and their colleagues contributed to the development of commercial wood fiber–plastic products by investigating the mechanism of adhesion between plastic and wood and quantifying the ability of certain coupling agents to improve the adhesion. [128, 133]

Adhesive Wood Product Development

Aircraft

Glued wood aircraft and glued wood propellers were the first glued wood products that the FPL had a hand in developing. The intricacy of a wooden wing frame, with glued wood box spars and wood ribs, contains approximately 1,000 glued joints.

◀ A complicated wing made of wood (with the exception of the aileron), including two wing spars and numerous ribs, containing more than 1,000 bonded joints. USDA Technical Bulletin No. 205. (1930)

In the 1910s (WWI) the FPL maintained a propeller laboratory to improve the manufacture of solid and glued-laminated wood propellers. [62]

The FPL's adhesives researchers were also heavily involved in research for aircraft in WWII, including wood-to-wood, wood-to-metal, and metal-to-metal bonding applications.

Lumber Joints and Laminates

The goal was to develop a wide variety of new products made by laminating veneer and small clear lumber cuttings from smaller trees and species not previously considered commercially viable, instead of using solid wood from large old-growth trees. The goal was more product from less raw material. A host of new products emerged, from shoe lasts and wagon hubs to bowling pins, baseball bats, tennis rackets, and laminated barn rafters and beams.

▲ FPL manufactured wood propellers. The three short propeller blades are likely for variable pitch applications and are made with compreg, a resin-impregnated veneer laminate. (1940s) {M 56957 F}

Laminated Beams

Laminated beams and arches are made by laying (or bending) the individual lumber pieces against a set of strongbacks and then clamping in place until the adhesive sets. The basic concepts were brought to the FPL from Germany. The first glued-laminated arches bonded with casein glue were made using FPL gluing techniques and the improved casein glue in 1935. Glulam beams using casein glue were installed in an FPL service building, as a demonstration and test, where they were in service for 75 years. [26] (See Packaging Laboratory in the Structures section.)

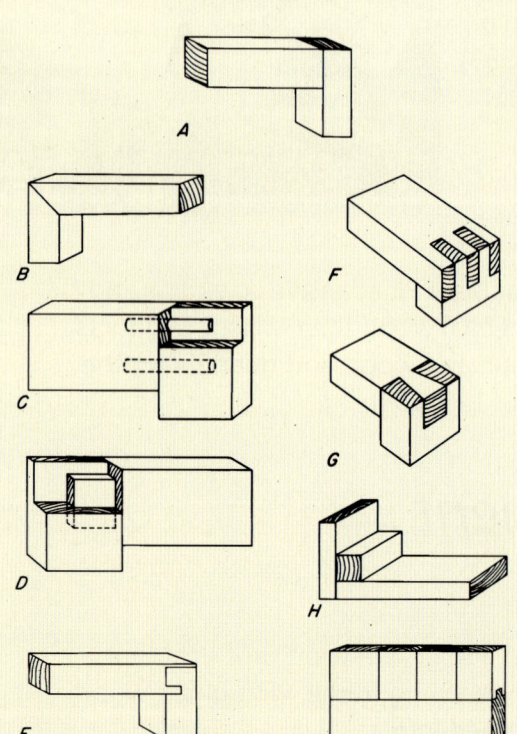

▲ Furniture joint specimens for determining joint strength. (1950s) {M 92376 F}

► Process of laying up individual plies in the manufacture of a designed laminated arch. (1960s) [44]

► Laminated beams in the new FPL Centennial Research Facility. (2009)

Gluing Treated Timbers

FPL scientists demonstrated the possibility of gluing lumber previously treated with certain preservatives to form strong, reliable, and durable glued-laminated structural timbers. Procedures were improved to a degree necessary to allow viable commercial production. [47, 78]

EGAR (Edge, Glue, and Rip)

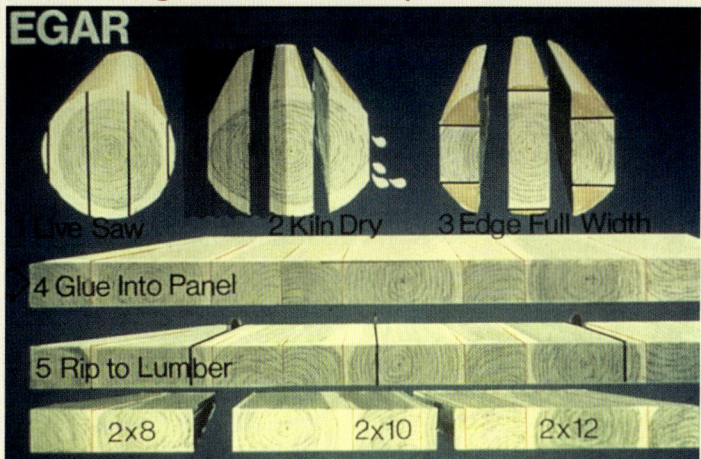

◄ FPL scientists developed the EGAR system, which can produce lumber of any width from small trees, resulting in a 10% increase in yield. With this method, wide dimension lumber can be produced from small logs by sawing the logs into full-width boards, kiln drying these boards or flitches, trimming the edges, gluing the boards into panels with waterproof glue, and then ripping the panels into lumber of desired widths. (1970s) [92]

◄ Ken Compton rip-sawing a panel that was made using the EGAR system. Results show that lumber recovery is increased over conventional sawing and that EGAR lumber is as strong as that produced by standard methods. (1970s)

Plywood

The gluing technique of synthetic resin adhesives developed at the Laboratory helped the fledgling Douglas-fir plywood industry in the 1930s. The FPL supported the growth of the structural grade plywood for exterior exposure by developing short-term, accelerated laboratory tests for durability and long-term outdoor test fence aging studies. [22, 25, 74]

In 1964, the FPL supported the development of a new structural plywood industry based on southern pine veneer with information about the suitability of existing glues and the potential for meeting strength and durability standards established previously for western species plywood. [71].

Stressed-Skin and Sandwich Panels

Stressed-skin panels can be used for walls, floors, and roofs. They combine framing and sheathing in one unit. Gluing the plywood skins to the lumber framing creates a panel structure that is stronger and more rigid than a conventionally nailed wall or floor system. The stiff, lightweight panels save materials, reduce weight, and speed onsite construction. [59, 98]

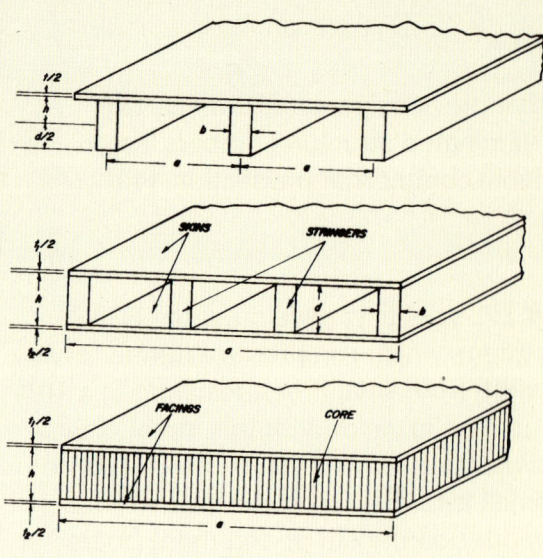

◄ Common types of stress-skin and sandwich panels that include insulation. (1970s) {M 141766}

Sandwich Panel

◄ Example of plywood-faced, lumber-framed, foam core, stressed-skin panel used for the walls of the FPL Wood Fiber Products Research building. The panel has resin-treated paper on the interior surface and Tedlar (Teflon) film on the exterior surface. {M 150689-17}

Through WWII and into the 1950s, and primarily at the behest of the U.S. Navy, the Laboratory continued to experiment with adhesives and develop techniques for bonding lightweight core materials, such as balsa wood, and paper or plywood skins to form strong, stiff, lightweight panels. These "sandwich" panels were used in wood aircraft construction and for lightweight prefabricated building panels used in housing for military personnel. [39]

FPL's expertise with sandwich panels led to its involvement in evaluating modified phenolic adhesives and epoxies for bonding wood-to-metal and metal-to-metal, including aluminum-skin, and aluminum honeycomb core sandwich panels. [48, 49, 52]

◄ Example of aluminum-faced, aluminum honey-comb-core adhesive-bonded sandwich. Note that the honeycomb core is itself formed by bonding thin aluminum foil coated with adhesive applied in a very specific pattern before bonding and expanded into hexagonal cells. (1950s) {M 107832}

When FPL became involved in the development of metal-faced, honeycomb-core sandwich panels for the military, FPL engineers developed the climbing drum peel test for determining the peel strength of the bond between the stiff faces and flexible cores. The test is particularly sensitive to the surface preparations and has been standardized by ASTM (test method D 1781). [58]

Two demonstration buildings of glued lumber-framed, plywood-skin panels, and one of sandwich panel construction, were built on the FPL campus in 1935 and 1937. These first of a kind buildings were still performing well after 31 years when they were taken down and tested. (See Housing section.) [98]

Glued-Laminated Beams for Exterior Service

The U.S. involvement in World War II again led to overwhelming demand for glued wood products, and again FPL was heavily involved in developing new products for the armed forces. The number of uses for glued wood products was much greater than in WWI due to the existence of the new synthetic adhesives including urea-, melamine-, phenol-, and later, resorcinol-formaldehyde adhesives. Magnus "Len" Selbo was a pioneer in the development of structural glued-laminated timber fabrication and evaluation, working closely with the Navy and private industry to develop structural laminated timbers for ship construction using phenolic and resorcinolic adhesives. [44, 64, 69, 73]

These water-resistant and waterproof adhesives broadened the range of glued wood end products available for the war effort, including plywood packing crates, aircraft trainers, gliders, and reconnaissance planes, landing craft, PT boats, large (nonmagnetic) mine-sweepers, truck bodies, aircraft hangers and factories, and other large, clear span structures and prefabricated buildings. [29]

The performance of glued-laminated timbers has been verified by long-term exposure tests and many years of actual service. Len Selbo conducted early experiments with the gluing of preservative-treated wood on the one hand, and treatment of laminated untreated wood on the other, which made possible large straight and curved timbers to be created. These could then be used in full exterior and marine structural applications such as in ships, rail-road trestles, bridges, and marine bulwarks. [34, 47, 75, 76, 78]

◀ Glued-laminated white oak ship frames used in construction of a navy minesweeper. (1950s) {M 94561 F}

Press-Lam

Press-Lam is manufactured by peeling or slicing a log into veneers from 1/4- to 1/2-inch thick. The veneers are then sawn or sliced to width and dried in a veneer dryer or in a heated press. Immediately upon removal from the dryer, the pieces are spread with adhesive, stacked to the desired thickness, and bonded in an unheated press using the residual heat from drying to set the adhesive. [84, 85]

◀ Samples of Press-Lam, thin boards that have been laminated to make thick boards. This process not only provides large boards but also randomly places defects in plies throughout the finished board, giving a stronger product. (1970s) {27 B}

▲ Samples of tongue-and-groove decking made from Press-Lam. (1970s) {27 A}

Nu-Frame House

The Nu-Frame house was a demonstration house designed and built in the late 1960s to demonstrate how adhesives and low-cost lumber and finishing materials could be used to build a sturdy, durable house. Adhesives were used throughout—from sandwich floor panels and roof trusses to interior and exterior wall panels, siding, and an innovative roof planking that eliminated sheathing and included a finished exterior surface film. Elastomer-based construction adhesives were used to connect many of the structural elements. The house stood on the FPL campus for several years until dismantled in about 1980 to make way for the new University of Wisconsin hospital next door. (See Housing section.) [81, 82]

Missile Nose Cone Fairing

One interesting project that FPL assisted in developing was the composite wood and aluminum nose fairing used to protect missiles when launched. The FPL evaluated the strength, creep resistance, and resistance to corrosion by salt water of several candidate adhesives. The fairings were made by bag-molding thin Sitka spruce veneers and aluminum strips to form a plywood cone with an aluminum ring for attachment to the missile body.

Particle Composites

One of the major accomplishments of the FPL has been their contribution to the development of wood composites, ranging from flakeboards, oriented strandboard, particleboard, and medium-density fiberboard to hardboard. The key to all these products is the adhesive. FPL has contributed to all these developments by evaluating the effects of many wood and processing variables and by evaluating the performance of the adhesives and the bonded composites derived from them.

The development of these products turned wood waste, such as sawdust, planer shavings, and board edgings, into saleable products and materially improved the air quality of towns and cities where the mills are located by eliminating wood waste burning. These products also opened new uses for wood species that were considered by many as useless. Probably one of the best examples is the use of aspen. For years, this species was considered a "weed tree" and efforts were made to replace it as soon as possible with "preferred species." Based on joint research between the Lake States Forest Experiment Station, Forest Service North Central Region, Universities of Minnesota and Michigan, Michigan State College, Michigan College of Mining and Technology, Superior Wood Products, North Western Railway System, and FPL, it was found that aspen has some very desirable properties not only for particle composites but also for paper, and it ceased being considered a "weed tree."

◄ Bryan River inspecting samples of particleboards after exposure to the elements.

Structural Composite Panels

◄ Structural composite panels made from wood particles. As the size of trees becomes smaller, the challenges are to fill the need for larger boards from smaller trees; or composite boards from wood chips, flakes, particles, or fiber, and to utilize forest residues such as thinnings and culled trees. (2000s) [155, 159]

◄ Composite wood products made possible by adhesives. (left to right) Plywood, oriented strandboard (OSB), particleboard, medium-density fiberboard MDF), and hardboard.

Test Methods and Standards Development

Strength

The strength of adhesives was not widely known before WWI, and methods for measuring strength were not widely available. Animal glue, for example, was selected by grade and grade was judged by "jelly strength" viscosity as well as other physical characteristics of the glue. [21]

The plywood shear and block shear tests were developed at FPL because actual strength values for adhesive bonds were required to assess gluing techniques and for aircraft design during WWI. [4, 15]

Both methods were used in Army and Navy standards, and were also standardized by ASTM test methods D 905 and D 906 respectively during WWII. These test methods have been used in tens of thousands of tests conducted at the Lab to evaluate adhesives and bonding methods for nearly 100 years. They are the standard methods for quality control strength tests in the laminate and plywood industries today. Military specifications for adhesives based on this work were critical to WWI aircraft production. [31]

Block Shear Test

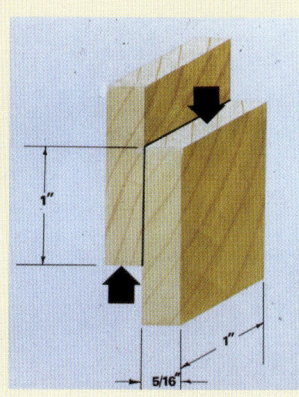

◄ Block shear test specimen.

► Early (probably the first) block shear compression tool developed at the FPL for testing the shear strength of lumber bonds for aircraft manu-facturing during WWI. (1918) {M 150836}

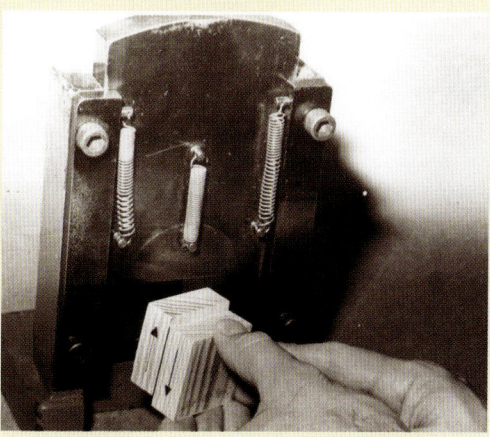

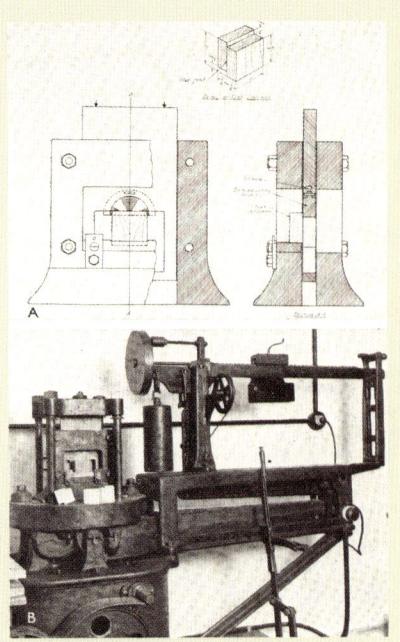

◀ Equipment used at FPL for the block shear joint test to evaluate adhesives.(1910s)

▶ Daniel Yelle conducting a block shear test of an experimental adhesive. (2000s)

Lap Shear Test

▲ Equipment used for the plywood-joint test to evaluate adhesives. (1910s)

◀ Lap shear test specimens ready for test.

▼ Failed lap shear test specimens. Note the mixture of wood and adhesive failure.

Scarf and Finger Joints

The early test method used to determine the strength of various scarf and finger joints was adapted from an FPL method for determining the tensile strength of clear lumber. [38]

In the 1950s, finger joints began to replace scarf joints because of the ease of manufacture and material savings. Selbo developed a strip tension test for finger joints that was adopted by the glulam industry (AITC Test 106) for quality control tests. [64, 65, 79]

▼ Fingers cut in the end of the board prior to adhesive application. (1960s) {7 Q}

◀ Various ways to bond the ends of boards starting with a scarf joint, two designs of finger joints, and a butt joint. {25 K}

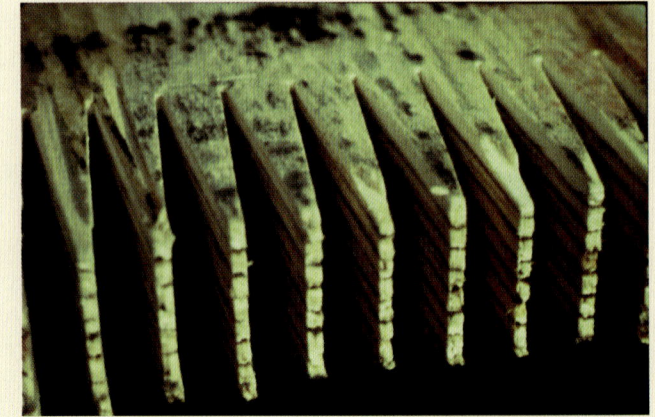

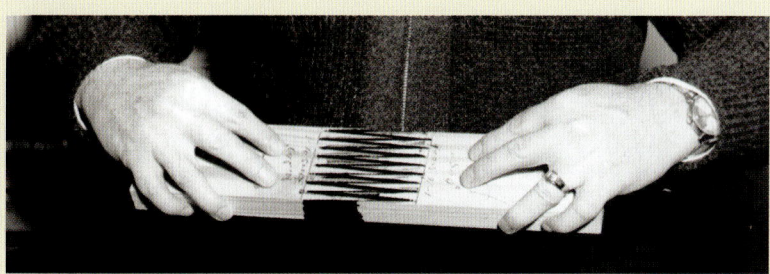

◄ Mating of the two sets of fingers after adhesive has been applied. (1960s) {M 101 216}

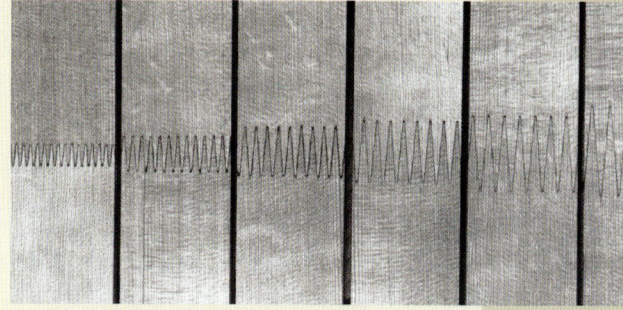

▶ Samples of boards that have been end-glued using different finger lengths. All six of these finger joints have the same slope (1 in 12) and tip thickness (0.045 inch), differing only in finger length. Strength increases with finger length because glue-joint area is enlarged and the ratio of total tip width to total joint width decreases. Using this method, long boards can be made out of previously wasted short pieces. (1960s) {M 120 739} [69]

Durability Exposures

Until the beginning of the 20th century, available wood adhesives were derived from natural resources such as vegetable starch and protein, animal protein, milk, and blood. Adhesives derived from these sources were mostly used for furniture solid parts and plywood. Because none of the commercially available adhesives was waterproof, test exposures were limited to determining how much humidity they could stand before they either softened or were degraded by mold. Simple exposures consisting of several levels of constant humidity or cycles between high and low humidity before strength testing sufficed to compare durability. WWI prompted the need for more water-resistant adhesives for aircraft manufacture. The FPL responded with improved versions of vegetable, animal, casein, and blood glues (described earlier) that required more rigorous tests of water resistance, including a minimum strength when tested wet after a defined soak treatment. [24]

Starting in the early 1930s, synthetic resin glues became available at reasonable cost. These were essentially waterproof adhesives and, as such, required more severe test exposures to evaluate bond quality. FPL developed the soak–dry and boil–dry exposures combining water, drying, and swell–shrink stress effects. Tests revealed that, although strength after exposure was an important indicator of original bond strength, a more revealing indicator of strength in service, or durability, is the amount of wood failure in the failed joint after testing the exposed specimens. These test methods, generalized as wet–dry–test–read % wood failure, eventually became the standards for exterior-grade plywood durability. Test plywood panels bonded with phenol-formaldehyde adhesives were also prepared and installed at the FPL outdoor exposure site in Madison, Wisconsin. Many of these panels were never removed for testing and remained on exposure for over 40 years with no failure of the adhesive. [25]

▲ Romie Klassy applying pressure to secure a satisfactory bond for laboratory testing of glued scarf joints. (1960s) {111 I}

Blood-Extended Phenol
3/8-inch, 3-ply
Douglas-Fir

▲ Plywood bonded with blood-extended phenol-formaldehyde resin has withstood over 40 years of full outdoor aging. Although the outer 1/8 inch of veneer has eroded, the bond integrity has endured. (1990s) {M 910051}

Similarly, in the early 1940s, FPL began to bond laminated joints with resorcinol-formaldehyde adhesive. Researchers needed more severe exposure treatments to evaluate bond quality than simple high humidity or soaking, which had been used with casein glue. The answer was a 180-day cyclic soak–dry treatment delamination test developed during WWII at the Laboratory. Under the pressure of wartime, the test duration was reduced to 21 days by using vacuum, pressure, and elevated temperature. [44]

Ultimately, through additional testing and experience, the test time was reduced to 7 days in ASTM test method D 1183, and finally to 3 days in ASTM specification D 2559. The pass–fail criterion, as set in the original 180-day test, was the amount of delamination visible during the drying phase. The cyclic delamination test became, and remains, the standard for evaluating the quality and durability of glues, and joints in beams for exterior–wet use service referenced by the American National Standard for Structural Glued Laminated Timber (ANSI/AITC A190.1).

Rate-Process Method of Accelerated Aging

The FPL now has a great deal of actual experience with long-term durability of wood products bonded with very durable thermosetting phenolic and resorcinolic adhesives in such severe environments as encountered in ship timbers, bridges, buildings, and other outdoor uses. We know from these many years of experience that the adhesives are at least as durable as the wood. But what happens when a new adhesive comes along without that history? How can we confidently predict how it will last without waiting for years of experience? How can we determine the chemical rate of aging of very durable adhesives?

In the 1960s, researchers from both adhesives and forest products industries, FPL, and the Canadian Forest Products Laboratory formed a task group (Steering Committee on Accelerated Testing of Adhesives, or SCATA) to develop accelerated test methods for aging adhesive-bonded wood products to predict the service life of durable adhesives. Working within this group, Robert Gillespie at FPL developed the "rate process" method for measuring the chemical aging effects of heat, moisture, and chemicals upon the rate of strength degradation. [74, 89]

Application of this method at several temperature levels and moisture levels permits the determination of rate curves that can be extrapolated from accelerated conditions of heat and moisture to non-accelerated conditions, or some other normal service conditions. The advantage is to gain an estimate of service life of the adhesive or the bond in a reasonably short time. Test specimens are also typically placed in outdoor exposure to determine service life of the bond under real conditions for comparison to the service life predicted by accelerated rate-process methods. The rate-process method was standardized by ASTM D 4502. [96]

◀ Fred Phelps inspecting plywood shear test specimens exposed to outdoor aging. (1960s) {M 129690}

Stress–Strain Behavior of Adhesives

The rigidity or stress–strain behavior of an adhesive is of importance in some cases. Ed Kuenzi pioneered the torsion ring test for measuring the strength and stress–strain behavior of rigid epoxy adhesives during the years when FPL was involved in metal bonding and stressed-skin panel development. The method was standardized by ASTM as test method E 229. [68]

In the 1970s, elastomer-based adhesives and cross-linked thermoplastic adhesives were developed by industry that had potential advantages in building construction, including gap-filling ability, ease of application, low bonding pressure, and tolerance of bonding conditions. However, in contrast to thermosetting adhesives, these new adhesives were less rigid than wood, and their mechanical properties were more affected by heat and moisture than wood. Knowledge of these properties was needed by building designers.

At about the same time the Department of Housing and Urban Development instituted a program, "Operation Breakthrough," to develop and promote the application of new technology to housing construction. Adhesives researchers at the FPL developed test methods and evaluated the strength and stress–strain behavior of adhesives under various conditions of heat, moisture, and time. The method for determining strength and shear modulus has been standardized as ASTM test method D 3983. [94, 99]

Fracture Toughness, Fatigue, and Long-Term Loading Effects

Block shear and plywood shear tests are the standards for determining quality of adhesive bonds, but unless the bond is defective, most of the failure is in the wood and the test results don't reveal much about the actual behavior of the adhesive. In the 1980s, a double cantilever-beam cleavage test for fracture toughness of wood bonds was developed at FPL and subsequently modified to improve its performance and utility. [100, 129, 130]

This method was found to be sensitive to formulating and bonding variables of adhesives as well as an understanding of the fracture mechanisms in joints. This method forces failure to occur in the adhesive layer, or very close to it, thus yielding more information about the adhesive performance. [136]

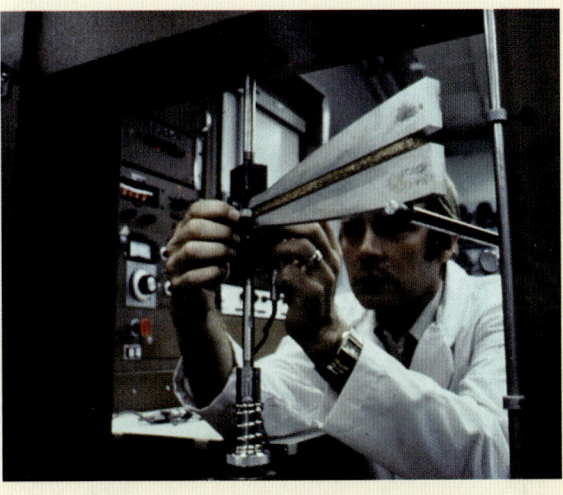

◀ Arnie Okkonen evaluating adhesive bond fracture toughness with an experimental contoured wood, double cantilever beam specimen developed at the Laboratory. (1970s) {112 E} [100]

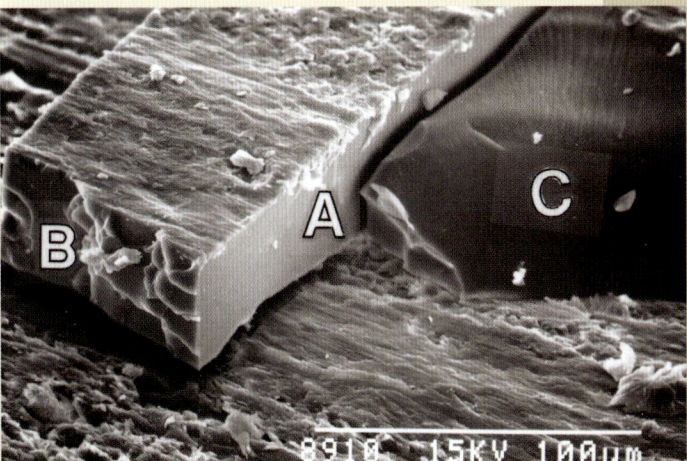

▲ Scanning electron micrograph of three different types of fracture surface in improved urea-formaldehyde adhesive layer. The three types of fracture result from (A) shrinkage during cure, (B) vacuum–pressure soak–dry shrinkage, and (C) fracture during cleavage testing. (1990s) {M 930027}

Warren Olson and colleagues tested the fatigue strength of several types of adhesives for wood aircraft [54].

Herb Eickner conducted extensive tests of the shear, fatigue, bending, impact, and long-time load strength properties of structural metal-to-metal adhesives for high-performance aircraft. [49]

Formaldehyde Emissions

The emission of formaldehyde from plywood and particleboard in housing became a national health problem in the 1970s. George Myers determined that formaldehyde was held in several forms within the board and that this resulted in different rates of emission from the board. He developed and evaluated several methods for measuring emissions from boards and effects of various material and environmental factors upon rates of emissions. This work provided an understanding of emission sources and processes and materially assisted in the development of industrial product standards for allowable formaldehyde emission from board products. [107, 110, 112]

◄ Melissa Baumann working with environmental chambers designed to measure off-gassing of chemicals from adhesive-bonded wood products. Both the adhesives and the wood itself can generate these gaseous emissions as a result of heat and moisture effects in the bonding process. (1990s) [146]

Reactivity

The potential scarcity of phenol for the critical wood adhesive phenol-formaldehyde prompted Tony Conner and colleagues to look for a way to model the reactivity of other phenolic compounds with formaldehyde. This would reduce the need to physically test a large number of possible compounds that could potentially be obtained from sources such as tannin, lignin, biomass pyrolysis, and coal gasification. They combined modeling techniques with experimental data collected from numerous phenolics and found good correlation between predicted and actual reactivity of several phenol substitutes with formaldehyde. Rate constants for the reactions of a variety of phenolic compounds with formaldehyde in an aqueous solution under alkaline conditions were obtained from experimental measurements. Their efforts resulted in the largest available database for the reactivity of different phenolic compounds with formaldehyde. [147, 150, 152]

► Tony Conner exploring the molecular structure of adhesives to better understand how to improve wood adhesives. (1990s)

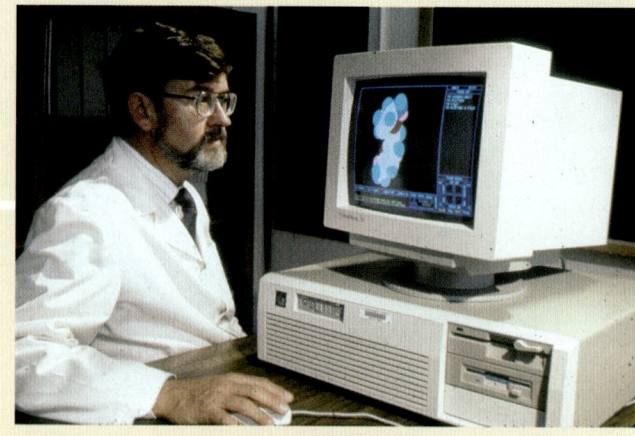

Technology Transfer

Research Reports and Industry Media

Technology transfer has been a primary mission of adhesives researchers at the FPL, starting with the gluing of wood for aircraft in 1917. After the wartime drain on timber supplies, the FPL played a major role in a national movement to conserve timber. [83]

The laboratory began a program to transfer the gluing technique developed for aircraft manufacturing to a broader group of wood products manufacturers. The goal was to foster the development of new, efficient glued wood products (alluded to in the quote from Tom Truax at the beginning of this section). FPL's adhesive scientists conveyed this type of information to industry through magazine articles, workshops for manufacturers, manuals, and handbooks. All these publications were aimed at supplying users with the best scientifically based technical information. In addition to the many research reports represented by those referenced in this section, the Laboratory's adhesive scientists have also published numerous articles over the years summarizing the information gained through research and the state-of-the-art in glues, gluing, and glued wood products. [10, 12, 35, 42, 61, 63, 67, 70, 77, 88, 97, 103, 115, 148]

Manuals and Handbooks

Manual for the inspection of aircraft wood and glue [12]
Gluing of wood in aircraft manufacture [18]
The gluing of wood [27]
Design of wood aircraft structures ANC–18 [31]
Army/Navy Civil Committee Bulletin ANC–19 [31]
Manual on wood construction for prefabricated houses [40]
Fabrication & design of glued laminated wood structural members [51]
Adhesive bonding of wood [88]
Adhesives in building construction [97]
Wood as an adherent [127]
Fracture of adhesive-bonded wood joints [137]
Adhesive bonding of wood materials [148]
Wood adhesion and adhesives [159]

Symposia

The pioneering national symposium on wood adhesives was sponsored by FPL, under the direction of Richard Blomquist, bringing together prominent government and industrial scientists to foster an interchange of ideas. [63]

This effort was revived in 1975 by Robert Gillespie and has continued every five years since then, attracting scientists and industry personnel from around the world. [90, 101, 111, 122, 141, 151, 162]

ASTM Test Methods Based on FPL Research

D 905, Standard test method for strength properties of adhesive bonds in shear by compression loading 1947

D 906, Standard test method for strength properties of adhesives in plywood type construction in shear by tension loading 1947

D 1002, Standard test method for apparent strength of single-lap-joint adhesively bonded metal specimens by tension loading (metal to metal) 1949

D 1101, Standard test methods for integrity of adhesive joints in structural laminated wood products for exterior use 1953

D 1781, Standard test method for climbing drum peel for adhesives 1960

D 2339, Standard test method for strength properties of adhesives in two-ply wood construction in shear by tension loading 1965

D 2559, Standard specification for adhesives for structural laminated wood products for use under exterior (wet use) exposure conditions 1966

D 2651, Standard guide for preparation of metal surfaces for adhesive bonding 1967

D 3632, Standard test method for accelerated aging of adhesive joints by the oxygen pressure method

D 3983, Standard test method for measuring strength and shear modulus of nonrigid adhesives by the thick adherend tensile-lap specimen 1981

D 4502, Standard test method for heat and moisture resistance of wood-adhesive joints 1985

D 4688, Standard test methods for evaluating structural adhesives for finger jointing lumber 1987

D 4896, Standard guide for use of adhesive-bonded single lap-joint specimen test results 1989

D 5266, Standard practice for estimating the percentage of wood failure in adhesive bonded wood joints 1993

D 5574, Standard test methods for establishing allowable mechanical properties of wood-bonding adhesives for design of structural joints 1994

E 229, Standard test method for shear strength and shear modulus of structural adhesives 1963

100 Years of Adhesive Research

1917	WWI created need to evaluate adhesives for aircraft plywood, wing spars, and propellers. Certain woods scarce in required sizes.	Animal glue Blood albumin Starch
1917–1918	Developed block shear and plywood shear tests. Developed water resistant casein glue.	Casein glue
1922	Modern block shear test developed.	Soybean protein
1925–1926	FPL involved in controversy of mechanical versus chemical bonding mechanisms.	
1934	Began evaluation of phenolic adhesives for exterior plywood.	Phenol-formaldehyde
1938	Wood failure criteria for plywood adhesive performance developed.	
1941	WWII created need for structural wood laminates in large and unusual shapes for use in marine and exterior service.	
1942	Developed 180-day cyclic delamination test to determine weathering resistance of laminated wood beams.	Urea, melamine, and resorcinol-formaldehyde
1942–1948	Developed techniques for bonding laminated ship timbers and beams, including bonding treated wood.	USDA Technical Bulletin 1069
1945–1958	Pioneered techniques for bonding metal-to-wood and metal-to-metal.	
1949	FPL block shear and plywood shear tests standardized.	ASTM D 905 ASTM D 906
1950	Developed the "FPL etch," a surface treatment to prepare aluminum for bonding. The "FPL etch" was the world standard for ensuring durable bonds in jet aircraft until the early 1980s.	Epoxy ASTM D 2651
1951	Began evaluating synthetic resins for furniture.	Polyvinyl acetate
1953	FPL cyclic delaminating test standardized.	ASTM D 1101

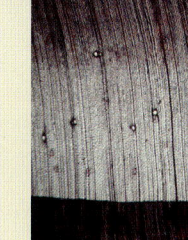

1956	Developed "climbing drum" test to determine the peel strength of honeycomb sandwich panel facings	
1960	FPL climbing drum peel test standardized.	ASTM D 1781
1963	FPL modified cyclic delaminating test for qualifying adhesives for laminating timbers for exterior use standardized.	ASTM D 2559
1965	Developed chemical rate process method to forecast bond service life.	
1969	Began evaluating adhesives for on-site building construction.	Elastomer-based mastics
1978	Developed thick adherend test to determine adhesive stress–strain behavior and cantilever beam test to determine fracture toughness of wood bonding adhesives.	
1979–1983	Developed methods for and defined variables affecting formaldehyde emissions from particleboard and hardwood plywood.	Urea-formaldehyde
1981	Thick adherend test standardized.	ASTM D 3983
1985	Durability rate-process method standardized.	ASTM D 4502
1987	Adhesives from Renewable Resources Conference.	
1975, 1980, 1985, 1990, 1995, 2000, 2005	Wood adhesives symposiums (FPL & Forest Products Society).	Technology transfer
1991–1995	Developed difurfuryl diamines as a new biobased adhesive for bonding wood.	
1997	Developed hydroxymethylated resorcinol primer for more durable bonds.	
2006	Developed soy flour and phenol formaldehyde adhesive.	

1910

Non-water resistant adhesives were available and poor outdoor performance of wood bonds was common. Furniture was the primary glued wood product. Practice of gluing was an art form.

2010

Adhesives research carried out by FPL, industry, and universities has resulted in many new products made from wood that in the past was just wasted. Better understanding of adhesive chemistry and mechanisms of bond formation has led to new highly improved durability with epoxy and other useful adhesives previously considered less durable, and with adhesives derived from renewable resources. Many new companies and thousands of jobs have resulted from the development of glued wood products.

Further Information

[1] Important extracts taken from letters written to the Laboratory in reference to glues in connection with questionnaire of December 20, 1917 / Forest Products Laboratory / Unpublished report of a poll of glue manufacturers regarding the availability of water proof or water resistant adhesives during the beginning of research on glues for aircraft construction / Project 157 (1917)

[2] The gluing of different woods / A.C. Lindauer / FPL / Unpublished Project L–157–6 report (1919)

[3] Memorandum on tests on Bakelite joints / A.C. Lindauer / FPL / Unpublished report of Project L–157 / 1919

[4] Method of testing strength of joint glues / FPL Tech. Note F–16 (unknown date, about 1920)

[5] Casein glues exceptionally durable in damp places / FPL Technical note No.157 (1921)

[6] Effect of structure on glue penetration / E. Gerry, T.R. Truax / Furniture manufacture and artisan / (n.s. 23) 84:49–51 (1922)

[7] Process of manufacturing waterproof adhesives / Samuel Butterman, Charles K. Cooperrider / U.S. Patent No. 1,456,842 (May 29,1923)

[8] Blood albumin glue / Alfred C. Lindauer / U.S. Patent No. 1,459,541 (June 19,1923)

[9] Conditions affecting the making of glued joints / T.R. Truax / FPL Report No. 216 (1924)

[10] Glues for manual training work / T.R. Truax / Industrial education magazine Vol. 25(9) (Mar. 1924)

[11] The gluing characteristics of different woods with animal glue / T.R. Truax, C.A. Harrison / Unpublished Forest Products Laboratory report / Project L–157–6J31 (1924)

[12] Manual for the inspection of aircraft wood and glue for the United States Navy / prepared by Forest Products Laboratory, Forest Service, Department of Agriculture, 1st ed., Washington, D.C.: U.S Navy Department, Bureau of Aeronautics (1928)

[13] Nature of adhesion between wood and glue / F.L. Browne, D. Brouse / Industrial & Engineering Chemistry (Jan. 1929)

[14] Blood albumin glue: their manufacture preparation and application. FPL report No. 281–2 (1929)

[15] The significance of mechanical wood joint tests for the selection of woodworking glues / T.R. Truax, F.L. Browne, D. Brouse / Industrial and engineering chemistry / (also FPL Report No. R825) January (1929)

[16] Water resistant glues / FPL Technical Note No. 4 (revised 1929)

[17] Water resistant animal glue / Clarance E. Hrubesky, Fredrick L. Browne / U.S. Patent No. 1,712,077 (May 7,1929)

[18] Gluing Wood in Aircraft Manufacture / T.R. Truax / USDA Technical bulletin No. 205 / Washington, D.C. (1930)

[19] Casein glues: their manufacture, preparation, and application / FPL Report No. 280 (1930)

[20] Controlling conditions in gluing different kinds of woods / T.R. Truax / Woodworking Industries Vol. 9(2) (Feb. 1931)

[21] Development of wood adhesives and gluing technic / T.R. Truax / FPL Report No. 947 (1932)

[22] Methods of increasing durability of plywood / D. Brouse / FPL Report No. R1125 (1932)

[23] Behavior of casein and blood glue joints under different conditions of exposure / D. Brouse / Furniture manufacturer 48(3) (Sept. 1934)

[24] Serviceability of glue joints / Don Brouse / FPL Report No. R1172 (1938)

[25] Contributions of synthetic resins to improvement of plywood properties / D. Brouse / FPL Report No. R1212 (1939)

[26] The glued laminated wooden arch / T.R.C. Wilson / USDA Technical bulletin No. 691 (1939)

[27] The gluing of wood / T.R. Truax / USDA Bulletin No. 1500 (1940)

[28] Glues for use in aircraft / FPL Report No. 1337 (1941)

[29] Wood goes to war / C.P. Winslow / FPL Report No. D1426 (1942)

[30] Sawed gluing surfaces / L.H. Reinecke / FPL unpublished wartime report (1943)

[31] Wood aircraft inspection and fabrication / Forest Products Laboratory / Army–Navy Civil Committee Bulletin, ANC–19 / (1944) (first issued in 1942) (Also the companion Design of wood aircraft structures / ANC–18 / (1944)

[32] Effect of high and low temperature on resin glue joints in birch plywood / R.F. Blomquist / FPL Report No. 1345 (revised 1944)

[33] Experiments with preservatives for soybean glue and soybean-glued plywood / FPL Report No. 1447 (1944)

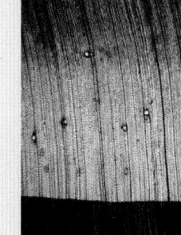

[34] Laminating lumber for extreme service conditions / C.D. Dosker, A.C. Knauss / American Society of Mechanical Engineers 66(12) (1944)

[35] Types and uses of adhesives / T.R. Truax, R.F. Blomquist / American Society for Testing Materials. Committee D–14 on Adhesives. Symposium on Adhesives (1946)

[36] A comparison of the durability of 23 urea resins in glue joints exposed to nearly saturated air at 75 °C (167 °F.) / A.E. Gabriel / Report prepared for the Army–Navy–Civil Committee on Aircraft Design Criteria, FPL Report No. 1355 (1946)

[37] Effect of thickness of glue line strength and durability of glued wood joints / R.A. Cockrell, H.D. Bruce / FPL Report No. 1616 (1946)

[38] End joints of various types in Douglas-fir and white oak compared for strength / R.F. Luxford, R.H. Krone / FPL Report No. 1622 (1946)

[39] Durability of glued joints between aluminum and end-grain balsa / H.W. Eickner / FPL Report No. 1566 (1947)

[40] Manual on wood construction for prefabricated houses / Prepared by Forest Products Laboratory for Office of the Housing Expediter, Housing and Home Finance Agency (1947)

[41] Polyvinyl-resin emulsion woodworking glues: A study of some of their properties / W.Z. Olson / FPL Report No. R1691 (1947)

[42] Postwar developments in woodworking glues / D. Brouse, R.F. Blomquist / Woodworking Digest Vol. 49(12) (Dec. 1947)

[43] Drying and conditioning glued joints / FPL Report No. 475 (1948)

[44] Laminating of structural wood products by gluing / A.C. Knauss, M.L. Selbo / FPL Report No. 1635 (1948)

[45] Results of accelerated tests and long-term exposures on glue joints in laminated beams / T.R. Truax, M.L. Selbo / FPL Report No. R1729 (1948)

[46] A study of methods for preparing clad 24S–T3 aluminum-alloy sheet surfaces for adhesive bonding / H.W. Eickner / FPL Report No. 1813 (1950)

[47] Summary of information on gluing treated wood / M.L. Selbo / FPL Report No. 1789 (1950)

[48] Adhesives for bonding wood to metal / H.W. Eickner, R.F. Blomquist / FPL Report No. 1768 (1951)

[49] The shear, fatigue, bend, impact, and long-time-load strength properties of structural metal-to-metal adhesives in bonds to 24S–T3 aluminum alloy / H.W. Eickner / FPL Report No. 1836 (1953)

[50] Resume of some of the newer products in wood utilization / FPL Report No. 1967 (1953)

[51] Fabrication and design of glued laminated wood structural members / A.D. Freas, M.L. Selbo / USDA Technical bulletin No.1069 (1954)

[52] Development of improved structural epoxy-resin adhesives and bonding processes for metal / J.M. Black, R.F. Blomquist, R. Frederick / FPL Report No. 2008 (1954)

[53] Drying and conditioning glue joints / FPL / FPL Report No. 485 (1955)

[54] Resistance of several types of glue in wood joints to fatigue stressing / W.Z. Olson, D.W. Bensend, H.D. Bruce / FPL Report No. 1539 (1955)

[55] Important factors in gluing with animal glue / FPL Report No. 869 (1956)

[56] Durability of water-resistant woodworking glues / FPL No. 1530 / (revised 1956)

[57] Rate of development of joint strength by four resin glues on eight species of wood / W.Z. Olson, H.D. Bruce, V.R. Soper / FPL Report No. 1547 (1956)

[58] Development and evaluation of the climbing peel method for testing adhesive bonds in sandwich and metal-to-metal constructions / H.W. Eickner, F. Werren / WADC Technical Report 56 –386 (ASTIR Document AD–11049) (Oct. 1956)

[59] Sandwich panels for building construction / L.W. Wood / FPL Report No. 2121 (1958)

[60] Wood gluing and laminating / R.F. Blomquist / Wood and wood products Vol. 64(3) (Mar. 1959)

[61] An international look at glues and gluing, 1959 / Forest products journal 10(2) (Feb. 1960)

[62] The U.S. Forest Products Laboratory / F.J. Champion / FPL Report No. 1698 (revised 1960)

[63] Symposium on adhesives for the wood industry / FPL Report No. 2183 (1960)

[64] Test for quality of glue bonds in end-jointed lumber / M.L. Selbo / FPL Report No. 2258 (1962)

[65] A new method for testing glue joints of laminated timbers in service / M.L. Selbo / Forest products journal 12(2) (1962)

[66] Epoxy-resin adhesives for gluing wood / W.Z. Olson and R.F. Blomquist / Forest products journal 12(2) (1962)

[67] Adhesives / R.F. Blomquist / Encyclopedia of chemical technology, 2nd ed. Vol. 1, Wiley & Sons, New York (1963)

[68] Determination of mechanical properties of adhesives for use in the design of bonded joints / Edward Kuenzi / FPL Research Report FPL–011 (1963)

[69] Effect of joint geometry on tensile strength of finger joints / M.L. Selbo / Forest products journal Vol. 13(9) (1963)

[70] Adhesives—Past, present, and future / R.F. Blomquist / Presented as the 1963 Marburg Lecture at the National meeting of the American Society for Testing and Materials, June 26, 1963 / ASTM (1964)

[71] Experiments in gluing southern pine veneer / R.F. Blomquist, W.Z. Olson / FPL Research Paper FPL–032 (1964)

[72] Improving the gluing characteristics of plywood surfaces by sanding / F.H. Kaufer / FPL Research Note FPL–051 (1964)

[73] Performance of melamine resin adhesives in various exposures / M.L. Selbo / Forest products journal Vol. 15(2) (1965)

[74] Accelerated aging of adhesives in plywood type joints / R.H. Gillespie / Forest products journal 15(9) (1965)

[75] After two decades of service—glulam timbers show good performance / M.L. Selbo , A.C. Knauss / Forest products journal Vol. 15(11) (1965)

[76] Evaluation and development of adhesives for marine assembly gluing / M.L. Selbo / Report prepared for Navy Department, Bureau of Ships, Order No. 1700S–757–65, Project Serial No. SR 007–03–02, Task 1004 / Forest Products Laboratory (1966)

[77] Synthetic resin glues / FPL Research Note FPL–0141 (1966)

[78] Long-term effect of preservatives on gluelines in laminated beams / M.L. Selbo / Forest products journal 17(5) (1967)

[79] Performance of three types of adhesives in finger joints / M.L. Selbo, R.A. Jokerst / Unpublished report for American Institute of Timber Construction (1967)

[80] A simplified test for adhesive behavior in fire exposure / E.L. Schaffer / FPL Research Note FPL–0175 (1968)

[81] Construction of Nu-Frame research house / L.O. Anderson / FPL Research Paper FPL–88 (1968)

[82] Adhesives and gluing methods used in the Nu-Frame house / C.G. Roessler, R.M. Lulling / FPL unpublished office report (1970)

[83] History of the Forest Products Laboratory / Charles A. Nelson / Forest Products Laboratory (1971)

[84] FPL Press-Lam process: Fast, efficient conversion of logs / E. Schaffer / Forest products journal 21(11) (1972)

[85] Residual heat of drying accelerates adhesive cure / R.W. Jokerst / FPL Research Paper FPL 179 (1972)

[86] Mastic construction adhesives in fire exposure / B.H. River / FPL Research Paper FPL–400 (1973)

[87] Surface damage before gluing—weak joints / B.H. River, V.P. Miniutti / Wood and wood products 80(7) (1975)

[88] Adhesive bonding of wood / M.L. Selbo / USDA, Forest Service Technical Bulletin 1512 (1975)

[89] Durability of adhesives in plywood / R.H. Gillespie, B.H. River / Forest products journal 26(10) (1976)

[90] Adhesives for products from wood / Proceedings of the 1975 symposium sponsored by the Forest Products Laboratory, Madison, WI. (1976)

[91] Knife- versus abrasive-planed wood: Quality of adhesive bonds / R.W. Jokerst, H.A. Stewart / Wood fiber 8(2) (1976)

[92] Yield and strength of softwood dimension lumber produced by EGAR system / K.C. Compton, H. Hallock, C.C. Gerhards, R. Jokerst / FPL Research Paper FPL–293 (1977)

[93] Accelerated aging of phenolic-bonded flakeboards / A.J. Baker, R.H. Gillespie / General Technical Report WO–5 / USDA Forest Service (1978)

[94] Measurement of shear modulus and shear strength of adhesives / B.H. River, R.H. Gillespie / Report prepared for Dept. of Housing and Urban Development / National Technical Information Service, PB80–121742 (1978)

[95] Strength and shear modulus of several construction adhesives as influenced by environment and loading conditions / B.H. River / NIST PB80–109887 (1978)

[96] Precision of the rate-process method for predicting bondline durability / M.A. Millett, R. Gillespie / NTIS PB80–121866 (1978)

[97] Adhesives in building construction / R.H. Gillespie, D. Countryman, R.F. Blomquist / USDA Forest Service Handbook No. 516 (1978)

[98] Structural sandwich performance after 31 years of service / J. Palms, G.E. Sherwood / FPL Res. Pap. FPL 342 (1979)

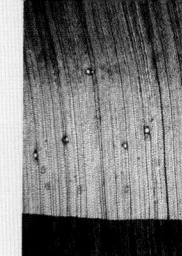

[99] Long-term load deformation behavior and strength of elastomer-based adhesives / B.H. River, R.H. Gillespie / Report prepared for Dept. of Housing and Urban Development / National Technical Information Service, PB82–13530 (1979)

[100] Tapered double cantilever beam fracture tests of phenolic-wood adhesive joints, Part 1. Development of specimen geometry; effects of bondline thickness, wood anisotrophy and cure time on fracture energy / R. Ebewele, B. River, J. Koutsky / Wood and fiber 11(3) (1979)

[101] Wood Adhesives—Research, Applications, and Needs / Proceedings of a symposium sponsored by Forest Products Laboratory and Washington State University. Madison, WI. September 23–25, 1980 / Forest Products Research Society, Madison, WI. (1981)

[102] Behavior of construction adhesives under long-term load / B.H. River / FPL Research Paper FPL 400 (1981)

[103] Adhesive bonding wood and other structural materials / R.F. Blomquist, ed. / Clark C. Heritage Memorial Workshop on Wood, Vol. 3 (1981), USDA Forest Service , Forest Products Laboratory, Madison, WI. / Materials Research Laboratory, Penn. State Univ., University Park, PA (1982)

[104] Steam injection pressing / R.L. Geimer / In: Maloney, Thomas M., ed. Proceedings of the 16th Washington State Univ., International Symposium on Particleboard (1982)

[105] Relationship between phenolic adhesive chemistry, cure, and joint performance. Part I, Effects of base resin constitution and hardener on fracture energy and thermal effects during cure / R.O. Ebewele, B.H. River, J.A. Koutsky / Journal of adhesion 14(3/4) (1982)

[106] Microscopy of abrasive-planed and knife-planed surfaces in wood-adhesive bonds / L. Murmanis, B.H. River, H. Stewart / Wood and fiber science, 15(2) (1983)

[107] Formaldehyde emission from particleboard and plywood paneling: measurement, mechanism and product standards / G.E. Myers / Forest products journal 33(5) (1983

[108] Accelerated, real-time aging of four construction adhesives / B.H. River / Adhesive age 27(2) (1984)

[109] Potential of carbohydrates for exterior adhesives / A.W. Christiansen, R.H. Gillespie / Forest products journal 36(7/8) (1986)

[110] Mechanisms of formaldehyde release from bonded wood products / G.E. Myers / In: B. Meyer, Andrews, B.A. Kottes: Formaldehyde release from wood products: ACS Symposium Series 316; American Chemical Society (1986)

[111] Wood Adhesives in 1985—Status and Needs / / Proceedings of the symposium sponsored by the Forest Products Laboratory and the Forest Products Research Society, Madison, WI. / Forest Products Research Society, Madison, WI. (1986)

[112] Resin hydrolysis and mechanisms of formaldehyde release from bonded wood products / G.E. Myers / In; Wood Adhesives in 1985: Status and Needs / Forest Products Research Society Proceedings No. 47344 (1986)

[113] Evaluation of flakeboard bonded with xylitol-modified alkaline phenolic resin / R.W. Jokerst, A.H. Conner / Forest products journal 38(2) (1988)

[114] Improving the fatigue resistance of adhesive joints in laminated wood structures / T.L. Laufenberg, B.H. River, L. Murmanis, A.W. Christiansen / Report No. NASA CR–182165. Report prepared for National Aeronautics and Space Administration, Lewis Research Center by Forest Products Laboratory / (1988)

[115] Adhesives from renewable resources: ACS Symposium 385 / R.W. Hemingway, A.C. Conner, S.J. Branham (eds.) / American Chemical Society, Washington, D.C. (1989)

[116] Wood adhesives from kraft lignin / R.H. Gillespie / Chapter 9, Adhesives from Renewable Resources, ACS Symposium Series 385, R.W. Hemingway, A.H. Conner (eds.), American Chemical Society, Washington, D.C. (1989)

[117] Carbohydrate-modified phenol-formaldehyde resins formulated at neutral conditions. / A.H. Conner, L.F. Lorenz, B.H. River / Chapter in Adhesives from Renewable Resources. ACS Symposium Series 385. R.W. Hemingway, A.H. Conner (eds.), American Chemical Society. Washington, D.C. / 1989

[118] How over drying wood reduces its bonding to phenol-formaldehdye adhesives: A critical review of the literature. Part I. Physical responses / A.W. Christiansen / Wood and fiber 22(4) (1990)

[119] Hydroxymethylated lignin bonded Douglas-fir flakeboard / R.W. Jokerst / Forest products journal 40(2) (1990)

[120] The inspiration for the FPL etch / Personal communication by H. Eickner to Bryan River / (about 1990)

[121] How overdrying wood reduces its bonding to phenol-formaldehdye adhesives: A critical review of the literature. Part II. Chemical reactions / A.W. Christiansen / Wood and fiber 23(1) (1991)

[122] Wood Adhesives 1990 / A.H. Conner, A.W. Christiansen, G.E. Myers [and others], eds./ Proceedings of a symposium sponsored by the Forest Products Laboratory and The Forest Products Society; 1990 May 16–18; Madison, WI. / Forest Products Research Society (1991)

[123] Phenol-formaldehyde resin curing and bonding in steam-injection pressing. I. resin synthesis, Characterization, and cure behavior / G.E. Myers, A.W. Christiansen, R.L. Geimer, R.A. Follensbee, J.A. Koutsky / Journal of applied polymer science, Vol. 43, 237–250 (1991)

[124] Wood–polymer bonding in extruded and nonwoven web composite panels / A.M. Krzysik, J.A. Youngquist, G.E. Myers, I.S. Chahyadi, P.C. Kolosick / In: Wood Adhesives 1990, Proceedings of a symposium, Madison, WI. / A.H. Conner, A.W. Christiansen, G.E. Myers [and others] eds. Forest Products Society (1991)

[125] One step method for the preparation of difurfuryl diamines / A.H. Conner, M.S. Holfinger, C.G. Hill, W.J. McKillup, R. Reimann / U.S. Patent 5,292,903 (1991)

[126] Polyamine-modified urea-formaldehyde resins. II. Resistance to stress induced by moisture cycling of solid wood joints and particleboard / R.O. Ebewele, B.H. River, G.E. Myers, J.A. Koutsky / Journal appled polymer science Vol. 43:1483–1490 (1991)

[127] Wood as an adherend / Bryan H. River, Charles B. Vick, Robert H. Gillespie / Treatise on adhesion and adhesives. Vol. 7. New York: Marcel Dekker (1991)

[128] Lignocellulosic-plastic composites from recycled materials / J. Youngquist, G.E. Myers, T. Harten / Chapter 4, In: Emerging Technologies for Materials and Chemicals from Biomass: Proceedings of symposium: 1990. ACS Symposium Series 476; American Chemical Society Washington D.C. (1992)

[129] Fracture testing wood adhesives with composite cantilever beams / C.T. Scott, B.H. River, J.A. Koutsky / Journal of testing and evaluation 20(4) (1992)

[130] Contoured beam specimen for determining the fracture toughness of wood-adhesive joints / B.H. River, E.A. Okkonen / Journal of testing and evaluation 21(1) (1993)

[131] Difurfuryl diisocyanates: New adhesives derived from renewable resources / M.S. Holfinger, A.C. Conner, L.F. Lorenz, C.G. Hill / Journal applied polymer science 49:337–344 (1993)

[132] Phenol-formaldehyde resin curing and bonding in steam-injection pressing Part II. Differences Between Rates of Chemical and Mechanical Responses to Resin Cure / A.W. Christiansen, R.A. Follensbee, R.L. Geimer, J.A. Koutsky, G.E. Myers / Holzforschung 47: 76–82 (1993)

[133] Bonding mechanisms between polypropylene and wood: coupling agent and crystallinity effects / P.C. Kolosick, G.E. Myers, and J.A. Koutsky / In: Wood Fiber/Polymer Composites, Fundamental concepts, processes, and material options: Proceedings of 1st Wood Fiber–Plastic Composite Conference; Madison, WI, and 45th annual meeting of the Forest Products Society, New Orleans, LA, Forest Products Society (1993)

[134] Polyamine-modified urea-formaldehyde-bonded wood joints. III. Fracture toughness and cyclic stress resistance and hydrolysis resistance / R.O. Ebewele, B.H. River, and G.E. Myers / Journal of applied polymer science 49:229–245 (1993)

[135] Improving durability of urea-formaldehyde-bonded wood joints / R.O. Ebewele, Bryan H. River, George E. Myers / Adhesives age Vol. 36(13) (Dec. 1993)

[136] Failure mechanisms in wood joints bonded with urea-formaldehyde adhesives / B.H. River, R.O. Ebewele, G.E. Myers / Holz als Roh- und Werkstoff 52:179–184 (1994)

[137] Fracture of adhesive-bonded wood joints / Bryan H. River / Handbook of adhesive technology. New York: Marcel Dekker (1994)

[138] Hydroxymethylated resorcinol coupling agent and method for bonding wood / C.B. Vick, K.H. Richter, B.H. River / U.S. Patent No. 5,543,487 (1996)

[139] Utilization of proteins and carbohydrates as adhesives to bond wood / A.H. Conner / Proc. Of the Ninth International Conference on Jojoba and Its Uses and of the Third International Conference on New Industrial Crops and Products / Association for the Advancement of Industrial Crops (1996)

[140] Durable wood adhesives from furfuryl diisocyanates / K.M. Coppock, D.R. Holm, A.H. Conner, C.G. Hill / In: Wood adhesives 1995, A.W. Christiansen, A.H. Conner (eds.), Forest Products Society (1996)

[141] Wood Adhesives 1995 / A.H. Conner, A.W. Christiansen, eds. Proceedings of a symposium sponsored by USDA Forest Service, Forest Products Laboratory and the Forest Products Society / Forest Products Society Proceedings No. 7296 / Forest Products Society, Madison WI. (1996)

[142] Critical variables in the rapid cure and bonding of phenolic resins / R.L. Geimer, A.W. Christiansen / Forest products journal 46(11/12) (1996)

[143] More durable epoxy bonds to wood with hydroxymethylated resorcinol coupling agent / C. B. Vick / Adhesives age 40(8) (1997)

[144] Utilization of black wattle bark and tannin liquefied in phenol in the preparation of resol-type adhesives / M.A.E. Santana, M.G.D. Baumann, A.H. Conner / Lignocellulosic plastics and composites / A.L. Leao, F.X. Carvalho, E. Frollini, eds. (1997)

[145] Bondability of salvaged yellow-cedar with phenol-resorcinol adhesive and hydroxymethylated resorcinol coupling agent / E.A. Okkonen, C.B. Vick / Forest products journal 48(11/12) (1998)

[146] Terpene emissions from particleboard and medium-density fiberboard products / Melissa G.D. Baumann, Stuart A. Batterman, Guo-Zheng Zhang / Forest products journal Vol. 49(1) (Jan. 1999)

[147] Predicting the reactivity of adhesive starting materials / A.H. Conner / International Contributions to Wood Adhesion Research / Forest Products Society. Proceedings No. 7267 Alfred W. Christiansen, Louis A. Pilato, eds. (1999)

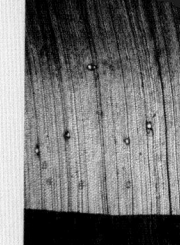

[148] Adhesive bonding of wood materials / revised by C.B. Vick / FPL–GTR– 113 (1999)

[149] Durability of one-part polyurethane bonds to wood improved by HMR coupling agent / C.B. Vick, E.A. Okkonen / Forest products journal 50(10) (2000)

[150] Predicting the reactivity of phenol compounds with formaldehyde under basic conditions: An Ab initio study / A.H. Conner / Journal applied polymer science 78:355–363 (2000)

[151] Wood Adhesives 2000 / A.H. Conner, A.W. Christiansen, eds. / Proceedings of the symposium sponsored by the Forest Products Laboratory and in cooperation with the Forest Products Society, the Adhesion Society, the International Union of Forestry Research Organizations (IUFRO), and the Japan Wood Research Society; Harvey's Resort Hotel and Casino, South Lake Tahoe, Nevada / Forest Products Society, Madison WI. (2001)

[152] Predicting the reactivity of phenolic compounds with formaldehyde. II. Continuation of Ab Initio Study / T. Mitsunaga, A.H. Conner, C.G. Hill Jr. / J. Applied Polymer Science 86:135–140 (2002)

[153] Development of a novolak-based hydroxymethylated resorcinol for wood adhesives / A.W. Christiansen, C.B. Vick, E.A. Okkonen / Forest products journal 53(2) (2003)

[154] Selectivity of bonding for modified wood / Charles R. Frihart, Rishawn Brandon, and Rebecca E. Ibach / Proceedings of the 27th Annual Meeting of the Adhesion Society, Inc "From molecules and mechanics to optimization and design of adhesive joints": February 15–18, 2004, Wilmington, NC. (Blacksburg, VA): Adhesion Society (2004}

[155] Competitive soybean flour/phenol-formaldehyde adhesives for oriented strandboard / J.M. Wescott, C.R. Frihart / Proc. 38th International Wood Composites Symposium / Washington State University (2004)

[156] Analysis of soy flour/phenol-formaldehyde adhesives for bonding wood / L.F. Lorenz, C.R. Frihart, J.M. Wescott / In: Wood adhesives 2005, San Diego, CA Nov. 2005 / Forest Products Society (2005)

[157] Method for quantifying percentage wood failure in block-shear specimens by a laser scanning profilometer / C.T. Scott, R Hernandez, C.R. Frihart, R Gleisner, T. Tice / Journal ASTM International 2(8) (2005)

[158] Wood adhesives prepared from lucerne fiber fermentation residues of *Ruminococcus albus* and *Clostridium thermocellum* / P.J. Weimer, Linda F. Lorenz, Charles R. Frihart, William R. Kenealy / Applied microbiology and biotechnology Vol. 66 (2005)

[159] Wood adhesion and adhesives / C.R. Frihart / Chapter 9 in: Handbook of wood chemistry and wood composites, R.M. Rowell, ed. / CRC Press, New York, NY (2005)

[160] Wood structure and adhesive bond strength / C.R. Frihart / Characterization of the cellulosic cell wall / Blackwell Pub. Co., Ames, IA (2006)

[161] Adhesive bonding of wood treated with ACQ and copper azole preservatives / L.F. Lorenz C.F. Frihart / Forest products journal 56(9) (2006)

[162] Wood Adhesives 2005 / C.R. Frihart, ed. / Proceedings of the symposium sponsored by the Forest Products Laboratory and in cooperation with the Forest Products Society, the Adhesion Society, the International Union of Forestry Research Organizations (IUFRO), the Japan Wood Research Society, The Adhesive and Sealant Council, Inc. and Adhesives and Sealants Industry Magazine; November 2–4, 2005, Holiday Inn on the Bay, San Diego, CA, USA / Forest Products Society, Madison, WI (2006)

[163] Characteristics of wood adhesion bonding mechanism using hydroxymethyl resorcinol / D.J. Gardner, C.F. Frazier, A.W. Christiansen / Proceedings of the symposium: Wood Adhesives 2005, C.R. Frihart, ed. Sponsored by USDA, Forest Service, Forest Products Laboratory, and in cooperation with the Forest Products Society, the Adhesion Society, the International Union of Forestry Research Organizations (IUFRO), the Japan Wood Research Society, The Adhesive and Sealant Council, Inc. and Adhesives and Sealants Industry Magazine; November 2–4, 2005, Holiday Inn on the Bay, San Diego, CA, USA / Proceedings No. 7230, Forest Products Society, Madison, WI. (2006)

[164] High-soy-containing water-durable adhesives / J.M. Wescott, C.R. Frihart A.E. Traska /Journal adhesion science and technology 20(8) (2006)

[165] Dispersion adhesives from soy flour and phenol-formaldehyde / C.R. Frihart, J.M. Wescott, A.E. Traska / Proc.: 30th annual meeting of the Adhesion Society / Feb. 2007, Tampa Bay FL (2007)

[166] Model for understanding the durability performance of wood adhesives / C.R. Frihart / Proc.: 30th annual meeting of the Adhesion Society / Feb. 2007, Tampa Bay FL (2007)

[167] Water-resistant vegetable protein adhesive dispersion / J.M. Wescott, C.R. Frihart / U.S. Patent No. 7,345,136 (2008)

[168] Why do some wood-adhesive bonds respond poorly to accelerated moisture-resistant test? / C.R. Frihart, J.M. Wescott / 9th Pacific Rim bio-based composites symposium, Rotorua, New Zeland (November 5–8, 2008)

[169] Adhesive groups and how they relate to the durability of bonded wood / C.R. Frihart / Accepted by journal of adhesion science and technology (2009)

Laminated Products

◄ Church in Middleton, Wisconsin, showing the beauty and functionality of large laminated arches.

Glued-Laminated Construction

FPL's pioneering work on the engineering design of glued-laminated construction helped launch the laminating industry in the United States. Many churches and other buildings have been constructed with laminated arches and beams. Spans as long as 250+ feet have been constructed. With modern waterproof glues, wood ships have been restored to important roles in our Navy as nonmagnetic minesweepers, torpedo boats, and other fast, light vessels. The laminated parts of these modern ships now can be constructed with wood treated prior to laminating. [1–3, 6, 7, 10]

Challenge: One of the major uses of wood is as a beam. Historically, large beams were simply a matter of selecting a large tree and sawing it to form the beam. However, with the decreased availability of large trees these large beams were less available. One way to replace them is by fabricating beams by laminating boards together.

FPL's Contributions: FPL researchers designed and evaluated various beams to determine how to economically fabricate beams to maximize strength, and they determined if underutilized species can be satisfactorily used.

Results: The results of research have eliminated the need to cut large trees to produce satisfactory beams. Also, many smaller, less utilized species of wood can now be assembled and used as large beams.

▲ Russell Moody evaluating the failure of the outer ply in a tested beam. (1970s) {28F}

◄ Billy Bohannan, foreground, and Freeman Drew, Structures Research Engineer of the Association of American Railroads, examine failures in a 50-foot laminated wood beam tested in bending. Results were used to develop new theory relating beam size to strength. (1960s) {M 130611} [13, 16, 17]

Laminated Construction (Glulam)

Glued-laminated timber (glulam) is a highly engineered wood product made up of several layers of solid lumber bonded together with adhesives. Widely used in wood-frame residential and commercial construction, glulam is primarily used for beams, arches, posts, and other heavy load applications. These members are the structural backbones of a building, and understanding their behavior is essential to ensuring occupant safety. Manufacturers and researchers have long recognized that when bonding lumber laminations into a glulam beam, the resulting strength of the beam is greater than the cumulative strength of the individual laminations. Understanding this "laminating effect" is important to optimizing glulam performance, achieving the most efficient material usage, and producing a product that performs safely. Research work led by FPL scientists has explained and quantified the theoretical and empirical link between constituent lumber, finger joint, and glulam beam properties to determine engineering performance. This research has provided American glulam manufacturers a direct means to compare and market their glulam products in the European Common Market. [32–34, 39–41]

FPL developed a procedure for manufacturing laminated products that makes it possible to utilize effectively the various grades of lumber in accordance with their inherent strength properties. With this procedure, it is possible to estimate statistically the frequency of occurrence, size, and distribution of knots in glued-laminated members, thus providing the foundation for the engineering use of such members on a sound, economical, and safe basis. [13, 36, 44, 45]

◀ Laminated beam showing high-quality boards placed as the top and bottom plies of the beam and lower quality boards placed toward the center of the beam. This design maximizes the strength of the beam with no increase in weight. (1980s) {78 K}

High Yield of Laminated Structural Products (Press-Lam)

FPL developed a process to improve the yield of structural products from trees at a reasonable cost using rotary cutting of up to 1/2-inch-thick veneer from logs, press-drying the thick veneer in less than 13 minutes, applying glue to the hot sheets, and laminating the sheets into thick structural materials. The process averaged a 90% yield of dry product from southern pine logs, compared with the 40% percent yield common at the time. Several plies vertically laminated together produce structural beams of desired width, thickness, and length that have greater uniformity, superior strength and equivalent stiffness relative to solid-sawn lumber. A bridge was built in the George Washington National Forest in Virginia using Press-Lam. [20]

Parallel-Laminated Veneer (PLV)

Parallel-laminated veneer (PLV) (now referred to as laminated veneer lumber, LVL) is made by feeding slices of veneer through a conventional glue spreader. The slices are layered on top of each other with the grain of each layer running parallel. Lumber-type products of desired length can be made. This enables wide dimension, thick structural material to be made from small-diameter logs. Advantages include virtual elimination of warp, more uniform mechanical properties, and increased product yield. FPL scientists studied the use of thick-sliced veneer and evaluated the strength properties of PLV. [14, 21, 23, 26]

◀ Testing of instrumented beam, made from parallel laminated veneers, to determine its bending stiffness and strength. (1970s)

Products from Parallel-Laminated Veneer: FPL demonstrated that it is economically feasible to use commercially developed and manufactured parallel-laminated lumber to produce high-quality, high-value products such as ladder rails, mobile home truss chords, joists, beams, and upholstered hardwood furniture frames. This process using lower grade material increases lumber yield, offers more uniform strength properties, and eliminates dependence of product dimensions on log dimensions. [23]

Wood I-joists

Wood I-joists are increasingly being used to support wood floors and roof areas in many houses. Many of these I-joists are made using laminated veneer lumber (LVL) for the flange components. [15, 28]

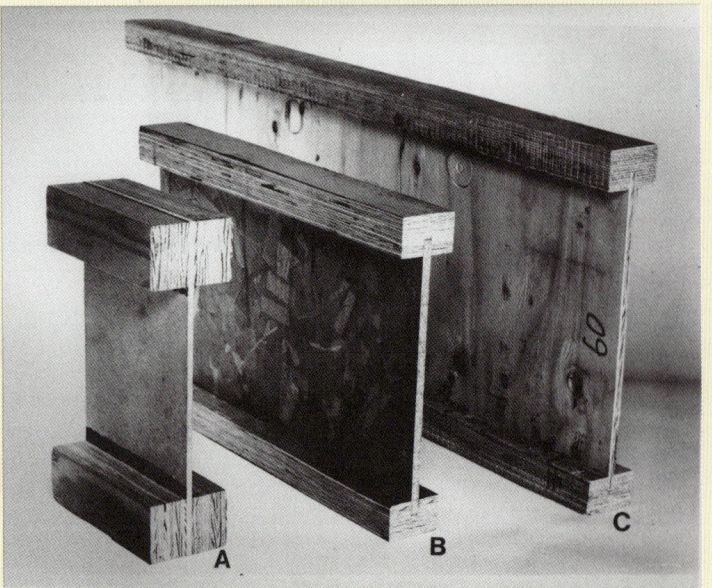

◀ Wood I-Joist made using parallel laminated veneer flanges with various web materials: A, hardboard; B, flakeboard; C, commercial plywood. (1970s)

Other Laminated Wood Products

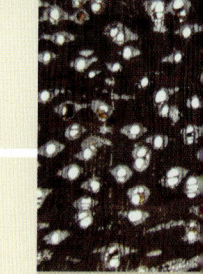

Nail-Laminated Posts: In cooperation with the University of Wisconsin, criteria were developed for design of spliced nail-laminated columns used in construction of post-frame buildings. [30]

Lam I-Joists: A new structural building product was made from small-diameter, fire-prone timbers. The focus was on the processing of small-diameter curved and culled timber into dimensional 2 by 4 studs and then converting that material into a value-added laminated I-beam, called LamLumber. [43, 45]

Baseball Bats: Laminated baseball bats are permitted by the National Collegiate Athletic Association and the major leagues and are being produced by some manufacturers as a result of FPL research. Use of these bats, with hickory cores and ash striking faces, eliminates the excessive breakage that occurs in the presently demanded slim-handled bats made of light-weight ash. [5]

Laminate Show-Through: In the production of furniture and other objects using particle-board and other interior constructions, the core often shows through the surface laminate. Production control is possible using an instrument developed by Bruce Heebink for measuring the amount of such show-through. [8]

Paper Overlays: Research resulted in a medium-density, resin-treated paper suitable for masking defects in plywood, thus giving superior surface characteristics to products made of lower grade veneers. Overlay papers are bonded to the plywood and widely used in pre-fabricated housing. Such papers can effectively mask defects in low-grade lumber, decrease its swelling and shrinking with moisture changes, and provide an excellent base for paint. [4, 9]

Large Crossties from Small Trees: Using the enormous inventory of smaller sized hardwood trees, a system was developed to make larger railroad crossties. By using steel dowels and glue, 7- by 9-inch mainline ties can be manufactured by laminating two smaller pieces cut from small logs. [31]

Composite Crossties: A system was developed to treat thin boards with preservatives and then laminate them to reach crosstie size, providing a more durable tie. Another system converted discarded crossties into flakes and then, by gluing the flakes together, into new, durable particle-composite crossties. [18, 22]

► Samples of laminated wood products along with a rafter and truss made by industry. Products like these were evaluated or improved by FPL's cooperative work with industry and universities to extend the timber supply by effectively using as much of the tree as possible. (2000s) [25, 27, 38, 43, 45]

1910

Virtually nonexistent in the United States.

2010

Laminated arches and other laminated products in common use. Companies employed 55,000 people and shipped products with a total value of $9,617,832,000. (2006 Annual Survey of Manufacturers)

Further Information

[1] The glued laminated wooden arch / T.R.C. Wilson / USDA technical bulletin No. 691 (1939)

[2] Strength of glued laminated Sitka spruce made up of rotary-cut veneer / R.F. Luxford / FPL No. 1512 (1944)

[3] Tests of glued laminated wood beams and columns and development of principles of design / T.R.C. Wilson, W.S. Cottingham / FPL No. 1687 (1947)

[4] Some properties of paper-overlaid veneer and plywood / FPL / Southern lumberman (July 15, 1949)

[5] Laminated baseball bats / George E. Heck / Wood (Oct 1950)

[6] Laminated southern pine / Alan D. Freas / Southern Lumberman Vol. 187(2345) (Dec. 15, 1953)

[7] Fabrication and design of glued laminated wood structural members / A.D. Freas, M.L. Selbo / (USDA Technical bulletin No. 1069 (1954)

[8] A new technique for evaluating show-through of particle board cores / B.G. Heebink / Forest products journal Vol. 10(8) (Aug. 1960)

[9] Paper overlaid lumber / Bruce G. Heebink / Forest Products Journal Vol. 11(4) (Apr. 1961)

[10] Effect of joint geometry on tensile strength of finger joints / M.L. Selbo / Forest products journal Vol. 13(9) (Sept 1963)

[11] Time-related flexural behavior of small Douglas-fir beams under prolonged loading / R.L. Youngs, H.C. Hilbrand / Forest Products Laboratory (1963)

[12] Prestressed laminated wood beams / B. Bohannan / FPL–RP–8 (1964)

[13] Beam strength as affected by placement of laminae / Peter Koch, Billy Bohannan / Forest products journal Vol. 15(7) (July 1965)

[14] Tensile strength of lumber laminated from 1/8-Inch-thick veneers / R.C. Moody / FPL–RP–181 (1972)

[15] Predicting performance of hardboard in I-Beams / T.J. Ramaker / FPL–RP–185 (1972)

[16] Evolution of Glulam strength criteria / Billy Bohannan, R.C. Moody / Forest products journal Vol. 23(6) (June 1973)

[17] Design criteria for large structural glued-laminated timber beams using mixed species of visually graded lumber / R.C. Moody / Research paper FPL 236 (1974)

[18] Recycling the wood crosstie: researchers develop and test laminated particle tie / B.G. Heebink / Railway track and structures Vol. 73(7) (July 1977)

[19] EGAR process makes wide-dimension lumber from small logs / G.B. Harpole, E. Wiliston, H.H. Hallock / Southern lumberman (December 15, 1977)

[20] Press-Lam: progress in technical development of laminated veneer structural products / R.W. Jokerst, R.C. Moody, C.C. Peters, E.L. Schaffer, J.L. Tschernitz, J.J. Zahn / Research paper FPL 279 (1977)

[21] Bending strength of vertically glued laminated beams with one to five plies / Ronald W. Wolfe, Russell C. Moody / FPL–RP–333 (1979)

[22] Hardwood Press-Lam crossties: processing and performance / J.L. Tschernitz, E.L. Schaffer, R.C. Moody, R.W. Jokerst, D.S. Gromala, C.C. Peters, W.T. Henry / FPL–RP–313) (1979)

[23] Parallel-laminated veneer: processing and performance research review / T.L. Laufenberg / Forest products journal Vol. 33(9) (Sep 1983)

[24] Economic feasibility of synthetic fiber reinforced laminated veneer lumber (LVL) / T.L. Laufenberg, R.E. Rowlands, G.P. Krueger / Forest products journal Vol. 34(4) (April 1984)

[25] Fiber-reinforced wood composites / R.E. Rowlands, R.P. VanDeweghe, T.L. Laufenberg, G.P. Krueger / Wood and fiber science Vol. 18(1) (1986)

[26] Butt joint reinforcement in parallel-laminated veneer (PLV) lumber / D.S. Larson, L.B. Sanberg, T.L. Laufenberg, G.P. Krueger, R.E. Rowlands / Wood and fiber science Vol. 19(4) (1987)

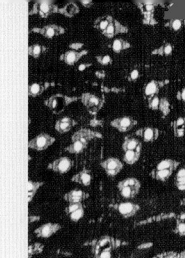

[27] Prefabricated wood composite I-beams: A literature review / R.J. Leichti, R.H. Falk, T.L. Laufenberg / Wood and fiber science Vol. 22(1) (1990)

[28] Prefabricated wood I-joists: an industry overview / R.J. Leichti, R.H. Falk, T.L. Laufenberg / Forest products journal Vol. 40(3) (March 1990)

[29] Designing, engineering, and testing wood structures / Thomas M. Gorman / Journal of materials education. Vol. 13(516) (1991)

[30] Bending properties of reinforced and unreinforced spliced nail-laminated posts / David R. Bohnhoff, Russell C. Moody, Steven P. Verrill, L.F. Shirek / FPL–RP–503 (1991)

[31] Technical practicality of dowel-laminating crossties before drying and treating / Richard A. Hale, John P. Howe / Forest products journal Vol. 42(6) (June 1992)

[32] Glued-laminated timber: Laminating effects / R. Falk, F. Colling / Conference proceedings: Pacific timber engineering conference, July 11–15, 1994, Gold Coast, Australia, Vol. 2. Fortitude Valley MAC, Qld.: Timber Research and Development Council, (1994)

[33] Fiber stress values for design of Glulam timber utility structures / Roland Hernandez, Russell C. Moody, Robert H. Falk / FPL–RP–532 (1995)

[34] Laminating effects in glued-laminated timber beams / Robert H. Falk, Francois Colling / Journal of structural engineering Vol. 121(12) (Dec. 1995)

[35] The Forest Products Laboratory—Giving its work away / Wood Vol. 100(4) (1896–1996) (Wood and wood products centennial, 100 years of American woodworking) (1996)

[36] Efficient hardwood glued-laminated beams / Harvey B. Manbeck, J.J. Janowiak, P.R. Blankenhorn, P. Labosky, Jr. R.C. Moody, R. Hernandez / Proceedings of the International Wood Engineering Conference, Sheraton New Orleans, New Orleans, Louisiana, USA, October 28–31, 1996. Volume 1. [S.l.]: International Wood Engineering Conference; Madison, WI: printed by Omnipress (1996)

[37] Prestressing wood beams with bonded tension elements / John Peterson / Thesis (Ph.D.) University of Wisconsin, 19646 (1996)

[38] ISO 9000: Issues for the structural composite lumber industry / Steve G. Winistorfer, Harold J. Steudel / Forest products journal Vol. 47(1): 43–47 (1997)

[39] Glued-laminated timber / Russell C. Moody, Roland Hernandez / In: Smulski, Stephen, Ed. / Engineered wood products—A guide for specifiers, designers and uses. Madison, WI: PFS Research Foundation: 1-1–1-39. Chap. 1. (1997)

[40] Design and performance aspects of United States and European glulam / Robert Falk / In: Proceedings, conference on research standardization applications; 1997 June 6; Technical University, Graz, Austria. Graz, Austria: Institute for Steel Construction, Wood Construction, and Industrial Construction: 1–21. Chap. 2. (1997)

[41] In-Place shear strength of wood beams / Douglas R. Rammer, David I. McLean, William F. Cofer / 5th world conference on timber engineering, August 17–20, 1998, Montreaux, Switzerland: proceedings, Volume 1. Lausanne: Presses Polytechniques et Universitaires Romandes, c1998. 1: 207–214 (1998)

[42] Localized notch reinforcement for wooden beams / Lawrence A. Soltis, Robert J. Ross, Douglas R. Rammer / U.S. Patent 5,852,909 (Dec. 29, 1998)

[43] Lam I-joist: a new structural building product from small-diameter, fire-prone timber / J.F. Hunt, J.E. Winandy / FPL–RN–0291 (2003)

[44] Improved utilization of small-diameter ponderosa pine in glued-laminated timber / Roland Hernandez / David W. Green, David William, David E. Kretschmann, Steven P. Verrill / FPL RP–625 (2005)

[45] Glue LamLum: A tool for evaluating the financial feasibility of laminated lumber plants / E.M. Bilek, J.F. Hunt / FPL–GTR–165 (2006)

Bridges

◄ The Alton Sylor Memorial Bridge, Joncy Gorge, in Angelica, New York, received first place for America's Best Timber Bridges 2004–2005 for highway bridges with main spans over 40 feet.

In 1988, the U.S. Congress passed legislation known as the Timber Bridge Initiative (TBI). Its objective was to establish a national program to provide effective and efficient utilization of wood as a structural material for highway bridges. Responsibility for the development, implementation, and administration of the timber bridge program was assigned to the USDA Forest Service. Within the program, the Forest Service established three primary program areas: demonstration bridges, technology transfer, and research. The demonstration bridge program, administered by the Forest Service National Wood in Transportation Information Center (NWITIC) in Morgantown, West Virginia, provided matching funds on a competitive basis to local governments to demonstrate timber bridge technology through the construction of demonstration bridges. The NWTIC also maintained a technology transfer program to provide assistance and state-of-the-art information related to timber bridges.

Responsibility for the research portion of the TBI program was assigned to the Forest Products Laboratory (FPL), the national wood utilization research laboratory of the U.S. Forest Service. The primary focus of TBI research was the development of new and improved technology for timber bridge materials and systems. In 1992, the FPL research program was expanded to include wood transportation structures such as noise barriers, marine facilities, retaining walls, and sign supports. At the same time, a substantial joint research program was initiated between the Federal Highway Administration (FHWA) and FPL to implement the FHWA timber bridge research program mandated under the Intermodal Surface Transportation Efficiency Act (ISTEA) of 1991. In subsequent years, the FHWA also initiated a substantial covered wood bridge rehabilitation and research program under the National Historic Covered Bridge Preservation Program mandated in 1999 by the Transportation Equity Act for the 21st Century (TEA21) and in 2004 by the Safe, Accountable, Flexible, and Efficient Transportation Act—A Legacy for Users (SAFETA–LU). To assist in implementing this program, the FHWA developed partnerships with FPL and the National Park Service (NPS).

Funding for the Forest Service NWITIC ended in fiscal year (FY) 2004, leaving significant voids in research capability, education, and technical assistance to governmental agencies, industry, and research institutions. In response, FPL moved to reestablish the National Center for Wood Transportation Structures (NCWTS) at Iowa State University (ISU), based on more than 25 years of cooperative research between ISU and FPL in the area of wood bridges as well as their extensive expertise in transportation structures. This Center was established and minimally funded in FY 2007 as a university–government–industry partnership to provide greater program efficiency and leverage federal funding to maximize public benefit at minimal cost. Given the longstanding cooperative relationship with FHWA, and the developing partnership with the NPS, both agencies were also included as partners in the new NCWTS at ISU. This partnership, joint national program, and creation of the NC-WTS are particularly timely and valuable because wood bridges represent more than 27% of the nation's bridges and afford an opportunity to efficiently utilize naturally sustainable forest resources. Additionally, more than 25% of our nation's bridges are structurally deficient or functionally obsolete. The problem is especially critical on rural road systems, where wood bridges offer many advantages due to their ease of construction.

Challenge: How to use hardwoods and nontraditional softwoods in design of wood bridges to maximize load-carrying capacity over a long span of both distance and time in an economical manner.

FPL's Contributions: FPL developed the knowledge base for design criteria, wood species evaluation, preservative treatments, and long-term evaluation of bridges. Early FPL research was on optimizing glulam technology for timber bridges, development of press-lam bridges, and technical guidance for inspecting engineers. More recent research included evaluation of stress-laminated deck technologies, crash-worthy rail and curb systems for bridges, nationwide field monitoring of timber bridges, inclusion of hardwood and nontraditional softwood species for bridges, environmental impacts of preservative chemicals in the bridge environment, and technical guidance in the use of nondestructive methods for inspecting bridges.

Results: The inspection and replacement of many wood bridges, especially on rural roads, resulted in improved designs and application of underutilized wood species, leading to better and safer transportation structures in our nations highway system.

Early Wood Bridges

Nature provided the first wood bridges in the form of fallen trees placed across streams that allowed people to cross without getting wet. The next step was to place the trees where one wanted to cross. With the advent of vehicles, groups of trees were placed across streams and planks nailed across the trees. From there, more and more elaborate bridges were designed to carry larger and heavier loads. FPL, working with Forest Service Alaska Region engineers, evaluated the strength of log bridge stringers after several years of use. [4, 5]

◀ Bridge in Alaska made using whole trees for bridge stringers and guard rails. (1970s)

▲ Forest Service and logging road bridge with abutment cribbing and span using trees. (1970s) {7 J}

▲ Overview of the test site for evaluating spruce and western hemlock tree bridge stringers in bending. (1970s)

◄ Failed bridge stringer. Forces up to 120,000 pounds were applied with reasonable control of the load rate. (1970s)

Nationwide Field Monitoring of Timber Bridges

Field monitoring, evaluations, and load testing have been completed on numerous stress-laminated decks constructed of various wood species and materials and exposed to different environmental conditions. A stress-laminated deck is made up of a series of lumber laminations that are stressed together with steel bars.

▲ Load testing of bridge using fully loaded trucks.

▲ Example of field monitoring a stress-laminated deck bridge. Note the ends of the steel bars that compress the laminated wood of the decks. Also note the remote data acquisition system box on upper center guardrail. (2000s) [25]

FPL conducted field evaluations of existing structures to determine the load distribution characteristics of stress-laminated timber deck bridge systems to refine procedures and design criteria for AASHTO (American Association of State Highway and Transportation Officials) acceptance. (2000s)

▲ Instrumentation of glulam girder railroad bridge for static and dynamic loading.

▲ Static (non-moving) load on the instrumented bridge.

◄ Dynamic (moving) load on the instrumented bridge.

System Development and Design

Design Criteria for New Bridge Systems and Components

Press-Lam

The FPL demonstrated that vertically laminated bridge decking fastened with adhesives could be used (a new system in the 1970s). Plant fabrication of glued-laminated bridge decking greatly reduces on-site labor and makes this construction advantageous for areas such as Alaska. [6]

PRODUCING VENEER

CLIPPING VENEER

PRESS DRYING VENEER

APPLYING ADHESIVE TO HOT VENEER

LAY-UP OF DIMENSION STOCK

LAMINATING DIMENSION MATERIAL

▲ Schematic of the process used at FPL to produce bridge press-lam components. (1970s) {M 147291}

◄ Manufacturing press-lam bridge deck panels. (1970s) {M 147294}

Glued-Laminated Bridge Components

▲ Glued-laminated bridge components being strength tested at FPL to determine bridge girder load distributions for use in evaluating differences between glulam and solid-sawn wood girders. (1960s) {M 121 619}

▲ Laminated arch bridge in Trough Creek State Park, Pennsylvania, demonstrating moderate span bridge design using glulam. (1990s)

Structural Glued-Laminated Timber Using Nontraditional Species

◄ Early example of a 32-foot-long stress-laminated bridge system consisting of composite glulam beams (southern pine and red pine) (Sawyer County, Wisconsin) (1991) [11]

FPL researchers also investigated the feasibility of gluing nontraditional species, such as hardwoods, in glulam and then treating them with waterborne preservatives. [12]

New System (1980–1990) Showing Stess-Laminated Box and Deck Design [6, 9]

◀ Example of a three-span stress-laminated 72-foot-long bridge using eastern hemlock. The 32-foot center section is stress-laminated box sections using southern pine and eastern hemlock. The bridge is in Baraga County, Upper Peninsula of Michigan. (1992) [15]

Structural Composite Lumber (1990s)

Field evaluations and load testing have been completed on six stress-laminated T-Beam bridges constructed of laminated veneer lumber (LVL). [16]

◀ Installing a pre-fabricated stress-lam T-beam section bridge super-structure. T-beam was made using structural compos-ite lumber (SCL) (laminated veneer lumber, LVL). [18]

▲ Placing a box beam with wood posts and rail on a wood bridge deck.

Wood Design Values and Properties

FPL scientists refined National Design Specification (NDS) engineering design properties for visually graded hardwood lumber through in-grade testing and added hardwood species and secondary softwood species to the NDS (information used in bridge design). [12, 28, 30]

Preservatives [20, 23]

◀ Underside of early glulam wood bridge showing preservative-treated laminated stringers, lateral supports, and tim-ber abutment wall. (2000s)

An example of preservative- (CCA-) treated red pine stress-laminated deck and creosote-treated rails and curbs. The bridge is near Mancelona, Michigan. (2000s)

Alternative Transportation System Timber Structures

Rural county wood bridge with steel running plates, wood curbs and side rails based on an early static design rail system. (1960s) {M 128 097}

Crashworthy Bridge Rails and Curbs [16]

Historically, the design of bridge railings, used to contain errant vehicles crossing the bridge, was based on static load criteria. It is now recognized that dynamic full-scale vehicle crash testing is necessary. [17]

Full-scale crash-testing of timber guardrail on a longitudinal timber deck. (1990s)

▲ A side view of a crash-test setup for a transverse deck bridge rail system (tests performed at the University of Nebraska–Lincoln). (1990s)

▼ Wood bridge with wood abutments and wood side rail posts with steel rails.

Inspection and In-Place Evaluation [19, 23, 26, 27, 29, 31, 32]

Wood Bridge Manual for Inspection and Maintenance

A comprehensive manual was published as a guide for bridge engineers and maintenance personnel in the inspection of wooden bridges for decay. The manual discusses methods for locating possible areas of decay and procedures for controlling or eradicating it. Agriculture Handbook 557, *Wood Bridges—Decay, Inspection, and Control*, is still used by Forest Service and highway engineers. [3]

▲ Inspecting a bridge for defects, wear, and decay. Common tools include picks, drills, and stress waves. (2000s)

▲ Robert Vatalaro, University of Minnesota–Duluth, using a resistance microdrilling tool to evaluate the condition of the wood members in the bridge substructure. (2000s)

Technology Transfer

In addition to numerous technical reports on specific studies. FPL engineers worked closely with local, state, and federal agencies to modify standardized bridge design plans, expand the number of wood species that can be used, and improve inspection and rehabilitation of bridges. [22]

FPL scientists worked with the Federal Highway Administration to develop standard design plans and crash-tested rail design details for use by bridge engineers. [10, 17, 22]

▲ Photo of the cover of *Timber Bridges: Design, Construction, Inspection and Maintenance.* (1990) [7]

◀ Early untreated wood Burr arch truss bridge in Pennsylvania. The bridge features a nailed laminated arch and is protected from the weather by roof and siding.

▲ An example of two wood highway bridges using laminated arches and beams. The larger one is a three-hinge glulam deck arch bridge at the Keystone Wye interchange off U.S. Highway 16, near Mount Rushmore, South Dakota. The arch spans 155 feet and supports a 26-foot-wide roadway. The lower bridge is a beam type (1968). {7 K}

▲ Keystone Wye interchange in 2008.

1910

Bridge design was limited, and bridges were constructed using large, untreated logs and timbers.

2010

Efficient designs for safety, and use of laminated treated wood is common.

Further Information

[1] Distribution of wheel loads on timber bridges / E.C.O. Erickson / Research paper FPL 44 (1965)

[2] Design procedure for glued-laminated bridge decks / R.L. Tuomi, W.J. McCutcheon, / Forest products journal Vol. 23(6) (June 1973)

[3] Wood bridges: decay inspection and control / Wallace E. Eslyn, Joe W. Clark / USDA Agriculture handbook No. 557 (1979)

[4] Bending strength of large Alaskan Sitka spruce and western hemlock log bridge stringers / R.L. Tuomi, R.W. Wolfe, R.C. Moody, F.W. Muchmore / FPL Research paper 341 (1979)

[5] Strength of log bridge stringers after several year's use in southeast Alaska / R.C. Moody, R.L. Tuomi, W.E. Eslyn, F.W. Muchmore / Research paper FPL 346 (1979)

[6] Design, fabrication, testing, and installation of a Press-Lam bridge / J.A. Youngquist, D.S. Gromala, R.W. Jokerst, R.C. Moody, J.L.Tschernitz / Research paper FPL 332 (1979)

[7] Timber bridges: design, construction, inspection, and maintenance / Michael A. Ritter / USDA, Forest Service, Engineering management EM–7700–8 (1990)

[8] Methods for assessing the field performance of stress-laminated timber bridges / Michael A. Ritter, E.A. Geske, W.J. McCutcheon, R.C. Moody, J.P. Wacker, L.E. Mason / Proceedings of the 1991 international timber engineering conference 1991 September 2–5, London, England: TRADA (1991)

[9] Behavior of stress-laminated parallel-chord timber bridge decks: experimental and analytical studies / Al G. Dimakis, Michael G. Oliva, Michael A. Ritter / FPL–RP–511 (1992)

[10] Standard plans for southern pine bridges / Paula D. Hilbrich Lee, Michael A. Ritter, Michael Triche / FPL–GTR–84 (1995)

[11] System stiffness for stress-laminated timber bridge decks / Julio F. Davalos, R.C. Moody, R. Hernandez / Proceedings of the International Wood Engineering Conference, Sheraton New Orleans, New Orleans, Louisiana, October 28–31, 1996. Vol. 1 / International Wood Engineering Conference; Madison, WI: printed by Omnipress, (1996)

[12] Yellow poplar glued-laminated timber: product development and use in timber bridge construction / Roland Hernandez, M.A. Ritter, R.C. Moody, P.D. Hilbrich Lee / FPL GTR–94 (1996)

[13] Timber bridges in the United States / Sheila Rimal Duwadi, Michael A. Ritter / Public roads Vol. 60(3) (winter 1997)

[14] Research accomplishments for wood transportation structures based on a national research needs assessment [microform] / Michael A. Ritter, Sheila Rimel Duwadl / General technical report FPL GTR–105 (1998)

[15] In situ performance of stress-laminated timber bridge decks / James A. Kainz, James P. Wacker, Michael A. Ritter / Proceedings of the SEM spring conference on experimental and applied mechanics and experimental/numerical mechanics in electronic packaging III: June 1–3, 1998, Houston, Texas. Society for Experimental Mechanics, Inc. (1998)

[16] Stress-laminated SCL bridges with prestressing strand / Michael A. Ritter, Matthew S. Smith, Terry J. Wipf / PTEC '99: Pacific timber engineering conference, 14–18 March 1999, Rotorua, New Zealand: proceedings Vol. 3. Rotorua: New Zealand Forest Research Institute, Forest research bulletin No. 212 (1999)

[17] Crashworthy railing for timber bridges / Michael A. Ritter, Ronald K. Faller, Sheila Rimal Duwadi / PTEC '99: Pacific timber engineering conference, 14–18 March 1999, Rotorua, New Zealand: proceedings Vol. 3. Rotorua: New Zealand Forest Research Institute, Forest research bulletin No. 212 (1999)

[18] T-section glulam timber bridge modules: modeling and performance / Paul A. Morgan, S.E. Taylor, M.A. Ritter, John M. Franklin / ASAE 1999 Annual international meeting: July 18–21, 1999, Toronto, Ontario, Canada. St. Joseph, MI: ASAE paper No. 99–4207 (1999)

[19] Inspection of timber bridges using stress wave timing nondestructive evaluation tools: a guide for use and interpretation / Robert J. Ross, Roy F. Pellerin, Norbert Volny, William W. Salsig, Robert H. Falk / FPL GTR–114 (1999)

[20] Wood in transportation program: an overview / Sheila Rimal Duwadi, Michael A. Ritter, Edward Cesa / Proceedings of the fifth international bridge engineering conference on bridges, other structures, and hydraulics and hydrology: Apr. 3–5, 2000, Tampa, FL. Washington, DC: National Academy Press, 2000: Transportation Research Record 1696, Paper No. 5B0105, Vol. 1, (2000)

[21] Effect of wood preservatives on stress-laminated southern pine bridge test decks / James A. Kainz, Nur Yazdani, Joy Kadnar / FPL RP–599 (2001)

[22] Standard plans for timber bridge superstructures / James P. Wacker, Matthew S. Smith / FPL–GTR–125 (2001)

[23] Ultrasonic inspection of large bridge timbers / Robert Emerson, Robert J. Ross / Forest products journal Vol. 52(9) (Sept. 2002)

[24] Extending service life of timber bridges with preservatives / James P. Wacker, Douglas M. Crawford / Proceedings of the 19th U.S.–Japan bridge engineering workshop: UJNR panel on wind and seismic effects: Tsukuba, Japan, October 27–30, 2003. Tsukuba-shi, Ibaraki-Ken, Japan: Public Works Research Institute, Technical memorandum PWRI No. 3920: p. 77–85 (2003)

[25] Field performance of stress-laminated highway bridges constructed with glued laminated timber / J.P. Wacker / Proceedings, 2004 structures congress: building on the past, securing the future, May 22–26, Nashville, TN. Washington, DC: American Society of Civil Engineers (2004)

[26] Wood and timber condition assessment manual / Robert J. Ross, Robert H. White, Roy F. Pellerin, Xiping. Wang, Brian K. Brashaw / Forest Products Society (2004)

[27] Nondestructive assessment of timber bridges using a vibration-based method / Xiping Wang, James P. Wacker, John R. Erickson / Proceedings of the 2005 structures congress: April 20–24, New York, NY: The Structural Engineering Institute of the American Society of Civil Engineers (2005).

[28] Structural products offer opportunities for lower value hardwoods / TechLine, Forest Products Laboratory (April 2005)

[29] Condition assessment of timber bridges / Brian K. Brashaw, Robert J. Vatalaro, James P. Wacker, Robert J. Ross / FPL GTR–159–160 (2005)

[30] Hardwoods for timber bridges: a national program emphasis by the USDA Forest Service / James P. Wacker, Ed Cesa / Wood design focus. Vol 15(2) (summer 2005)

[31] Estimating bridge stiffness using a forced-vibration technique for timber bridge health monitoring / James P. Wacker, Xiping Wang, Robert J. Ross / Proceedings : NDE conference on civil engineering: a joint conference of the 7th structural materials technology: NDE/NDT for highways and bridges and the 6th international symposium on NDT in civil engineering, August 14–18, St. Louis, MO. Columbus, OH: American Society for Nondestructive Testing, Inc.s ISBN: 1571171487 (2006)

[32] New inspection technologies improve bridge safety / Newsline, Forest Products Laboratory / Vol. 6(4) (Fall 2007)

Veneer

◄ Veneer being laid on top of another layer of veneer that has had adhesive applied. Multiple layers of veneer are bonded together to make plywood, with each layer oriented at 90° to the next to form a panel with crossbanded layers.

Veneer is wood that has been cut from a log using a sharp knife rather than a saw. Veneer can be "peeled" or "sliced." When peeling, like a paper towel off a roll, the log (bolt) is turned against a knife and a thin layer of wood is cut (peeled). When slicing, the log is affixed and the knife moves up and down, slicing parallel to the length of the log. The wood properties of these thin sheets are essentially the same as sawn wood, but the surface of the wood—because of cutting, drying, and laminating into plywood—can significantly change the surface characteristics of the veneer. [21, 22]

Challenge: How to make flat sheets out of round logs. One method is to peel the wood from the log, another is to slice it with a knife. This peeled or sliced material is called veneer and is the basic component of plywood or laminated veneer lumber.

FPL's Contributions: FPL scientists improved the methods to efficiently peel the wood so that more of the log can be converted into veneer, including development of specialized pre-heating "cooking" schedules for softening the wood prior to veneer peeling.

FPL research on processing variables resulted in reduction of thickness variations caused from uneven peeler crushing because of the large difference in density between earlywood and latewood. This significantly improved the processing of Douglas-fir and southern pine into veneer. FPL developed information for making veneer from other little-used species. A veneer improvement program was developed to transfer the technology.

Results: Significantly improved utilization of wood through the development of new wood processing methods. An important benefit of peeling or slicing wood is the elimination of sawdust. Bonding the veneer sheets together makes possible large plywood panels that have unique properties. (See Plywood section.)

▲ Bill Nelson and Marshall Begel evaluating veneer using nondestructive test methods.

Peeled Wood

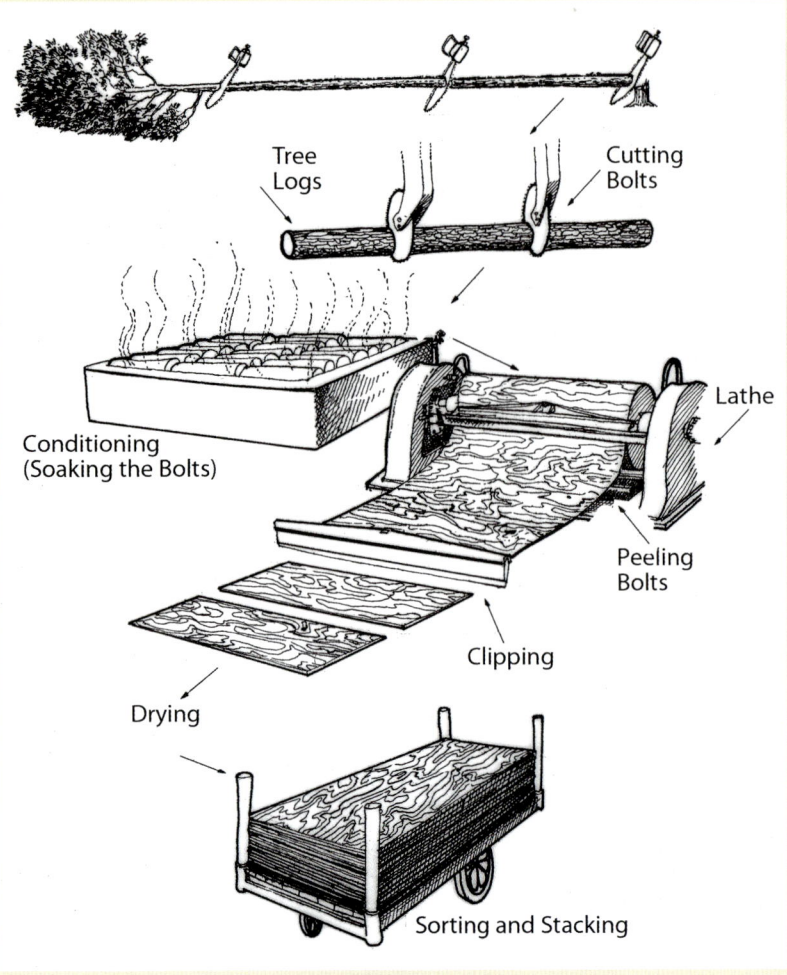

Tree Logs

Cutting Bolts

Conditioning (Soaking the Bolts)

Lathe

Peeling Bolts

Clipping

Drying

Sorting and Stacking

◀ Veneer manufacturing is generally done by cutting the tree into short-length logs (bolts), soaking the bolts in hot water for hours (heating schedule determined by species), mounting the bolts on the lathe, peeling thin layers of veneer, clipping the veneer to specified lengths, drying the veneer, and sorting and stacking. This process does not create sawdust but is time consuming because of the requirement to soak the logs. (1983) {ML 835 454} [15, 17]

Conditioning the Wood

The conditioning of the wood prior to peeling is controversial. However, research has shown that moisture content, permeability (largely inherent in the wood species), and temperature of wood can have a marked effect on veneer cutting. [17] Most conditioning is done in large vats with steam or hot water; however, some experiments in 1953 were done using electrical heating. Approach is not presently commercially used.

◀ Test setup to determine possible use of electrical conductivity to speed heating of the veneer bolts. (1950s) {M 94142 F} [4]

Veneer Cutting

▲ Veneer lathe with bolt in position for peeling. (1970s) {42 C}

FPL research provided information on the influence of machine settings on veneer quality. [2, 17]

▲ Robert Patzer, Lucy Ebisch, and John Hunt handling peeled veneer off the FPL experimental veneer lathe. (1970s) {101 J}

▲ John Lutz and Edward Locke, FPL Director, observing the cutting action taking place during peeling of southern pine to make veneer. (1960s) {M 120 647}

▶Clipper used to remove sections with holes and other defects in the veneer prior to drying. (1970s)

▶Blowup of the knife cutting veneer and the distortions that occur. (1960s) {30 L}

◀ Roger Russell taking high-speed photos of the cutting action to aid in understanding what is happening as the knife slices through the wood. (1960s) {64 C}

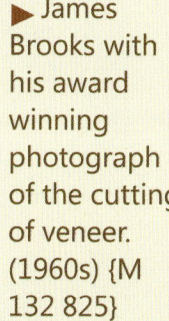

▶James Brooks with his award winning photograph of the cutting of veneer. (1960s) {M 132 825}

Based on this FPL research, instruments were developed for aligning the knife and the nosebar on the veneer lathe. The use of these instruments has made it possible for veneer cutters, with a minimum of experience, to consistently produce well-cut veneer. [5]

Powered Backup Roll

One of the challenges in cutting veneer with a lathe was the problem of spin-out of the bolt. This spin-out occurs when the force required for cutting exceeds the force that can be delivered by the chucks that grip the bolt ends to turn the bolt. The chucks keep spinning but the bolt quits turning. Because the bolt can no longer be turned by the chucks, the remaining wood in the bolt cannot be peeled and is lost with the removal of the bolt from the lathe. [18]

◄ Experimental power backup roll developed by Frank Fronczak, foreground. Robert Patzer is in background. This major development allowed the veneer lathe to continue the peeling action to the minimum diameter of the bolt and improved the quality of the veneer. Without the power backup roll, less veneer is obtained from the bolt. Power backup rolls are now commercially used. (1980s) {98 C} [18]

Veneer Drying

◄ Bruce Heebink inspecting a sheet of veneer after drying in an experimental press dryer. (1960s) {M 123 259}

▲ Conventional veneer dryer being used to dry thick-sliced wood samples. (M 100 565)

FPL researchers developed factors that affected drying of veneer and provided industry with suggestions for operating mechanical dryers. [17]

► James Muehl operating a second generation veneer press dryer. To date, this approach has not been commercialized. (1980s) {6 Q} [20]

Hemlock Veneer

In the 1950s, working with the Douglas Fir Plywood Association, the Laboratory conducted research on how best to utilize hemlock, which was in abundant supply when there was a shortage of Douglas-fir. Because of the success of this research, needs of the country for plywood were met and mills were able to continue manufacturing plywood. [3]

Southern Pine Veneer

Prior to 1964, there were no structural softwood plywood plants operating in the South, primarily because southern pine then posed technical problems in plywood manufacturing. Based on years of collaborative research by the Plywood Association, the FPL, and the Forest Service Southern Research Station prior to 1964, a means to work with the heavily resined, small-diameter, southern pine timber was developed. Three plants, two in Texas and one in Arkansas, were in production by late 1964. [7]

At the same time, equipment manufacturers began making the special machinery used in southern pine plywood manufacturing. Seventeen plants were operating in seven states of the southern pine region less than two years after southern pine plywood production was proved commercially feasible. [9, 10, 16]

Veneer Species Guides

FPL scientists developed two guides covering important production aspects of the veneer industry:

"Wood and Log Characteristics Affecting Veneer Production" [12]
"Veneer Species that Grow in the United States" [13]

Sliced Wood

◀ Harry Panzer operating the Laboratory's veneer slicer. Slicing is another commercial means of veneer production but does not produce a continuous sheet. Though slicing is usually used for hardwoods, in this research it was used to study the cutting of Douglas-fir white pocket wood, a waste material, for use as wall paneling. (1960s) {42 J} [8]

Thick Slicewood

FPL pioneered the production of thick material (up to 1/2 inch) cut from logs and bolts by the use of a knife instead of a saw. This sliced material enabled much larger recovery from the original raw material. It has been successfully used for making such items as fences and bin pallets for fruits and vegetables. [14]

◀ Experimental thick wood slicer, R.H. McAlister at the controls and Alfred Mergen on top of the slicer. (1960s) {M 132 769}

▲ 1/2-inch-thick boards produced by the thick wood slicer. Yellow poplar, Douglas-fir, white fir, and pine (top to bottom). Note the fractures (called lathe checks) in the boards. These fractures lower some strength properties but enhance the treatment of the wood with preservatives. (1960s) {M 119 151}

Veneer Mill Improvement Program (VIP)

To help transfer FPL research to industry, the Laboratory set up a Veneer Mill Improvement Program (VIP). This program measured veneer mill raw-material conversion efficiency and, with aid of a computer simulation model, could predict gains from process improvements. The VIP analyzed log bucking, block centering in the lathe, and veneer peeling and clipping. This information was used to identify areas where improvements were feasible and quantified their effects. This program could result in higher productivity, decreased wood waste, and improved veneer quality. (1986) [19]

1910

Little softwood veneer produced. Hardwood veneer used for furniture.

2010

Major veneer production in both the northwest (using Douglas-fir and hemlock) and the south (using southern pine).

Further Information

[1] Thin plywood / Forest Products Laboratory F29 (1919)

[2] Experiments in rotary veneer cutting / H.O. Fleischer / Wood working digest Vol. 52(4) (Apr. 1950)

[3] Veneer cutting and drying properties: western hemlock / FPL No. 1766–7 (1951)

[4] Heating veneer logs electrically / H.O. Fleischer, L.E. Downs / FPL No. 1958 (1953)

[5] Instruments for aligning the knife and nosebar of the veneer lathe and slicer / H.O. Fleischer / Forest products journal Vol. 6(1) (Jan. 1956)

[6] Beech for veneer and plywood, and the gluing of beech / H.O. Fleischer / Northeastern logger Vol. 7(8) (Feb. 1959)

[7] Southern pine area regarded ready for establishing plywood industry / H.O. Fleischer, E.G. Locke / Lumber journal Vol. 66(11) (Nov. 1962)

[8] Slicewood—a promising new wood product / J.F. Lutz, H.H. Haskell, R.H. McAlister / Forest products journal Vol. 12(4) (May 1962)

[9] Southern pine plywood / H.O. Fleischer, John F. Lutz / Forest products journal Vol. 13(1) (Jan. 1963)

[10] Timber quality aspects of the southern pine plywood development / H.O. Fleischer / Proceedings, Society of American Foresters annual meeting (SAF) (1963)

[11] A survey of some developments in slicing and veneer cutting / John F. Lutz. / Plywood & panel (Feb. 1967)

[12] Wood and log characteristics affecting veneer production / J.F. Lutz / Research paper FPL 150 (1971)

[13] Veneer species that grow in the United States / J.F. Lutz / Research paper FPL 167 (1972)

[14] Thick slicing of wood: effects of wood and knife inclination angle / C.C. Peters, A.F. Mergen, R.A. Patzer / Forest products journal Vol. 22(9) (Sept. 1972)

[15] Techniques for peeling, slicing, and drying veneer / J.F. Lutz / Research paper FPL 228 (1974)

[16] Manufacture of veneer and plywood from United States hardwoods with special reference to the south / John F. Lutz / Research paper FPL 255 (1975)

[17] Wood veneer: log selection, cutting and drying / John F. Lutz / Forest Service Technical bulletin No. 1577 (1978)

[18] Powered back-up roll: new technology for peeling veneer / Frank J. Fronczak, Stephen P. Loehnertz / Research paper FPL 428 (1982) (Patent No. 4,381,023) (1983)

[19] Tighten up your mill with veneer improvement program / Jeanne Danielson, William VonSegen, Tim Donivan / Plywood & panel world Vol. 27(3) (June/July 1986)

[20] A continuous [sic] press dryer for veneer / Stephen P. Loehnertz / Forest products journal. Vol. 38(9) (Sept. 1988)

[21] Wood handbook: wood as an engineering material / FPL–GTR–113 (1999)

[22] Diagnostic guide for evaluating surface distortions in veneered furniture and cabinetry / A.W. Christiansen, M. Knaebe / FPL–GTR–143 (2004)

Plywood

◀ Plywood in various forms. Flat panel in center has been deconstructed to show the seven layers of cross-banded veneers.

Plywood consists of a glued combination of three or more sheets of veneer, usually with the grain of the alternate plies at right angles. Compared with solid wood, the chief advantages of plywood are an equalization of strength properties in length and width, greater resistance to splitting, and less dimensional change from warping because of the gain or loss of moisture. [14, 16]

Although plywood is most widely known as a construction material appearing in the form of panels, it is extremely versatile. If the plies are laminated with the wood grains parallel, the panels can be sawn into laminated veneer lumber (LVL) (see Laminated Products section). Plywood panels may also be molded into various shapes for special purposes.

Challenge: A major challenge was how to make large panels of wood that have more uniform properties and are resistant to splitting. One answer was plywood. It is so common and versatile that it is used in hundreds of application. It is strong and split resistant, comes in large sizes, and can be cut to any shape.

FPL's Contributions: FPL was a major contributor in the development of the southern pine plywood industry and in specialized pre-heating "cooking" schedules for softening logs prior to peeling. This later accomplishment decreased thickness variations resulting from uneven peeler crushing because of the large differences in density between southern yellow pine earlywood and latewood. Scientists helped develop water-resistant adhesives so plywood could be successfully used outdoors. They also developed test methods and used the results to evaluate design formulas for plywood. This research assisted in the development of plywood specifications.

Results: Using wood veneer, it is possible to make large panels of plywood that have many uses. FPL research in composites and engineering has been critical to the development of many types of commercial composite products, from plywood to fiberboard, and the national standards that ensure quality in the marketplace.

Adhesives Are Critical

Without a good adhesive, plywood is useless, which is true of any glue-laminated product. Over the years, the FPL has conducted research on adhesives that has improved the performance of laminated wood products.

Research emphasis has been on improving the moisture resistance of adhesives, improving surface bond quality and strength, determining how adhesives and wood interface to create a bonded wood surface, developing test methods to evaluate bond quality and durability, and helping to set specifications for effective bonding of wood. (See Adhesives section.)

► Plywood samples being exposed to the elements at the FPL test fence to determine the durability of new glues. (1930s) {M 29823 F}

▼ Condition of plywood panels made of the same woods and thickness combinations, but glued with different adhesives, after three years exposure. Panels on the left were glued with phenol-formaldehyde adhesive; those on the right were glued with casein adhesive. (1930s) {M 36235 F} {M 36242 F} [1]

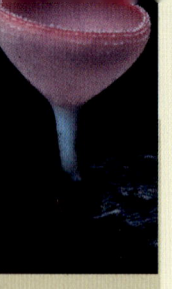

Phenolic resin glues have been found suitable for fabricated products that must resist continued exposure to moisture, high humidity, and extremes of temperature. [13] More recently FPL has made significant advances in bio-based adhesives based on denatured plant proteins to replace existing commercial resins made from petroleum. (See Adhesive section.)

Hemlock Plywood

In cooperation with the Douglas Fir Plywood Association, FPL investigated the manufacture of veneer and plywood from western hemlock logs, worked out recommendations for best practice, and assisted in the preparation of commercial specifications. [2]

Southern Pine Plywood

One of the major contributions to the better utilization of wood was the FPL's research on the proper processing of southern pine so that it could be made into plywood. Then, working with industry and the plywood associations, FPL developed standards for its use. [6, 7]

Plywood Grading

▶ Early grade stamp from APA–The Engineered Wood Association certifying the quality of the plywood the public buys. C-D is the quality rating of the facing plies; 32/16 is the span rating; "Interior" indicates the plywood is for uses not permanently exposed to weather; PS-1-74 is the Product Standard (the new one is PS-1-07); "Exterior Glue" is the type of water-resistant glue used to make the plywood. [13]

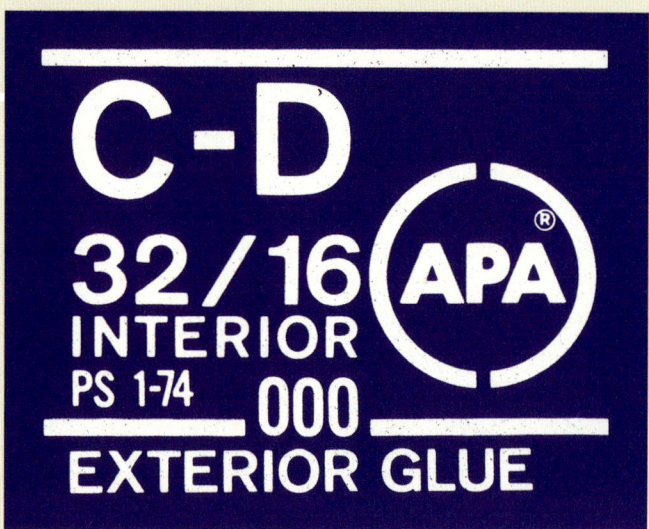

Plywood Properties, Design Guides, Product Standards

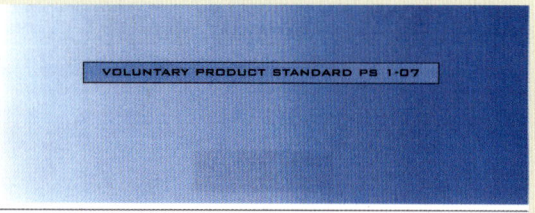

◀ FPL contributed to the development of plywood properties, design standards, and product standards. (2007) [13]

Surface Profiler

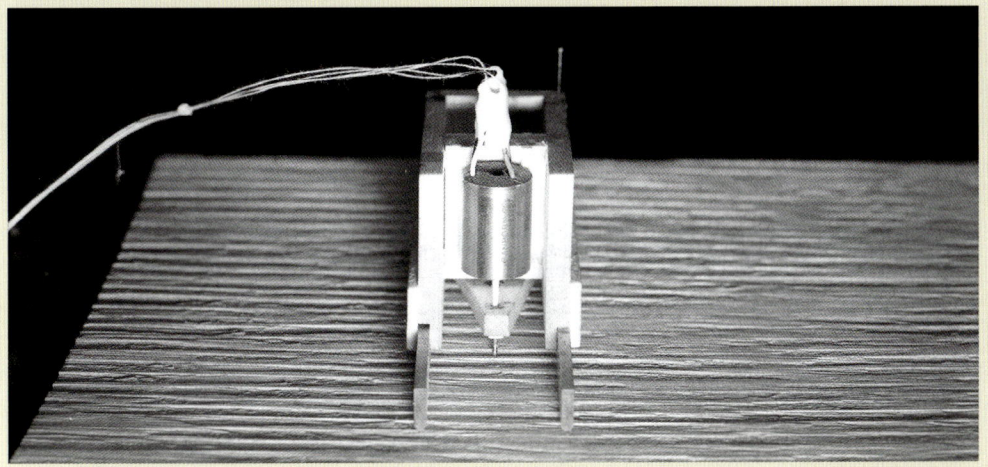

◀ FPL developed a gage that rapidly indicates the surface condition of machined wood and the feasibility of its use in glued joints, finishing situations, and in other instances where the surface of the material is a controlling factor. (1970s) {M 139 938} [15]

Plywood Composites

▶ Piece of softwood plywood showing the different layers of cross-banded veneer. Plywood can be made in as few as three layers or many more. The plies can be different wood species or thicknesses. Two pieces of plywood can be laminated to another material (such as the foam shown in the picture) to make a sandwich material. Some plywoods are made with water-resistant adhesives, others with non-water-resistant adhesives. Plywood can even be pressure-treated with preservatives or fire-retardant chemicals. (108 E)

Softwood plywood

Engineering Design of Products Made from Plywood

◀ One of the many uses of plywood has long been complex engineered structures, such as high-speed watercraft and aircraft. Based on research at FPL, the strength properties of plywood were determined, allowing development of design formulas that are used in complex engineered structures. (1940s) {M 48675} [4, 14]

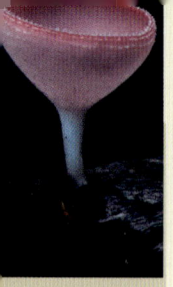

▲ Howard Hughes's Flying Boat H–4 (commonly known as the "Spruce Goose"). It is an example of a complex engineering product made primarily from spruce lumber and birch plywood.

Formed Plywood Beam

▶ Laminated plywood designed as a decorative I-Beam.

1910

Virtually nonexistent in the United States.

2010

Major industry developed, with many new products and national standards that ensure quality in the marketplace.

Companies employed 38,489 people and shipped products with a total value of $8,259,210,000. (2006 Annual Survey of Manufacturers)

Further Information

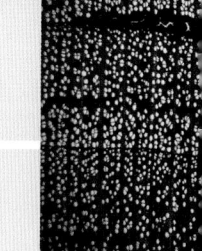

[1] Artificial resin glues for plywood / FPL No. 1055 (1939)

[2] Comparative strength of western hemlock and Douglas-fir plywood / R.F. Luxford / Forest Products Laboratory (1941)

[3] Design of plywood webs in box beams: supplement: the effect of repeated buckling on the ultimate strengths of box beams with shear webs in the inelastic buckle range / W.C. Lewis, T.B. Heebink, W.S. Cottingham / FPL No. 1318E (1944)

[4] Durability of resorcinol glue bonds in gusset-type assembly joints similar to those used in wood boats / M.L. Selbo / FPL No. 1714 (1956)

[5] Design of plywood webs in box beams / FPL No. 1318 (1958)

[6] Forest Products Laboratory experiments making veneer from southern pine logs / John F. Lutz / Plywood Vol. 12(1) (June 1961)

[7] Southern pine plywood / H.O. Fleischer, John F. Lutz / Forest products journal Vol. 13(1) (Jan. 1963)

[8] Bending strength and stiffness of plywood / FPL Research note FPL 059 (1964)

[9] Manufacture and general characteristics of flat plywood / FPL Research note FPL 064 (1964)

[10] If we need it: construction plywood from hardwoods is feasible / J.F. Lutz, / Plywood & panel magazine Vol. 14(9) (Feb. 1974)

[11] Manufacture of veneer and plywood from United States hardwoods with special reference to the South / Research paper FPL 255 (1975)

[12] Economic evaluation of process innovations in plywood manufacturing / Henry N. Spelter / Executive summaries: 42nd annual meeting, June 19–22, 1988, Quebec, Canada. Madison, Wis.: Forest Products Research Society (1988)

[13] Performance standards and policies for structural-use panels / American Plywood Association (1997)

[14] Wood handbook: wood as an engineering material / FPL–GTR–113 (1999)

[15] Diagnostic guide for evaluating surface distortions in veneered furniture and cabinetry / A.W. Christiansen, M. Knaebe / FPL–GTR–143 (2004)

[16] Encyclopedia of forest sciences (4 Vol.) / Jeffery Burley, Julian Evans, John A. Youngquist / Elsevier Academic Press, San Diego, CA (2004)

Flooring & Paneling

◄ Experimental mixed wood flooring in John Kulp's FPL office. The wood strips are held in blocks by splines, and the blocks are glued directly to a concrete subfloor. Janice Loder is taking notes. (1950s) {M 100 323 F} [3]

Challenge: How to use small pieces of wood in flooring.

FPL's Contributions: In the manufacturing of flooring, much of the wood is too short to be used in conventional flooring. Research at FPL demonstrated combining short pieces into squares that can be installed in decorative ways, just like tiles. FPL developed paperplywood flooring and press-dried wood paneling. FPL demonstrated the use of little-used species and suppressed-growth trees for flooring.

Results: New uses for wood that was previously wasted. Wood flooring is a major wood industry.

Mixed Species and Veneer Flooring

◄ Bruce Heebink discussing the use of new experimental mixed species wood flooring with Sharon Royston in FPL office. (1950s) {M 100 33

► Bruce Heebink explaining the use of veneer in wood flooring with Nancy Lane. (1950s) {M 100 457} [2]

▲ Short lengths, not suitable for paneling pieces, make handsome parquet flooring. (1960s) {M 129 357} [3, 5]

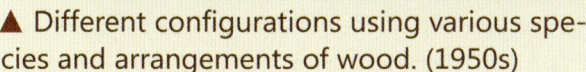

▲ Different configurations using various species and arrangements of wood. (1950s)

Suppressed, Small-Diameter Douglas-fir Flooring

◄ Suppressed, small-diameter Douglas-fir used for flooring. This use provides an outlet for material that is otherwise considered to be waste. (2000s)

Suppressed, Small-Diameter Western Larch Flooring

◄ Suppressed, small-diameter western larch used as flooring. (2000s) [14]

NCAA Basketball Flooring

◀ Special hard-wood basketball floor at Wake Forest University.

Many basketball floors are temporary and must be taken up and put down depending on what event is being featured in the arena. Meeting quality standards for this kind of portable flooring is difficult. FPL scientists, working with the University of Minnesota–Duluth's Natural Resources Research Institute (NRRI) and industry, evaluated the substrate structure (which is usually plywood or oriented strandboard), finishing and drying schedules, moisture absorption, vibration characteristics, and connectors that hold the floor together. Based on this evaluation, they developed an improved product that was easier to manufacture, set up, and take down, and still meet the demanding standards for high-performance basketball. (2000s) [13]

Paper-Plywood Flooring

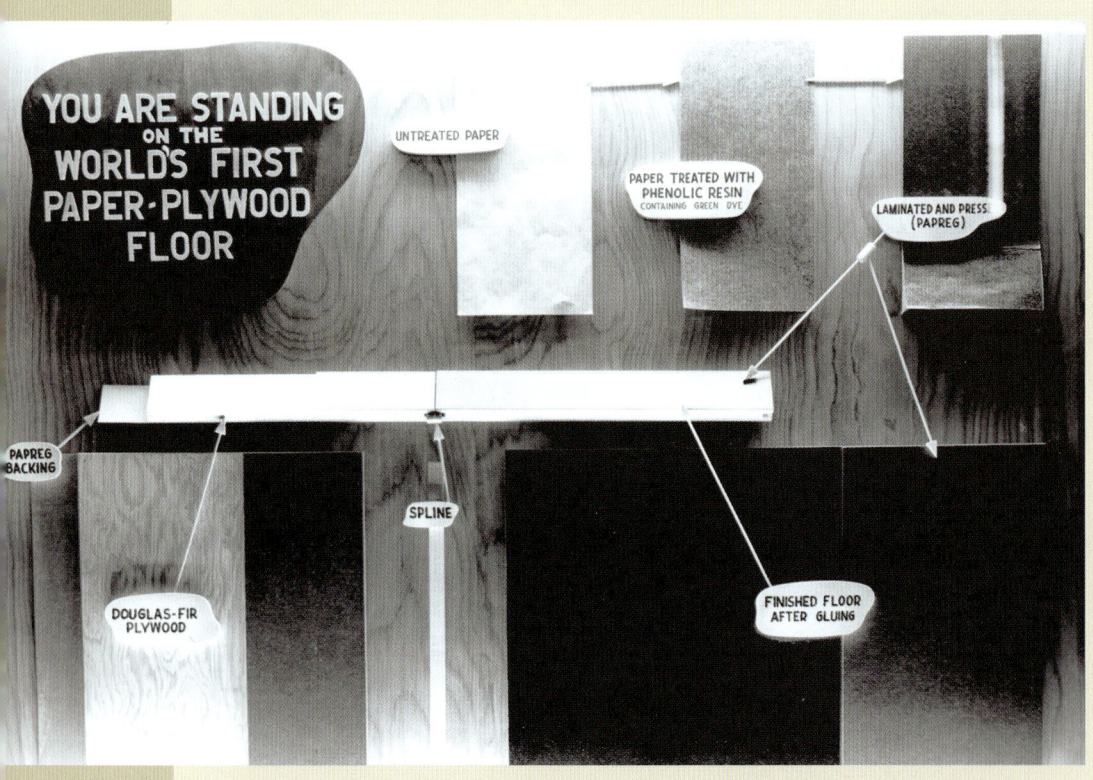

▲ Display showing how paper-plywood is made. The resin-treated paper (papreg) was developed during WW II. (1950s) {M 92088 F} [1]

▲ Floor in the lobby of the Forest Products Laboratory that was made with papreg laminated to plywood. Papreg is paper that is saturated with resin and then pressed with heat, which cures the resin. Pat Palmer on the left and Joyce Sobek on the right. (1950s) {M 92090 F}

Hardwood Floors Last Longer

FPL scientists and industry cooperators developed technology that increases the hardness, stain resistance, and fire retardancy of wood used in flooring. This technology is based on combining wood with organic monomers to fill the void space in wood, thereby forming a hard polymer network. The technology is patented by Shannon Fuller (Nashville, Tennessee), Dale Ellis (FPL), and Roger Rowell (FPL) and has been commercialized. The technology can be applied to any wood to improve its hardness, which increases its lifetime in use. The increased product lifetime results in conservation of resources, and the technology can be applied to wood species not now used by the flooring industry. [10–12]

Press-Dried Wood Paneling

◀ Untrimmed boards are held flat between steam-heated platens of this press while drying. Slight over-drying gives brown tints to red oak suggestive of chestnut or walnut. This press-dried material can be used for high-quality paneling. (1960s) {M 128 315} [8]

▶ Celeste Kirk adjusting a table in front of press-dried wall paneling. (1960s) {M 136 500-9} [8]

◀ Director H.O. Fleischer inspecting oak wall paneling with Oluwadare Awe, Nigeria's Assistant Conservator of Forests.

1910

Common woods used.

2010

Developed treated wood to improve wear resistance and methods to utilize waste material in flooring and paneling.

Companies employed 36,139 people and shipped products with a total value of $6,224,636,000. (2006 Annual Survey of Manufacturers)

Further Information

[1] Durability of papreg-to-papreg and papreg-to-birch glue joints / Herbert W. Eickner / FPL No. 1538 (1950)

[2] Veneer flooring / Bruce G. Heebink / Journal of the Forest Products Research Society (Sept 1952)

[3] New block flooring uses low grades / Southern lumberman Vol. 191(2388) (Oct. 1955)

[4] New veneer-lumber flooring developed for concrete slabs / Douglas A. Zischke / Southern lumberman Vol. 191(2393) (Dec. 15, 1955)

[5] Develop flooring blocks from low quality hardwood / Wood and wood products Vol. 60(12) (December 1955)

[6] Wood floors for dwellings / E.M. Davis, F.L. Browne, D. Brouse, H.W. Eickner / USDA Agriculture handbook No. 204, Supersedes USDA Circular No. 489 (July 1938) (1961)

[7] Paneling and flooring from low-grade hardwood logs / B.G. Heebink / Research Note FPL–RN–0122 (1966)

[8] Press-dried paneling shows promise: FPL research on Appalachian program produces commercial possibility / B.G. Heebink / Forest products journal Vol. 16(1) (Jan. 1966)

[9] Wood / E.H. Bulgrin / McGraw–Hill yearbook of science and technology New York: McGraw-Hill (1974)

[10] Finishing and maintaining wood floors / D.L. Cassens, W.C. Feist / USDA, North Central Regional Extension Publication No. 136 (1988)

[11] Chemical modification of wood / Rowell, R.M. / International Academy of Wood Science No. 1 (1996)

[12] Hardened and fire retardant products / B. Shannon Fuller, W. Dale Ellis, Roger M. Rowell / US Patent No. 5,683,820 (November 4, 1997)

[13] Checking out the action underneath the action at this year's NCAA Tournament / Gordie Blum / Forest Products Laboratory NewsLine Vol. 4, Issue 2 (2005)

[14] Small-diameter success stories II / Jean Livingston / FPL–GTR–168 (2006)

Overview of Wood Composites and Their Technologies

◀ Examples of wood composites that show simple panels to complex three-dimensional (3D) engineered wood products—all made from material that had been considered waste.

Wood composites are made from any fibrous or particulate wood materials that are bonded together using either natural bonding (no resin) or a thermoset resin or a thermoplastic or inorganic binder. This product mix ranges from fiberboard to laminated beams. Composites are used for a number of structural and nonstructural applications. The product lines range from panels for interior covering purposes to panels for exterior uses, and in furniture and architectural trim materials used in many different types of buildings. Lignocellulosic fibers and particles other than wood (such as straw) can many times be readily substituted for wood to produce other biocomposites with engineering properties similar to those of the wood composites discussed below.

Wood composite (and other biocomposite) materials can be engineered to meet a range of specific properties. When wood materials and processing variables are properly selected, the result can provide high performance and reliable service. With solid wood, properties are determined at the cellular level, and properties can be highly variable for pieces of solid wood both within and between wood species. With composite wood materials, properties are determined at the fiber, particle, flake, or veneer level, and properties are less variable. A key determinant of composite properties is the type of woody element used. These elements are available in a great variety of sizes and shapes and can be used alone or in combination. Wood and biocomposites fall into three general categories: engineered wood composites, wood–inorganic composites, and wood–plastic composites.

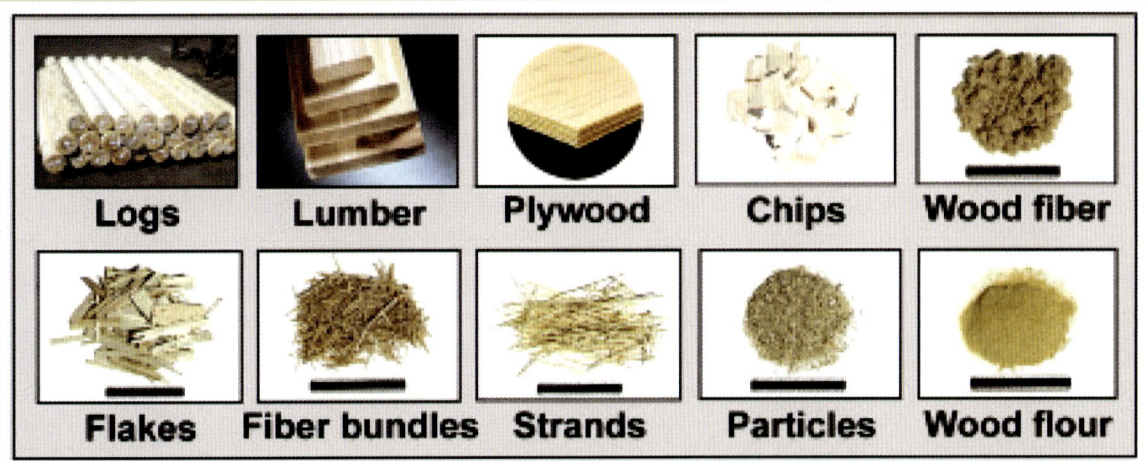

| Logs | Lumber | Plywood | Chips | Wood fiber |
| Flakes | Fiber bundles | Strands | Particles | Wood flour |

▲ Overview of the range of sizes and elements used in commercial wood composite technology.

Engineered Wood Composites

Engineered wood composites use a thermoset or heat-curing resin binder and can be grouped into three sub-categories based on the physical configuration of the wood element used to make the products: laminated, particle- or flake-based, and fiber-based composites. Within limits, the manufacturing processes are variants of that for oriented strandboard (OSB).

The performance of composites can be tailored to the end-use application of the product by varying the physical configuration of the wood material, adjusting the density of the composites, varying the resin type and amount, and incorporating additives to increase water, decay, or fire resistance.

Adhesives

Commonly used thermoset resin-binder systems include phenol-formaldehyde, urea-formaldehyde, melamine-formaldehyde, and isocyanate (diphenylmethane di-isocyanate, or MDI). These adhesives have been chosen based upon their suitability for bonding bio-based materials; the selection of one from this group is based on desired composite strength, durability requirements, and cost.

Laminated Composites

Laminated composites consist of wood veneers bonded with a resin binder and fabricated with either parallel- or cross-banded veneers. When laminae are laid parallel, the resulting product has higher performance properties parallel to the grain and is often used as a lumber substitute. When cross-banded, the composite product is moderately strong but has higher dimensional stability, which is critical when used as a panel product such as plywood.

Particle Composites

Particle-, flake-, strand-, or fiberboard composites are normally classified by density, element size, and process. Each is made with a dry woody element, except for fiberboard, which can be made by either dry or wet processes. Fiberboard is broadly classified into three groups: insulating board, medium-density fiberboard, and hardboard. Insulating board is also sometimes referred to as cellulosic fiber insulating board.

Particleboard panel products typically are made from small lignocellulosic particles and flakes, rather than fiber. The particles are bonded together with a synthetic adhesive under heat and pressure. The density levels for particleboard are the same as those for medium-density fiberboard.

In manufacturing **fiberboard**, lignocellulosic materials are first reduced to fibers or fiber bundles and then put back together by special forms of manufacture into fiberboard panels. Dry processes are used to make boards with high density (hardboard) and medium density (medium-density fiberboard, or MDF). Wet processes are used to make both high-density hardboard and low-density insulation fiberboard. Wet-process hardboards differ from dry-process fiberboards in that water is used as the distribution medium for forming the fibers into a mat. As such, wet-form technology is really an extension of paper manufacturing technology. Wet-process boards can also sometimes be made without additional resin-type binders.

Insulating board, also known as cellulosic fiber insulating board, is a generic term for a low-density homogeneous panel that is made from interfelted lignocellulosic fibers and that has been consolidated under heat to a density range between 160 and 500 kg/m³.

Medium-density fiberboard (MDF) is made from lignocellulosic fibers combined with a synthetic resin. The dry-process technology utilized to manufacture MDF is a combination of that used in the particleboard industry and that used in the hardboard industry. There are three density levels for MDF: low, <640 kg/m³; medium, 640–800 kg/m³; and high, >800 kg/m³.

Hardboard is a generic term for a homogeneous panel that is made from interfelted lignocellulosic fibers and has been consolidated under heat and pressure.

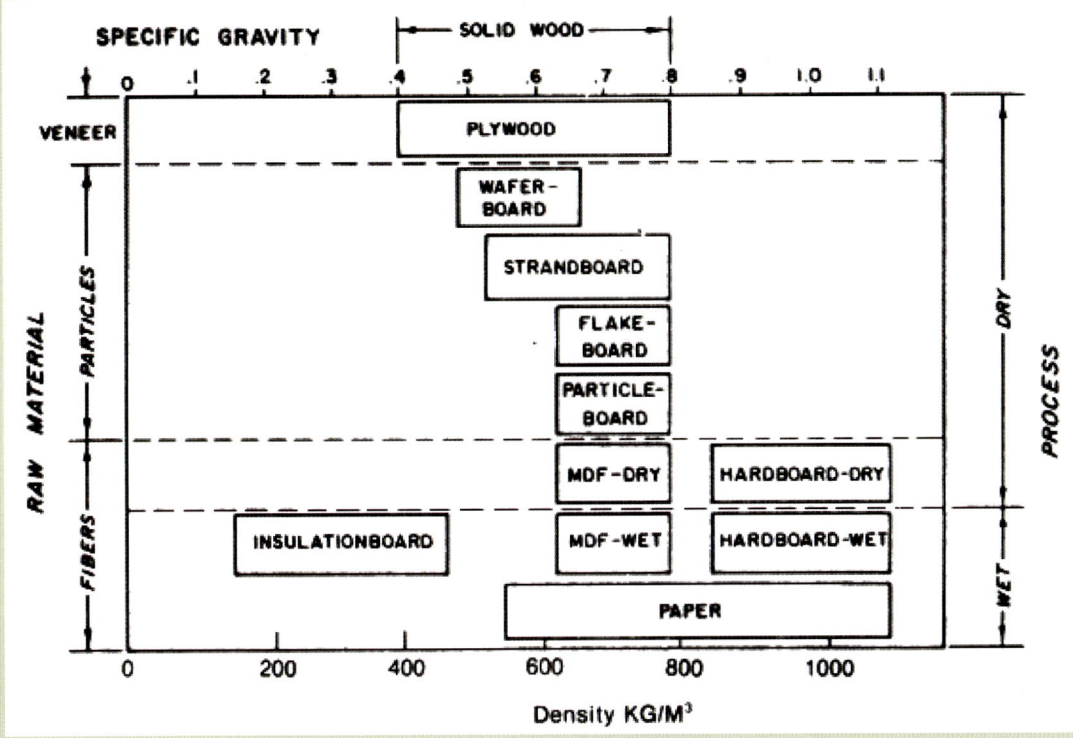

◄ Overview of potential options in wood composites relative to variations in density, raw material, and process. (Fiberboard manufacturing practices in the United States / O. Suchsland, G.E. Woodson / USDA Agriculture Handbook 640 (1986).) Note that insulation board is now known as cellulosic fiber insulating board.

Particle Composites

◄ Oriented strandboard (OSB) siding on a house under construction. OSB and other particle composites are higher valued products made from wood that in the past would have been burned or not utilized. (2000s)

Wood composites encompass a large number of products. (To help understand the differences, see the Overview of Wood Composites.) In this book, plywood is treated in a separate section. Composites made from pieces such as wafers, strands, flakes, or particles are covered in this section. Fiberboards—made from much smaller particles (such as fibers) that are combined using adhesive (dry formed) or combined wet using moisture (wet-formed)—are another similar family of fiber composites and are covered in the section on Fiberboards. [21, 31]

Challenge: How can more of a tree be converted into useful products? Much of a tree may not be usable as boards because it is too small, crooked, or split or has other defects. One method for utilizing the material, other than converting it to pulp or burning it, is to cut it into flakes or other particles and reconstitute it by gluing the particles together.

FPL's Contributions: FPL developed test procedures to determine the physical and mechanical properties of various composites. The FPL also evaluated the response of those properties to changes in particle geometry, species, binder type and amount, manufacturing techniques, and additives. Initial product development efforts resulted in panels and composite lumber elements, but research has shown that composites can be formed into complex engineered systems such as I-joists, wall structures, and flooring. [7, 25]

Results: Today, high-value products can be made from low-value wood material that was often considered waste and burned. FPL research contributed significantly to the development of oriented strandboard (OSB). OSB is a successful commercial product and finds many uses wherever structural panel material is desired, such as siding, roof decking, and sub-flooring.

Structural Particleboard

Particleboard is a wood product manufactured from wood particles, such as wood chips, sawmill shavings, or saw dust, and a synthetic resin or other binder (adhesive), which is pressed or extruded into panels. Depending upon components and manufacturing processes, the panels can be used as sub-flooring.

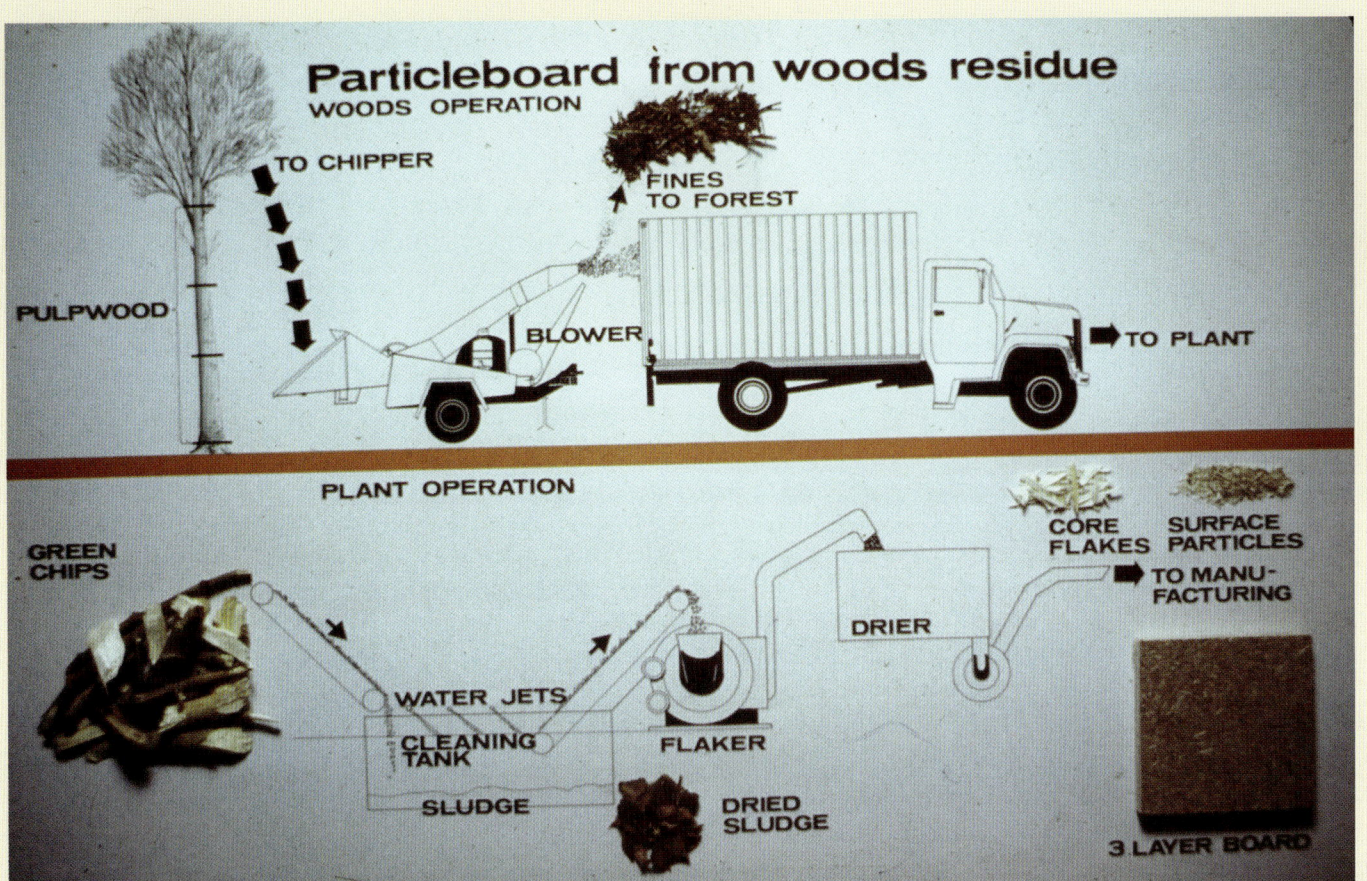

Particleboard from woods residue

WOODS OPERATION

TO CHIPPER

FINES TO FOREST

PULPWOOD

BLOWER

TO PLANT

PLANT OPERATION

GREEN CHIPS

CORE FLAKES

SURFACE PARTICLES

TO MANU-FACTURING

DRIER

WATER JETS

CLEANING TANK

FLAKER

SLUDGE

DRIED SLUDGE

3 LAYER BOARD

▲ Process for converting wood residue to usable panel products. (1960s) {10 N} [2]

STRUCTURAL PARTICLEBOARD FROM FOREST RESIDUES

◄ Early FPL research assisted in the use of forest residues to make useful products. (1970s) [4, 5, 9]

Some Products Made from Particle Composites

Particleboard can be used alone, or other materials such as veneer, plywood, impregnated paper, plastic, or metal can be bonded to it to form a sandwich structure, giving the assembly unique properties.

▶ Particleboard being combined with veneer to form a composite structure. (1970s)

▲ Wood particles can be molded into various shapes such as this bowl. (1960s) {M 101 003}

▲ Samples of structural composites. Left to right, flakeboard, particleboard, press lam, and comply. (1970s) [4, 6, 21]

Complex Three-Dimensional Engineered Floor and Wall Sections from Particle Composites

▲ Use of particle composite for flooring. It can also be used as complex three-dimensional (3D) engineered wall sections. {12 F}

▲ Bob Geimer showing a 3D wall section made of a particle composite using a press molding technique. (1970s) {52 N}

A wood-based building component of cellulosic particles and/or fibers and adhesive binder is, in a single pressing operation, molded into an integral product of sheathing and support members for use in constructing wood-frame buildings. The dies utilized during press-molding form a flat panel containing many evenly spaced channels that serve as support members to replace conventional framing such as studs, joists, and rafters. The integral product may serve as roof, wall, or flooring components in the usual wood-frame building applications. [10]

Flakeboard

The use of wood particles in the form of flakes, strands, or wafers provides another outlet for little used or waste wood and can result in commercial structural composite panels such as oriented strandboard (OSB)

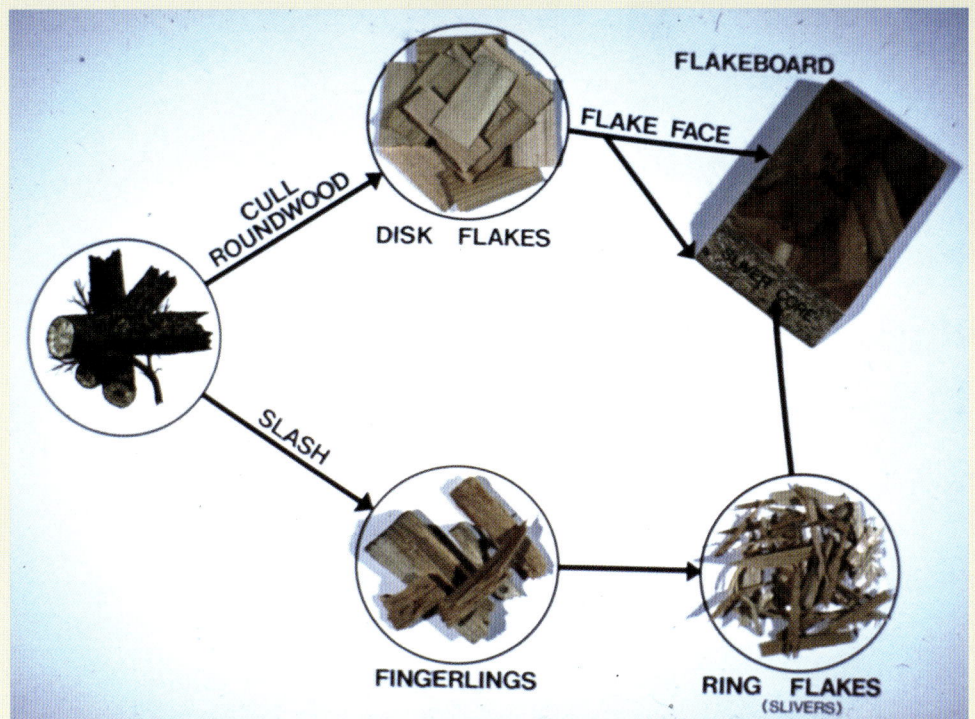

◀ Process for utilizing wood in flakeboard, a form of particle composite. (1960s) {10 M} [1, 2]

▲ Typical size of flakes used in flakeboard. (1960s) [1]

Applying Adhesive to the Wood Flakes

◀ Bill Rogers (left) and Copeland Francis handling flakes in preparation to apply adhesive to the flakes. (1980s) {76 H} [8, 12, 19, 21]

Lay-up of Adhesive-Treated Flakes for Pressing

After wood particles and adhesive have been mixed, they are formed into a thick pad (lay-up), placed in a heated press, and compressed into a panel.

◀ Laboratory lay-up of a mat of adhesive-treated flakes for pressing into a panel. {21 H}

Computer-Controlled Presses

Virtually every laboratory and industrial press is computer controlled today. The basic concepts were laid down by Bob Geimer and others in 1982 and 1990 after years of work developing some of the earliest computerized hot-press control technologies. [14, 16]

Steam Injection Pressing

By injecting steam into the mat, the press time for a 1/2-inch-thick structural flakeboard can be decreased from 4-1/2 minutes to about 90 seconds. However, the major advantage of the system is making possible the pressing of very thick mats (those greater than 1-1/2 inches). Also, smaller equipment is used, resulting in less energy consumption and lower costs. In addition, the system can incorporate additives for fire resistance and greater durability into the board. [13]

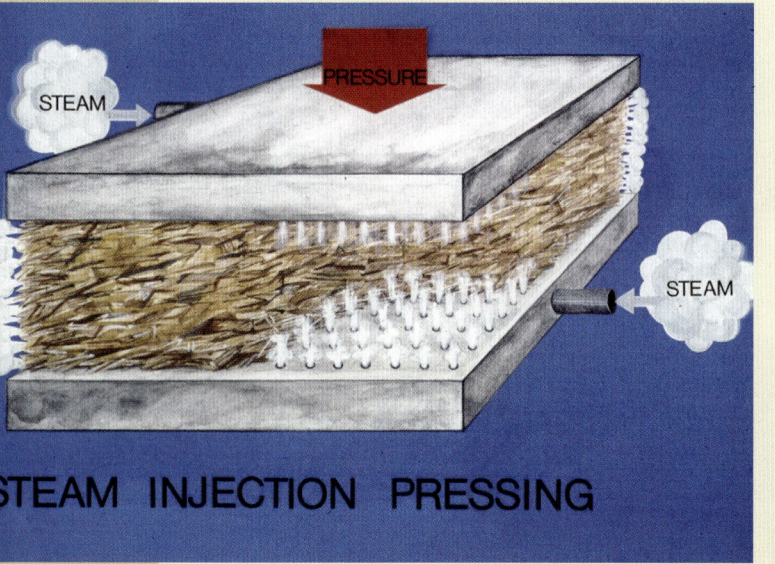

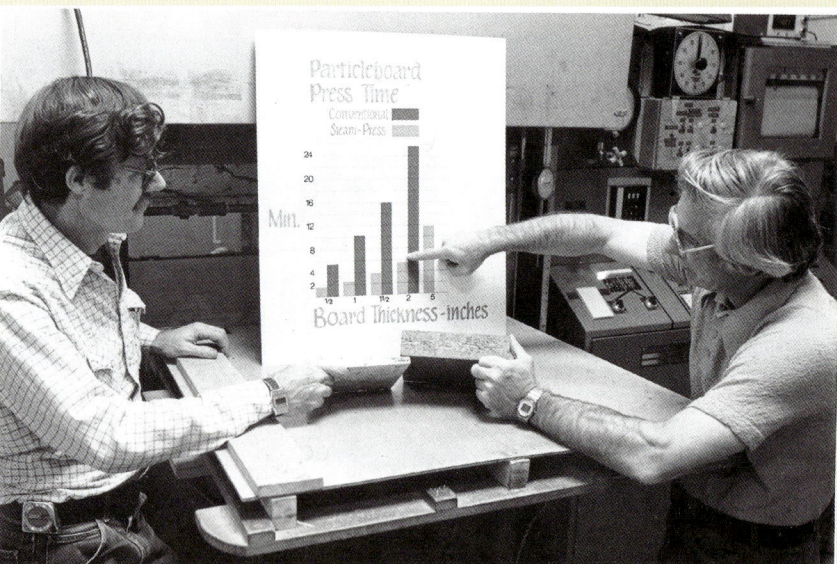

▲ Bob Geimer developed the concept of steam injection pressing, which significantly decreases pressing time and improves the adhesive bonding. He was granted U.S. Patent No. 4,393,019 (July 12, 1983) {113 D}

▲ Jim Wood and Bob Geimer discussing the test results of conventional particle composite press time versus steam-injection press time. (1980s) {M 150 635-3}

◀ The FPL's research with wafers, strands, and flakes contributed significantly to industry's development of oriented strandboard (OSB), a commercially popular particle composite that is in general use today. (Left to right, Gordy Stevens, Copeland Francis, and Bill Kruel.) (1970s) {3 C} [8]

Structural Flakeboard

The Forest Service Structural Flakeboard program showed the technical and economic feasibility, from harvesting through marketing, of producing a structural grade sheathing board from hardwood and softwood residues. This opened a new, high-quality use for forest residues, which can serve as a substitute for softwood plywood. (1975) [8, 11, 12, 17, 20, 22, 26, 29]

▶ Tongue-and-groove flooring made as a three-layer flakeboard with the face flakes oriented. (1970s) {134 M}

Structural Flakeboard Roof Decking from Residue

Tests indicate that structural flakeboard roof decking made from unmerchantable hardwoods would be suitable for industrial and commercial buildings. This roof decking can economically meet engineering requirements and provide better thermal insulation than some alternative materials.

◀ Compressed flakes were used to make a flakeboard roof decking material. (1970s) {101 D}

Structural Flakeboard I-Joist

◄ Bob Geimer holding an I-joist made from wood flakes bonded together. (1970s) {12C} [7]

Stabilized Particle Composites

Because of the extreme pressure used in producing particle composites, the densified wood is subject to an expansion phenomenon known as springback. This effect has restricted the use of particle composites for certain applications but they can now be stabilized by a steaming process developed by FPL. [19]

Engineered Wood Fiber Surfaces

The Americans with Disabilities Act (ADA) requires firm and stable surfaces for use of mobility devices such as crutches, wheel chairs, and walkers. Scientists at FPL developed a novel wood-based playground surface that is resilient and soft enough to cushion falls and impacts, but also firm and stable enough to support wheel chairs and walkers.

◄ Test set up to evaluate the engineered wood surface for its cushioning capacity. The silver colored ball at the top of the picture is an instrumented loadcell with similar weight, density, and size to a child's head. It is used to measure the effectiveness of the wood surface to slow the ball when dropped onto the wood surface. (Left to right) Ted Laufenberg, Mr. Zeager, Ted Illjes. (2000s) [23, 24]

The goal is to make a wood composite surface that improves the safety of playground surfaces and make trails accessible to the disabled.

▲ Jane Schmieding using a wheelchair on a stabilized engineered wood fiber trail in Governor Dodge State Park, Wisconsin (2000s) [28]

◀ Jane Schmieding manuvering a wheel-chair on an engineered wood fiber surface at a playground in Governor Nelson State Park, Wisconsin. (2000s) [27]

Emission Standards for Formaldehyde

Since 1980, the FPL has been working with other government agencies and industry to develop better ways to predict and measure the amount of formaldehyde that escapes into the air from various types of panel products. This work has led the Department of Housing and Urban Development to incorporate a formaldehyde emission standard for interior panels in its proposed product standard for mobile homes. By meeting this standard, panel manufacturers can decrease formaldehyde contamination levels in all living spaces. (1997) [18]

Coatings

Vapors from volatile organic additives, resins, and treatments may create a possible health hazard if the wood composite is used indoors. Scientists at FPL and Mississippi State University have evaluated commercially available coatings that may minimize the vapors. These include paints, varnishes, and other coatings. A user-oriented bulletin and a technical publication explain the limited effectiveness of both barrier and non-film-forming coatings. (1983) [15]

1910

Particle composites were virtually nonexistent in the United States.

2010

Flakeboard, oriented strandboard, particleboard industries established.

Companies employed 21,220 people and shipped products with a total value of $7,868,726,000. (2006 Annual Survey of Manufacturers)

Further Information

[1] How the physical properties of flake-type particle boards are affected by the species of wood, flake dimensions, binder content, and density / H.H. Haskell, B.G. Heebink / Forest Products Laboratory (1962)

[2] Board materials from wood residues / Lewis, W.C. / Research note FPL 045 (1964)

[3] Steam post-treatments to reduce thickness swelling of particleboard (exploratory study) / B.G. Heebink, F.V. Hefty / Research note FPL 0187 (1968)

[4] Basic engineering properties of particleboard / J.D. McNatt / Research paper FPL 206 (1973)

[5] Weathering characteristics of particleboard / R.L. Geimer / Research paper FPL 212 (1973)

[6] Properties of structural particleboards / W.F. Lehmann / Forest products journal Vol. 24(1) (Jan. 1974)

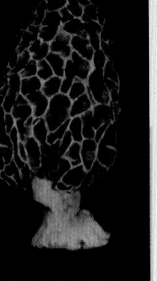

[7] Product and process variables associated with a shaped particle beam / R.L. Geimer / Forest products journal Vol. 25(9) (Sept. 1975)

[8] Flake alignment in particleboard as affected by machine variables and particle geometry / R.L. Geimer / Research paper FPL 275 (1975)

[9] Particleboards from lower grade hardwoods / Bruce G. Heebink, William F. Lehmann / Research paper FPL 297 (1977)

[10] Combination sheathing support-member building product / Robert L. Geimer, William F. Lehman / United States Patent 4,061,813 (December 6, 1977)

[11] Cyclic moisture conditions and their effect on strength and stability of structural flakeboards / W.F. Lehmann / Forest products journal Vol. 28(6) (Jun 1978)

[12] Flakeboard properties as affected by flake cutting techniques / Eddie W. Price, William F. Lehmann / Forest products journal Vol. 29(3) (Mar. 1979)

[13] Steam injection pressing / Robert L. Geimer / Proceedings of the 16th Washington State University International Symposium on Particleboard (Mar. 30–Apr. 1, 1982)

[14] Automation of a laboratory particleboard press / R.L. Geimer, G.H. Stevens, R.E. Kinney, Forest products journal Vol. 32(4) (Apr. 1982)

[15] Effect of coating systems on the vaporization of pentachlorophenol from treated wood / L.L. Ingram, G.D. McGinnis, P.M. Pope, W.C. Feist / American Wood-Preservers' Association. Proceedings of the 79th annual meeting Stevensville, MD: AWPA (1983)

[16] User-friendly programming for a computerized laboratory press / Robert L. Geimer, Richard Kinney, Mark Podlipec / Forest products journal Vol. 40(3) (Mar. 1990)

[17] Creep and creep–rupture of plywood and oriented strandboard / J. Dobbin McNatt, Theodore L. Laufenberg / Proceedings of the International timber engineering conference, 1991 September 2–5, London, Vol. 3. London: TRADA (1991)

[18] Volatile organic chemical emissions from composite wood products: A review / Melissa G. D. Baumann / Technical summary. In: The fibril angle; newsletter. Washington, DC: Cellulose, Paper, and Textile Division of the American Chemical Society: 1–12 (Spring) (1997)

[19] Impact of steam pressing variables on the dimensional stabilization of flakeboard / Jin Heon Kwon, Robert L. Geimer / Forest products journal 48(4): 55–61. (1998)

[20] Creep and creep–rupture behavior of wood-based structural panels / Theodore L. Laufenberg, L.C. Palka, J. Dobbin McNatt / FPL–RP–574 (1999)

[21] Wood Handbook: wood as an engineering material / GTR–113 (1999)

[22] Providing moisture and fungal protection to wood-based composites / Janet K. Baileys, Brian M. Marks, Alan S. Ross, Douglas M.;Crawford, Andrzej M. Krzysik, James H. Muehl, John A. Youngquist / Forest Products journal 53(1): 76–81 (2003)

[23] Improving engineered wood fiber surfaces for accessible playgrounds / Theodore L Laufenberg, Andzej Krzysik, Jerrold Winandy / FPL–GTR–135 (2003)

[24] Field performance testing of improved engineered wood fiber (EWF) surfaces for accessible playground areas / Theodore L. Laufenberg, Jerrold E. Winandy / FPL–GTR–138 (2003)

[25] Fundamentals of composite processing—proceedings of a workshop / Jerrold E. Winandy, Frederick A. Kamke, Eds. / FPL–GTR–149 (2004)

[26] Competitive soybean flour/phenol-formaldehyde adhesives for oriented strandboard / James M. Wescott, Charles R. Frihart / In: 38th International Wood Composites Symposium Proceedings, April 6–8, 2004, Washington State University, Pullman, Washington: 199–206. (2004)

[27] Stabilized engineered wood fiber for accessible playground surfaces installation and serviceability results: Governor Nelson State Park, Wisconsin / Theodore L. Laufenberg, Jerrold E. Winandy / FPL–GTR–154 (2004)

[28] Stabilized engineered wood fiber for accessible trails installation and serviceability results: Governor Dodge State Park, Wisconsin / Theodore L. Laufenberg / FPL–GTR–155 (2004)

[29] Physical and mechanical properties and fire, decay, and termite resistance of treated oriented strandboard / Nadir Ayrilmis, S. Nami Kartal, Theodore L. Laufenberg, Jerrold E. Winandy, Robert H. White, Forest products journal Vol. 55(5): 74–81 (2005)

[30] Using wood composites as a tool for sustainable forestry: proceedings of scientific session 90, XXII IUFRO world congress / edited Jerrold E. Winandy, Robert W. Wellwood, Salim Hiziroglu / FPL–GTR–163 (2005)

[31] Composite materials from forest biomass: a review of current practices, science, and technology / Roger M. Rowell / American Chemical Society: Distributed by Oxford University Press, c2007. ACS symposium series 954 (2007)

Fiberboards

◄ Hardboard exterior application siding. (1950s) {134 K}

Fiberboard includes insulating board, medium-density fiberboard (MDF), and hardboard. Fiberboards differ from particleboards mainly in the physical configuration of the prepared fiber. Because of its fibrous nature, fiberboard exploits the inherent strength of wood to a greater extent than does particleboard. [12, 13, 19]

Challenge: A major challenge is what to do with forest and manufacturing residues. In the past, they were burned in slash piles in the forest or tepee burners at the sawmill. This resulted in loss of material and air pollution.

FPL's Contributions: Research on adhesives, processing, and testing has contributed to a number of new products. One general type of product is fiberboard. Working with industry, FPL helped establish standards for various types and grades of fiberboards.

Results: This research has resulted in a significant reduction in wood waste and new products for the consumer and improves sustainability of our nation's forest resources.

◄ Various types of fiberboard. (1950s) {134 G} [1, 2, 4, 11]

► Sample of table top made from wood fibers. (Veneer overlaid medium density fiberboard.) Because fiberboard can be machined throughout its thickness, to give a smooth surface, it is preferred to particleboard for complex edge formations. (1980s) {10 E}

Fundamental Properties of Fiberboard and Hardboard

FPL scientists helped established much-needed procedures for evaluating the properties of fiberboard that were adopted as standards by ASTM International (ASTM). These standards are of great importance in establishing North American Building Codes and international standards by the Food and Agricultural Organization Committee on Mechanical Wood Technology. [5, 9, 11, 14, 18]

Low-Density Fiberboard

Working with the fiberboard industry, the FPL researched methods for evaluating the structural properties, fire resistance (developed the 8-foot funnel test to determine flame spread resistance), vapor movement, and decay resistance. [3, 4, 11, 14]

◄ Earl Geske testing an engineered wall system constructed of 2 by 4 lumber with fiberboard sheathing. (1960s) [11, 14]

Hardboard Lap Siding

FPL scientists were selected by the Department of Housing and Urban Development and the hardboard industry to perform two studies of the performance of hardboard lap siding. It was found that well-manufactured siding would remain at low in-service moisture contents in the (hot and humid) south Florida climate, provided that there was not significant water leakage past it. The second finding was that the siding would dissipate modest levels of leakage by drainage at the laps, provided that the laps were not painted shut. The third finding was that sealant joints associated with window and door casings would commonly leak, even though they appeared to be intact and functional. [20, 23]

Hardboard I-Beams Conserve Wood and Decrease Cost (Complex Systems)

A study on the design of built-up I-beams incorporating wood flanges glued to a tempered hardboard web has provided information valuable to architects, engineers, and home builders. [6, 10]

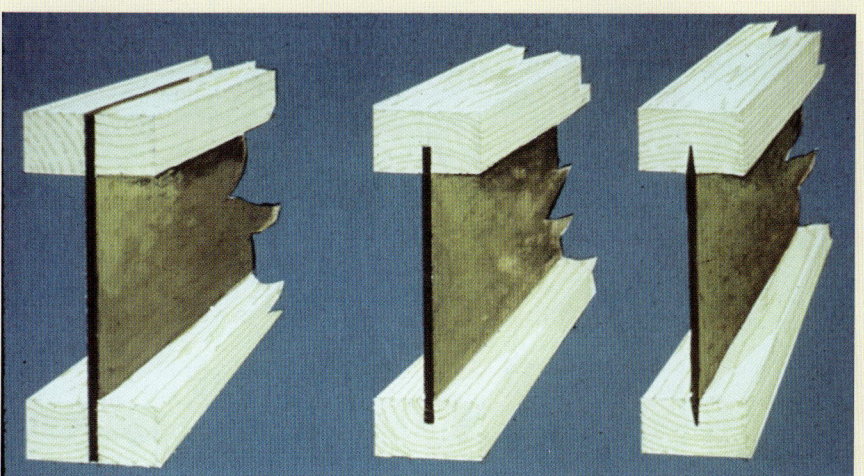

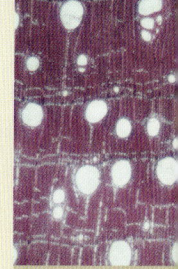

Better Hardboard Produced with Less Pollution

FPL scientists developed a process to produce dry-formed hardboard with better physical and mechanical properties by regulating wood fiber pH more closely. Dry-formed hardboard requires less water use in production, which will decrease water pollution, control costs, and produce a better product. [7, 8]

Acetylated Fiberboard

Wood products are affected by moisture and biological organisms. Research on improving the resistance of wood was conducted by FPL scientists.

◄ Roger Rowell holding an acetylated, three-dimensional pressed fiberboard. Acetylating improves the fiberboard's dimensional stability and resistance to biological attack. (1990s) [16, 24]

Wood Fiber–Inorganic Bonded Composite Panels

► Wood–inorganic bonded composite panels are made from wood fiber or particles and cement. The panels are durable, require little advanced technology, are inexpensive, but are slow to cure. FPL scientists helped develop the use of carbon dioxide injection to speed up the cure rate. (1990s) (14 I) [15, 17]

◀ Large wood fiber–cement composite panels used to decrease road noise. (2000s) {59 C} [21]

1910

Virtually nonexistent in the United States.

2010

Siding and paneling are commercial products, utilizing wood fiber that in the past was burned.

Further Information

[1] Hardboard: processes, properties, potentials / P.K. Baird, S.L. Schwartz / FPL No. 1928 (1952)

[2] Building fiberboards / FPL No. 1903–9 (1953)

[3] Properties of insulating fiberboard sheathing / R.F. Luxford / FPL No. 2032 (1960)

[4] Insulating board, hardboard, and other structural fiberboards / W.C. Lewis / FPL–RN–077 (Aug. 1965)

[5] Acoustical absorption properties of wood-base panel materials / W.D. Godshall / FPL–RP–104 (1969)

[6] Predicting performance of hardboard in I-Beams / T.J. Ramaker / FPL–RP–185 (1972)

[7] The role of phenolic resin in imparting properties to dry-formed hardboards / D.J. Fahey, D.S. Pierce / Tappi Vol. 56(3) (Mar. 1973)

[8] Effects of wood and pulp properties on medium-density, dry-formed hardboard / Neil D. Nelson / Forest products journal Vol. 23(9) (Sept. 1973)

[9] Dielectric properties of wood and hardboard: variation with temperature, frequency, moisture content, and grain orientation / W.L. James / FPL–RP–245 (1975)

[10] Hardboard-webbed beams: research and application / J.D. McNatt / Forest products journal Vol. 30(10) (Oct. 1980)

[11] Fiberboard and hardboard research at the Forest Products Laboratory: a 50-year summary / Gary C. Myers, J. Dobbin McNatt / FPL–GTR–47 (1985)

[12] A comparison of hardboards manufactured by semidry-, dry-, and wet-formed processes / Gary C. Myers / Forest products journal Vol. 36(7/8) (July/Aug. 1986)

[13] Fiberboard manufacturing practices in the United States / O. Suchsland, G.E. Woodson / USDA Agriculture handbook 640 (1986)

[14] Properties of composite panels / John A. Youngquist, Andrzej M. Krzyski, Poo Chow, Roger, Meimban / In: Rowell, Roger M.; Young, Raymond A.; Rowell, Judith, K. / Paper and composites from agro-based resources. Boca Raton, FL: CRC Lewis Publishers: 301–336. Chap. 9. (1997)

[15] Cement-bonded wood composites as an engineering material (PDF 539 KB) / Ronald W. Wolfe, Agron Gjinolli / In: Proceedings: The use of recycled wood and paper in building applications. Proceedings No. 7286. Madison, WI: Forest Products Society: 84–91 (1997)

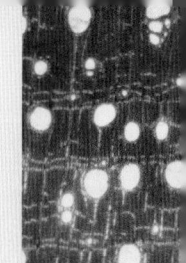

[16] Worldwide in-ground stake test of acetylated composite boards / R.M. Rowell, B.S. Dawson, Y.S. Hadi, D.D. Nicholas, T. Nilsson, D.V. Plackett, R. Simonson, M. Westin / In: Proceedings of the the international research group on wood preservation, 28 th annual meeting; 1997 May 25–30; Whistler, Canada. Sec. 4, Processes. The research group on wood preservation. Document IRG/WP/ 97–40088 (1997)

[17] Accelerated aging of low-density cement-bonded wood composites made conventionally and with carbon dioxide injection = Ubrzano starenje lakih cementnih drvnih ploca proizvedenih konvencionalno i uz injekciju ugljik-dioksida / Robert L. Geimer, Mario Rabelo de Souza, Ali A. Moslemi / Drvna industrija Vol. 47(2) (1997)

[18] The comparative performance of wood fiber–plastic and wood-based panels / Robert H. Falk, Dan Vos, Steven M. Cramer / In: Proceedings, 5th international conference on wood fiber–plastic composites; 1999 May 26–27; Madison, WI. Madison, WI: Forest Products Society: 269–274 (1999)

[19] Wood Handbook: wood as an engineering material / FPL GTR–113 (1999)

20] Performance of back-primed and factory-finished hardboard lap siding in southern Florida / Charles Carll, Mark Knaebe, Vyto Malinauskas, Peter Sotos, Anton TenWolde / FPL–RP–581 (2000)

[21] Design of wood highway sound barriers / Thomas E. Boothby, Courtney B. Burroughs, Craig A. Bernecker, Harvey B. Manbeck, Michael A. Ritter, Stefan Grgurevich, Stephen Cegelka, Paula D. Hillbrich Lee / FPL–RP–596 (2001)

22] Fundamentals of composite processing--proceedings of a workshop / Jerrold E. Winandy, Frederick A. Kamke, Eds. / FPL–GTR–149 (2004)

[23] Durability of hardboard lap siding-determination of performance criteria / Charles Carll, Anton TenWolde / FPL–RP–622 (2004)

[24] Acetylation / Roger M. Rowell / Forest products journal. Vol. 56(9) (Sept. 2006)

Molded Products

◄ Sample of a molded pulp panel and a tri-axial pressure molding device. (1980s) {84 K}

One of the earliest molded pulp products was the egg carton. The primary purpose is to separate and help cushion the eggs. Molded egg cartons are still used today.

Challenge: How to utilized recycled fibers in high-strength panels and other products.

FPL's Contributions: FPL developed the method of tri-axial pressure molding of wood pulp slurries and studied factors affecting processing. FPL also developed methods for mixing plastic and wood particles and extruding or thermoforming the material into usable shapes.

Results: The patented FPL technology product—Spaceboard—was licensed to two companies for commercialization. Molded fiber–plastic products are commercial products today.

◄ Examples of different molded pulp products made from recycled fiber. These examples range from 0.05 to 3.0 inches thick. (1990s) {114 B} [12]

FPL Spaceboard

A totally new type of fiberboard was made by Vance Setterholm by molding fibers into a three-dimensional integrated rib and face panel and then gluing two panels, rib-to-rib, to form a cellular sandwich structure or laminate. When joined with adhesive, the cells are sealed cubes. This provides high strength in every direction. "Spaceboard" can be manufactured from a fraction of an inch thick to several inches thick. It can be produced from a variety of recovered wastepapers and doesn't require cleaning of the wastepaper-derived fibers to the same degree as recycling wastepaper back into paper. Therefore, Spaceboard can use a wide variety of recovered post-consumer papers, many of which have little use today. (1980s) [1, 4, 7]

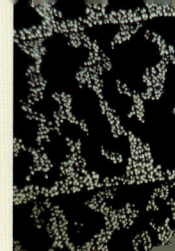

◀ Building panels made by tri-axial pressure molding, with some of the cells filled with insulation. {15 I}

Three-Dimensional Engineered Fiberboard (3DEF): A New Structural Building Product

▲ Many of our national forests contain an abundance of small, tightly spaced trees and underbrush that can substantially contribute to catastrophic forest fires. Existing stands of dead or dying trees can also add fuel to forest fires. (2000s)

▲ Two- to five-inch tree tops of suppressed-growth lodgepole pine on the Big Horn National Forest in Wyoming. In an effort to help reduce this excessive growth and utilize the dead and dying trees, the FPL has developed a process to produce a three-dimensional structural fiberboard product using a wide range of wood fibers. These new engineered wood products will help address sustainable forest management issues and help promote economically viable utilization of hazardous fuels such as this presently unmerchantable small-diameter lodgepole pine. (2000s) [20]

◀ John Hunt (right) and Hongmei Gu (visiting scientist) with a wet-formed three-dimensional engineered fiberboard product consisting of a resin-free, pulp-molded core that is made from a wide range of inexpensive, underutilized fiber sources. (2000s) [3, 16, 19]

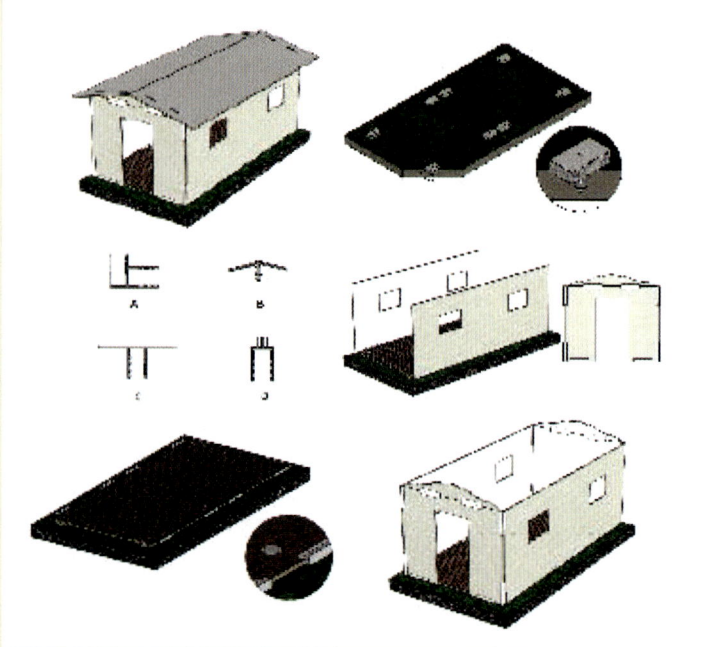

◀ Emergency housing systems designed using three-dimensional engineered fiberboard.

Temporary building systems for lightweight, portable, easy-to-assemble, reusable, recyclable, and biodegradable structures are possible. (2000s) [27]

Wastepaper–Plastic Composite Panels

Wood fiber has been used as a filler for plastic products with great success. FPL researchers working with a consortium of 15 fiber suppliers and plastics manufacturers, tackled the problems of effectively utilizing the wood fiber from wastepaper. [2] (See Recycle–Reuse section.)

Wood–Plastic Composites

◀ 2002 review of research on wood–plastic composites in the United States. [21]

Extruded Wood–Plastic Products

▼ Craig Clemons observing a wood–plastic composite being extruded.[21]

Possible uses for extruded building products include decking, railing, and siding. The fiber used in this product is from salt cedar, an invasive species. In 2005, the Bureau of Land Management (BLM) and FPL worked together to establish if several problem species could be used in wood–plastic composites in exterior applications. Salt cedar from the Colorado River basin in southwest Arizona and one-seed juniper from Utah were harvested by the BLM (Bureau of Land Management) and sent to FPL for evaluation for use in wood product composites. (2000s) [30] (See Sustainability–Using Problematic Natural Resources).

▲ Examples of commercial wood–plastic material used for decking. (2000s) [15, 21, 22, 29]

▲ Sample of a deck made with commercial wood–plastic composite material. (2000s)

Thermoformed Complex Engineered Wood–Plastic Products

◀ Early FPL research with industry resulted in scissor handles made from thermoformed wood pulp and plastic. (1990s) [5, 25, 26]

◀ Sue Paulson (left) and Eunice Tshabalala (right) discuss experimental wood–plastic composite chairs.

▼ John Youngquist explaining the use of wood particles in complex thermoformed molded products. This opens another avenue for using previously wasted wood particles. (2000s) [14]

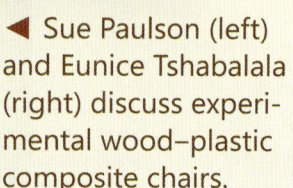

▲ Samples of car parts made from wood fiber–plastic composites. (2000s)

▶ Nicole Stark shows a wood–plastic composite panel for use in the automotive industry.

Wood–Plastic Shingles

◀ Experimental house at FPL with roof shingles made primarily from sawdust and recycled soda bottles (2000s)

▲ FPL test structures for evaluating three-year Wisconsin temperature histories for roof systems using wood, wood–thermo-plastic composite, and fiberglass shingles. (2000s) [6, 17, 22, 28]

1910

Virtually nonexistent in the United States.

2010

Wood fiber and plastic composites are now used to make commercially available products.

In 1995, about 0.5% of the wood deck market was wood–plastic composite material. In 2008, over 20% of decking was wood–plastic composite material, and the new composite material is moving into shingles and siding.

New three-dimensional engineered fiberboards are being developed that can be made, using either wet- or dry-formed processes, from virtually any virgin or recycled biofiber resource, including wood residues, agricultural biomass, recycled newsprint, small-diameter timber, and mixed wood and agricultural residues. As engineered biocomposites, these products can be manufactured to provide a range of specific levels of structural performance, durability, moisture resistance, and insulation. These new products could have significant implications on resource sustainability and economic development for rural communities.

Further Information

[1] FPL Spaceboard development / J.F. Hunt, D.E. Gunderson / TAPPI proceedings of the 1988 Corrugated Containers Conference, October 24–27, Orlando, FL. Atlanta, GA TAPPI Press, (1988)

[2] Wastepaper fiber in plastic composites made by melt blending: demonstration of commercial feasibility / George E. Myers, Craig M. Clemons / Forest Products Laboratory. Final report for Solid Waste Reduction and Recycling Demonstration Grant Program, Project No. 91–5, Wisconsin Department of Natural Resources (1993)

[3] 3-D fiber forming: a step toward engineered sandwich construction / J.F. Hunt, R.L. Noble, R.M. Smyrski / 3rd international conference on sandwich construction, September 12–15, 1995, Southampton, UK. (1995)

[4] Spaceboard: today's structural recycled fiber products for the future /John F. Hunt, Rose Smyrski / Meeting society's needs through forest products technology marketing: proceedings, technology transfer national workshop, November 7–10, 1994, Madison, WI: USDA Forest Service, Forest Products Laboratory (1995)

[5] Waste-wood-derived fillers for plastics / Brent English, Craig M. Clemons, Nicole Stark, James P. Schneider / FPL–GTR–91 (1996)

[6] Factors that affect the application of woodfiber–plastic composites / Brent W. English, Robert H. Falk / Wood-fiber–plastic composites: virgin and recycled wood fiber and polymers for composites. Madison, WI: Forest Products Society Proceedings, Forest Products Society No.7293 (1996)

[7] Mechanical properties of Spaceboard panels and pallets made from recycled linerboard mill sludge / John F. Hunt, Tim C. Scott, Don Jenkins, Kermit Hovey / Tappi environmental conference 481–488. Book 1 (1997)

[8] Agro-Fiber thermoplastic composites / Anand R. Sanadi, Daniel F. Caulfield, Rodney E. Jacobson / In: Rowell, R.M.; Young, R.A.; Rowell, J.K.; comp. ed. / Paper and composites from agro-based resources. Boca Raton, FL: CRC Lewis Publishers: 378–401. Chap. 12. (1997)

[9] Effect of species and particle size on properties of wood-flour-filled polypropylene composites / Nicole Stark, Mark J. Berger / In: Proceedings, Functional fillers for thermoplastics and thermosets; 1997 December 8–10; Sand Diego, CA. Portland, ME: Intertech Conferences (1997)

[10] Utilization of natural fibers in plastic composites: problems and opportunities / Roger M. Rowell, Anand R. Sanadi, Daniel F. Caulfield, Rodney E. Jacobson / In: Leão, Alcides L.; Carvalho, Francisco X.; Frollini, Elisabete. Lignocellulosic–plastics composites. Sao Paulo, Brazil: Universidade de Sao Paulo Press: 23–52 (1997)

[11] Injection-molded composites from kenaf and recycled plastic / Poo Chow, Dilpreet S. Bajwa, Wen-da Lu, John A. Youngquist, Nicole M. Stark, Qiang. Li, Brent English, Charles G. Cook / Proceedings of 1st Annual American Kenaf Society Meeting: Feb. 1998, San Antonio, TX. Vernon, TX: American Kenaf Society, (1998)

[12] Know your fibers: process and properties or (A material science approach to designing pulp molded products) / John F. Hunt / In: Proceedings, 4th international molded pulp packaging seminar; 1999 November 19; Chicago, Il. Mequon, WI: International Molded Pulp Environmental Packaging Association: 16 p. (1999)

[13] Pulp extrusion for recycling wastepapers and paper mill sludge's / Stefan Zauscher, C. Tim Scott, J.L Willett, Daniel J Klingenberg / Tappi journal Vol. 83(6) (June) (2000)

[14] Wood/polymer composites: a state-of-the art review / John Youngquist / International Institute of Forest Research Organizations Vol. II, p. 170. (2000)

[15] Outdoor durability of wood–plastic composite lumber / Nicole Stark, Craig Clemons / Advanced Housing Research Center (Oct. 2000)

[16] Apparatus for molding three-dimensional objects / John F. Hunt / US Patent No. 6,190,151 (Feb. 20, 2001)

[17] Effects of weathering on color loss of natural fiber thermoplastic composites / R.H. Falk, C.Felton, T. Lundin / Composites in manufacturing Vol. 17(4) (Fourth Quarter 2001)

[18] Pulp extrusion at ultra-high consistencies: selection of water soluble polymers for process optimization / C. Tim Scott / 2002 TAPPI Fall technical conference and trade fair. Atlanta, GA (2002)

[19] 3D engineered fiberboard: a new structural building product / John F. Hunt, Jerrold E. Winandy / In: Proceedings of the sixth panel products symposium; 1–11 October 2002. Llandudno, Wales, UK. The BioComposites Centre, UWB, Bangor, Gwynedd, LL57 2UW, UK. 106–117 (2002)

[20] Using wood-based structural products as forest management tools to improve forest health, sustainability and reduce forest fuel: A research program of the USDA Forest Service under the National Fire Plan / J.F. Hunt, J.E. Winandy / Proc. of the 6th pacific rim bio-based composites symposium, Nov 10–13, 2002. Portland OR. Vol. 1: 316–322. (2002)

[21] Wood–plastic composites in the United States: the interfacing of two industries / Craig Clemons / Forest products journal Vol. 52(6) (2002)

[22] Accelerated weathering of natural fiber-filled polyethylene composites / Thomas Lundin, Robert H. Falk / Journal of materials in civil engineering (Nov./Dec. 2004)

[23] Considerations in recycling of wood–plastic composites / J.E. Winandy, N.M. Stark, C.M. Clemons / 5th global wood and natural fibre composites symposium: April 27–28, 2004 in Kassel, Germany: p. A6.1–A6.9 (2004)

[24] Effect of processing method on surface and weathering characteristics of wood-flour/HDPE composites / Nicole M. Stark, Laurent M. Matuana, Craig M. Clemons / Journal of applied polymer science Vol. 93 (2004)

[25] Natural fibers / Craig M. Clemons and Daniel F. Caulfield / Functional fillers for plastics / Weinheim: Wiley-VCH, (2005)

[26] Wood thermoplastic composites / Daniel F. Caulfield, Craig Merrill Clemons, Rodney E. Jacobson, Roger M. Rowell, / Handbook of wood chemistry and wood composites. Boca Raton, Fla.: CRC Press (2005)

[27] Emergency housing systems from three-dimensional engineered fiberboard / J.E. Winandy, J.F. Hunt, C. Turk, J.R. Anderson / FPL–GRT–166 (2006)

[28] Analysis of three-year Wisconsin temperature histories for roof systems using wood, wood–thermoplastic composite, and fiberglass shingles / J.E. Winandy, C.H. Hatfield / Forest products journal Vol. 57(9) (2007)

[29] Cone calorimeter tests of wood-based decking materials. R.H. White, M.A. Dietenberger, N.M. Stark / in proceedings, 18th BCC Conference on Flame Retardancy of Polymeric Materials, Stamford, CT (May 20–23, 2007)

[30] Use of saltcedar and Utah juniper as fillers in wood–plastic composites / Craig Clemons, Nicole Stark / FPL–RP–641 (2007)

Housing

◀ Straddling a fissure from an earthquake, this wood-frame house remained in good condition except for a front entryway. A dramatic example of the rigidity of wood-frame construction. (Alaskan earthquake 1964)

Starting with log cabins, wood has been a major material for building houses in the United States. [34, 50, 64]

Challenge: How to help provide satisfactory housing for the public.

FPL's Contributions: The FPL has had, and continues to have, a major impact on housing. Much of this impact has been through the development of test methods, building codes, and the generation of basic engineering design properties for using wood efficiently.

Results: More efficient use of wood and safer housing.

◀ LeRoy Anderson (left) and Joe Liska make notes on damage done to a wood-frame house that four days before had been torn from its foundation by an upthrust of the earthquake that struck Anchorage, Alaska, on Good Friday. (1964) {M 126 426}

Earthquakes

Based on observations and analysis, FPL established design and performance requirements for buildings of wood diaphragm construction that adequately resist forces imposed by earthquake, wind, or blast loadings. In California, this information is being used in the construction of wood-frame schools that are safer than those built in the past, in that they will not collapse as a result of earthquake shock. [26, 48, 53, 56]

Strength and Rigidity of Frame Walls

Early (1920s) research on wood structures dealt with how to construct rigid wall sections using wood efficiently and sparingly. In 1983, the contribution of gypsum wallboard to racking resistance of light-frame walls was evaluated. [45].

▲ Test house built to evaluate moisture distribution and temperature gradient in walls. The arrow shows where small test panels were inserted in the building. (1930s) {M 26955}

Moisture Barriers

In 1937, FPL proposed the use of moisture barriers (now referred to as vapor retarders) to prevent the accumulation of moisture in the walls of buildings. Materials made for this purpose are in common use today, particularly in northern climates where, when properly installed, they help prevent cold-weather accumulation of moisture within house walls and subsequent paint failure. [3, 21, 39, 41, 52, 58, 60, 63] (For further information on moisture, see the Environment section.)

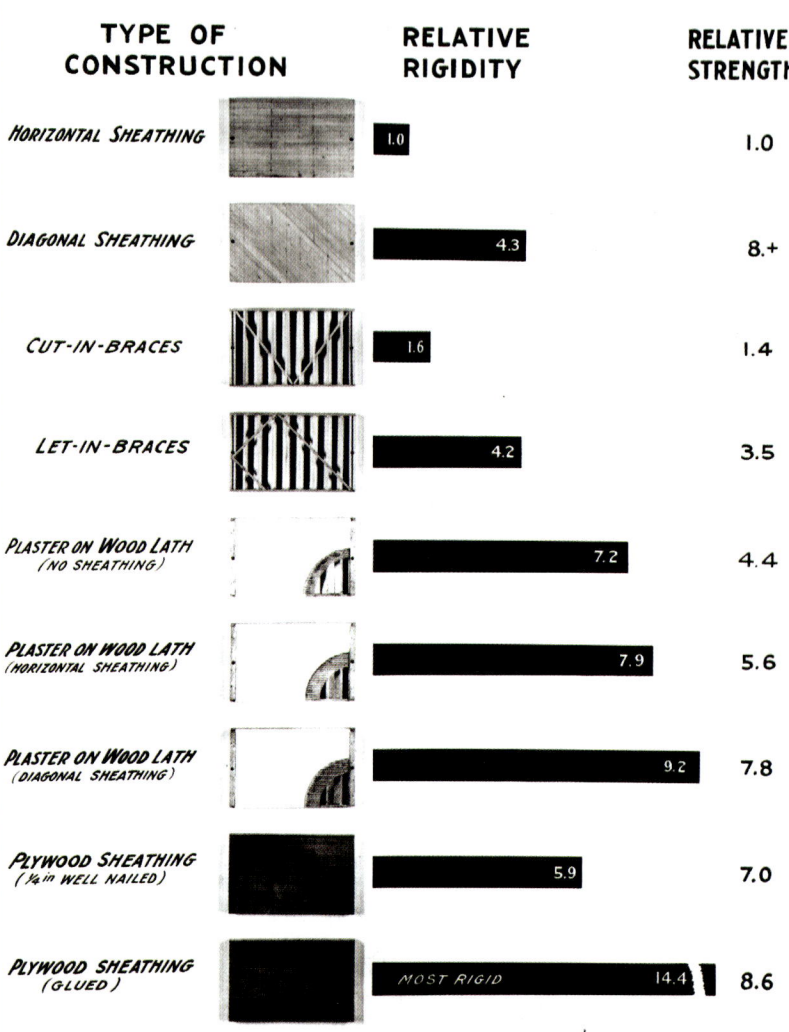

STRENGTH AND RIGIDITY OF FRAME WALLS
RESULTS OF TESTS OF 9FT. BY 14FT. WALL PANELS

TYPE OF CONSTRUCTION	RELATIVE RIGIDITY	RELATIVE STRENGTH
HORIZONTAL SHEATHING	1.0	1.0
DIAGONAL SHEATHING	4.3	8.+
CUT-IN-BRACES	1.6	1.4
LET-IN-BRACES	4.2	3.5
PLASTER ON WOOD LATH (NO SHEATHING)	7.2	4.4
PLASTER ON WOOD LATH (HORIZONTAL SHEATHING)	7.9	5.6
PLASTER ON WOOD LATH (DIAGONAL SHEATHING)	9.2	7.8
PLYWOOD SHEATHING (¼in WELL NAILED)	5.9	7.0
PLYWOOD SHEATHING (GLUED)	MOST RIGID 14.4	8.6

▲ Strength and rigidity of frame walls. (1920s) {M 23958 F}

Prefabricated Panelized Housing

During the early 1930s, the FPL laid the research base for most of today's prefabricated houses by developing the principles of stress-cover panelized construction. For the first time, the covering materials of walls and roofs were used as strength members rather than just to keep out wind, rain, and snow. The space between the facings of the panels was filled with mineral-wool insulation to increase fire resistance in addition to giving heat and sound insulation. The panel system proved well adapted to factory assembly methods. This research led to the development of structural insulated panels (SIPs). [2, 4, 5, 8, 10, 57]

◄ First all-wood prefabricated house being erected on FPL grounds. It was completed in November 1937, just in time for a visit by Mrs. Eleanor Roosevelt, who was very interested in low-cost housing. This research contributed significantly to the realization of prefabricated housing, a major manufacturing industry today. {M 32420 F}

► Completed prefabricated house with garage. For a number of years, the FPL drafting group used the house for their office. (1940) {M 34210 F}

◄ Before tearing down this two-story experimental house, FPL engineers subjected it to a final series of tests. After 25 years, the house easily withstood simulated winds up to 125 miles per hour. (1960s) {M 124 574}

Conventional Wood-Frame Housing

► Much of the early research on properties of wood and fasteners was directly applicable to improving efficient house design. (1950s) {4 C} [54]

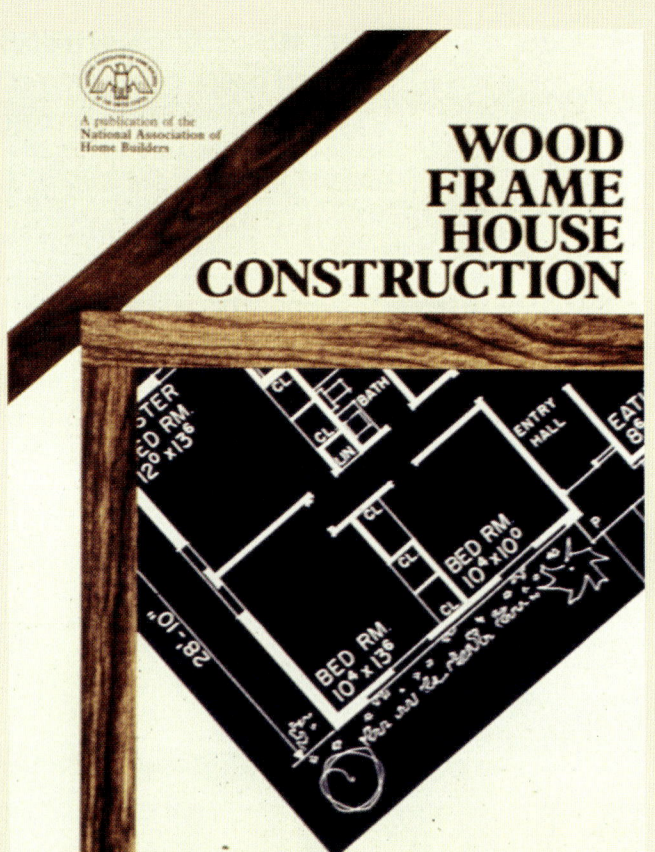

One of the most important contributions to housing was the FPL's manual on wood-frame construction, USDA Agriculture Handbook 73. [54]

◄ Much of the vast amount of wood research was compiled in numerous publications and manuals. For example, in 1947 a manual on wood construction for prefabricated houses was published. In 1955, a manual on wood-frame house construction was published, and in 1989 it was updated and republished. Written in simplified language and illustrated with numerous drawings, it makes the results of the Laboratory's housing research easily available and understandable to the prospective home owner, carpenter, and contractor. {38 L} [5, 32, 42, 54]

Housing Performance

FPL furnished the technical knowledge needed to ensure that houses financed by the Federal Housing Administration meet certain performance standards and will not require excessive maintenance. Included over the years have been laboratory-developed data on such things as roof trusses, sheathing, sandwich house construction, stiffness of various floor systems, properties of insulating fiberboards, vapor barriers, racking strength of wall construction, fire resistance, and paint performance. [7, 61, 65]

Nu-Frame Experimental House

▲ Sharon Peterson pointing out the double-chord roof truss in the model of the Nu-Frame experimental house. (1960s) {M 132 220} [15]

Unique features of the Nu-Frame house:

- Double-chord roof trusses use low-grade material.

- Single struts and gussets require fewer pieces and fasteners.

- Truss spacing is twice that for conventional single-chord trusses.

- Exterior sheathing is low-grade, random-width pine edge-glued to 12-inch widths; glued units are tongued and grooved to allow interlocking courses.

- Siding, redwood or cedar, is glued to sheathing—no face nailing is required.

- Unit is fastened to studs with mastic adhesive and blind nailing.

◄ Finished Nu-Frame house. (1968) It was moved to an open area on the UW campus, adjacent to the FPL. {M 134 773-5} [15]

Houses of Sandwich Construction

The theory of sandwich construction, much of which was developed at the FPL, was applied to structural sandwich panels. The practicability and durability of these panels was demonstrated by building a demonstration house with floor, wall, and ceiling panels made using sandwich construction. Many commercial applications such as exterior and interior curtain walls, doors, furniture, ship construction, and trailers stem from this research. [6, 9, 11–13, 25]

► Twenty-one years after it was first erected at FPL, to demonstrate a new concept called sandwich construction, this structure was dismantled. Floor, wall, and roof panels consisted of paper-honeycomb cores and plywood, hardboard, or metal facings. Some panels were tested and found fully as strong as when the building was originally erected as a long-term exposure experiment. (1960s) {M 135 468-5}

HUD's Operation Breakthrough

The technical expertise provided by FPL for the Department of Housing and Urban Development's "Operation Breakthrough" is an example of a payoff from research. There was extensive use of wood and wood-based materials and glued systems in 16 of the 24 contract housing systems. The Laboratory provided the needed in-depth review on the structural engineering, glue technology, and over-the-road transport of these systems. (1970s) [24]

Low-Cost Housing

In 1969, the FPL designed low-cost housing that was safe and efficient. Two of the five designs are shown on the next page. Detailed building plans and a construction manual were prepared. Prototype houses were built and plans sent to anyone who requested them. Hundreds of homes were built using these basic plans. This research was one of the most successful accomplishments of the Laboratory in the building research area. [17, 18, 20, 22]

▶ Building plan for a single-story home.

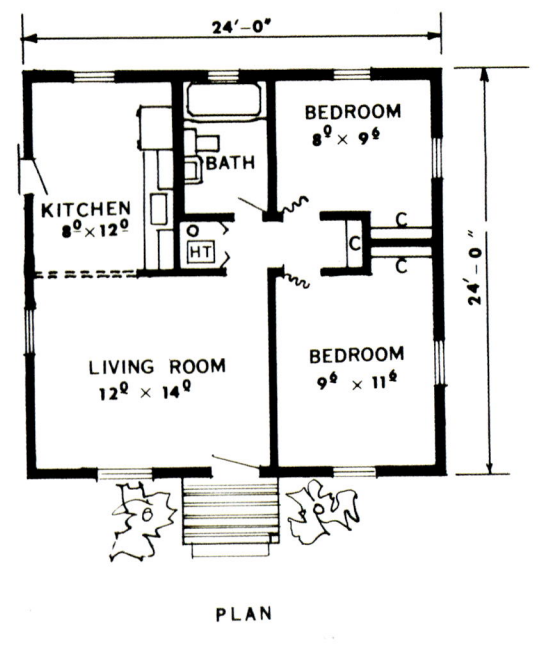

◀ Low-cost house under construction. (1970s) {4 G}

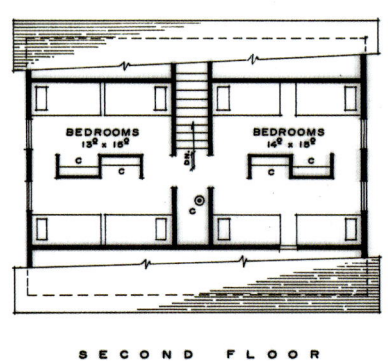

◀ Building plan for a two-story home. {M 136 216}

USDA Handbook No. 364, *Low-Cost Wood Homes for Rural America—Construction Manual*, has over 100 pages covering important details on building low-cost wood homes. Construction techniques are emphasized, and all phases from foundation to final finishing are covered. (Unfortunately the building plans have not been updated to current building standards.) [17]

▲ The team of FPL staff who worked with LeRoy Anderson to produce award-wining low-cost housing plans, specifications, and manuals gathered to congratulate the housing research leader just before he and Mrs. Anderson departed for Washington, D.C., to receive a superior service award from Secretary of Agriculture Clifford M. Hardin. Anderson (fourth from left, front row) spearheaded the year-long program to produce five sets of low-cost house plans with specifications and a manual on low-cost house construction. The team consisted of some 60 scientists, engineers, craftsmen and support staff: Front row, left to right: Fred Stenge, James Brooks, Berry Hendricks, LeRoy O. Anderson, Mrs. Anderson, Mrs. H.O. Fleischer, FPL Director H.O. Fleischer, Stella Webb, Mary Haukereid, Judy Broge, Karl Kanvik, and Alexander Pirola. Second row: Alan Freas, Raymond O'Neill, Gerald Campbell, Gerald Sherwood, Ann Gallagher, Elizabeth Caygill, Winifred Statz, Harold Mitchell, Harley Davidson, Charles Oliver, Warren Leadley, and Leif Ersland. Third row: Walter Sievers, Robert Edlund, Joseph Liska, Dorothy Sutton, Virginia Stehr, Lilian Marks, George Paulson, Fred Werren, Jerrold Moen, A.J. Motelet, Earl Geske, and Spence Munson. Fourth row: Robert Gillespie, Robert Lulling, Patricia Palmer, Walter Last, Norman Tessman, Magnus Selbo, Mary Douros, Esther Vetter, Walter Burns, Bruce Lythjohan, Edward Mraz, Francis Hefty, and Roman Baltes. Fifth row: Richard Strasbaugh, Wally Youngquist, Gordon Logan, Bruce Heebink, Billy Bohannan, Wayne Lewis, Reuben Simon, John Black, Lauris Anderson, Chester Roessler, William Quinn, Laverne Becker, and Ole Weum. (May 16, 1969) Note: Though most studies don't involve this many people, this photo shows that when reference is made to "FPL scientists" it usually means a number of FPL employees working together to get the job done.

Wind Damage

► Tornado damage. The problem of designing a building to resist high wind conditions is emphasized from the great damage wrought by tornados throughout the country. To understand how a structure resists wind, it is necessary to prove theories with actual test data. Tests of this nature provide understanding of how the many aspects of a structural design work together or against each other in reacting to wind forces. To assist in this research, a test frame was designed and built to test complete structures. {89 M}

Test frame for evaluating complete structures. This particular structural design was shown to withstand a wind of 220 miles per hour and a snow load of up to 20 feet of new snow. This test frame is also used to help determine where savings in material might be realized without sacrificing the structure's integrity, thus extending our timber supply. (1980s) {8 H} [14, 23, 31, 47, 48]

New Ways to Analyze Structural Performance of Light-Frame Floors and Walls

FPL researchers, working in cooperation with several universities, have developed new methods to analyze floors and walls that use light-frame components. They have also developed a computer program for wood truss analysis. This research makes it possible to predict stiffness and strength of various components under various types of loads. [35, 43, 44, 55]

Truss-Framed House

Cross section of a house in which all the components are tied together. The truss-framed house was designed by Roger L. Tuomi of the FPL staff. It was the outgrowth of the study of the effect of wind loads that indicated the importance of securely tying the roof to the side walls and the side walls to the floor and foundation of the structure. The truss-framing system is directly applicable to residential or light-frame construction. It consists of a special floor system, trussed rafters, and conventional wall studs all tied together with rigid joints into a unitized frame. Superior performance in both strength and stiffness have been demonstrated by full-scale structural tests. Savings of up to 30% in framing material are possible. Another important fact was that the whole house was made using wood no bigger than a 2 by 4. Using engineering-designed trusses, it is no longer necessary to use large trees to provide wide lumber like 2 by 10's. This has a direct impact on forest management from the standpoint of how long it is necessary to grow a tree if size is not critical and opens the possibility of using small-diameter thinnings and many other species in construction. (1970s) {121 J} [33, 40, 51]

◄ Truss-framed house under construction. Because the house sections can be made in a plant under controlled conditions, erection on the site is much quicker and the quality of construction improved. The frames are placed on 24-inch centers. Because the trusses span the width of most homes, supports are not needed in the basement, and load-bearing walls are not necessary on the first floor. {58 D}

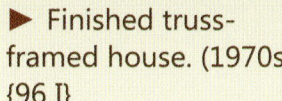

► Finished truss-framed house. (1970s) {96 I}

Sound Insulation

▲ Dunc Godshall adjusts impedance tube specimen holder in early tests of sound absorption by wood-fiber hardboard. An audio speaker sends a single tone through the tube. The sound properties are then determined based on the measured effect of the reflected sound wave on the sound sent from the speaker. (1960s) {M 136 655-9} [16]

▲ John Hillis using a portable system for determining the acoustic performance of buildings. It consists of a noise source and a sound level meter shown on the tripod. This system should permit builders to pinpoint acoustic deficiencies during construction and correct them immediately. (1970s) {M 142 767-14} Robert Jones reported on the results. [27–29, 36, 49]

Insulation Economy and Energy Improvement

FPL engineers analyzed heat loss and total cost effectiveness over the useful life of a home to demonstrate that there is an optimum insulation thickness for any climate. The optimum thickness results in least total cost of insulating and operation for the life of the building and reduction in energy use. [38, 46]

Saving Energy in Older Homes

A report on saving energy in the older home describes many ways to reduce energy waste and save fuel in older wood-frame houses. Research-based recommendations are given for weather stripping, passive use of shade and sun, insulating, use of vapor barriers, attic venting, and stopping air leaks. [37]

Rehabilitation Manual

A guide has been written for appraising the condition of older wood-frame dwellings, with information on building examination, esthetic values, improvement planning, and practical details for accomplishing the job. Many homes can be remodeled or rehabilitated at a lower cost than new construction. [30, 59]

Advanced Housing Research Center

America needs durable, affordable, and energy-efficient housing. In response to this need, the Advanced Housing Research Center (AHRC) was established at the FPL in 2001. [62]

Housing research at the AHRC is facilitated and coordinated through both internal and external partners. The Center relies heavily on partnerships and alliances with universities, industry, consumers, associations, and other government agencies, including the following:

- Coalition for Advanced Wood Structures (CAWS)
- Federal Agency Housing Partnership (FAHP)
- Partnership for Advancing Technology in Housing (PATH)
- American Society of Heating, Refrigerating and Air-Conditioning Engineers (ASHRAE)
- ASTM International (formerly American Society for Testing and Materials)
- American Forest & Paper Association (AF&PA)
- Forest Products Society (FPS)
- Society of Wood Science and Technology (SWST)
- The Minerals, Metals, and Materials Society (TMS)
- National Association of Corrosion Engineers (NACE)

▶ Latest demonstration house located at FPL as part of the Advanced Housing Research Center. Working with industry and the latest in new technology and materials, this house showcases opportunities to use new ideas in design and materials. For example, this house has wood–plastic roof shingles made from recycled plastic, four different types of siding, high-density wood flooring, and cellulosic insulation. The house is instrumented to evaluate temperature and moisture movement in the structure. (2000s)

◀ High-density flooring made from suppressed-growth Douglas-fir used in the demonstration house located at FPL. Sample of the small-diameter of a section of the suppressed tree, which is usually considered as waste material, is shown in the right corner with a pencil to show size. (2000s)

◄ Instrumentation wiring in the demonstration house located at FPL used to evaluate moisture and temperature profiles in the walls and attic. (2000s)

Structural Insulated Panels (SIPs) [57]

▲ Stacks of structural insulated panels, made from oriented strandboard (OSB) and styrofoam, on the Washington, D.C., mall, waiting to be used in the construction of a house. (2000s)

▲ House built on the Washington, D.C., mall as part of the 2005 Folklife Festival Celebration, to demonstrate the use of structural insulated panel (SIP) construction. Following the celebration, the house was dismantled and reassembled for use by Habitat for Humanity. (2000s)

◄ From left to right viewing the inside of the SIP house: Marnette Colborne, Executive Director, Habitat for Humanity, Haywood NC Chapter; Dennis Hardman, President, APA - The Engineered Wood Association; Karen Martinson, and Mike Ritter, FPL staff, helped explain the housing materials to the thousands of people that visited the SIP house. (2005)

Structural Insulated Panel House

This Structural Insulated Panel house, called the "Sustainable Resource House," was designed to reduce long-term maintenance and energy costs while demonstrating sustainable forestry, modern efficient wood products, and "green building."

The house showcased sustainable wood products that the FPL developed or played a role in developing. The 1,200-square-foot house was displayed on the National Mall in Washington,

D.C., during the Smithsonian Folklife Festival in 2005 to commemorate the 100th anniversary of the U.S. Forest Service and sustainable forestry. Following the festival, the house was donated to the Haywood Habitat for Humanity for reconstruction in Canton, North Carolina.

In addition to FPL, sponsoring organizations included APA–The Engineered Wood Association, the Structural Insulated Panel Association, Southern Pine Council, and the Metal Roofing Alliance.

1910

Use of large-sized solid lumber was common. Factory-built and prefabricated buildings were virtually nonexistent.

2010

Wood structures are significantly improved in terms of strength, durability, safety, and energy efficiency.

Through design, small-sized lumber is used more efficiently, thus providing a positive impact on forest management.

Companies employed 69,866 people and shipped products with a total value of $11,293,000,000 just in manufactured homes (mobile homes) and prefabricated wood buildings. (2006 Annual Survey of Manufacturers)

Further Information

[1] The rigidity and strength of frame walls / G.W. Trayer / FPL No. 896 (1929)

[2] Progress report on prefabricated house system under development by the Forest Products Laboratory, Madison, Wisconsin / R.F. Luxford / FPL No. 1165 (1937)

[3] Condensation in walls and attics during cold weather / L.V. Teesdale / Society of Architects monthly bulletin Vol. 21(10–11) (Apr.–May 1937)

[4] Prefabricated housing / R.F. Luxford / Wisconsin engineer Vol. 42(7) (Apr. 1938)

[5] Manual on wood construction for prefabricated houses / Prepared by the Forest Products Laboratory, Forest Service, USDA, in collaboration with the technical staff of the Housing and Home Finance Agency / U.S. Government Printing Office (1947)

[6] Physical properties and fabrication details of experimental honeycomb-core sandwich house panels / U.S. Housing and Home Finance Agency technical paper No. 7 (1948)

[7] Manual of test methods for housing materials and constructions / W.C. Lewis / In cooperation with the Housing and Home Finance Agency. Part I. Materials -- Part II. Assemblies -- Part III. Structures (1948)

[8] Prefabricated house system developed by the Forest Products Laboratory / R.F. Luxford / FPL No. 1165 (1958)

[9] Durability of resin-treated paper honeycomb core / K.H. Boller / FPL No. 2158 (1959)

[10] Fabricated wall panels with plywood coverings / R.F. Luxford / FPL No. 1099 (1959)

[11] Case studies of four sandwich panels houses / L.J. Markwardt, Lyman W. Wood. / Presented at Building Research Institute conference on sandwich panel design criteria, Washington, D.C. (Nov. 17–19, 1959)

[12] Mechanical properties of several honeycomb cores / Gordon H. Stevens, Edward W. Kuenzi / FPL No. 1887 (1962)

[13] Performance of sandwich panels in FPL experimental unit / Anderson, L.O. / Research paper FPL 12 (1964)

[14] Houses can resist hurricanes / L.O. Anderson, Walton R. Smith / Research paper FPL 16 (1965)

[15] Construction of Nu-frame research house: utilizing new wood-frame system / L.O. Anderson / Research paper FPL 88 (1968)

[16] Acoustical absorption properties of wood-base panel materials / W.D. Godshall, J.H. Davis / Research paper FPL 104 (1969)

[17] Low-cost wood homes for rural America—construction manual / L.O. Anderson, / USDA Agriculture handbook No. 364 (1969)

[18] Designs for low-cost wood homes / L.O. Anderson, Harold F. Zornig / USDA, Forest Service (For sale by the Supt. of Docs. U.S. Government Printing Office) (1969)

[19] Longtime performance of sandwich panels in Forest Products Laboratory experimental unit / G.E. Sherwood / Research paper FPL 144 (1970)

[20] Lower housing costs through improved design and wood utilization / J.A. Liska / Journal of forestry Vol. 68(7) (July 1970)

[21] Condensation problems in your house: prevention and solution / L.O. Anderson, G. Sherwood / USDA Agriculture information bulletin No. 373 (Sep 1974)

[22] FPL designs meet family housing needs / G.E. Sherwood / Research Paper FPL 173 (1972)

[23] Wood structures can resist hurricanes / G.E. Sherwood / Civil engineering Vol. 42(9) (Sep. 1972)

[24] Design and development of housing systems for operation breakthrough / U.S. Dept. of Housing and Urban Development / For sale by the Supt. of Docs., U.S. Government Printing Office (1973)

[25] Characteristics of load-bearing sandwich panels for housing: state-of-the-art report. Prepared for the U.S. Dept. of Housing and Urban Development. NTIS, Springfield, VA Order No. PB 220899 (1973)

[26] Performance of wood construction in disaster areas / J.A. Liska / American Society of Civil Engineers. Journal of the structural division Vol. 99, No. ST12 (Dec. 1973)

[27] Improved acoustical privacy in multifamily dwellings / R.E. Jones / Sound and vibration Vol. 7(9) (Sep. 1973)

[28] Lab-field correlations for airborne sound transmission through party walls / R.E. Jones / Research paper FPL 240 (1975)

[29] Sound insulation evaluations of several single-row-of-wood-stud party walls under laboratory and field conditions / R.E. Jones / Research paper FPL 241 (1975)

[30] New life for old dwellings: appraisal and rehabilitation / Gerald E. Sherwood / USDA Agriculture handbook No. 481) (1975)

[31] A conventional house challenges simulated forces of nature / Roger L. Tuomi, William J. McCutcheon / Forest products journal Vol. 25(6) (June 1975)

[32] Review of the Housing and Urban Development minimum property standards as they relate to protection of wood in use / Forest Products Laboratory. Committee for review of HUD minimum property standards, Chair. Rodney C. De Groot. (1977)

[33] Lightweight truss-framed house for safety and energy efficiency / Roger L. Tuomi / Agricultural engineering Vol. 58(5) (May 1977)

[34] Long run housing demand by type of unit and region / Thomas C. Marcin / Research paper FPL 308 (1978)

[35] Racking strength of light frame nailed walls / R.L. Tuomi / American Society of Civil Engineers, Journal of the structural division Vol. 104, No. ST7 (Jul 1978)

[36] How to design walls for desired STC ratings / R.E. Jones / Sound and vibration Vol. 12(8) (Aug. 1978)

[37] Saving energy in the older home / Alfred E. Oviatt, Jr.,Theodore J. Brevik / University of Wisconsin–Extension G2911 (1978)

[38] Energy efficiency in light-frame wood construction / Gerald E. Sherwood, Gunard E. Hans / Research paper FPL 317 (1979)

[39] Moisture interactions in light-frame housing: a review / E.L. Schaffer / Building air change rate and infiltration measurements. Philadelphia: American Society for Testing and Materials ASTM STP 719 (1980)

[40] The truss-framed system for residential and light commercial buildings / Roger L. Tuomi / Southern lumberman Vol. 241(2991) (Dec. 15, 1980)

[41] Movement and management of moisture in light-frame structures / G.E. Sherwood / Conference sponsored by Forest Products Laboratory and Forest Products Research Society in cooperation with the National Association of Home Builders, National Forest Products Association and Truss Plate Institute, and held in Denver, Colorado, (Sep. 22–24, 1981)

[42] Wall and floor systems: design and performance of light-frame structures / Doris Robertson, proceedings coordinator / Kendall/Hunt Publishing, Forest Products Research Society Proceedings No. 7317 (1983)

[43] Methodology to evaluate racking resistance of nailed walls / R.Y. Itani, R.L. Tuomi, W. McCutcheon / Forest products journal Vol. 32(1) (Jan 1982)

[44] Design approaches for light-frame racking walls / L.A. Soltis, R.W. Wolfe, R.L. Tuomi / Wall and floor systems: design and performance of light-frame structures, Sept. 22–24, 1981, Denver, Colorado. Kendall/Hunt Publishing Company, Forest Products Research Society Proceedings 7317 (1983)

[45] Contribution of gypsum wallboard to racking resistance of light-frame walls / R.W. Wolfe / Research paper FPL 439 (December 1983)

[46] Thermal insulation, materials, and systems for energy conservation in the '80s: a conference / sponsored by ASTM Committee C-16 on Thermal Insulation and DOE–ORNL, Clearwater Beach, Fla., 8–11 Dec., 1981 / F.A. Govan, D.M. Greason, J.D. McAllister, editors. Philadelphia, PA ASTM special technical publication 789 (1983)

[47] Structural analysis of light-frame subassemblies / D.S. Gromala / State-of-the-art and research needs: proceedings of the workshop / Rafik Y. Itani, Keith F. Faherty, editors. New York: American Society of Civil Engineers (1984)

[48] Low-rise buildings subjected to seismic, wind, and snow loads / L.A. Soltis / Journal of structural engineering Vol. 110, ST4 (Apr. 1984)

[49} Airborne sound transmission loss characteristics of wood-frame construction / Fred F. Rudder, Jr. / FPL–GTR–43 (March 1985)

[50] Regional variations in housing characteristics and wood products consumption for residential construction in the United States. The blue and the gray: proceedings of the 1987 joint meeting of the southern forest economist workers and the mid-west forest economists; Asheville, N.C., Apr. 8–10, 1987. Raleigh, N.C., North Carolina State University (1987)

[51] Wooden building system with flange interlocks, and beams for use in the system / R.L. Tuomi / Arlington, VA: U.S. Patent No. 4,677,806 (July 7,1987)

[52] Moisture in building envelopes / Stephen L. Quarles, Anton TenWolde / Executive summaries: 43rd annual meeting, June, 1988 Reno, Nevada. Madison, Wis.: Forest Products Research Society (1988)

[53] Seismic behavior of low-rise wood-framed buildings / R.H. Falk, L.A. Soltis / Shock and vibration digest Vol. 20(2) (Dec. 1988)

[54] Wood-frame house construction / O. C. Heyer, L.O. Anderson / USDA Agriculture handbook 73 (1955) Latest revision / Gerald E. Sherwood, Robert C. Stroh / (1989)

[55] Light-frame wall and floor systems: analysis and performance / G. Sherwood, R.C. Moody / FPL–GTR–59 (1989)

[56] Seismic performance of low-rise wood buildings / Lawrence A. Soltis, Robert H. Falk / Shock and vibration digest Vol. 24(12) (1992)

[57] Time is ripe for structural insulated panels: 60 years of research, durability testing now bearing fruit for builders / Gerald E. Sherwood, Henry Spelter / Automated builder. Vol. 31(10) (Oct. 1994)

[58] FPL roof temperature and moisture model--description and verification / Anton TenWolde / FPL–RP–561 (1997)

[59] Rehabilitation of wood-frame houses. Agriculture handbook / A. TenWolde / USDA Agriculture handbook No. 704) (1998) This publication was produced in cooperation with the National Association of Home Builders, National Research Center. This publication is a revision of Gerald E. Sherwood's "New life for old dwellings: appraisal and rehabilitation" (USDA Agriculture handbook No. 481) (1975) (1998)

[60] Roof temperature histories in matched attics in Mississippi and Wisconsin / Jerrold E. Winandy, H. Michael Barnes, Cherilyn A. Hatfield / FPL–RP–589 (2000)

[61] Performance of back-primed and factory-finished hardboard lap siding in southern Florida / C. Carll, M. Knaebe, V. Malinauskas, P. Sotos, A. Tenwolde / FPL–RP–581 (2000)

[62] Partnership for advancing technology in housing / Elizabeth Burdock, Michael Ritter, Jean Livingston, Stephanie Carnes / Forest products journal Vol. 51(3) (March 2001)

[63] Attic and crawl space ventilation: implications for homes located in the urban–wildland interface / Stephen L. Quarles, Anton TenWolde / Woodframe housing durability and disaster issues, Forest Products Society (Oct. 4–6, 2004)

[64] Improved grading system for structural logs for log homes / D.W. Green, T.M. Gorman, J.W. Evans, J.F. Murphy / Forest products journal 54(9) (2004)

[65] Durability of hardboard siding: Determination of performance criteria / C. Carll, A. TenWolde / FPL–RP–622 (2004)

Structures

► Baird Creek Bridge, an early masterpiece in timber trestle construction. It is 1,130 feet long and 235 feet high. Each brace has a split ring connector at each end connection. With a knowledge of the strength properties of wood, determined by FPL research, and split ring connectors, structures like this are possible. (1940s) {M 40816 F}

A century ago the construction of large structures was limited by the size of the wood cut from a tree and methods of connecting the members. With a knowledge of wood properties, the use of adhesives, and wood connectors, the use of wood in structures has been greatly expanded. {1, 2, 22, 23, 25–27, 30–34, 40, 45]

Large Engineered Wood Structure

◄ Construction of a wood structure to be used for storing lighter-than-air (LTA) airships like dirigibles for the U.S. Navy. Hanger is approximately 1,058 feet long, 174 feet high, with a 234-foot clear span. It required over 3,000,000 feet of fire-retardant lumber. (1940s) {M 47507 F}

► Internal wood structure of a lighter-than-air hanger—an engineer-designed combination of wood braces and connectors. (2007)

Challenge: With the diminishing supply of large trees, how can wood be used in the construction of large structures?

FPL's Contributions: FPL helped develop design data based on wood properties and new connector designs so that large structures can be built. FPL helped develop principles for design of glued-laminated materials such as laminated arches and beams. By laminating smaller pieces of lumber with the appropriate glue, large laminated aches and beams can be fabricated. FPL developed tests of glulam, finger joints, efficient lay-ups placing stronger materials at locations of greater stress, assessing effects of beam size, moisture conditions, lamination quality, and radial stresses in glulam arches.

Results: Large structures are possible. A new industry developed in the laminated arch and beam area, providing outlets for lower grade lumber, increasing the size of wood structures, and creating hundreds of jobs.

Laminated Arches and Beams

Early research at FPL evaluated the use of small boards to fabricate large laminated arches and beams. [1–6, 12, 13, 15, 18, 23, 38]

◄ Curved laminated beam being set up for compression test. (1935) {M 25188 F}

► Failure of the curved laminated beam. Test results are used to adjust beam design to prevent failures of this type. (1935) {M 25193 F}

FPL Laminated Arch Building (Packaging Laboratory)

In 1934, a new packaging laboratory was built. However, it was no ordinary building—it was built using laminated arches and also included other arches made from wood. The purpose was to provide a useful building but also one in which a visitor could observe different types of arches and see the advantages of design to decrease material and improve aesthetics. The building suffered a major fire in later years, but when the firefighters learned the building was built with wood beams, they were able to save the structure. (1930s) [3, 4, 6, 9, 10]

◄ Second laminated arch building being built in the United States. The first was built in Peshtigo, Wisconsin. The building to the left of the new building is the FPL Fire Test Laboratory. (1934) {M 25088 F}

▲ Erecting the first arch for the new building. (1934) {M 25072 F}

▲ Arches in place. (1934) {M 25074 F}

▲ Completed structure. Note the truss arch near the end of the building. (1935) {M 25270}

◄ Placing roof sheathing in place. (1935) {M 25139}

▶ Close-up of glued-laminated arch. (1960s)

▶ One of the charred laminated arches still holding up the roof. (1995)

▲ Fire broke out in the laminated arch packaging laboratory on December 6, 1995. When firefighters found out that the building had wood arches, they entered it and suppressed the fire. Following the fire the building was used for composite products research. In 2010 it was dismantled and the arches saved for testing.

Barns

In the early 1930s, small curved laminated beams were used in barn construction. [7]

▲ Finished barn. (1930s) {M 23802}

In the early 1950s, FPL established methods for designing and constructing floors that provided barns with greater resistance to wind and other shock loadings. [11]

▲ Interior of a barn made with curved laminated beams. (1930s) {M 23801 F}

Bowed Laminated Beams

▲ Bowed laminated beams used for supporting a roof on FPL garage. (1930s) {M 24708 F}

▲ Close-up of curved laminated beam. (1930s) [16, 21]

FPL's Fiber Products Pilot Plant (Laminated Arch) and Chemistry Building

▶ In the mid-1960s, these three interconnected buildings were erected at the FPL to house research in wood fiber products and wood chemistry. The 460-foot-long fiber products pilot plant (foreground) is built mainly of wood, with glued-laminated southern pine arches, a southern pine roof deck, and walls of sandwich panels with plywood faces. Roomy, well-equipped laboratories facilitate research on new environmentally preferable pulping processes, the recycling of used paper, lumber, and other wood fiber wastes for new products, the conversion of sawdust to livestock feed, and basic research on the chemistry of wood and bark. (1971) {M 138 643-2}

Fiber Products Pilot Plant Construction

◀ (Left to right) Gordon Logan, Governor Gaylord Nelson, and Wally Youngquist discussing the new pilot plant construction.

▲ Completed construction of the arch sections and roof being applied to new pilot plant.

◀ Curved laminated arch being swung into position.

▲ Applying the sandwich panels to the sides of the pilot plant.

▲ Interior of the new pilot plant. Note the open space design that allows for major changes in equipment without having to tear out walls.

▶ Dedication program held November 8, 1967, on the upper floor of the south end of the new pilot plant. This $4 million FPL addition uses FRTW (fire-retardant-treated wood). [19]

▶ Standing beside a massive laminated arch of the new pulp and paper pilot plant are Herbert O. Fleischer, Director, with Mrs. Fleischer; Mrs. McGarvey Cline, widow of the Laboratory's first director who held office from 1910 to 1912; Mrs. Howard Weiss, widow of the second director who held office from 1912 to 1917; Mrs. Edward G. Locke, whose late husband was director form 1959 to 1966; and Mrs. George M. Hunt, whose husband held the directorial reins from 1946 to 1951.

Building description:

1st floor—full building length; generally experimental pulping equipment

2nd floor—about half building length; experimental paper making equipment

Building size: 460 feet long, 60 feet wide, 50 feet high, 47,652 square feet gross area

Objectives of construction:

- To provide the greatest volume of unimpeded space possible within fixed cost limits.
- To demonstrate the most up-to-date and economical structural concepts employing wood and wood-base materials. These objectives were achieved through use of glued-laminated wood arches and supplemental framework, stressed-skin lumber-plywood panels, lumber and fiberboard roof deck, and redwood trim.

Arch framework:

A typical two-story arch consisting of the two legs and connecting beam that weighs 20,200 pounds and contains 10,600 board feet of southern pine lumber.

Wall enclosure:

- Walls enclosed with stressed-skin panels.
- Nominal 2 by 4 lumber frame adhesive-bonded to 3/8-inch-thick Douglas-fir plywood covers. Internal voids filled with polystyrene foam insulation.
- Exterior panel face covered with beige-colored polyvinyl fluoride film; interior panel face covered with Kraft paper treated with phenolic resin and painted.
- More than 800 panels, most of them 4 by 8 feet in size, were required.

Pole Barns

◄ Pole barn under construction showing the preservative-treated poles and curved laminated beams used for roof supports. (1950s)

▼ Completed pole barn structure.

◀ Pole barn showing the use of laminated beams and preservative-treated poles. (1950s) {M 100 385} [14, 35]

▲ Experimental buildings: Nu-Frame house in the background and pole barn with slant-leg rigid frame showing various design features in the foreground. (1960s) [17, 20]

Wood Laminated Beams Used in a House

◀ Example of a room constructed using laminated beams. (1960s) {M 116 377}

Large-Diameter Timbers

▶ McKenzie River Ranger Station on the Willamette National Forest in McKenzie Bridge, Oregon. Example of use of large-diameter timbers. (2007)

Small-Diameter Timbers

◀ Structures like this were used as information kiosks at the 2002 Winter Olympic Games in Utah. They were built to demonstrate the valuable use of small-diameter timber that has been removed from the forest, helping to return our forests—and the economies of forest-dependent communities—to a healthy state. [41–45]

1910

Use of wood limited to small structures.

2010

Through development of timber connectors, adhesives, design, and knowledge of wood properties, large structures are presently built.

Further Information

[1] Built-up wood / O.M. Butler American forestry magazine Vol. 25(310) (Oct 1919)

[2] Built-up wood columns conserve lumber / John A. Scholten / Engineering news-record Vol. 107(9) (Aug. 27, 1931)

[3] Glued laminated arches developed at the Laboratory / L.V. Teesdale / U.S. Forest Service, Forest Products Laboratory (1934)

[4] Latest developments in glued arch construction: Forest Products Laboratory tests new service building: commercial projects under way / T.R.C. Wilson. American lumberman No. 3063 (Dec 21, 1935)

[5] The glued laminated wooden arch / T.R.C. Wilson / USDA Technical bulletin No. 691 (1939)

[6] Formulas for columns with side loads and eccentricity / J.A. Newlin / Building standards monthly Vol. 9(12) (Dec. 1940)

[7] Durability of glued laminated barn rafters / FPL No. 1232 (1945)

[8] Report of conference on maintenance and repair of timber structures / FPL No. 1626 (1946)

[9] Tests of glued laminated wood beams and columns and development of principles of design / T.R.C. Wilson, W.S. Cottingham / FPL No. 1687 (1947)

[10] Formulas for columns with side loads and eccentricity / Lyman W. Wood / FPL No. 1782 (1950)

[11] The diaphragm action of haymow floors / D.V. Doyle / Presented at the winter meeting of the American Society of Agricultural Engineers, Chicago, Ill. (Dec. 17, 1952)

[12] Laminated southern pine / Alan D. Freas / Southern lumberman Vol. 187(2345) (Dec. 15, 1953)

[13] Fabrication and design of glued laminated wood structural members / A.D. Freas, M.L. Selbo / USDA Technical bulletin No. 1069 (1954)

[14] Performance of structural pole frames under test load / D.V. Doyle / FPL No. 2111 (1958)

[15] Factors affecting strength and design principles of glued laminated construction / Alan D. Freas / FPL No. 2061 (1962)

[16] Stresses within curved laminated beams of Douglas-Fir / Charles B. Norris / FPL Research note FPL 020 (1963)

[17] Study of wind stresses in three typical pole building frames / J.E. Parker / Research note FPL 049 (1964)

[18] After two decades of service—Glulam timbers show good performance / M.L. Selbo, A.C. Knauss, H.E. Worth / Forest products journal Vol. 15(11) (Nov 1965)

[19] $4 million FPL addition uses FRTW (fire retardant treated wood) / Wood preserving news (February 1967)

[20] Experimental pole-type structure: initial evaluation / D.V. Doyle / FPL–RP–115 (1969)

[21] Residual stresses in curved laminated wood beams / John J. Zahn / Journal of the structural division, Proceedings of the American Society of Civil Engineers Vol. 95, No. ST12 (Dec 1969)

[22] Wood products and their use in construction / A.D. Freas / Unasylva Vol. 25(2–3–4), No. 101–102–103 (1971)

[23] Strength criteria of glued-laminated timber / B. Bohannan / Performance concept in buildings, proceedings of the joint Rilem–ASTM–CIB symposium. Washington: National Bureau of Standards, 1972 NBS special publication 361 (1972)

[24] Evolution of Glulam strength criteria / Billy Bohannan, R.C. Moody / Forest products journal Vol. 23(6) (June 1973)

[25] Racking strength of walls: let-in corner bracing, sheet materials, and effect of loading rate / R.L. Tuomi / Research paper FPL 301 (1977)

[26] Improved utilization of lumber in glued laminated beams / Russell C. Moody / Research paper FPL 292 (1977)

[27] Longtime performance of trussed rafters with different connection systems: 10-year evaluation / T.L. Wilkinson / Research paper FPL 204 (1978)

[28] Design, performance, and installation of a Press-Lam basement beam in a factory-built house / J.A. Youngquist, D.S. Gromala, R.W. Jokerst, R.C. Moody, J.L. Tschernitz / Research paper FPL 316 (1978)

[29] Purdue plane structures analyzer II: a computerized wood engineering system / Stanley K. Suddarth, Ronald W. Wolfe / General technical report FPL 40 (1984)

[30] Wood as a building material / A.D. Freas / Encyclopedia of materials science and engineering / Michael B. Bever, editor Elmsford, NY: Pergamon Press (1986)

[31] Timber structures / Russell C. Moody, Alan D. Freas / Handbook of architectural technology. New York, NY: Van Nostrand Reinhold (1991)

[32] Designing, engineering, and testing wood structures / Thomas M. Gorman / Journal of materials education Vol. 13(516) (1991)

[33] Strength of bolted timber connections with steel side members / T.L. Wilkinson / FPL RP–513 (1992)

[34] Fiber stress values for design of Glulam timber utility structures / Roland Hernandez, Russell C. Moody, Robert H. Falk / FPL–RP–532 (1995)

[35] Repetitive member adjustment for wood structural design / Ron Wolfe, Steve Cramer / Building an international community of structural engineers: proceedings of Structures Congress XIV. Volume 2. (New York): American Society of Civil Engineers, (1996)

[36] Structural performance of light-frame truss-roof assemblies / Ronald Wolfe / Proceedings of the International Wood Engineering Conference, Sheraton New Orleans, New Orleans, Louisiana, USA, October 28–31, 1996. Volume 3. [S.l.]: International Wood Engineering Conference; Madison, WI: printed by Omnipress (1996)

[37] ISO 9000: Issues for the structural composite lumber industry / Steve G. Winistorfer, Harold J. Steudel, / Forest products journal 47(1): 43–47 (1997)

[38] Glued-laminated timber / Russell C. Moody, Roland Hernandez / In: Smulski, Stephen, Ed., / Engineered wood products—A guide for specifiers, designers and uses. Madison, WI: PFS Research Foundation: 1-1–1-39. Chap. 1. (1997)

[39] Wind resistance of light-frame structures / Ronald W. Wolfe / In: Proceedings, third wood building/architecture technical seminar; 1996 November 14; Seoul, Korea. Korean Wooden Architecture Association 1–15 (1998)

[40] Wood products used in new nonresidential building construction, 1995 / David B. McKeever, Craig Adair / APA– The Engineered Wood Association (1998)

[41] Dowel-nut connection in Douglas-fir peeler cores / Ronald W. Wolfe, John R. King, Agron Gjinolli / FPL–RP–586 (2002)

[42] Characterizing wood properties of small diameter Northwest trees / Thomas M. Gorman, David W. Green / Small Diameter Timber Symposium Proceedings, 2002: Resource Management, Manufacturing and Markets, 2002 February 25–27, Spokane, WA. Pullman, WA: Washington State University (2002)

[43] Assessing the market potential of roundwood recreational buildings / Dorothy Paun, Randall Cantrell, Susan L. LeVan-Green / FPL–GTR–144 (2004)

[44] Strength of small-diameter round and tapered bending members / Ron Wolfe, Joe Murphy / Forest products journal Vol. 55(3) (Mar. 2005)

[45] Small-diameter roundwood, strong-post W-beam guardrail systems / David Kretschmann, Ronald Faller, John Reid, Jason Hascall, Dean Sicking, John Rohde, / WCTE 2006: 9th World Conference on Timber Engineering, August 6–10, 2006, Portland, OR, USA: program and abstracts. Portland, OR: World Conference on Timber Engineering (2006)

Pulping/Bleaching

◄ Experimental stone grinder for making groundwood pulp. (1935) {111 G}

One hundred years ago, spruce pulpwood supplied more than 60% of our paper needs, and the remainder came mostly from balsam fir and hemlock pulpwood. Dependence on the spruce–fir–hemlock type pulpwood resulted in the depletion of these species, and by 1915 the industry was dependent primarily on Canada for more than half the pulpwood needed to meet American requirements. [1–4]

Challenge: How to economically and environmentally increase the yield of pulp from wood and how to pulp other underutilized species, thus extending the timber supply and enhance sustainable forest management.

FPL's Contributions: FPL developed the neutral sulfite semichemical (NSSC) pulping process, a process that significantly increased pulp yield and allowed use of many underutilized hardwoods. FPL research significantly improved the sulfate (kraft) pulping process so that many softwoods could be used in paper making. FPL critically evaluated and suggested improvements in mechanical, thermomechanical, chemithermomechanical, and chemical pulping processes. Laboratory scientists evaluated various species of wood using the sulfite and prehydrolysis kraft processes for use in making rayon. FPL evaluated causes and solutions to pulp-chip storage losses.

Results: Almost all species of wood can now be converted into paper making fibers by some type of pulping process. Fundamental research has led to improvements in many of the pulping processes used today. FPL's development of the neutral sulfite semichemical pulping (NSSC) process greatly advanced the corrugated container industry, creating thousands of jobs and greatly assisting forest management by providing economic outlets for underutilized wood species. By 1953, thirty-one pulp mills used the NSSC pulping process. (Corrugating medium is one of the major grades of paperboard: it is the "wavy" fluting material found in corrugated fiberboard containers.)

Mechanical Pulping

◄ Axel Hyttinen placing a wood block into the grinder pocket. Pressure is applied to the blocks forcing them against a rotating stone. The abraded fibers are retrieved in a water slurry in the pit of the grinder. (1935)

The FPL was equipped with a pulpwood grinder that had a "synthetic stone" that was the same diameter but not as wide as industrial equipment. This facilitated research on the effect of grinding conditions on the quality and production of mechanical pulp that related more directly to the industry. Grinders are still used today in the industry but are being replaced with refiners that use wood chips. One of the primary products made out of groundwood and refiner pulp is newsprint. An advantage of these mechanical pulps is that the yield of pulp for a given amount of wood is over 90%, so most of the wood is used in the product. {M 100 673} [5, 23, 26, 29, 44]

Groundwood Pulp Bleaching

FPL investigations showed that the brightness of groundwood pulp could be improved considerably by the use of sodium or hydrogen peroxide. These chemicals are now being used to increase the utility of thousands of tons of mechanical pulp for newsprint, magazine, book, and specialty papers. [22, 35, 75]

Process for Making Pulp

The following five pictures depict some of the major steps in making pulp. Because the pulp is primarily made from wood fibers, which is discussed later, the first step is to harvest the trees and convert them to wood chips.

▶ One method of preparing wood for pulping or composite products is the use of a mobile chipper that can handle tree lengths and convert the trees to chips. (1990s)

◀ FPL laboratory set-up for chipping logs. Erwin Elert ready to place a log in the chipper. The chips travel up the conveyor and into a screen box to remove the oversized chips and fine material. (1960s)

▲ For paper made from chemical pulp, the chips are placed in a digester for chemical processing. This is a small FPL laboratory rotating digester used to study new pulping processes. (1960s) {97G PU}

◄ Early laboratory beater for processing chips from the digester. The laboratory and modern paper mills now use high-speed refiners. (1935) {M 22508 F}

▲ Resulting pulp.

Chemical Pulping

The two major chemical pulping processes, sulfite and sulfate (kraft), were invented prior to the establishment of the FPL. However, the laboratory conducted research on the improvement of these processes and commercial application of them to previously unused wood species. Most of this work was done in conjunction with the pulp and paper industry and helped the growth of the industry in the United States. [8, 9, 11, 33, 34, 45–47]

▶ One early major accomplishment of the FPL and the Forest Service Southern Experiment Station was the cooperative research done with industry to help develop a way to economically convert southern pine softwood to make newsprint and bleached paper. Later a polysulfide process to give higher yields in kraft pulping of southern pine and other softwoods was developed. Today the use of southern pine pulp is a multibillion industry. (1930s) [7, 24, 25, 27, 30, 39]

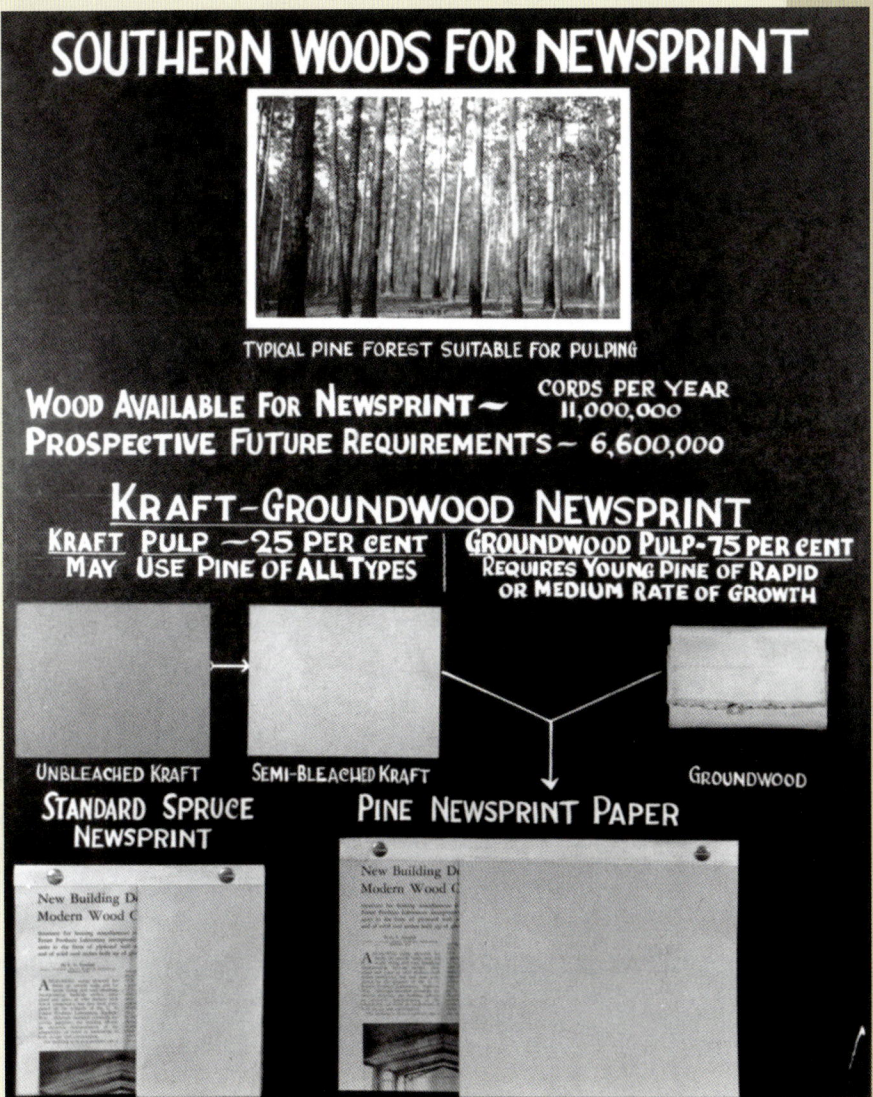

▲ Brown wrapping paper (center) and semi-bleached (top) and bleached paper made from southern pine wood using the sulfate pulping process. (1930s) {M 19224 F}

Southern Pine Bleaching

FPL pioneered the development of multistage bleaching methods for southern pine sulfate pulp. The commercial production of bleached sulfate pulp suitable for bond, writing, wrapping, printing, and specialty papers now amounts to millions of tons annually. Results have helped keep the cost of paper low. [14, 19]

Neutral Sulfite Semichemical Pulping (NSSC)

In 1925, the FPL developed the "neutral sulfite semichemical" (NSSC) pulping process. This process was successfully commercialized and led to the utilization of previously underutilized hardwood species and waste wood. (Details of that development can be found in the article *Development and Mill Transfer of the FPL Semichemical Pulping Process* by John McGovern found in The Related Articles Section.)

Initially, wasted extracted chestnut chips from the tannin industry were successfully converted to corrugating medium using the NSSC process, solving a major problem. Later research showed that many other hardwood species could also be successfully pulped. It is used with particular success in pulping mixtures of hardwoods. This research has significantly improved forest management by providing an economic outlet for underutilized wood species. Today millions of tons of NSSC pulp are commercially made. The growth of the corrugated container market provided the outlet for the use of this coarse-fiber pulp. (See the Packaging section.) [15, 37, 51, 68, 73]

◀ Corrugated fiberboard made from blackjack oak using the FPL-developed "neutral sulfite semichemical" (NSSC) pulping process. (1920s) {M 34257 F}

Anthraquione in Pulping

Larry Landucci helped explain the mechanism of increased rate of alkaline delignification using anthraquinone. Anthraquinone is presently used in some mills to improve pulp operations. [72]

Aqueous Alcohols and Amines in Alkaline Pulping

Jesse Green and Necmi Sanyer found that "additions of alcohols and amines during alkaline pulping increased the rate of selectivity of delignification above that of conventional soda and kraft processes. ... The preferential alkali adsorption and solvent effects influenced the structure of wood polymers and reaction kinetics in these systems, providing a unique but complex tool for studying the mechanism of base-catalyzed delignification of wood." [70]

◀ Pressurized digester used to pulp chips or for acid hydrolysis to produce sugars for fermentation. For pulp, the chips are placed in the digester, chemicals added and the time, temperature, and pressure are adjusted to break down the lignin that holds the wood fibers together. At the completion of the cook, the pressure is released and the chips removed from the digester for washing and further processing, depending on the type of paper product desired.

▲ James Minor working on new methods of pulping wood. (1970s) {33 G}

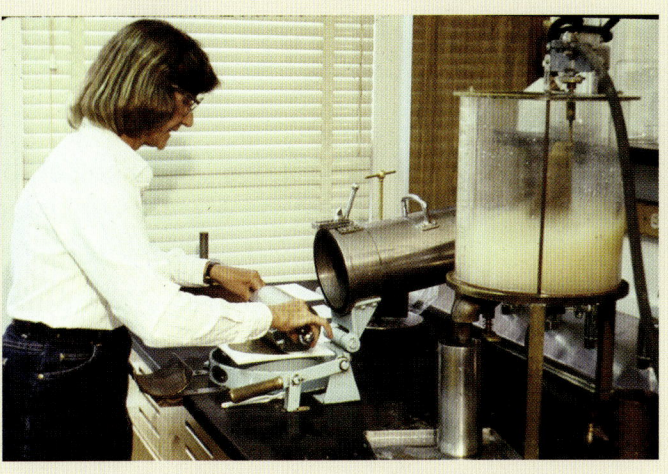

▲ Marguerite Sykes, using experimental pulps, makes handsheets of paper for evaluation in the paper test laboratory. (1980s){142 A}

Cold Soda Pulping

FPL scientists developed a high-yield cold soda pulping process for use in making corrugated board and printing paper. The process is relatively low in cost and is especially applicable to hardwoods. The pulp has more potential uses than groundwood pulp and is made at a lower cost than is semichemical pulp. Cold soda pulping was used in several mills but has been largely replaced by other processes, such as alkaline-peroxide, which was derived from the cold soda process. [40, 79]

Pulping of Douglas-fir

FPL published results of research on the sulfite and sulfate (kraft) pulping processes and bleaching of Douglas-fir. Stimulated by this and subsequent investigations, Douglas-fir is now the principal wood used in the Northwest for the manufacture of sulfate pulp. [12, 28, 32]

Pulping of Aspen

FPL's research helped expand the use of aspen for pulp and paper manufacture. This wood is now one of the leading hardwood species for this purpose. Aspen pulp is valued in the manufacture of printing papers and is also used in tissues, toweling, blotting, and glassine papers. Prior to this research, aspen was considered a "weed" tree in the United States, and it still is in some parts of the world. [31, 78, 81, 97]

Pulping of Forest Thinnings

Removing the excessive small trees found on national forests is necessary to decrease the incidence and extent of catastrophic wildfires. Junyong Zhu and co-workers conducted research to determine the properties of this available resource. This research indicated that thinnings are good quality material for thermomechanical and kraft pulps. A major problem is the cost of transporting this material to mills that have an adequate water supply. [87, 94]

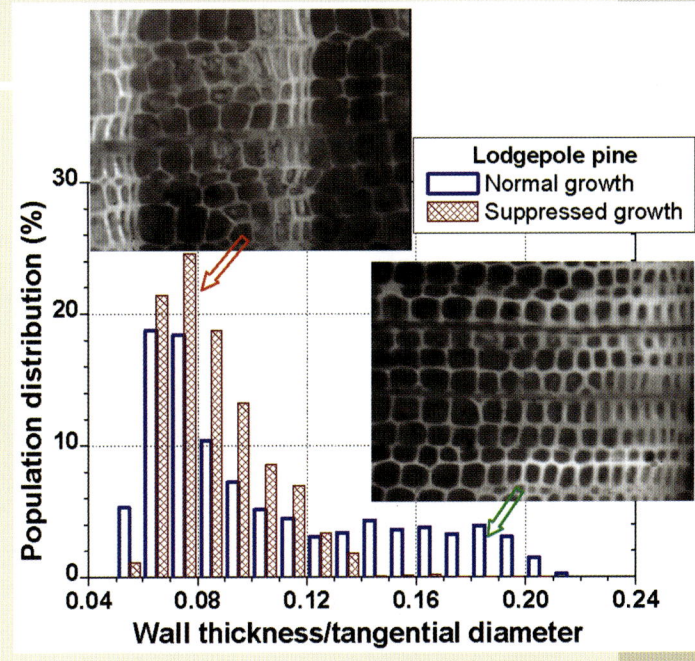

▶ Effect of tree growth conditions on wood tracheid geometry. Normal growth produced a bimodal distribution with a clear distinction between earlywood (thin wall) and latewood (thick wall), whereas suppressed growth produced a narrower distribution with little distinction between earlywood and latewood. Photographs were taken from wood core samples. (2000s) [105]

Pulp Yield from Unbarked Wood

FPL scientists showed that the pulp yield per cord of wood could be increased about 4% by eliminating the de-barking operation because of the elimination of loss of wood during de-barking. This allowed use of small tree branches, tops, and other wood wastes, and the bark provided some additional fiber. However, more pulp cleaning and chemicals are required. [59, 64, 66]

Supporting Tropical Forestry

Jim Laundrie and co-workers developed correlations between wood density and tropical climatic life zones. This research indicated that most species from within a climatic zone have similar processing characteristics. Pulp, paper, and panel products have now been made using mixed-species samples from various Philippine, Colombian, and Ghanan climatic zones. [68]

Pulp Analysis

FPL scientists developed and published techniques for the determination of wood and pulp constituents by quantitative paper chromatography. These early techniques have been replaced by newer techniques, such as high-pressure liquid chromatography (HPLC). Modern methods, built off the principles of paper chromatography, and the main analytical process used today with HPLC are directly related. [20, 21, 36, 42, 43, 53, 56, 57, 74, 88, 99]

Treatment Controls Pulpwood Chip Deterioration

Pulpwood chips stored in outside piles are subject to serious losses of wood substance because of fungal attack and even spontaneous combustion. Laboratory-scale studies have demonstrated that chemicals recovered from the kraft pulping process, when applied to the fresh pulpwood chips, effectively prevent such losses. [6, 10, 52, 54, 55, 61, 67]

In addition to chemical treatments, a new "standby storage" approach has been evaluated by Edward Springer. Setting aside a volume of chips needed for a safe reserve and keeping them in reserve over extended periods permits the use of fresh chips on a day-by-day basis. Storage losses are limited to the reserve pile, resulting in savings in the million-dollar-per-year range per pulpmill without the cost, and environmental disadvantages of applying chemicals. Today most mills apply "first-in-first-out" and "just-in-time" logistical purchasing. [69]

▲ Three piles of wood chips being studied for the effects of time, fungal action, and two chemical treatments on the loss of wood substance. (1960's)

▲ Steam in a wood chip pile because of spontaneous heating caused by biological action of organisms in the chips.

Bleach Plant Washer Mineral Scale

One of the problems in some processes for bleaching of pulp is a buildup of mineral scale in the pipes. Alan Rudie and Peter Hart (Mead–Westvaco) have conducted research on how to minimize the scale buildup. [101–104]

▶ Buildup of calcium oxalate scale in bleach plant pipe. (2000s)

Pulping and Bleaching Technologies Decrease Environmental Impact

About 50% of wood pulp produced in the United States (29 of 58 million tons) is bleached to improve the brightness of paper products. Over time, FPL's collective contributions to pulp bleaching research has helped the industry to eliminate the use of elemental chlorine. FPL scientists, through the efforts of Rajai Atalla, Ira Weinstock, Richard Reiner, Carl Houtman, Craig Hill, and others, developed a totally chlorine-free bleaching process using polyoxometalates (POMs). Early research results indicated that using a closed-loop process, POM's could substantially decrease effluents, in that it uses only 1% as much water as present processes. Furthermore, it doesn't produce any organic by-products, because the organic matter removed during the bleaching is oxidized to carbon dioxide and water. Also, because POMs cause little damage to cellulose fibers, they may be used to bleach recycled paper, with little loss in the strength of the fibers, thus reducing the need for virgin wood fiber. This process has yet to be commercialized. [85, 89, 90, 92, 93, 95]

Results of FPL research indicated that pulping and bleaching of wood, using oxygen in place of sulfur and chlorine compounds can reduce the undesirable environmental impacts associated with current pulping and bleaching methods. While some of these processes work effectively, pulp quality and cost do not compete well with the kraft pulping process. [62, 63, 76, 80, 82, 84]

▶ FPL's modern pilot plant used for pulping and bleaching studies and processing recycled fibers. (2000s) {10 A}

1910

No southern pine pulp industry existed, and limited use of hardwoods in pulping.

2010

The pulping industry is now found in many parts of the United States. Companies employed 6,736 people and shipped products with a total value $4,257,431,000. (2006 Annual Survey of Manufacturers)

Further Information

[1] The effect of variable grinding conditions on the quality and production of mechanical pulp / McGarvey Cline / Prepared for the International Congress of Applied Chemistry, New York, (September 1912)

[2] Paper pulps from various forest woods: experimental data and specimens of soda and sulphite pulps / compiled by Henry E. Surface / FPL Library has manuscripts at B6781, published report at 1 F768P (1912)

[3] Some experiments on the conversion of longleaf pine to paper pulp by the soda and sulphate processes / Sidney D. Wells / Paper trade journal Vol. 57(18) (Oct. 16, 1913)

[4] Wood waste and other pulpwood's used in 1914 by United States mills / Henry E. Surface / Paper Vol. 18(14) (June 14, 1916)

[5] Factors in the quality of groundwood: notes on the control of the groundwood process / G.C. McNaughton / Paper Vol. 21(4) (Oct. 3,1917)

[6] Reduce pulpwood decay by proper storage / FPL Technical note No. 123 (1920)

[7] Book paper from southern pines and gums / Sidney D. Wells / Paper Vol. 27(12) (Nov. 24, 1920)

[8] Uniformity of digester chip charges / FPL Technical note No. 126 (1921)

[9] American pulpwoods: the suitability of various species of American woods for pulp and paper production / Sidney D. Wells / Paper Vol. 32(14) (July 25, 1923), Vol. 32(15) (Aug. 1, 1923)

[10] The control of decay and molding of wood pulp / C. Audrey Richards / Paper trade journal Vol. 78(18) (May 1, 1924)

[11] The relation between cooking conditions and yield and quality of sulphite woodpulp / R.N. Miller / Paper trade journal 81(23): 55–59 (Dec. 3, 1925)

[12] The determination of the pulping characteristics of Douglas fir by the sulphate process / C.E. Curran, J.A. Staidl / FPL (1927)

[13] Semi-sulfite pulping studies: preliminary experiments / W.H. Monsson, G.H. Chidester / FPL No. 834 (1929)

[14] Production of a commercial white paper from southern pine pulp / M.W. Bray, R.H. Doughty / FPL No. 862 (1929)

[15] Utilization of wood waste and "waste" woods through the semichemical pulping process / C.E. Curran / FPL No. 897 (1929)

[16] Can technical development help the paperboard industry / C.C. Heritage / Paper Industry 11(4): 658, (July 1929)

[17] The effect of varying cooking time and temperature on the yield and chemical compositions of sulphite pulp using a soda base / John Neilson McGovern. Thesis (M.S.)—University of Wisconsin–Madison (1930)

[18] The utilization of fibrous plants for pulp and paper: a digest of the results made available at the Forest Products Laboratory by the investigations of the United States Department of Agriculture / compiled by Earl R. Schafer / FPL No. 1002 (1931)

[19] White papers from southern pines III, Pulping longleaf pine for strong, easy-bleaching pulp / Mark W. Bray, C.E. Curran / Paper trade journal Vol. 96(6) (Feb. 9, 1933)

[20] Application of elementary statistical methods in the testing of pulp and paper / F.A. Simmonds, R.H. Doughty / Paper trade journal 97(25):32–37 (Dec. 21, 1933)

[21] A comparison of sheet machines for pulp evaluation / R.H. Doughty, C.E. Curran / Paper trade journal 97(25):38–44 (Dec. 21, 1933)

[22] A study of the color principle in Western hemlock groundwood pulp / C.E. Curran, E.R. Schafer, J.C. Pew / Paper trade journal 101(8): 35–41 (Aug. 22, 1935)

[23] Effect of temperature and consistency in mechanical pulping / Earl R. Schafer, J.C. Pew / Paper trade journal, (September 26, 1935)

[24] Pulpwood quality of southern pine as related to the requirements of newsprint production / C.E. Curran / Journal of forestry Vol. 34(3) (Mar. 1936)

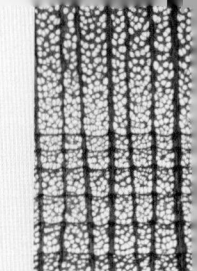

[25] Frontiers of the southern pulping industry / C.E. Curran / FPL No. 1141 (1937)

[26] Grinding of loblolly pine: relation of wood properties and grinding conditions to pulp and paper quality / E.R. Schafer, J.C. Pew, C.E. Curran / FPL No. 1163 (1937)

[27] Contribution of Forest Products Laboratory research to southern pulp and paper development / Carlile P. Winslow / FPL No. 1206 (1939)

[28] Sulphite pulp from Douglas fir / G.H. Chidester, J.N. McGovern / TAPPI technical association papers series 24(1) (June 1941)

[29] The grinding of Engelmann spruce for newsprint and magazine quality mechanical pulps / E.R. Schafer, J.C. Pew / FPL No. 1407 (1942)

[30] Development, problems, and possibilities of the southern pulping industry / Mark W. Bray / Southern pulp and paper journal Vol. 4(10) (Mar. 1942)

[31] Suitability of birch, aspen, and sugarberry for rayon pulp: results of certain sulfite pulping and bleaching experiments / Hugo Nihlen, J.N. McGovern / FPL No. 1441 (1943)

[32] Sulfate pulping, fiber measurements and bleaching of woods- and sawmill-waste from Douglas-fir, western hemlock, Pacific silver fir, and western redcedar I, Production of kraft and bleachable sulfate pulps and the evaluation of the woods and pulps / Mark W. Bray, J. Stanley Martin / Forest Products Laboratory (1945)

[33] Southern hardwoods for the pulp and paper industry / Mark W. Bray, Lloyd N. Lang / Southern lumberman Vol. 173(2177) (Dec. 15, 1946)

[34] Sulphite pulps and papers from sawdust and chip mixtures / E.L. Keller, J.N. McGovern / Pulp and paper magazine of Canada Vol. 48(7) (June 1947)

[35] Observations on bleaching groundwood pulps / Ralph M. Kingsbury, Earle S. Lewis, Forrest A. Simmond / Paper trade journal Vol. 126, TAPPI section (June 10, 1948)

[36] Comparison of several freeness testers on board stock / C.E. Hrubesky / TAPPI Vol. 32(7):315–318, (July 1949)

[37] Relations between yield and properties of spruce and aspen sulphite and sulphite semichemical pulps / J.N. McGovern / TAPPI Vol. 33(10) (Oct. 1950)

[38] New sources of paper pulps / E.R. Schafer / Crops in peace and war, the yearbook of agriculture 1950–1951 Washington, DC: U.S. Dept. of Agriculture (1951)

[39] Effect of bark on yield and quality of sulphate pulp from southern pine / J.S. Martin, K.J. Brown / TAPPI Vol. 35(1) (Jan. 1952)

[40] High-yield cold soda pulps and products from several woods / K.J. Brown, J.N. McGovern / The paper industry (Apr. 1953)

[41] Commercial processes of pulping woods / FPL Technical note No. 204 (1953)

[42] Techniques for the determination of pulp constituents by quantitative paper chromatography / Jerome F. Saeman, W.E. Moore, R.L. Mitchell, M.A. Millett / TAPPI Vol. 37(8) (Aug. 1954)

[43] Pulping characteristics of Lake States and Northeastern woods / E.R. Schafer, J.S. Martin, E.L. Keller / FPL No. 1675 (1955)

[44] Grinding pretreated hardwoods: experiments on quaking aspen, sweet gum, red alder, black tupelo, sugar maple, red oak, and cottonwood / Axel Hyttinen, E.R. Schafer / FPL No. 2015 (1955)

[45] Pulping sawdust chips made by a coarse-feed saw / J.S. Martin / Forest products journal Vol. 9(10) (Oct. 1959)

[46] Preparation and evaluation of Insignis pine sulfite, bisulfite, and groundwood pulps for use in newsprint / Necmi Sanyer, E.L. Keller, D.J. Fahey / Forest Products Laboratory (1961)

[47] Bleached kraft pulp from several southern hardwoods / F.A. Simmonds, E.L. Keller, G.H. Chidester / Southern pulp and paper manufacturer Vol. 26(8) (Aug. 10, 1963)

[48] Possibilities for the nonchlorination bleaching of softwood sodium bisulfite pulps / F.A. Simmonds / Paper trade journal Vol. 148(17) (Apr. 27, 1964)

[49] Factors affecting yield increase and fiber quality in polysulfide pulping of loblolly pine, other softwoods, and red oak / Necmi Sanyer, James F. Laundrie / TAPPI Vol. 47(10) (Oct. 1964)

[50] Pulp yields for various processes and wood species / FPL–RN–031 (1964) (Reissued May 1980)

[51] Magnesium-base sulfite semichemical pulps for corrugating boards / Necmi Sanyer, Eugene L. Keller / TAPPI Vol. 48(2) (Feb. 1965)

[52] Outside storage of pulpwood chips—a review and bibliography / George J. Hajny / Tappi Vol. 49(10) (Oct. 1966)

[53] Swelling of prehydrolysis-kraft pulp fibers in cadmium ethlyenediamene / F.A. Simmonds, R.A. Horn / TAPPI Vol. 52(5) (May 1969)

[54] Spontaneous heating in piled wood chips: I. Initial Mechanism / E.L. Springer, G.J. Hajny / Tappi Vol. 53(1) (Jan. 1970)

[55] Spontaneous heating in piled wood chips: II. Effect of temperature / E.L. Springer, G.J. Hajny, W.C. Feist / Tappi Vol. 54(4) (Apr. 1971)

[56] Effect of pulping on cellulose structure. Part I, A hypothesis of transformation of fibrils / Volker E. Stockmann / Tappi Vol. 54(12) (Dec. 1971)

[57] Effect of pulping on cellulose structure. Part II, Fibrils contract longitudinally / Volker E. Stockmann / Tappi Vol. 54(12) (Dec. 1971)

[58] High yields of kraft pulp from rapid-growth hybrid poplar trees / J.F. Laundrie, J.G. Berbee / Research paper FPL 186 (1972)

[59] Kraft pulping of pulpwood chips containing bark / R.A. Horn / Paper trade journal Vol. 156(46) (Nov. 6, 1972)

[60] Assessment of a rapid-growth hybrid poplar for kraft pulping / J.N. McGovern / Forestry Research notes No. 180 (1973)

[61] Spontaneous heating in piled wood chips: contribution of bacteria / William C. Feist, Edward L. Springer, George J. Hajny / Tappi Vol. 56(4) (Apr. 1973)

[62] Carbohydrate stabilization with iodide in oxygen bleaching of kraft pulps / J.L. Minor / Tappi Vol. 57(2) (Feb. 1974)

[63] Oxygen pulping of shortleaf pine with sodium carbonate / J.L Minor / Tappi Vol. 58(3) (Mar 1975)

[64] Corrugated fiberboard containers from high-yield roughwood kraft linerboard pulp / R.A. Horn, J.W. Koning, Jr. / Tappi Vol. 59(2) (Feb. 1976)

[65] Southern pine thermomechanical pulp in newsprint and corrugating medium furnishes / Donald J. Fahey, Wm. James Frederick, Jr. / Southern pulp and paper manufacturer Vol. 39(5) (May 1976)

[66] Kraft pulping of roughwood chips / R.A. Horn / Complete-tree utilization and biosynthesis and structure of cellulose: proceedings of the 8th Cellulose Conference, Part II / edited by T.E. Timell New York: Wiley, 1976. Applied polymer symposia No. 28 (1976)

[67] Evaluation of the sodium N-methyldithiocarbamate + sodium 2, 4-dinitrophenol treatment in an outside chip pile (1–74–3): final report / E.L. Springer, M. Benjamin, W.C. Feist, L.L. Zoch, G.J. Hajny, This report was later revised and published in TAPPI, vol. 60(2) February 1977, as "Chemical treatment of chips for outdoor storage" (1977)

[68] Kraft and NSSC pulping of mixed tropical hardwoods / J.F. Laundrie / Proceedings of conference on improved utilization of tropical forests. Madison, WI: USDA Forest Service Forest Products Laboratory, (1978)

[69] Losses during storage of southern pine chips—the case for standby storage / Edward L. Springer / Tappi Vol. 61(5):69–72 (May 1978)

[70] Alkaline pulping in aqueous alcohols and amines / J.D. Green, N. Sanyer, / Tappi Vol. 65(5) (May 1982)

[71] Improved strength in high-yield pulps through chemical treatment / T.H. Wegner / Tappi Vol. 65(8) (1982)

[72] Anthraquinone losses during alkaline pulping / L.L. Landucci / Presented in part at the Canadian Wood Chemistry Symposium, Niagara Falls, Ontario, Sep 1982 and at the 185th American Chemical Society meeting, Seattle, Washington, (Mar 1984.) J. Ralph, Journal of wood chemistry and technology Vol. 4(2) (1984)

[73] Development and mill transfer of the FPL semichemcial pulping process / J.N, McGovern / Tappi J. Vol. 67(6): 30–34 (June 1984)

[74] An analysis of the wood sugar assay using HPLC: a comparison with paper chromatography / Roger C. Pettersen, Virgil H. Schwandt, Marilyn J. Effland / Journal of chromatographic science Vol. 22(11) (Nov. 1984)

[75] Bleaching groundwood and kraft pulps with potassium peroxymonosulfate, comparison with hydrogen peroxide / Edward L. Springer / TAPPI proceedings: 1986 Pulping Conference, 1986 October 26–30, Toronto, ON, Book 3. Atlanta, GA: TAPPI Press (1986)

[76] Use of calcium sulfite and air to bleach a delignified aspen kraft pulp / Edward L. Springer, James D. McSweeny / Tappi journal Vol. 69(4) (Apr. 1986)

[77] Increasing hardwood use in kraft linerboard / P.J. Ince / Southern pulp & paper Vol. 49(12) (Nov 1986)

[78] Strength and optical properties of chemically pretreated aspen chip groundwood. Tappi journal. Vol. 70(5) (May 1987)

[79] History of FPL cold soda CMP process, 1950–present / J.N. McGovern, E.L. Springer / TAPPI proceedings of the 1988 pulping conference: 1988 October 30–November 2, New Orleans, Louisiana. Atlanta, GA: TAPPI Press, (1988)

[80] Progress in bleaching pulps with the sulfite-air system / Edward L. Springer, James D. McSweeny / TAPPI proceedings of the 1988 pulping conference: October 30–November 2, New Orleans, Louisiana. Atlanta, Ga: TAPPI Press, (1988)

[81] Biomechanical pulping of aspen chips by Phanerochaete chrysosporium: fungal growth pattern and effects on wood cell walls / Irving B. Sachs, Gary F. Leatham, Gary C. Myers / Wood and fiber science Vol. 21(4) (Oct. 1989)

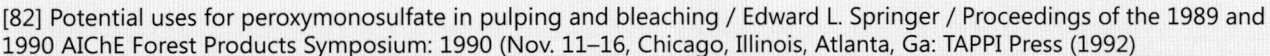
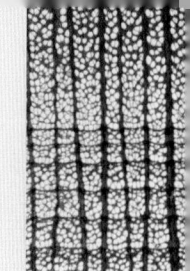

[82] Potential uses for peroxymonosulfate in pulping and bleaching / Edward L. Springer / Proceedings of the 1989 and 1990 AIChE Forest Products Symposium: 1990 (Nov. 11–16, Chicago, Illinois, Atlanta, Ga: TAPPI Press (1992)

[83] Dissolving pulp industry: market trends / Irene Durbak / FPL–GTR–77 (1993)

[84] Treatment of softwood kraft pulps with peroxymonosulfate prior to oxygen delignification / Edward L. Springer, James D. McSweeny / Tappi journal Vol. 76(8) (Aug. 1993)

[85] A New environmentally benign technology and approach to bleaching kraft pulp. polyoxometalates for selective delignification and waste mineralization / Ira A. Weinstock, Rajai H. Atalla, Richard S. Reiner, Mark A. Moen, Kenneth E. Hammel, Carl J. Houtman, Craig L. Hill / New J. Chem. 20(2): 269–275 (1996)

[86] Papermachine runnability of never dried, dried, and enzymatically treated dried pulp / S. Abubakr, K. Rutledge-Cropsey, J.H. Klungness / In: Srebotnik Ewald; Messner, Kurt, eds. Biotechnology in the pulp and paper industry—recent advances in applied and fundamental research: Proceedings, 6th international conference on biotechnology in the pulp and paper industry; Vienna, Austria: Facultas-Universitätsverlag: p. 151–156. (1996)

[87] Pulp quality from small-diameter trees / Gary C. Myers, S. Kumar, R.R. Gustafson, R.J. Barbour, S.M. Abubakr / FPL–GTR–100 (1997)

[88] A rapid modified method for compositional carbohydrate analysis of lignocellulosics by high pH anion-exchange chromatography with pulsed amperometric detection (HPAEC/PAD) / Mark W. Davis / Journal of wood chemistry and technology Vol. 18(2) (1998)

[89] Selective transition-metal catalysis of oxygen delignification using water-soluble salts of polyoxometalate (POM) anions—Part I. Chemical principles and process concepts / Ira A. Weinstock, Rajai H. Atalla, Richard S. Reiner, Carl J. Houtman, Craig L. Hill / Holzforschung. 52(3): 304–310 (1998)

[90] Selective transition-metal catalysis of oxygen delignification using water-soluble salts of polyoxometalate (POM) anions—Part II. Reactions of a -[SiVW11010]5- with phenolic lignin-model compounds / Ira A. Weinstock, Kenneth E. Hammel, Mark A. Moen, Lawrence L. Landucci, Sally Ralph, Cindy E. Sullivan, Richard S. Reiner / Holzforschung Vol. 52(3): 311–318. (1998)

[91] Pilot testing and recycling evaluation of newly developed environmentally benign pressure sensitive adhesives / Said Abubakr, David Bormett / Proceedings of the 1998 recycling symposium, March 8–12, 1998, New Orleans, LA. Atlanta, GA: TAPPI Press, (1998)

[92] Oxidative delignification of wood pulp or fibers using transition metal-substitute polyoxometalates / Ira A. Weinstock, Craig L. Hill / US Patent No. 5,824,189 (Oct. 20,1998)

[93] Delignification of wood and kraft pulp with polyoxometalates / Edward L.Springer, Richard S. Reiner, Ira A. Weinstock, Rajai H. Atalla, Michael W. Wemple, Elena M.G. Barbuzzi, / Pulping conference: Oct. 25–29,1998, Queen Elizabeth Hotel, Montreal, Quebec, Canada. Atlanta, GA TAPPI Press, c1998: Book 3, p. 1571–1583 (1998)

[94] Small-diameter trees used for thermomechanical pulps / Gary C. Myers, R. James Barbour, Said Abubakr / Tappi journal Vol. 82(10) (Oct. 1999)

[95] Progress in the development and optimization of polyoxometalate delignification systems / Rajai H. Atalla, Ira A. Weinstock, Edward L. Springer, Richard S. Reiner / TAPPI fall technical conference: engineering, pulping & PCE & I: Atlanta, GA : TAPPI Press (2003)

[96] Comparison of macrostickies measurement methods / Mahendra R. Doshi, Carl Houtman, Freya Tan, Lisa Davie, Gregg Sauve / Progress in paper recycling Vol. 12(3) (May 2003)

[97] Development of unbonded and bonded areas in relation to Populus species wood characteristics in grinding / L.K. Lehtonen, J.H. Lehto, A.W. Rudie / Journal of pulp and paper science Vol. 30(10) (Oct. 2004)

[98] Modeling of bleach plant washer mineral scale / Alan Rudie, Peter Hart / Proceedings of the 2004 TAPPI fall technical conference, October 31–November 3, 2004, Atlanta, GA. Atlanta, GA TAPPI Press (2004)

[99] Comparison of fiber length analyzers / Don Guay, Nancy Ross-Sutherland, Nicole Malandri, Aimee Stephens / TAPPI Practical papermaking conference, May 22–26, 2005, Milwaukee, Wisconsin. Atlanta, GA: Tappi Press (2005)

[100] Kraft pulp from budworm-infested jack pine / J.Y. Zhu, Gary C. Myers / Holzforschung Vol. 60 (2006)

[101] Mineral scale management. Part I, Case studies / Peter W. Hart. Alan W. Rudie / Tappi journal Vol. 5(6) (June 2006)

[102] Mineral scale management. Part II, Fundamental chemistry / Alan W. Rudie, Peter W. Hart / Tappi journal Vol. 5(7) (July 2006)

[103] Mineral scale management. Part III, Nonprocess elements in the paper industry / Alan W. Rudie, Peter W. Hart / Tappi journal Vol. 5(8) (Aug. 2006)

[104] Modeling and minimization of barium sulfate scale / Alan W. Rudie, Peter W. Hart. / Engineering, pulping, and environmental conference, Nov. 5–8, 2006, Atlanta, GA. Atlanta, GA TAPPI (2006)

[105] Forest thinnings: for integrated lumber and paper production / J.Y. Zhu, C. Tim. Scott, Roland Gleisner, Doreen Mann, D.W. Vahey, Dennis P. Dykstra, Forest products journal Vol. 57(11) (Nov. 2007)

Paper

◀ Structural variations in chemically swollen kraft pulp fibers. (1960s) {1 A}

In 1910, only a few preferred tree species, such as spruce, fir, and hemlock were utilized in a fledgling paper industry. Waste in lumber manufacturing was huge. [9, 27, 29, 31, 75]

Challenge: What can be done to better utilize the underutilized wood species, forest residues, manufacturing waste, and sawdust?

FPL's Contributions: FPL scientists conducted evaluations of new pulping and bleaching processes in terms of the resulting pulp's runnability on a paper machine. FPL research of paper machine variables, chemical additives, and wood species has helped to maximized paper and paperboard performance. FPL demonstrated the use of mixed species in making usable paper. FPL also developed new techniques to use recycled fiber and extend the timber supply.

Results: Because of ongoing research by FPL, industry, universities, institutes, and associations, almost any species of wood or wood waste can be pulped and converted into paper products. Major limitations to further growth are the economic availability of new fiber and the cleanliness and bonding characteristics of recycled fiber. [15, 62, 85]

Comparative Size and Forms of Wood Fibers Used in Paper

▶ Different sizes and forms of wood fibers that may be present in pulp, depending on the wood species used. [25, 37, 91–93]

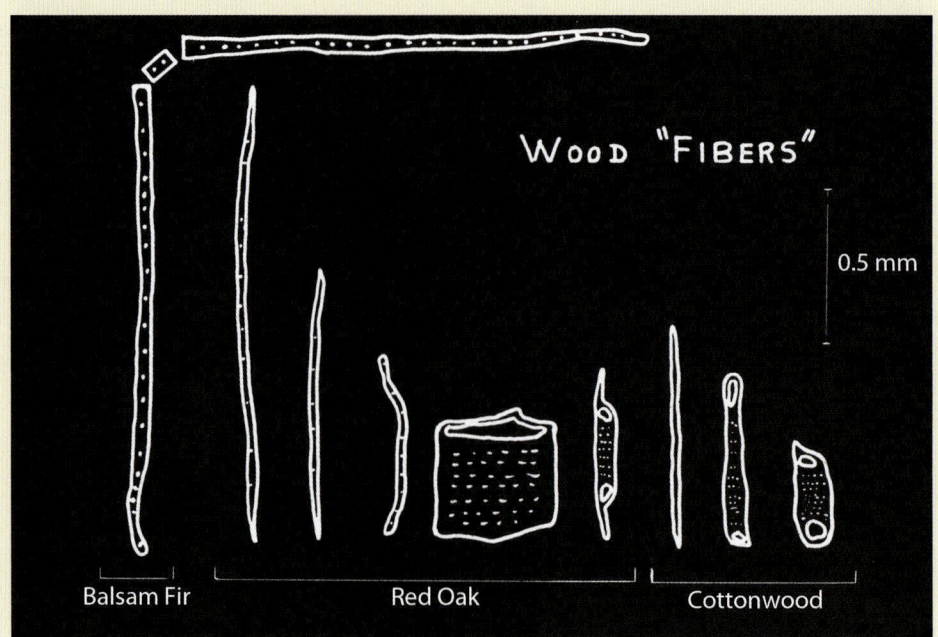

WOOD "FIBERS"

0.5 mm

Balsam Fir   Red Oak   Cottonwood

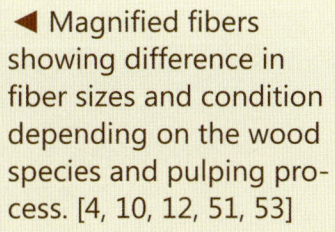

◀ Magnified fibers showing difference in fiber sizes and condition depending on the wood species and pulping process. [4, 10, 12, 51, 53]

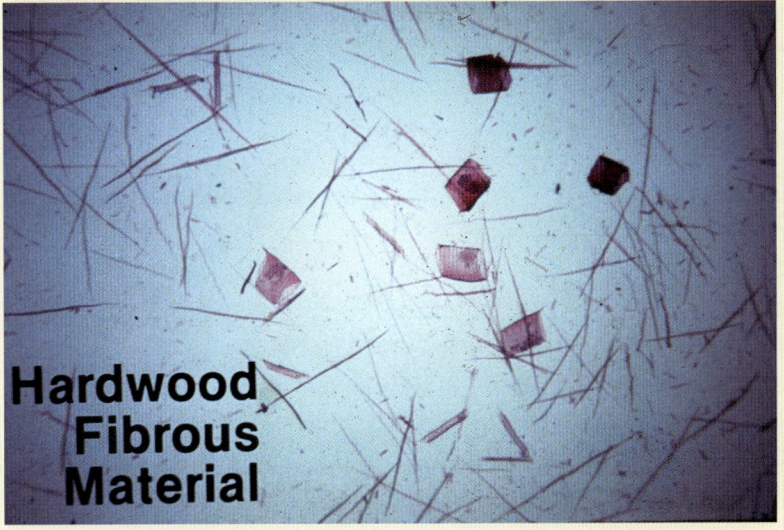

Hardwood Fibrous Material

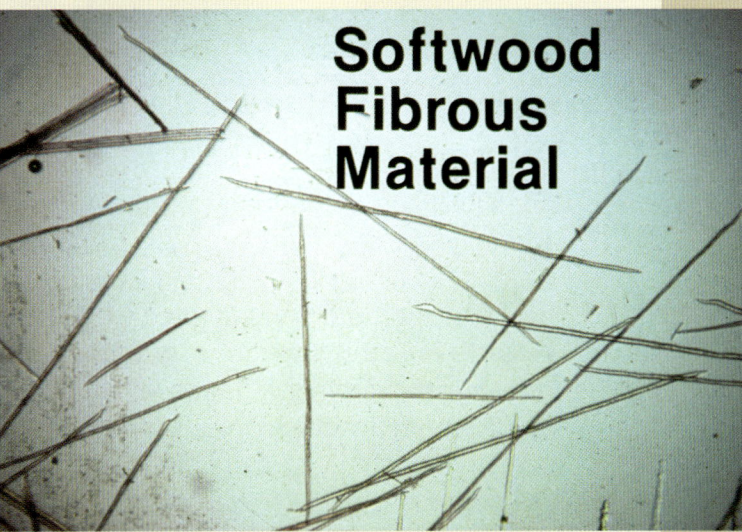

Softwood Fibrous Material

▲ Note the variety of fiber types, including the large vessel elements typically found in oak. {82 B}

▲ Note the long length of the fibers {110 F}

Though fiber lengths vary, they are all quite short, especially after processing.

FPL Paper Machines

▲ First paper machine at FPL. (1926) {5074 M} [6]

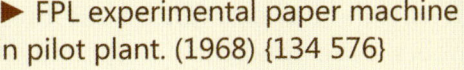

▶ FPL experimental paper machine in pilot plant. (1968) {134 576}

◀ Pulp is diluted with water and flowed onto the paper machine wire where the water is removed by gravity, pressure, and heat. This is the FPL experimental paper machine. Robert Hilton, Donald Fahey, and Charles Polly (front to back) observe the run that is being made to study the effect of expander rolls on paper properties. This paper machine is used to evaluate hundreds of different pulps, chemicals, and mechanical treatments to improve the papers and paperboards used today. (1960s) {M 101 199} [40]

Treated Papers

The FPL inaugurated many treatments for paper that have led to expanded uses in situations where paper would not normally be considered serviceable. For example, paper has been treated for flexibility by adding latex compositions to make satisfactory gaskets, sandpaper backing, and other uses; treated with chemicals to resist mold formation and become serviceable in damp areas; treated with chemicals for copy-sensitive office paper; processed for dimensional stability so that it will not change shape in such vital applications as maps and charts; treated with chemicals for grease proofing so it can be used for wrapping greased parts and prohibit contamination of adjacent materials; and treated with chemicals to make the paper less sensitive to moisture vapor and liquids. [20, 24, 35, 39, 41–43]

Fiber Picking on Printing Press

◀ Seen through the microscope, fiber "picking" on a printing press may be caused by large vessel elements so weakly bonded to surrounding fibers that ink easily picks them off. In the photo on the right, a large flat vessel element is clearly visible on the surface of a sheet made of conventionally refined kraft pulp high in oak fiber content. Photo on the left shows surface of the paper made of a kraft pulp similarly high in oak content, but specially refined to fibrillate vessel elements. (1960s) {M 134 555} [46]

Engineering Properties of Paper

One major accomplishment of the Laboratory was helping to convince industry to look at paper as an engineering material. This required the development of accurate methods of evaluating paper in terms of compression, tension, and shear. Vance Setterholm, working with Edward W. Kuenzi and other staff members, developed the devices shown on the next page. [33, 34, 36, 38, 45, 47, 49, 52, 54, 57]

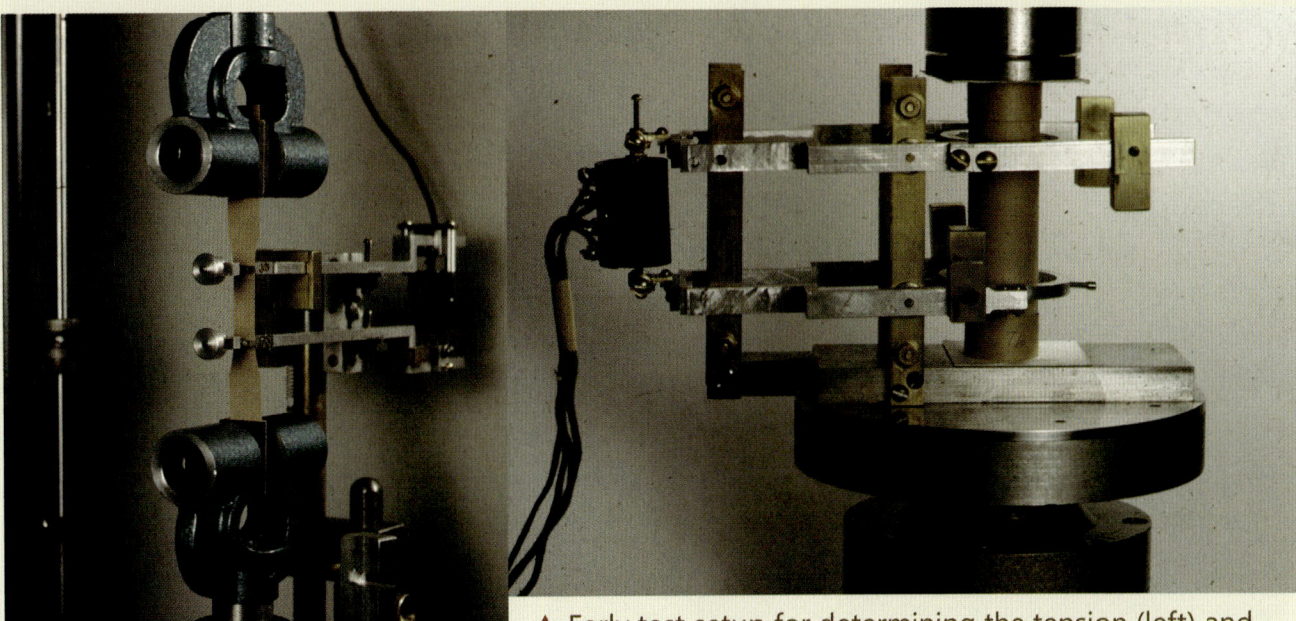

▲ Early test setup for determining the tension (left) and compression (right) stress–strain properties of paper. (1960s)

▼ Results of research on the effect of cyclic humidity changes on the strength of paperboard. (1970s) [3, 50, 66, 68, 69, 73]

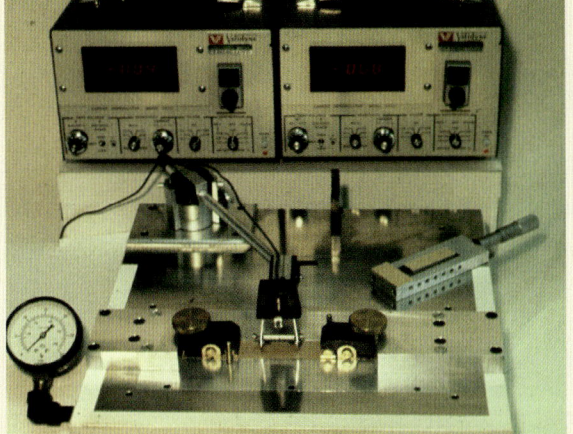

▲ A test device developed by Dennis Gunderson, John Considine, and staff to determine tensile or compression properties of paper. This device eliminates the need to form tubes for the compression tests. It also was modified so that moisture content in the specimen could be varied while the test is being run. Ultimately this device was packaged into a commercial machine for sale to industry. Using the results obtained from this device and the engineering design formulas developed by Thomas Urbanik and Millard Johnson (University of Wisconsin Engineering Mechanics Department Chairman), it is now possible to predict the compressive strength of corrugated fiberboard containers and determine how best to make the paper to maximize the strength using the minimum amount of fiber. (1980s) {59 N} [63, 64, 67]

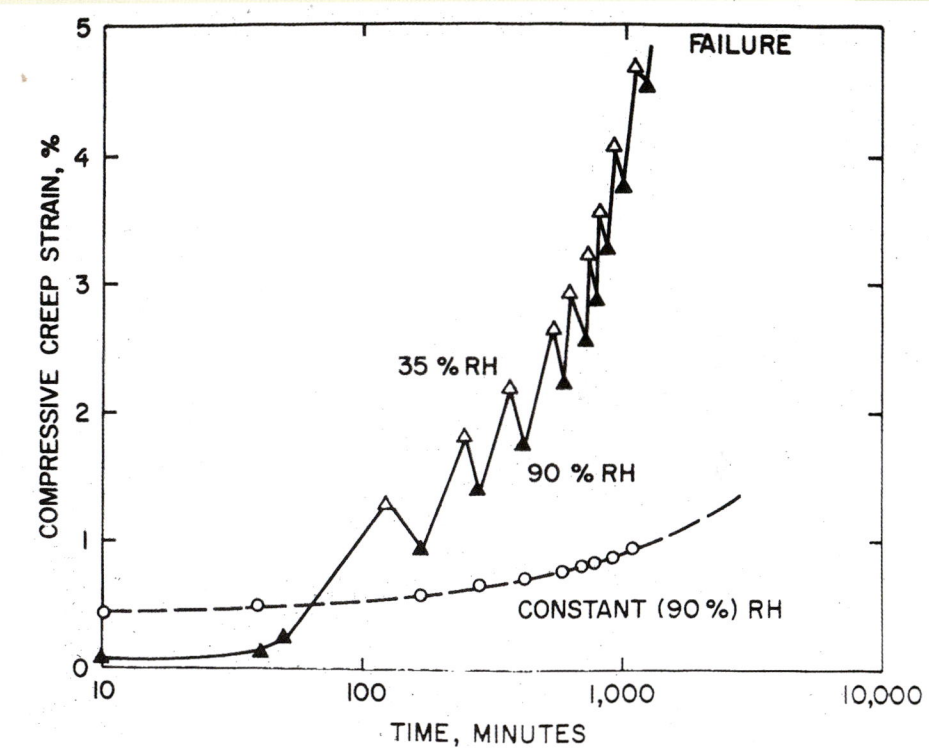

▶ John Considine (left) and David Vahey setting up an experimental test to determine full-field local displacement of two-sided paperboard. (2000s) [90, 95]

Structural Paper Products

◄ Paper products that benefit from treating paper as an engineering material. {17 P}

Honeycomb Core Panels

FPL developed a low-density honeycomb core made of resin-treated paper for use in aircraft sandwich panels. This was followed by production of a lower cost paper core for use in building panels. Paper honeycomb core has been used in partition panels, doors, spandrel-type wall panels, and other construction in houses, shelter buildings, warehouses, farm buildings, and lightweight shipping containers. Untreated paper honeycomb core material has been used as cushioning in packaging. [17, 21, 23, 28]

◄ Paper honeycomb material used as the core material in a sandwich structure. Engineering properties for paper are used to predict performance. (1940s) {36 A}

Press-Dry Papermaking

A method to produce strong paper from 100% hardwood pulp was developed. The process can also be used with softwood pulps to improve paper strength and smoothness. The process simultaneously applies heat and pressure to a web of wood fibers. Data have been collected from high-speed simulation tests for the subsequent design of high-speed commercial press-dry equipment. A commercial prototype was built, but the paper industry needs to develop improved methods to accomplish press drying at modern production speeds. [56, 61, 65]

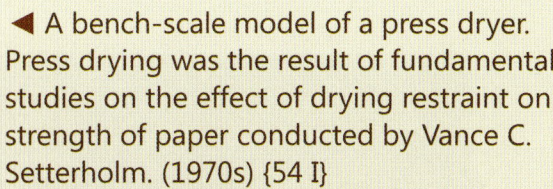

◀ A bench-scale model of a press dryer. Press drying was the result of fundamental studies on the effect of drying restraint on strength of paper conducted by Vance C. Setterholm. (1970s) {54 I}

▶ Scaled-up press dryer that was used in line with the FPL's experimental paper machine. (1970s) {75 B}

◀ Vance Setterholm at the reel end of the paper machine looking at a corrugated container made from press-dried paper. The FPL had the capability of starting with a log and ending up with a finished corrugated container, thus being able to study all the processing effects on the end product. (1980s) {91 L}

Wet Paper Stiffness

One major challenge in the use of paperboard in structural applications is its loss of stiffness when wet. Richard Weatherwax and Dan Caulfield developed the SOFORM (sulfur dioxide and formaldehyde) process to solve this problem.

◀ Chamber used to treat experimental corrugated fiberboard using the SOFORM process to make it less sensitive to moisture and free water so that it can be used in outdoor structures. Unfortunately, as effective as this treatment is, the use of sulfur dioxide and formaldehyde, even at the low levels studied, are of concern. Also, the material would be difficult to recycle. (1970s) {126 C} [18, 19, 32, 55, 59]

◄ Upper photo shows untreated doublewall corrugated fiberboard after soaking for 24 hours in water. The bottom photo shows SOFORM-treated doublewall corrugated fiberboard after soaking for 24 hours in water. (1970s)

Fiber Loading

FPL developed a new technology—fiber loading—for precipitating calcium carbonate filler into the fiber lumens and walls of both recycled and virgin fiber used in papermaking. This patented new technology (1) decreases the amount of sludge arising from recycling filled printing and writing papers by allowing the filler to also be recycled (currently fillers are removed in recycling operations and landfilled) and (2) improves the opacity and light scattering of printing papers, allowing basis weights to be decreased, which produces more paper from less fiber. Fiber loading can also significantly decrease the heat energy required to dry the paper. [72, 76, 79, 84]

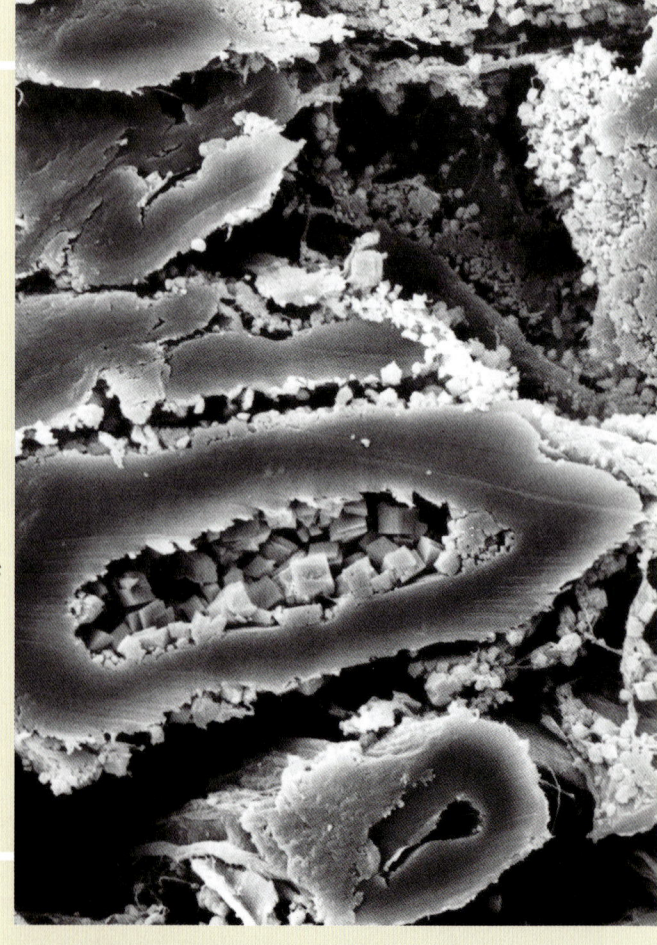

► Cross section of wood fibers that have been filled with calcium carbonate. (1990s)

Paper Permanence

Paper permanence—the stability of paper and maintenance of its initial properties over a lengthy time—can be measured by accelerated aging tests in the laboratory. Such aging can then be correlated with long-term natural aging. Unfortunately, a high-quality test to predict paper stability over time was not available, and ASTM-International (ASTM) was interested in developing a paper aging standard. The FPL and the Finnish Pulp and Paper Research Institute (FPPRI), Espoo, Finland, were chosen to do research on an accelerated light aging test method. Others were chosen for research on pollutant and thermal aging. A test method was developed, accepted by ASTM, and published as ASTM D 6789–02.

◄ Paper permanence study showing aging of papers in daylight, north-facing window. This study was used in the development of the ASTM Paper Aging Standard. The area of paper placement was 7 feet high and 14 feet long. These papers were made from furnishes whose composition varied from pure cotton linters to 100% chemithermomechanical pulp (CTMP). The paper exposure followed the normal day–night cycle. (2000s)

Under this paper permanence study, and for the next 100 years, all 15 papers selected for study will be stored in 10 North American libraries in a variety of climates. The libraries will periodically submit a storage condition report to the U.S. Library of Congress and the National Archives and Records Administration. At 10 intervals during the century-long storage, paper sheets will be removed and tested by FPL and FPPRI for optical and physical properties. Results will be shared worldwide. [74, 78, 81, 83, 86–89]

To assist scientists in the evaluation of new pulps and paper treatments, the FPL has a humidity-controlled test laboratory.

Paper Test Laboratory [7, 8, 82]

▲ Nancy Ross-Sutherland adding a diluted pulp sample to a fiber length analyzer. (2000s)

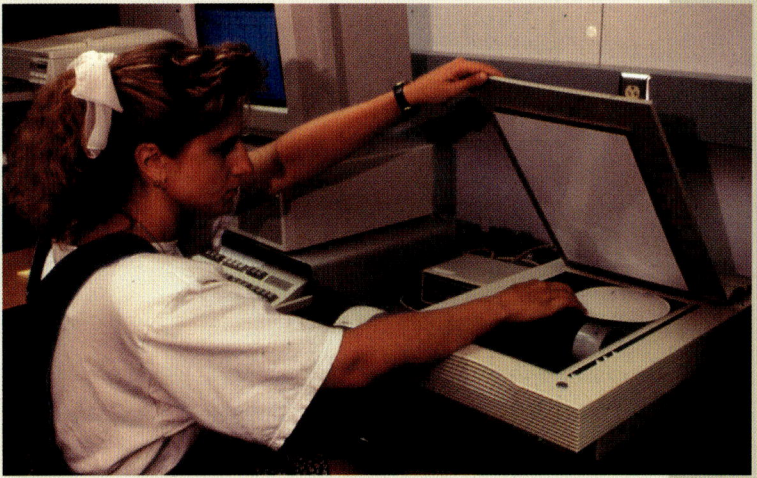

▲ Sara Fishwild scanning a test sample handsheet of paper to measure contaminant levels. (2000s)

◄ Experimental clothing made from wood fiber. (1920s) {M 10388 F} [1]

1910

Production limited to a few wood species.

2010

Most wood species can now be converted to paper. Companies employed 189,003 people and shipped products with a total value of $97,511,459,000 (paper, paperboard, and sanitary paper). (2006 Annual Survey of Manufacturers)

Further Information

[1] Spun paper and some of its uses / R.Thelen / Paper Vol. 18(20) (July 26, 1916)

[2] The use of bark for paper specialties / Otto Kress / Paper trade journal 63(15):36,38 (Oct. 12, 1916)

[3] How paper is affected by humidity / Otto Kress, Philip Silverstein / Paper 19(25):13–17,22, (Feb 28, 1917)

[4] Book paper from southern pines and gums / Sidney D. Wells / Paper trade journal Vol. 71(22) (Nov. 25, 1920)

[5] The relation of sheet properties and fiber properties in paper / R.H. Doughty / Paper trade journal in three parts: I. (July 9, 1931); II. (Oct. 8, 1931); III. (March 3, 1932)

[6] Forest Products Laboratory research on paper machine variables / Warren A. Chilson, P.K. Baird / Paper trade journal Vol. 97(14) (Oct. 5, 1933)

[7] The establishment of paper specifications / C.C. Heritage / Paper trade journal Vol. 97(19) (Nov. 9, 1933)

[8] Application of elementary statistical methods in the testing of pulp and paper / F.A. Simmonds, R.H. Doughty / Paper trade journal Vol. 97(25):32–37 (Dec. 21, 1933)

[9] Twenty-five years of research / C.E. Curran / Paper industry Vol. 17(4) (July 1935)

[10] The morphology of cellulose fibers as related to the manufacture of paper / George J. Ritter / Paper trade journal 101(18):93–100 (October 31, 1935)

[11] The use of bark for paper specialities / Otto Kress / FPL No. 889 (1938)

[12] Methods used at the Forest Products Laboratory for the chemical analysis of pulps and pulpwoods / compiled by Mark W. Bray / Chemical analysis of pulps and pulp woods: methods used at the Forest products Laboratory / Paper trade journal 87(25):59 (Dec. 20, 1928) (Revised ed. 1939)

[13] Cross-sectional dimensions of fibers in relation to paper-making properties of loblolly pine / J.C. Pew, R.G. Knechtges / Paper trade journal Vol. 109(15) (Oct. 12, 1939)

[14] Holocellulose research in cooperation with the Technical Association of the Pulp and Paper Industry / George J. Hajny, George J. Ritter / FPL No. 1497 (1941)

[15] The re-use of waste paper / FPL Technical note No. 179 (1943)

[16] The effect of temperature on certain mechanical properties of Forest Products Laboratory, high-strength, cross-laminated paper plastic (papreg) / FPL No. 1321 (1943)

[17] An investigation of mechanical properties of honeycomb structures made of resin-impregnated paper / C.B. Norris, G.E. Mackin / National Advisory Committee for Aeronautics Technical note No. 1529 (1948)

[18] The swelling and shrinking of wood, paper, and cotton textiles and their control / H. Tarkow / Tappi Vol. 32(5) (May 1949)

[19] Bound water and hydration / Alfred J. Stamm / Pulp and paper magazine of Canada Vol. 51(10) (Sept. 1950)

[20] Size-press treatment of paper with protein or latex solutions for nonporous package wrappers / D.J. Fahey, R.D. Hilton / Forest Products Laboratory (1952)

[21] Paper honeycomb as a core for structural sandwich construction / Edward W. Kuenzi / American Society for Testing Materials, Special technical publication No. 118 (1952)

[22] Potentialities of paper-base laminates as compared with other laminates / Alfred J. Stamm / FPL No. 1452 (1952)

[23] Paper-honeycomb cores for structural sandwich panels / Robert J. Seidl / FPL No. 1918 (1952)

[24] Experiments on use of a mixed southern gum cold soda pulp and starch coating for stiffening southern pine newsprint paper / D.J. Fahey, K.J. Brown, R.J. Seidl. Forest Products Laboratory (1953)

[25] American woods for papermaking / FPL Technical note No. 212 (1953)

[26] Paper birch utilization in the Lake States / E.M. Davis / FPL No. 1953 (1953)

[27] The U.S. Forest products Laboratory and its pulp and paper division / F.J. Champion / Papermaker Vol. 22(1) (Feb 1953)

[28] Thermal conductivity of paper honeycomb cores and sound absorption of sandwich panels / D.J. Fahey, M.E. Dunlap, R.J. Seidl / Southern pulp & paper manufacturing Vol. 16(9) (Sep 10, 1953)

[29] Pulp and paper: tools for hardwood utilization / J.N. McGovern / Forest Products Research Society Journal Vol. 8(5) (Dec. 1953)

[30] Some potentialities of overlaid lumber / B.G. Heebink / Forest products journal Vol. 5(4) (Apr. 1955)

[31] Fiber products as a factor in utilization / Edward G. Locke, Kenneth G. Johnson / Forest products journal Vol. 5(6) (June 1955)

[32] Dimensional stabilization of paper by crosslinking with formaldehyde / W.E. Cohen, A.J. Stamm, D.J. Fahey / TAPPI Vol. 42(12) (Dec. 1959)

[33] A partial list of articles relating to paper stiffness, 1925–1962 / Vance Setterholm, Roland Gertjejansen, James Swartout / Forest Products Laboratory WFPR–124 (1962)

[34] Effect of temperature and restraint during drying on the tensile properties of handsheets / Vance C. Setterholm, Warren A. Chilson, A.T. Luey / FPL No. 2265 (1962)

[35] Use of chemical compounds to improve the stiffness of container board at high moisture conditions / D.J. Fahey / Tappi 45(9):192A–(202A) (Sept. 1962)

[36] Electrical strain gage for the tensile testing of paper / Douglas M. Jewett / Research note FPL 03 (1963)

[37] Elements of wood fiber structure and fiber bonding / F.A. Simmonds, G.H. Chidester / Research paper FPL 5 (1963)

[38] Method for measuring the edgewise compressive properties of paper / V.C. Setterholm, R.O. Gertjejansen / TAPPI Vol. 48(5) (May 1965)

[39] How surface applications of starch affect hardwood–softwood papers / Warren A. Chilson, Donald J. Fahey / American paper industry Vol. 48(3) (Mar. 1966)

[40] How expander rolls widen and improve sheet properties / Warren A. Chilson / American paper industry Vol. 48(3) (Mar. 1966)

[41] Improvements in cylinder board from inter-ply chemical additions / D.J. Fahey / Paper trade journal Vol. 151(3) (Aug. 21, 1967)

[42] Chemical treatments for improving compressive strength of linerboard at high moisture conditions / D.J. Fahey / Research note FPL–084 (Dec. 1967)

[43] Application of chemicals to wet webs of paper and linerboard using the smoothing press / D.J. Fahey / Forest Products Laboratory (1968)

[44] The story of southern pine newsprint: Part II / Cranston Williams, New York: American Newspaper Publishers Association (1968)

[45] Method for measuring edgewise shear properties of paper / Vance C. Setterholm, Roy Benson, Edward W. Kuenzi / TAPPI Vol. 51(5) (May 1968)

[46] How to reduce vessel element picking in printing papers containing oak / Von L. Byrd, D.J. Fahey / Paper trade journal (Nov. 24, 1969)

[47] Effects of relative humidity and temperature on tensile stress–strain properties of kraft linerboard / Roy E. Benson / Tappi Vol. 54(5) (May 1971)

[48] Resistance of resin-impregnated paper overlays to accelerated weathering / D.J. Fahey, D.S. Pierce / Forest products journal Vol. 21(11) (Nov. 1971)

[49] Fiber bonding and tensile stress–strain properties of earlywood and latewood handsheets / V.L. Byrd / Research paper FPL 193 (1972)

[50] Effect of relative humidity changes during creep on handsheet paper properties / Von L. Byrd / Tappi Vol. 55(2) (Feb. 1972)

[51] How fiber morphology affects pulp characteristics and properties of paper / R.A. Horn / Chem 26 paper processing Vol. 8(5) (May 1972)

[52] Development of basic information for the design of paper shipping sacks / J. Chern, E.W. Kuenzi / Tappi Vol. 55(10) (Oct. 1972)

[53 Small-angle X-ray scattering by paper: a new method for investigating inter-fiber bonding / D.F. Caulfield / Tappi Vol. 56(3) (Mar. 1973)

[54] New concept in paper thickness measurement / V.C. Setterholm / Tappi Vol. 57(3) (Mar 1974)

[55] Soform paper: vapor phase cross-linked paper for uses requiring high wet stiffness / D.F. Caulfield / Proceedings of the International Paper Physics Conference, Atlanta, GA: Tappi Press (1975)

[56] Z-direction restraint, a new approach to papermaking / Vance Setterholm, R.E. Benson, J.F. Wichmann, R.J. Auchter / Research paper FPL 256 (1975)

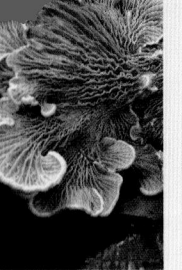

[57] Method for measuring the interlaminar shear properties of paper / Von L. Byrd, Vance C. Setterholm, John F. Wichmann / Tappi Vol. 58(10) (Oct. 1975)

[58] Corrugated fiberboard containers from high-yield roughwood kraft linerboard pulp / R.A. Horn / Tappi Vol. 59(2) (Feb. 1976)

[59] Cross-link wet-stiffening of paper: the mechanism / D.F. Caulfield, R.C. Weatherwax / Tappi Vol. 59(7) (July 1976)

[60] Papermaking factors that influence the strength of linerboard weight handsheets / John W. Koning, Jr., James H. Haskell / Research paper FPL 323 (1979)

[61] FPL press drying process: wood savings in linerboard manufacture / Tappi Vol. 64(4) (Apr. 1981)

[62] Computer model for economic study of unbleached kraft paperboard production / Peter J. Ince / General technical report FPL 42 (1984)

[63] Measuring the mechanical behavior of paperboard in a changing humidity environment / D.E. Gunderson / Proceedings of the 1986 International Process and Materials Quality Evaluation Conference. Atlanta: Tappi Press (1986)

[64] The compressive load–strain curve of paperboard: rate of load and humidity effects / D.E. Gunderson, J.M. Considine, C.T. Scott / Journal of pulp and paper science. Vol. 14(2) (Mar. 1988)

[65] Press drying: a way to use hardwood CTMP for high-strength paperboard / Richard A. Horn, David W. Bormett, Vance C. Setterholm / Tappi journal Vol. 71(3) (Mar. 1988)

[66] Compressive creep behavior of paperboard in a cyclic humidity environment: exploratory experiment / J.M. Considine, D.E. Gunderson, P. Thelin, C.V. Fellers / Tappi journal Vol. 72(11) (Nov. 1989)

[67] Method for measuring mechanosorptive properties / D.E. Gunderson / Journal of pulp and paper science Vol. 17(2) (Mar. 1991)

[68] Literature review of cyclic humidity effects on paperboard packaging / John Considine, Theodore L. Laufenberg / Cyclic humidity effects on paperboard packaging proceedings, September 14–15, 1992, Forest Products Laboratory, Madison, Wis. (1992)

[69] Cyclic humidity effects on paperboard packaging proceedings / editors; Theodore L. Laufenberg, Craig H. Leake / FPL Symposium (September 14–15, 1992)

[70] Raman spectroscopy / R.H. Atalla, U.P. Agarwal, J.S. Bond / Methods in lignin chemistry.New York: Springer–Verlag (1992)

[71] Method for forming structural components from dry wood fiber furnish / Dennis E. Gunderson, Roland L. Gleisner / US Patent No. 5,198,236 (March 30, 1993)

[72] Method for fiber loading a chemical compound / John H. Klungness, Daniel F. Caulfield, Irving B. Sachs, Marguerite S. Sykes, Feya Tan, Richard W. Shilts / US Patent No. 5,223,090 (June 29, 1993): US Patent No. Re. 35,460 (Feb. 25, 1997)

[73] Compressive creep behavior of corrugating components affected by humid environment / John M. Considine, D.L. Stoker, T.L. Laufenberg, J.W. Evans / Tappi journal Vol. 77(1) (Jan. 1994)

[74] Raman spectral features associated with chromophores in high-yield pulps / Umesh P. Agarwal, Rajai H. Atalla / Journal of wood chemistry and technology Vol. 14(2) (1994)

[75] Paper and composites from agro-based resources / R.M. Rowell, R.A. Young, J.K. Rowell / CRC Lewis Publishers, Boca Raton, FL (1996)

[76] Effect of fiber loading on paper properties / John H. Klungness, M.S. Sykes, F. Tan, S.M. Abubakr, J.D. Eisenwasser / Tappi journal Vol. 79(3) (Mar. 1996).

[77] Heterogeneity of lignin concentration in cell corner middle lamellae appeared in Wood science and technology Vol. 30 (1996) 99–104: response to the "Letter to the Editor" by D.A.I. Goring appeared in Wood science and technology Vol. 30 (1996) 282 / V.C. Tirumalai, U.P. Agarwal, J.R. Obst / Wood science and technology Vol. 30(6) (Nov. 1996)

[78] Photo yellowing of thermomechanical pulps: Looking beyond-carbonly and ethylenic groups as the initiating structures / Umesh P Agarwal, James D. McSweeny / Journal of wood chemistry and technology Vol. 17(1&2):1–26 (1997)

[79] Fiber loading: theory and applications / John H. Klungness, Marguerite S. Sykes, Freya Tan, Said Abubakr / In: Proceedings of the 4th international refining conference; 1997 March 18–20; Fiuggi, Italy. United Kingdom, England: Pira International: 1–10. (1997)

[80] FT–Raman spectroscopy of wood: identifying contributions of lignin and carbohydrate polymers in the spectrum of black spruce (*Picea mariana*) / Umesh P. Agarwal, Sally A. Ralph / Applied spectroscopy Vol. 51(11) (1997)

[81] Assignment of the photo yellowing-related 1675 cm–1 Raman/IR band to p-Quinones and its implications to the mechanism of color reversion in mechanical pulps / Umesh P. Agarwal / Journal of wood chemistry and technology Vol. 18(4): 381–402 (1998)

[82] Paper friction—Influence of measurement conditions / Anna Johansson, Christer Fellers, Dennis Gunderson, Urban Haugen / Tappi journal Vol. 81(5): 175–183. (1998)

[83] The aging of lignin rich papers upon exposure to light: Its quantification and prediction / James S. Bond, Rajai H. Atalla, Umesh P. Agarwal, Chris G. Hunt / 10th International Symposium on wood and pulping chemistry, Main Symposium: Jun. 7–10, 1999, Yokohama, Japan. Atlanta, GA: TAPPI Press, 1999: Vol. III, p. 500–504. (1999)

[84] Lightweight, high-opacity paper by fiber loading: filler comparison / John H. Klungness, Aziz Ahmed, Nancy Ross-Sutherland, Said AbuBakr / Nordic pulp & paper research journal Vol. 15:5: 345–350 (2000)

[85] United States paper, paperboard, and market pulp capacity trends by process and location, 1970–2000 / Peter J. Ince, Xiaolei Li, Mo Zhou, Joseph Buongiorno, Mary Reuter / FPL–RP–602 (2001)

[86] The aging of printing and writing papers upon exposure to light: part 1—optical and chemical changes due to long-term light exposure / James S. Bond, Xiaochun Yu, Umesh P. Agarwal, Rajai H. Atalla, Chris G. Hunt / Proceedings of the 11th ISWPC International Symposium on Wood and Pulping Chemistry: June 11–14, 2001, Nice, France. Saint Martin d'Heres, France: Ecole Francaise de Papeterle et des Industries Graphiques, 2001: Vol. II, p. 209–213 (2001)

[87] Aging of printing and writing paper upon exposure to light. Part 2—mechanical and chemical properties / Chris Hunt, Xiaochun Yu, James S. Bond, Umesh P. Agarwal, Rajai H. Atalla, / 12th ISWPC International Symposium on Wood and Pulping Chemistry : Madison, Wisconsin, USA, June 9–12, 2003: proceedings, volume II (2003)

[88] Raman spectra of lignin model compounds / Umesh P. Agarwal, Richard S. Reiner, Sally A. Ralph, Colby C. Hirth, Rajai H. Atalla / Proceedings of the 59th APPITA Annual Conference and Exhibition incorporating the 13th ISWEPC (International symposium on wood, fibre, and pulping chemistry), held in Auckland, New Zealand (May 16–19, 2005). (Carlton, Victoria, Australia) APPITA, (2005)

[89] Predicting photoyellowing behaviour of mechanical pulp containing papers / Umesh P. Agarwal / Proceedings of the 59th APPITA Annual Conference and Exhibition incorporating the 13th ISWEPC (International symposium on wood, fibre, and pulping chemistry), held in Auckland, New Zealand (May 16–19, 2005) (Carlton, Victoria, Australia) APPITA, (2005)

[90] Use of digital image correlation to study the local deformation field of paper and paperboard / J.M. Considine, Tim C. Scott, Roland Gleisner, Junyong Zhu / Proceedings of the advances in paper science and technology, 13th Fundamental research symposium, 2005 September 11–16, Cambridge, UK. The Pulp and Paper Fundamental Research Society (2005)

[91] Evaluation of forest thinning materials for TMP production / John H.Klungness, Roland Gleisner, Doreen Mann, Karen L. Scallon, Junyong Zhu / TAPPI journal Vol. 5(4) (Apr. 2006)

[92] Effect of plantation density on kraft pulp production from red pine (Pinus resinosa Ait.) / J.Y. Zhu, G.C. Myers / Journal of pulp and paper science Vol. 32(3) (July/Aug. 2006)

[93] Wood density and anatomical properties in suppressed-growth trees: comparison of two methods / David W. Vahey, J.Y. Zhu, C. Tim Scott / Wood and fiber science Vol. 39(3) (2007)

[94] Revealing organization of cellulose in wood cell walls by Raman imaging / Umesh P. Agarwal, Sally A. Ralph / Proceedings of the 14th International Symposium on Wood Fibre and Pulping Chemistry, held in Durban, South Africa, Technical Association of the Pulp and Paper Industry, TAPPSA, 6/25–6/28 (2007)

[95] Full-field local displacement analysis of two-sided paperboard / J.M. Considine, D.W. Vahey / 61st Appita annual conference and exhibition: Gold Coast, Australia, 6–9 May 2007: proceedings. Carlton, VIC: Appita, ISBN 9780975746928: p. 223–227 (2007)

Packaging

▲ Display of previous areas of packaging research at FPL, including wood boxes, crates, pallets, cushioning, fasteners, and corrugated fiberboard. (1980s) [40]

Early in its history, the FPL was engaged in packaging research. Huge quantities of wood were used to ship the products of farming and manufacturing as the country grew. Packaging research expanded in WWI and greatly expanded in WWII. (See the National Defense section.)

Challenge: One major challenge was how to package food and other goods to be shipped to people located in all parts of the world.

FPL's Contributions: FPL scientists analyzed the shipping environment, developed meaningful tests to simulate the environment, and developed new materials and new container designs. For heavy items, wood boxes and crates are still used, but based on FPL research, boxes and crates are now designed and made with less wood, and many more species of wood can be used. FPL research on solid and corrugated fiberboard has contributed to the fact that today over 90% of all items are shipped in corrugated fiberboard containers, resulting in significantly more efficient use of wood, reduced product damage, reduced cost, and containers that can be reused and recycled more than once. Based on years of research, a new standard for evaluating containers (ASTM D 4169) is now available. All this has resulted in savings to consumers through lower shipping costs and less damage to products they buy.

Results: When the Laboratory was established, most product shipments were in wood boxes, crates, and barrels, with few corrugated fiberboard containers used. The use of corrugated fiberboard in world trade is increasing and helping to reduce the cost of food and other products.

Container Evaluation

▲ FPL packaging test laboratory, located in the FPL experimental laminated arch building. (1940s) (See Structures section.) Sample wood crates are being made for testing. Prior to moving into the packaging laboratory, packaging research in the 1920s and 1930s was conducted in the FPL laboratory located on the University of Wisconsin campus. [5, 7, 8, 23]

▲ Kenneth Skidmore observing the reaction of a container subjected to the drum test that was developed in the 1910s by the FPL to simulate rough handling in shipping. (1940s) {M 46104 F} [1, 4]

▶ Raymond Seborg conducting a pressure test on a keg. (1930s)

◀ Evaluation of wirebound box designs was done as early as 1920 to assist in the preparation of standards [2, 6, 12]

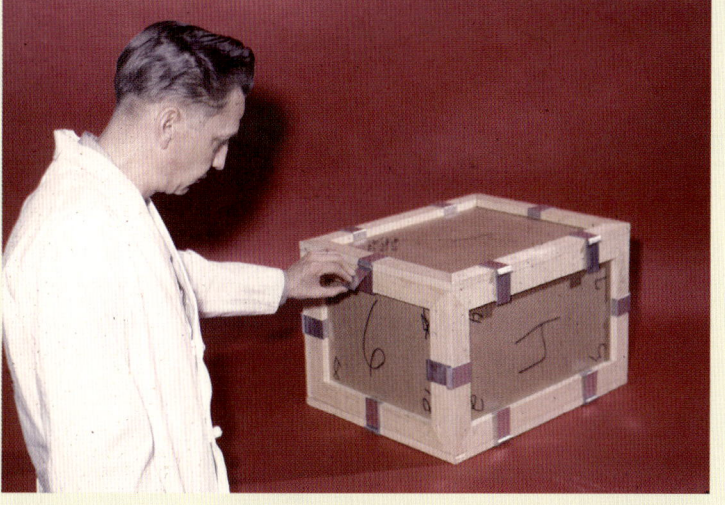

▲ FPL evaluated various bushel baskets to improve design and possible utilization of other species of wood. (1960s) [3, 27]

▲ Milo Schimming checking a clip designed by Robert Kurtenacker, Charles Roe, and Harry Wetzel to hold together a cleated panel box. The clips allow for quick assembly and disassembly to enhance reusability. (1960s) [26]

Corrugated Fiberboard

Corrugated fiberboard containers were developed in the late 1890s and experienced rapid growth after approval for use in shipping products by rail in the early 1900s. This expansion in use of corrugated fiberboard led to a boom in papermill construction and an outlet for poor quality trees. Through the development of high-strength paperboards combined with fluted designed cores, corrugated fiberboard (erroneously, but commonly called cardboard), a lightweight, high-stiffness, low-cost, sandwich material has replaced most solid wood used in containers. This has resulted in greatly extending the timber supply.

▼ Forrest Simmonds, Richard Auchter, and Conrad Marking investigating the runnability of various kinds of corrugating medium furnished by manufacturers in the United States, Canada, and Europe on the newer FPL experimental corrugator. (1970s) {M 136 064-4} [29]

▲ Early FPL corrugator used to flute corrugating medium and bond paperboard facings to the fluted medium to form corrugated fiberboard. (1930s) {M 26068 F}

FIBER AND CORRUGATED BOX STUDY
INFLUENCE OF MACHINE VARIABLES

VARIATION IN GLUING PRESSURE

BOXES A AND B IDENTICAL IN DESIGN, MADE FROM SAME ROLLS OF PAPER WITHOUT STOPPING MACHINE BUT WITH DIFFERENT PRESSURES FOR GLUING FIRST SHEET TO CORRUGATIONS.

LOW PRESSURE BOARD

HIGH PRESSURE BOARD

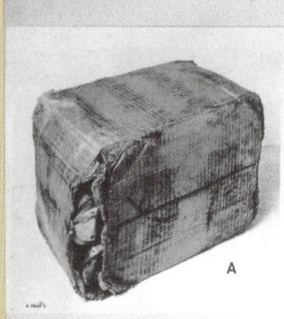

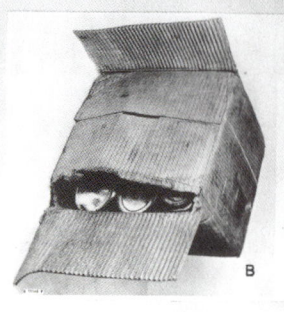

◄ Results of an early study of corrugator machine variables. Boxes A and B were identical in design and made from the same rolls of paper. Without stopping the machine, a different pressure for gluing the first sheet to the corrugations was applied. After subjecting the boxes to the revolving drum test, which simulates rough handling, it was found that box A made with low pressure withstood twice as many drops as box B, which was made with high pressure. This showed the importance of proper manufacturing methods to improve the performance of a corrugated container. (1930s) {M 29361}

Evaluation of Corrugated Fiberboard and Containers

In 1919, research on paperboard containers was conducted by Andrew MacKenzie. Research expanded to include corrugated fiberboard containers, with Thorwald (Tac) Carlson applying engineering principles to help understand how a corrugated structure performs under load and thus design better containers. Keith Kellicutt and Eugene Landt continued this line of research and in 1953 developed a box strength calculator that was successfully used for estimating the performance of a corrugated container from the properties of the paperboard components. FPL staff continued the research by improving the method for determining the edgewise compressive strength of the combined board, so having the paperboard components was unnecessary to use the calculator. Finally, Thomas Urbanik, working with Millard Johnson (University of Wisconsin), developed fundamental relationships between the paperboard components and the combined corrugated structure so that the maximum compressive strength of a container can be achieved utilizing the least amount of fiber. This research resulted in formulas for predicting container compressive strength, and the approach was commercialized. [1, 11, 15, 19, 24, 33, 36, 42, 44, 48, 49, 52]

► Tac Carlson showing how a corrugated container reacts in multiple panel waves to compression loading. (1930s) {M 36757} [11]

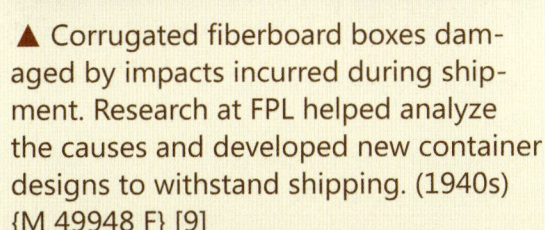

▲ Corrugated fiberboard boxes damaged by impacts incurred during shipment. Research at FPL helped analyze the causes and developed new container designs to withstand shipping. (1940s) {M 49948 F} [9]

◄ Short-column test of corrugated fiberboard. This test was developed to determine the edgewise compressive strength of corrugated fiberboard. It is now one of the methods used to determine if corrugated fiberboard meets requirements for use in shipping a product by common carrier. (1960s) [24]

▲ Four-point bending test used to evaluate stiffness of corrugated fiberboard. Results from the short-column test and bending stiffness test are used to estimate the compressive strength of corrugated containers. (1960s) [48]

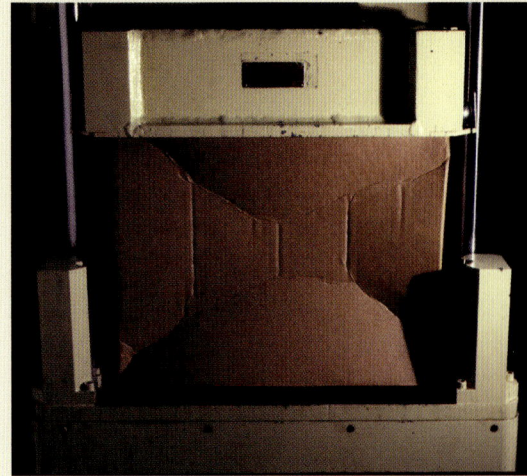

▲ Container failing a compression test. (1960s) [19, 20]

▲ Stacks of corrugated containers being exposed to a continuous dead load that simulates the stacking of containers in a warehouse. FPL developed formulas for estimating the length of time a container can be expected to support a load in the warehouse. (1950s) {M 147 609-5} [14, 31, 32, 45, 47, 51]

◀ (Left to right) Milo Schimming, Ed Kinney, Dave Schroeder, and Dunc Godshall observing the compression testing of a pallet load of containers. (1970s)

Fiber Distribution in Corrugated Structures

Mathematical analysis of the complex interaction of paperboard properties in a sandwich structure resulted in design curves for optimum distribution of fiber in a corrugated structure. FPL developed a test to measure the interlaminar shear stress–strain properties to determine and understand the causes of edgewise compression failure. [33, 42, 44]

Fundamental research by Thomas Urbanik on the engineering mechanics of corrugated fiberboard helped explain how it fails and how to make paperboard and corrugated fiberboard to maximize the strength of a box. FPL research also identified and developed methods to accurately evaluate the performance of paperboard components, corrugated fiberboard, and corrugated containers under cyclic humidity conditions that are experienced in the shipping and warehousing environment. This block of pioneering research now allows corrugated performance to be based upon end-use performance requirements. Another result of this research was assistance in the development of a new railroad standard based on a compression test of the corrugated fiberboard. This has enabled manufacturers to make more efficient use of packaging materials with a 10% to 20% savings in package weight and costs. [36, 51]

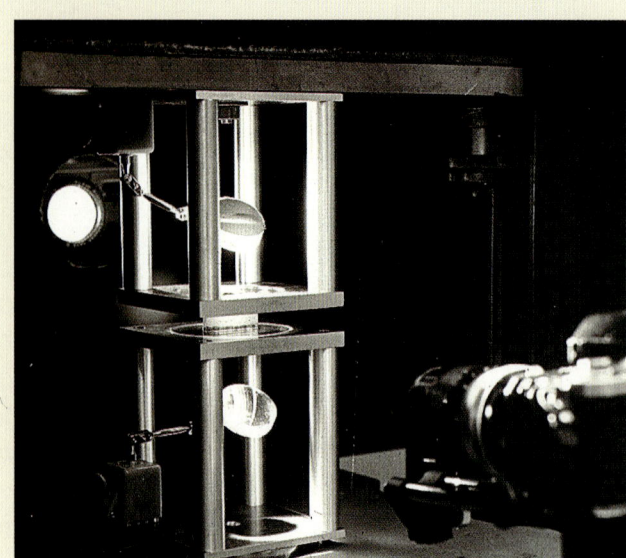

▲ Camera setup used to study the buckling of the paperboard components of corrugated fiberboard when loaded in compression. (2000s)

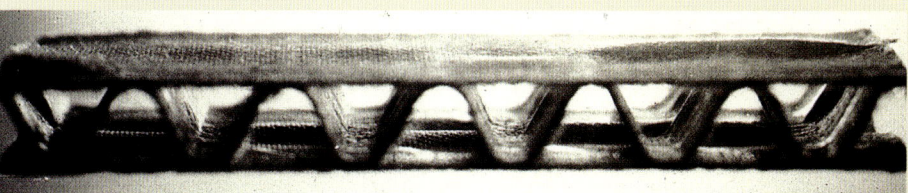

◄ Buckling of the paperboard facings. (2000s)

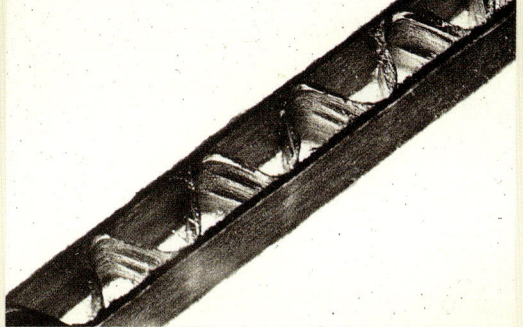

► Buckling of the fluted corrugating medium. Knowledge of the failure mechanism in the facings and corrugating medium of singlewall aids in the proper design of multiwall corrugated fiberboard, such as doublewall and triplewall. (2000s) [48, 49]

Ventilation Holes

◄ Method of shipping produce. Keith Kellicutt studied the locations of vent holes and their relationship to container strength. Results indicated that vent holes should be located away from the box corners. (1990s) [18]

Reuse of Packaging

◄ A common practice in the 1970s was for stores to save used corrugated containers for reuse by customers or to discard them as waste. Today, those not reused are crushed, baled, and returned to the pulp mill for recycling. Based on research by FPL and industry, over 70% of corrugated containers are recycled.

Shipping Environment

▲ Modes of transporting goods. (Left to right and top to bottom) Truck (being loaded with fresh produce), rail (truck trailer being loaded on railroad flat car), ship (corrugated fiberboard containers being loaded), and airplane.

FPL, working with industry, evaluated the effects of each of these methods of transportation on shipping containers. From this basic knowledge it was possible to develop new test methods that simulated the forces encountered by containers in shipment. This resulted in improved container design and materials to minimize weight and volume, thus significantly reducing shipping costs and conserving raw materials. (1970s–1980s) [10, 13, 21, 34, 35]

▲ Types of damage that can occur during shipping, such as crushing of the corners of the containers from netting, slings, and mishandling.

▲ Damage due to excessive strapping pressure.

Impact Testing

▶ Instrumented incline-impact tester used to simulate the effect of lateral shock and compression on containers, the type of damage that can occur by cargo shifting in a truck or railcar when the vehicle stops too suddenly. Research of this nature led to improved methods of blocking and bracing cargo. (1970s)

Vibration Testing

▶ Standing on the platform of the FPL-designed transportation simulator are Dunc Godshall (left) and Thomas Urbanik demonstrating the evaluation of corrugated fiberboard containers subjected to vibrations that are experienced in transportation. This research explained a significant amount of the damage that occurs in shipping, resulting in large savings. (1970s) [28, 30, 43]

Cushioning Testing

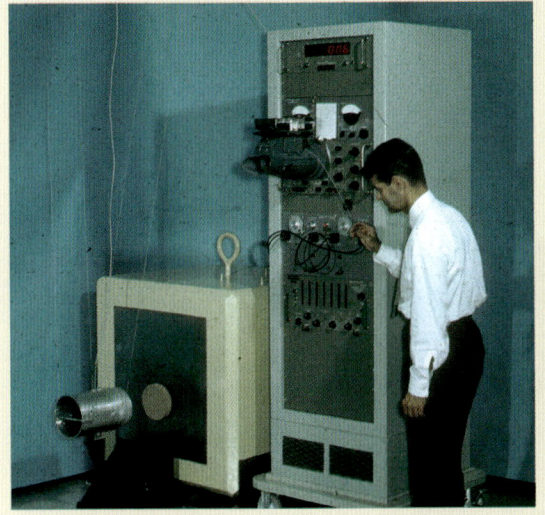

▲ Dunc Godshall conducting a dynamic cushioning test on corrugated fiberboard using the FPL-designed pendulum impact tester. Results of tests were used by Robert Stern in the development of the "Package Cushioning Design Manual." (1960s) [22]

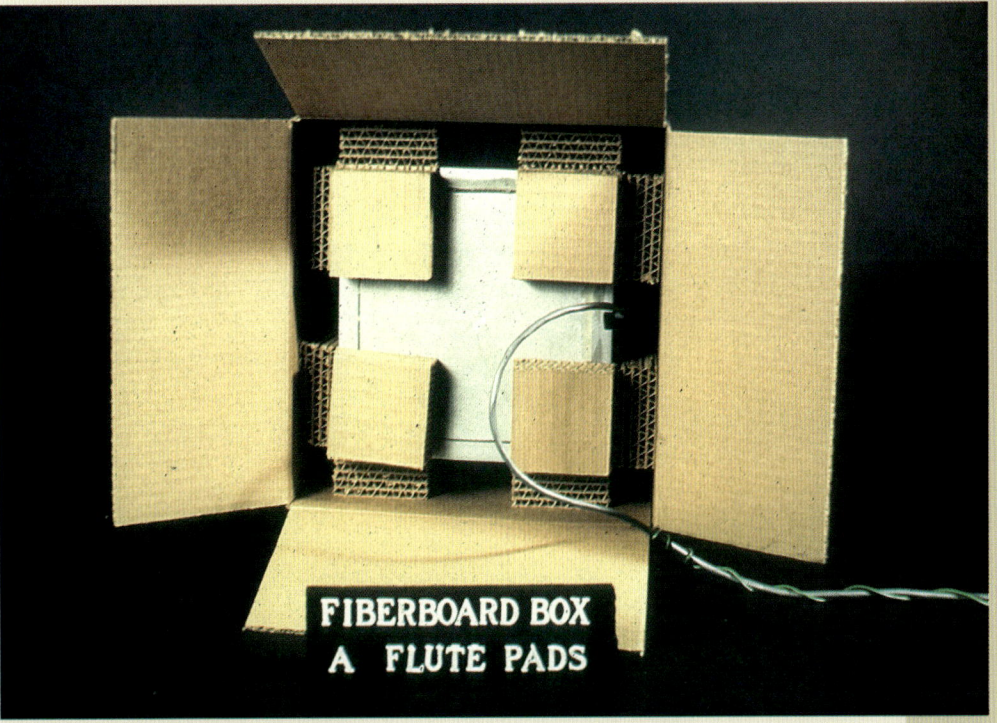

FIBERBOARD BOX
A FLUTE PADS

▲ Instrumented corrugated fiberboard container showing interior corrugated fiberboard cushioning pads. (1980s) [41]

Reduced Shipping Costs and Damage through Performance Evaluation (ASTM D 4169)

Utilizing information from the evaluation of shipping environments, development of meaningful methods for evaluating containers, and improved design principles, the FPL, working with industry and ASTM committee D–10, developed test standard ASTM D 4169. This performance standard takes into consideration new modes of transportation, new materials, and new materials-handling equipment. Tests include simulation of hazards such as shock, vibration, and compression that may occur in actual handling and shipment. A major aspect of D 4169 is that the tests are run in a sequence of expected exposures of the packaged item to the shipping environment as the product moves from manufacturer to consumer. Utilizing this standard allows manufacturers to minimize the packaging material necessary to provide cost-effective packaging and reduce excessive packaging and product damage. (1980s) [38, 39]

Thermal Properties of Corrugated Fiberboard Containers

► Test setup to study thermal properties of corrugated fiberboard containers. (1980s) [37]

Triplewall Corrugated Fiberboard

FPL developed design criteria for corrugated containers using a triple wall of paperboard in flat and corrugated form, which produces a strong container. Such containers can be used for heavy loads and extends the use of corrugated fiberboard. (1955) [16]

▲ Triplewall corrugated fiberboard. (1960s)

◄ Keith Kellicutt showing the strength of a triplewall corrugated fiberboard container. The holes in the container were so observers could assure themselves that there were no internal supports in the container. (1960s)

TRIPLE WALL CORRUGATED BOX
• HAS A COMPRESSIVE STRENGTH VALUE OF ABOUT 9,000 POUNDS
• MADE FROM SEVEN SHEETS OF PAPERBOARD

Fiberneer

◀ Use of a car to show the strength of a FPL product—called fiberneer—developed by Robert Kurtenacker. It resembles single-wall corrugated fiberboard but consists of paper-overlaid veneer facings on both sides of the corrugated (fluted) sheet. (1960s) [25]

Paper Honeycomb Core Material

◀ Kenneth Kreuger explaining honeycomb core material to visitors. Plywood facings were glued to the honeycomb core, making a lightweight sandwich material that is very stiff and strong. The sandwich material was used to make the large container shown in the picture. (1950s) {M 100 819} [17]

Controlling Spread of Invasive Species from Packaging

Researchers at the FPL have been working to prevent the spread of invasive species by focusing on the wood-based packaging materials used for international shipping. Their research aided in the development of the first international commodity quarantine standard for invasive species.

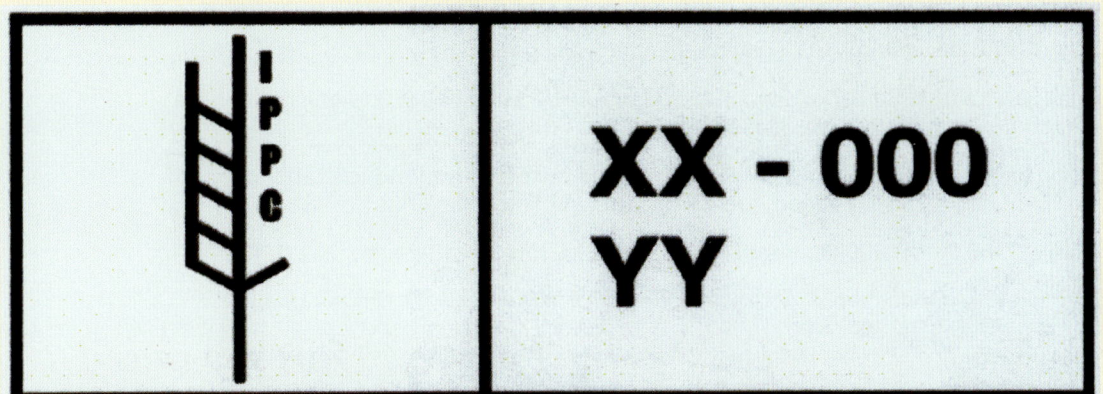

◀ Official stamp of the International Standard for Phytosanitary Measures No. 15 (ISPM15) to be applied to treated packaging materials used in international trade.

Barbara Illman and William Simpson developed heat treatment protocols to kill pests in packaging materials before they are shipped around the world. These protocols became the scientific basis to support new international quarantine measures for wood packaging materials. (2000s) [50]

1910

Wood boxes and crates were primarily used.

2010

Corrugated fiberboard containers are primarily used. Companies employed 218,310 people and shipped products with a total value of $68,591,660,000. (2006 Annual Survey of Manufacturers)

Further Information

[1] Tests on solid fibre board shipping containers for the improvement of designs and specifications / A.H. MacKenzie / FPL No. 309 (1919)

[2] Comparative strength of various thicknesses and species of sheet material for wirebound boxes / FPL No. 642 (1920)

[3] Drum tests of bushel baskets / J.G. Woodburn / Forest Products Laboratory (1921)

[4] A new box-testing machine. Forest products research in pictures No. 2 (1922)

[5] Effect of moisture content of lumber on strength of boxes / T.A. Carlson. / Timberman Vol. 27(2) (Dec. 1925)

[6] Tests on wirebound boxes: first progress report / I.B. Lanphier / Barrel and Box, Vol. 31(11) (Jan. 1927)

[7] Suitability of little-used species for containers / T.A. Carlson / American lumberman (2810):38–40, (Mar. 23, 1929)

[8] Principles of box and crate construction / C.A. Plaskett / USDA Technical bulletin 171 (1930)

[9] Determining the cause of failures in shipping containers / C.A. Plaskett / Fibre containers Vol. 15(4) (Apr. 1930)

[10] Packing for export shipment / C.A. Plaskett / Traffic world Vol. 48(17) (Oct. 24, 1931)

[11] A study of corrugated fibreboard and its component parts as engineering materials / T.A. Carlson / Fibre containers Vol. 24(7) (July 1939)

[12] Cleated plywood and wirebound boxes / H.B. McKean / American Management Association. Packaging series No. 6 (1943)

[13] Solid and corrugated fiberboard boxes for overseas shipment / T.A. Carlson / American Management Association. Packaging series No. 6 (1943)

[14] Safe stacking life of corrugated boxes / K.Q. Kellicutt, E.F. Landt / Fibre Containers (Sept. 1951)

[15] Basic design data for the use of fiberboard in shipping containers: Box strength calculator / K.Q. Kellicutt, E.F. Landt, V.C. Setterholm / Fibre Containers 38(2) (Feb. 1953)

[16] Evaluation of triple-wall corrugated fiberboard shipping containers / K.Q. Kellicutt / Technical note WADC 55–346 (1955)

[17] Paper-honeycomb cores for structural sandwich panels / Robert J.Seidl / FPL No. 1918 (1956)

[18] Effect of ventilating and handholes on compressive strength of fiberboard boxes / C.C. Peters, K.Q. Kellicutt / FPL No. 2152 (1959)

[19] Structural design notes for corrugated containers Note No. 12, Compressive strength of boxes—Part II / K.Q. Kellicutt / Paperboard packaging Vol. 45(3) (Mar.1960)

[20] Effect of age on paperboard and corrugated / K.E. Skidmore / Paperboard packaging Vol. 47(9) (Sept. 1962)

[21] Wanted—better simulated service tests for packaging / C.A. Jordan / ASTM Symposium on simulated service testing of packaging; Special technical publication No. 324 (1963)

[22] Package cushioning design handbook / R.K. Stern, C.W. Roe / Prepared by the Forest Products Laboratory in cooperation with the U.S. Air Force Packaging Evaluation Agency, Brookley Air Force Base, Ala. (1964)

[23] Wood crates: design manual / L.O. Anderson, T.B. Heebink / USDA Agriculture handbook No. 252 (1964)

[24] A short column crush test of corrugated fiberboard / John W. Koning, Jr. / TAPPI Vol. 47(3) (Mar. 1964)

[25] Fiberneer: development, production, and evaluation / R.S. Kurtenacker / Research paper FPL 52 (1966)

[26] Container Fastener / Robert S. Kurtenacker, Charles W. Roe, Harry J. Wetzel / U.S. Patent No. 3,279,829 (Oct. 18, 1966)

[27] Evaluation of 1-bushel pea baskets to determine design criteria / R.S. Kurtenacker / USDA Marketing research report No. 772 (1967)

[28] Effects of vertical dynamic loading on corrugated fiberboard containers / W.D. Godshall / Research paper FPL 94 (1968)

[29] Corrugating furnishes and mediums: their microscopy, sheet properties and runnability / Forrest A. Simmonds / Paperboard packaging Vol. 55(1) (Jan. 1970)

[30] Frequency response, damping, and transmissibility characteristics of top-loaded corrugated containers / W.D. Godshall / Research paper FPL 160 (1971)

[31] Long-term creep in corrugated fiberboard containers / J.W. Koning Jr., R.K. Stern / Tappi Vol. 60(12) (Dec 1977)

[32] Corrugated fiberboards: edgewise compression creep in cyclic relative humidity environments / Von L. Byrd, John W. Koning, Jr. / Tappi Vol. 61(6) (June 1978)

[33] Optimum fiber distribution in singlewall corrugated fiberboard / Millard W. Johnson, Jr., Thomas J. Urbanik, William E. Denniston / Research paper FPL 348 (1979)

[34] An assessment of the common carrier shipping environment / Fred E. Ostrem and W.D. Godshall / General technical report FPL 22 (1979)

[35] A new proposal for the performance testing of shipping containers / W.D. Godshall / Package development and systems (Sept./Oct. 1979)

[36] Principle of load-sharing corrugated fiberboard. Paperboard packaging Vol. 66(11) (Nov 1981)

[37] Overall effective thermal resistance of corrugated fiberboard containers / David W. Bormett / Research paper FPL 406 (1981)

[38] Package performance testing / ASTM standardization news Vol. 10(10) (Oct 1982)

[39] Shipping container performance testing / W. Duncan Godshall / SPHE Journal (Fall 1983)

[40] Packaging perspective, 1910–1985 / John W. Koning, Jr., James F. Laundrie / General technical report FPL 51) (1985)

[41] Cushioning properties of corrugated pads in edge and corner drop tests / J.Y. Liu, J.F. Laundrie / Boxboard containers Vol. 94(10) (May 1987)

[42] Analysis of the localized buckling in composite plate structures with application to determining the strength of corrugated fiberboard / Millard W. Johnson, Thomas J. Urbanik / Journal of composites technology and research Vol. II, No. 4 (winter 1989)

[43] Forced vibration response of nonlinear top-loaded corrugated fiberboard containers / T.J. Urbanik / Proceedings, 61st Shock and Vibration Symposium, Pasadena, CA, Jet Propulsion Laboratory, Vol. 1, October 16–18, pp: 253–274. (1990)

[44] Effect of in-plane shear modulus of elasticity on buckling strength of paperboard plates / T.J. Urbanik / Wood and fiber science Vol. 24(4): 381–384 (1992)

[45] Swept sine humidity schedule for testing cycle period effects on creep / T.J. Urbanik, S.K. Lee / Wood and fiber science Vol. 27(1): 68–78 (1995)

[46] Corrugated crossroads / John W. Koning, Jr. / Atlanta, GA: TAPPI Press, (1995)

[47] Strength criterion for corrugated fiberboard under long-term stress / T.J. Urbanik / Tappi journal, 81(3): 33–37 (1998)

[48] Effect of corrugated flute shape on fiberboard edgewise crush strength and bending stiffness / T.J. Urbanik / Journal of pulp and paper science Vol. 27(10) (October 2001)

[49] Finite element corroboration of buckling phenomena observed in corrugated boxes / Thomas J. Urbanik, Edmond P. Saliklis / Wood fiber science 35(3): 322–333 (2003)

[50] Heat sterilization times of red pine boards / William T. Simpson, Barbara L. Illman / Forest products journal, Vol. 54(12) (Dec. 2004)

[51] FE analysis of creep and hygroexpansion response of a corrugated fiberboard to a moisture flow: a transient nonlinear analysis / A.A. Rahman, T.J. Urbanik, M. Mahamid / Wood and fiber science Vol. 38(2) (2006)

[52] Box compression analysis of world-wide data spanning 46 years / T.J. Urbanik, B. Frank / Wood and fiber science Vol. 38(3) (2006)

Chemicals

◄ Longleaf pine wood that is saturated with resin because of treatment with the herbicide paraquat (bottom), and during normal gum resin production using sulfonic acid paste (top). [84]

Wood is a complex mixture of natural polymers: cellulose, hemicelluloses, and lignin, plus many smaller molecules termed "extractives." A myriad of chemicals can be derived from wood simply by dissolving them out as extractives (which vary widely among wood species), and many more are obtained from the structural polymers by physical, chemical, biochemical, and microbial technologies. One of the earliest chemical uses of wood was to obtain the gum (pitch) from wounded pine trees. Pine tar was used to seal wooden boats and coat ropes to slow decay. Pine tar was an important commercial incentive for England to invest in colonizing North America because pine tar was critical to building and maintaining the British Navy. Later it was found that by heating wood, a process referred to as "destructive distillation," it was possible to drive off gases that contained other chemicals, such as methanol (wood alcohol), and produce charcoal. [17, 19, 37, 40, 74]

Challenge: What chemicals can be derived from wood and how can they be utilized?

FPL's Contributions: Early work on wood chemicals included the traditional naval stores (rosin, pine tar, and turpentine) and destructive distillation chemicals (methanol, acetate of lime, and charcoal). The FPL began to develop analytical methods to characterize and quantify the major components of wood. By the 1930s, FPL was involved in wood hydrolysis to produce ethanol and other fermentation products. The research culminated in the Madison Wood Sugar process patented by the FPL, in association with the University of Wisconsin, and implemented by the U.S. Government in a wood hydrolysis plant constructed in Springfield, Oregon, during WWII. More recent research has emphasized the study of chemicals from wood such as sterols from tall oil for conversion to steroid drugs, continuing research on lignin structures, and defining the nanostructure of wood cells. (See Biochemistry section.)

Results: It is technically possible to make a large number of useful chemicals from wood. Chemicals such as ethanol, 1,2-propanediol, 1,3-propanediol, butanol, furfural, acetone, xylitol, acetic acid, and many others are possible, but the major limiting factor has usually been cost because of competition with the petrochemical industry or other sources of sugar, such as corn starch.

Trees Are the Source of Many Chemicals

A 1954 report by Edward G. Locke and Kenneth G. Johnson included the following example. [40] A more recent report (2004) by the U.S. Department of Energy largely replicates the list of fermentation products. [114] Other chemicals include osage orange as a dyestuff and conifer leaf oil. [4, 6, 98, 101]

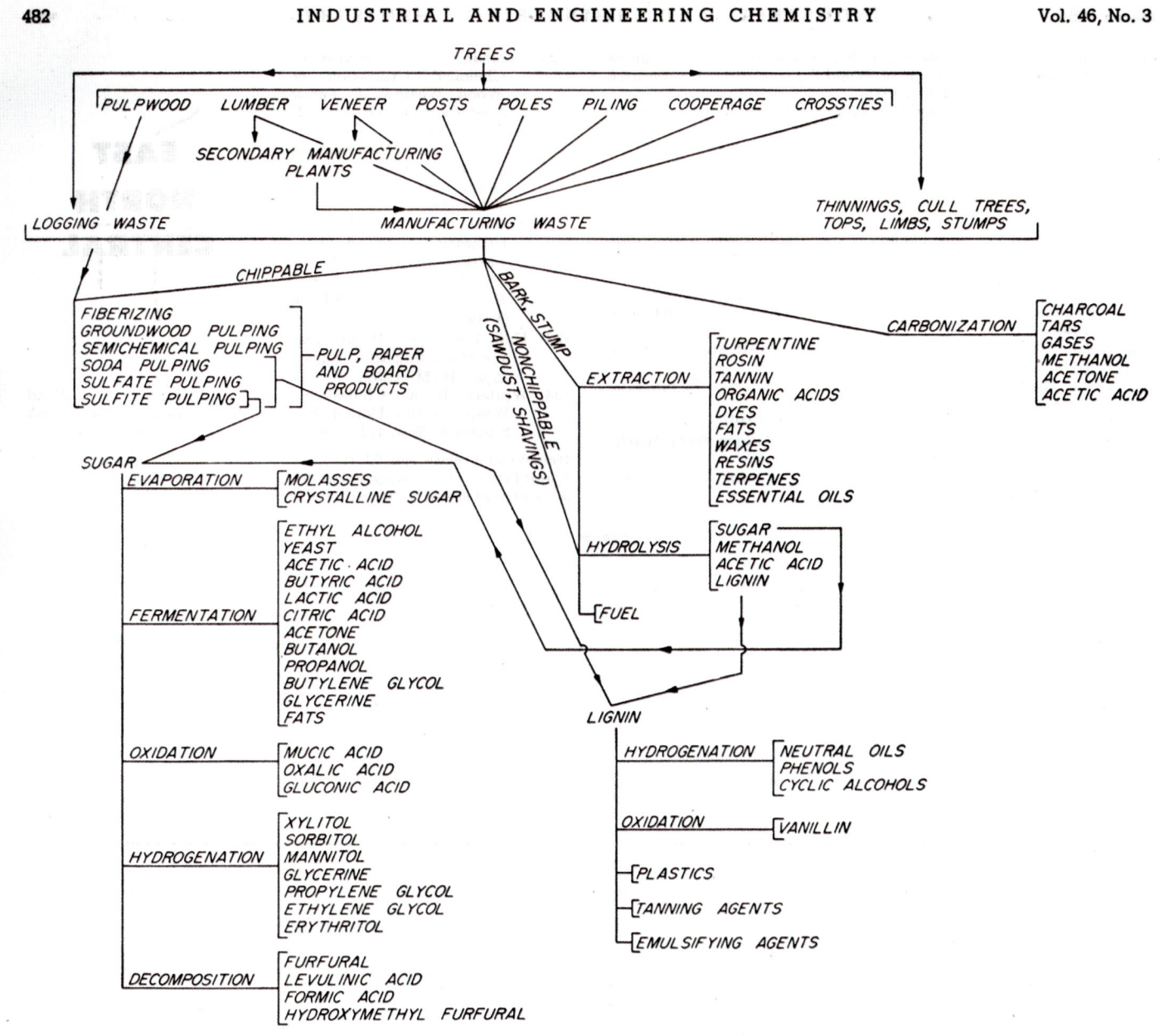

Figure 4. Chemical Utilization of Wood

Destructive Distillation

In 1910, the use of retorts to distill chemicals from wood resulted in recovery of pyroligneous acid, which was neutralized with lime and by evaporating the solution provided acetate of lime. Also produced and marketed was methanol (also called wood spirits, wood naphtha, or wood alcohol), tar, and charcoal. Advanced refining could also make acetic acid, acetone, and formaldehyde. [13, 30, 43]

FPL scientists determined the yields of destructive distillation products from several southern hardwood species in comparison with northern hardwoods. This led to the establishment of the hardwood distillation industry in the south around 1915 (four plants, capacity 500 cords daily). [1]

Temperature control procedures were established, giving 5% to 10% increases in yields of wood alcohol (methanol) and acetic acid from distillation of hardwood. (Estimated savings: $300,000/year in 1937 dollars, $3,750,000 in 2002 dollars.) [5]

Naval Stores

"Naval stores" is an old term for denoting chemicals derived from the gum (pitch or more technically, oleoresin) extracted from pine. The gum is a source for a number of products such as turpentine, rosin, dyes, fats, resins, essential oils, waxes, and organic acids. [18, 77, 94, 96]

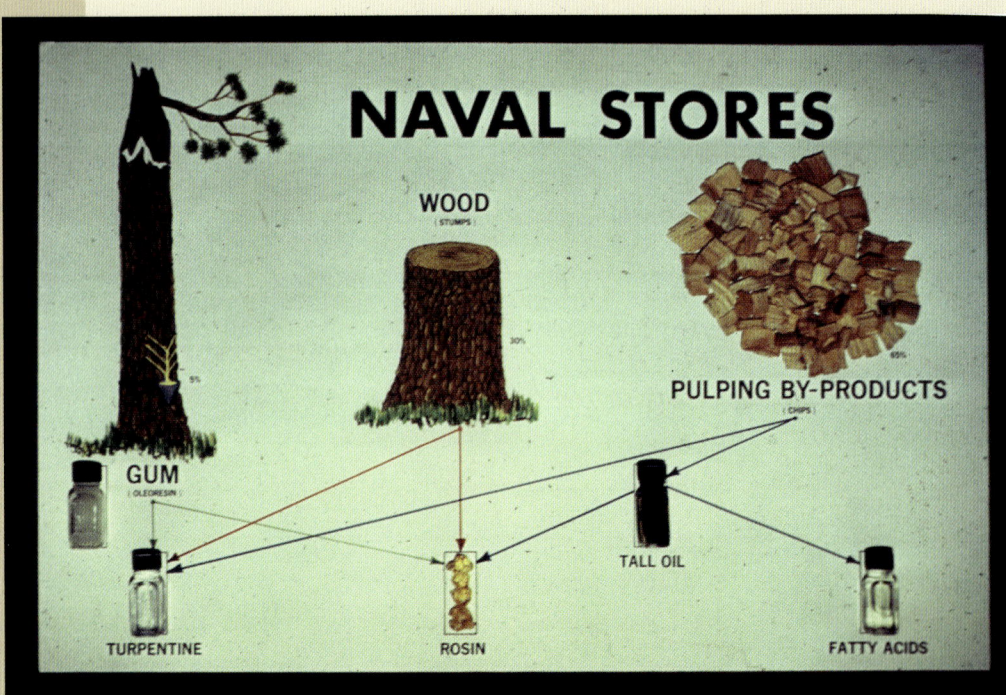

◄ Diagram of where naval stores come from. (1970s) {136 N}

Microbial Conversion of Tall Oil Sterols to C$_{19}$ Steroids

In addition to the traditional naval stores, extractions of wood and pulping wastes contain neutral (i.e., non-acidic) biologically active components. FPL also studied these neutral components. [75] Sterols, one portion of those neutrals, were studied in cooperation with the University of Wisconsin Department of Pharmacy as precursors for steroid drugs. [78] One wood-based sterol is used today in products to decrease the buildup of cholesterol in human blood.

► Early naval stores research by Eloise Gerry (left) on the effect of chipping or tree scoring to increase the gum flow and decrease damage to the tree. (A) After one year, left, 1/4-inch-high chip; middle, 1/2-inch chip; right, 3/4-inch chip. (B) Three years. (C) Five years. Note that after five years the person doing the 3/4-inch chipping would have to reach over 9 feet from the ground to make the next chip. The 1/4-inch chipping allowed for five more years of gum production and less damage to the tree. (1930s) {M 15754 F} [3, 8, 15]

Turpentines

Pioneering work on the analysis and composition of wood turpentines, and on their refining in a continuous-column still, resulted in greatly improved quality. [2]

◄ Optical properties of a resin acid extracted from pine wood are checked in a polarimeter by Duane Zinkel. (1960s) {M 123 594}

Improved analytical procedures were developed for assaying precursors of tall oil in wood, for determining tall oil constituents in pulp mill waste streams, and for determining the purity of products, all adding value to product quality. [99]

Paraquat Treatment of Pine for Naval Stores

Natural oleoresins (oils and resins) found in southern pine wood are now recovered during the kraft pulping process and are used in the production of turpentines, soaps, detergents, and resins. FPL research indicated that oleoresin yield can be increased several-fold when pine is stimulated by treatment with paraquat. In a cooperative study, paraquat-treated slash pine was pulped and information was obtained on the yields of tall oil (resin acids and fatty acids), turpentines, and pulp. Resin-soaked wood can be pulped without difficulty and without a loss in yield or quality of the pulp. Relative composition of the components of the resin acid, turpentine, and fatty acid fractions did not change significantly after paraquat treatment. (1977) [79, 82]

Chemicals in Jack Pine Bark

More than 100 chemical compounds were isolated from jack pine bark and fully identified. They include fats, waxes, resins, and sterols. Nineteen of the chemicals found are natural products that had not been previously reported. This comprehensive study of pine bark chemistry provided a broader technical basis for evaluating the feasibility of bark chemical utilization and contributed to a better understanding of the disease resistance, insect resistance, chemistry, and biosynthetic processes of wood and other natural products. [64] This was one of many FPL studies of chemicals from both wood and bark of various species.

Extractives from Eastern Hardwoods

A thorough review was completed on the chemical extractives in eastern hardwoods. This included both wood and bark. These extractives affect wood performance, color, odor, decay resistance, and other properties. [86]

These are just examples of FPL research on extractives. The results of other extractives research is found in reference [97].

Cancer Treatment

◄ Rajai Attala and John Obst examine the dark heartwood of a 180-year-old Pacific yew log in preparation for taxol extraction studies. The anti-cancer drug taxol was being extracted from the bark of Pacific yew trees. Attala and Obst were seeking to confirm the presence and quantity of this anticancer drug in the heartwood of yew. The compound previously identified by others as taxol turned out to be a related compound that was not biologically active. (1990s) {M 910 168-9} [100, 102]

FPL continues the effort to discover and identify new chemicals from wood. A recent discovery is an extractive common in willow named Salucci. (2000s) [112]

Wood Conversion

Instead of converting wood by destructive distillation, other physical, chemical, biochemical, and/or microbial technologies can be used. These processes start with a fundamental understanding of the composition of the wood cells (fibers, including ray cells), which contain cellulose, hemicelluloses, and lignin in different locations, concentrations, mixtures, and structures. Much FPL research has been devoted to understanding the complexity of wood and how to obtain the possible chemicals. [14, 17, 19, 36, 87, 116]

Cellulose

Cellulose constitutes approximately 50% of the wood substance by weight. Research showed that a number of chemicals can be derived from cellulose. [56, 71, 95, 105, 107–109]

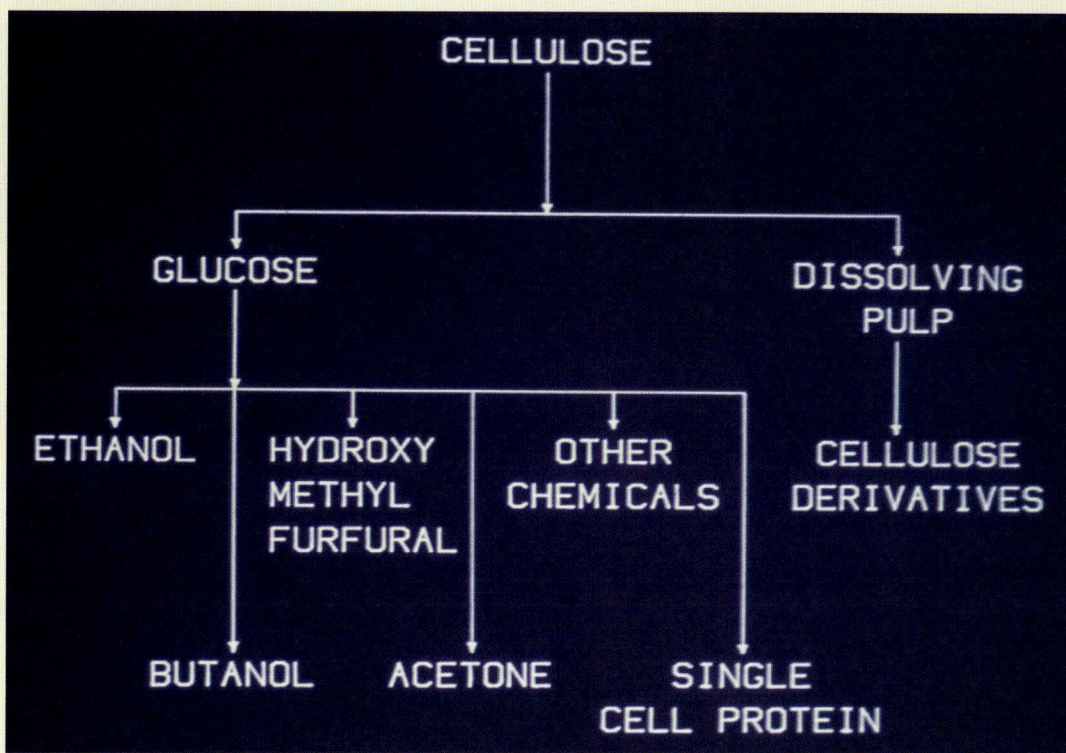

◄ Cellulose is a major component of wood. This diagram shows some of the many chemicals that can be made from cellulose.

Cellulose Derivatives

A technique was developed to enable the use of low-cost, hardwood pulps in the manufacture of cellulose derivatives such as rayon, acetate, carboxymethyl cellulose (CMC), cellulose nitrate, and cellulose ethers. This technique is based on maintaining wood pulp in a "never-dried" condition throughout the manufacturing process, which enhances chemical reactivity. [9, 31, 41, 44]

Hemicelluloses

Hemicelluloses are another major component of wood. The hemicelluloses are associated with the cellulose and the challenge is economically separating them and also separating both from the third major component, lignin. [42, 45, 87, 89, 104]

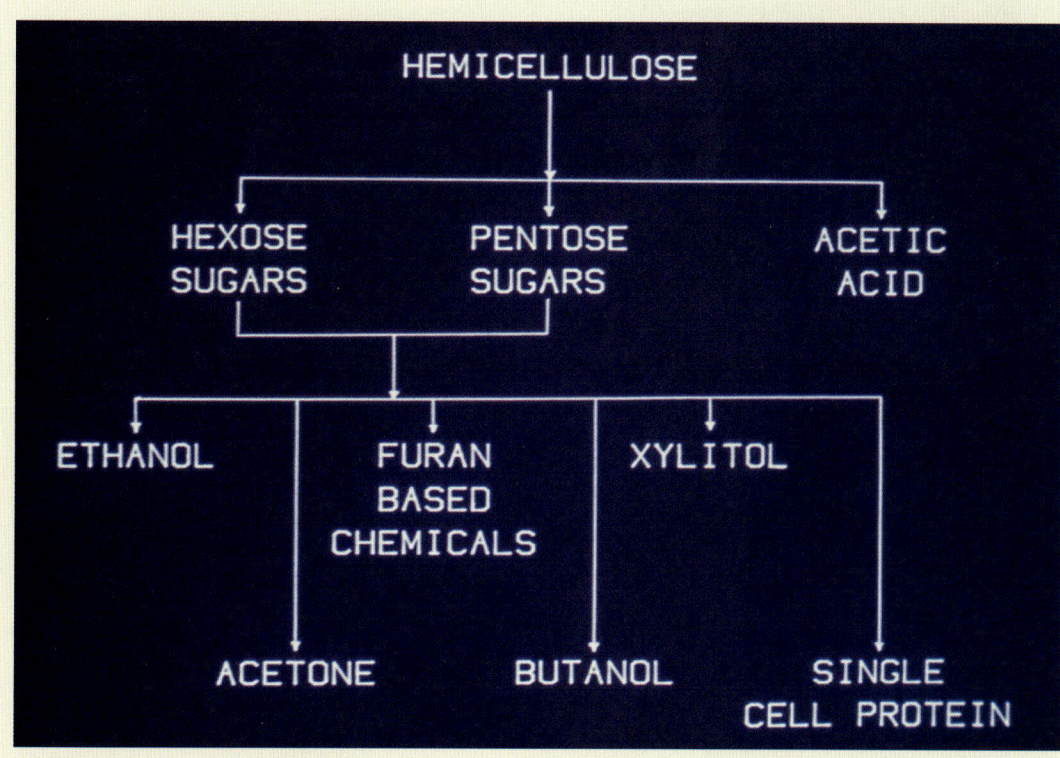

◄ Hemicelluloses are a major component of wood. This diagram shows some of the many chemicals that can be made from hemicelluloses.

Chemicals from Waste Liquors

Waste liquor from the acid sulfite pulping of hardwoods commonly contains 2% to 6% dissolved xylose sugar. Stream or river disposal of this liquor results in undesirable environmental impacts to the water and is a waste of a potentially valuable raw material. Furfural was produced from the xylose in samples of waste sulfite pulping liquors in 1969. [65]

Fodder Yeast

FPL scientists perfected the production of fodder yeast from wood sugar solutions. The process has been used for the production of feed yeast from formerly wasted sulfite pulping liquor. [29]

Ethanol

Two-Stage Process for Hydrolyzing Hardwoods

FPL modified a two-stage process for hydrolyzing hardwoods that recovers both pentose (five-carbon) and hexose (six-carbon) sugars from wood particles. Initially only six-carbon sugars were fermentable into ethyl alcohol. The two-stage process not only permits more efficient processing of wood to ethyl alcohol (and recovery of valuable by-products), but also requires approximately 40% less energy than a one-stage process. Both sugars may be used to make a molasses product for animal feed plus furfural and acetic acid. [7, 25–27, 33, 71, 73, 83, 87, 89–91, 93] Tom Jeffries's work showed how to convert five-carbon sugars to ethyl alcohol. [110, 111] (See Biochemistry section.) For a more in-depth discussion of early ethanol research at FPL see related article by George Hajny on page 494.

Pilot Plant

Some of the early wood hydrolysis pilot plant equipment used for the study of ethanol made from wood.

▲ Edward Beglinger filling the percolator. (1940s) {M 56706 F}

▲ Release valve for the percolator. (1940s) {M 56705 F}

▲ Fermentation tanks, centrifuge, and autoclave. (1940s) {M 56707 F}

▶ Jerry Saeman using a centrifuge in the study of wood sugars. (1940s)

Lignin

A third major component of wood is lignin. It constitutes 23% to 33% of the wood substance by weight in softwoods and 16% to 25% in hardwoods. Lignin is found throughout the cell wall but is concentrated between the cell walls. Lignin is a three-dimensional phenylpropanol polymer. FPL made many contributions to identifying its structure and distribution in wood, but lignin is still not fully understood. [17, 54, 92, 106, 113, 115, 116]

◀ Lignin is treated in a hydrogenation reactor to produce chemicals, which are analyzed in an endeavor to find higher valued uses. (1940s) {M 54880 F} [24, 35]

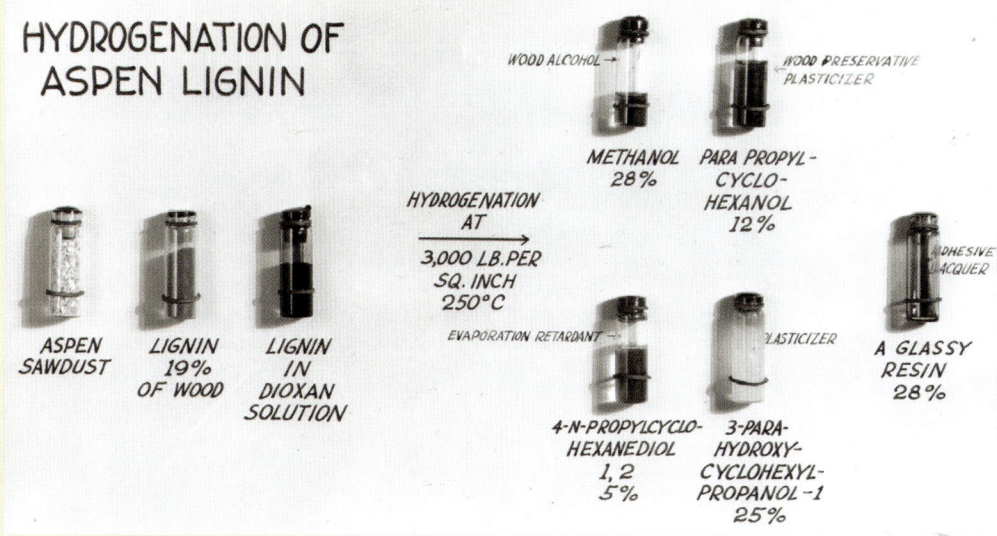

HYDROGENATION OF ASPEN LIGNIN

ASPEN SAWDUST — LIGNIN 19% OF WOOD — LIGNIN IN DIOXAN SOLUTION

HYDROGENATION AT 3,000 LB. PER SQ. INCH 250°C

WOOD ALCOHOL → METHANOL 28%

WOOD PRESERVATIVE PLASTICIZER → PARA PROPYL-CYCLO-HEXANOL 12%

EVAPORATION RETARDANT → 4-N-PROPYLCYCLO-HEXANEDIOL 1, 2 5%

PLASTICIZER → 3-PARA-HYDROXY-CYCLOHEXYL-PROPANOL-1 25%

ADHESIVE LACQUER → A GLASSY RESIN 28%

▲ Five products developed from lignin at the FPL. (1940s) {M 34256 F}

◀ John Harkin and Clarence Pew (background) studying the characteristics of lignin. (1960s) {M 131 351} [51, 61]

▲ Larry Landucci, Sally Ralph, and John Ralph discussing mechanisms of lignin degradation. (1980s) {M 150 557-9}

Lignin Uses

The main use of lignin is as fuel in the pulp and paper industry. However, some other uses include storage batteries, tanning agents, boiler-water treatment, adhesives, core binders, dispersing agent in cement, extenders for rubber, lignin-filled laminates, and resins. (Although some of these products and uses are available commercially, a saying in the industry is that you can make anything out of black liquor (lignin) except money.)

Technology Transfer

◀ John Rowe describing the potential wealth of chemicals in the wood and bark of trees to the Forestry Research Advisory Committee of the U.S. Department of Agriculture at its review meeting in Madison, Wisconsin. Listening to John Rowe are (from left) Ralph H. Bescher, Assistant Vice President, Forest Products Division of Koppers Co.; John A. Zivnuska, Dean of the School of Forestry, University of California, Berkeley; John R. Woodworth, Director of the Idaho Fish and Game Department; George M. Jemison, Deputy Chief of the Forest Service, who was chairman of the meeting; Frederick H. Claridge, North Carolina State Forester; Reynold E. Carlson, Professor of Recreation and Park Administration, University of Indiana; George F. Dow, Director of the Maine Agricultural Experiment Station, Orono; and Kel M. Fox, Grand Verde, Arizona cattleman. Standing beside John Rowe is Margaret K. Seikel of the Laboratory staff. (1960) {M 131 880}

Analytical Procedures for Wood and Wood Components

Test Methods

Improving the understanding of wood and wood chemicals has often required improved analytical procedures to evaluate wood components and changes in those components with processing.

To accomplish this, FPL developed numerous test methods and procedures to determine characteristics of wood and the types and amounts of its constituents (cellulose, hemicellulose, lignin, and extractives). Many of these methods have been incorporated in standard test procedures used in other laboratories. [49, 52, 58, 60, 62, 67, 69, 70, 103, 105, 109] Results of some of these tests are incorporated in the *Handbook of Wood Chemistry and Wood Composites.* [116]

FPL has been a leader in the development of solution NMR (nuclear magnetic resonance) for studying lignin, and Raman spectroscopy for understanding both lignin and carbohydrates. [106, 115]

▶ Larry Landucci at the console of an early ^{13}C (carbon-13)NMR instrument used in studying wood polymers and extractives. (1990s) [106]

Holocellulose

Among the methods developed was a procedure to isolate the total carbohydrate fraction from wood. The name holocellulose was given to this fraction to indicate that it contains all the cellulose and hemicelluloses. [16, 20, 70]

Microstructure of Wood

FPL scientists continue to probe the nanostructure of the wood cell wall. Current efforts focus on determining the degree of kinetic and environmental control over the lignin structure and the nature and purpose of the physical association between cellulose, hemicelluloses, and lignin in wood.

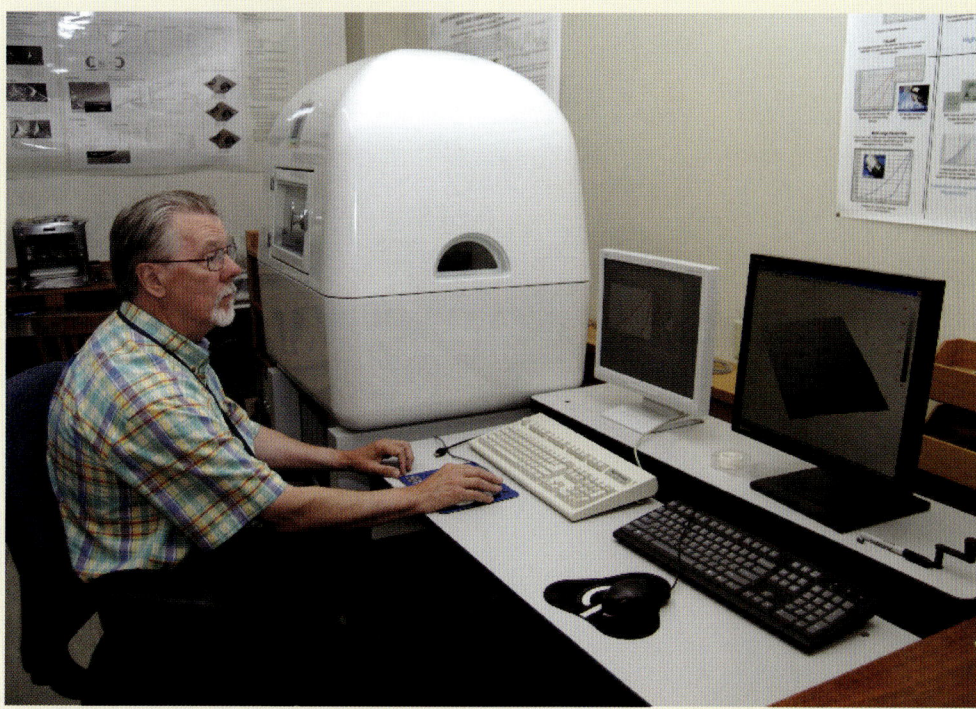

◀ James Beecher operating the nanoindenter used to help understand the nanostructure of the wood cell. (2000s)

Wood Sugar and Element Analysis

◀ Fred Matt recording data from an ion chromatograph for sugar analysis. (2000s)

▼ Dan Foster preparing a specimen for element analysis of a wood preservative using an inductively coupled plasma atomic emissions spectroscope. (2000s)

Pressure Digestion

An early example of using wood waste was to expose the waste to pressure digestion. Some of the wood components separated by this process can then be recombined to make a useful plastic-like product. [21]

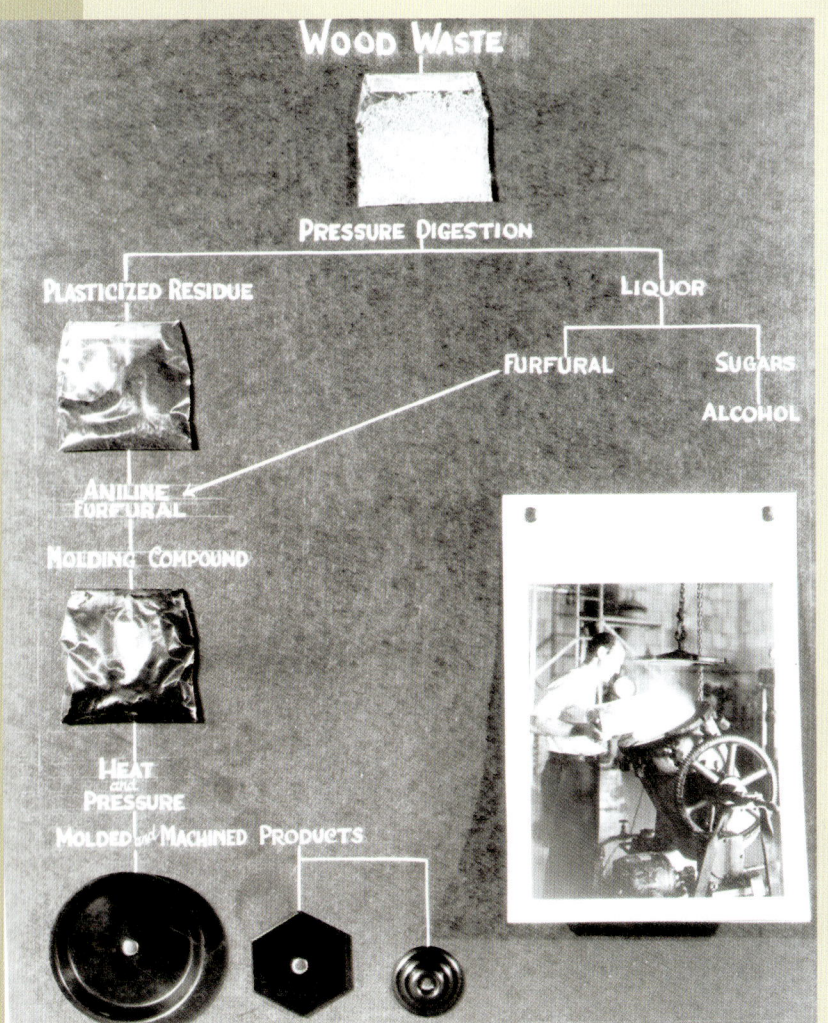

◀ Diagram of the use of pressure digestion to produce a plasticized residue and liquor from wood waste. With further processing, useful products can be made. (1930s)

▲ Various products made from wood waste converted using pressure digestion. (1940s) [23]

Charcoal

The FPL worked cooperatively with other Forest Service experiment stations and industry to improve the production of charcoal. [28, 47, 63, 80]

FPL developed a method for making a dense charcoal from wood as a substitute for imported coconut-shell charcoal used for gas masks. This charcoal was used to help protect soldiers from poison gases in WWI. [10, 11, 12, 53]

Al Stamm developed a method to continuously make charcoal. (1949) [30]

◄ Scale model of a seven-cord, masonry block charcoal kiln. FPL pioneered the design and engineering of numerous small charcoal kilns, which permit the production of charcoal by farmers and other rural residents without investment in a large industrial plant. Some of the uses for charcoal besides barbecuing include soil treatments, filters, foundry molds, black powder, electrodes, graphite, pharmaceuticals, plastics, rubber, poultry and stock feeds, molding resins, brake linings, and activated carbon. Charcoal for many of these products is made in commercial retorts. (1950s) {ZM 111 376} [46]

Animal Feeds

New combinations of wood-processing wastes are valuable as animal feed components. Short fibers in pulp mill waste provide digestible carbohydrates. Aspen sawdust, when properly supplemented, is equivalent to medium-quality hay and as a roughage replacement in feeding value. In addition, it appears suitable as an ingredient in maintenance rations in feedlots. (1960s–1970s) [66, 72, 76, 87]

◄ Andy Baker opening the discharge valve of a pressurized digester. Here, aspen sawdust pours from a digester after having been treated with anhydrous ammonia to improve its digestibility for livestock. (1970s) {M 137 451}

▶ An experiment, in cooperation with the University of Wisconsin, in feeding 30 sheep for a 90-day period on a maintenance diet of 77% mill waste from a tissue mill, 9% soybean meal, 9% cracked shelled corn, 2% urea (nitrogen source), 1% calcium phosphate, 1% mineral salt and vitamins. The sheep took to the diet and thrived on it. (1970s) {M 139 404-2}

Pine Pollen

▶ Linda Feldman, student, and Ariadne Freeman (right), a UCLA senior majoring in organic chemistry, conducting an experiment using a column chromatograph to separate complex phenols extracted from pine pollen for identification. (1960s) {M 127 152} [57]

Dimensional Stabilization of Wood and Wood Fibers

The dimensions of wood and wood fibers are affected by changes in moisture. These changes may cause problems in their use. FPL researchers have studied these effects and how chemical treatments can modify them. This research resulted in the development of compreg and other wood and wood fiber treatments. [32, 34, 38, 39, 48, 59, 85]

▶ *Wood Chemistry and Wood Composites Handbook*, edited by Roger Rowell, covers in detail many of the accomplishments of FPL research on the chemistry of wood and wood composites. (2005) [116] Chapters of the handbook are (1) Wood and Society, (2) Structure and Function of Wood, (3) Cell Wall Chemistry, (4) Moisture Properties, (5) Biological Properties, (6) Thermal Properties, (7) Weathering of Wood, (8) Surface Characterization, (9) Wood Adhesion and Adhesives, (10) Wood Composites, (11) Chemistry of Wood Strength, (12) Fiber Webs, (13) Wood Thermoplastic Composites, (14) Chemical Modification of Wood, (15) Lumen Modifications, and (16) Plasm Treatment of Wood.

HANDBOOK OF
WOOD CHEMISTRY
AND WOOD
COMPOSITES

Edited by
Roger M. Rowell

1910

Limited number of chemicals were obtained through destructive distillation.

2010

Hundreds of new chemicals have been identified and obtained by many methods, including carbonization, fermentation, evaporation, oxidation, extraction, hydrolysis, hydrogenation, and decomposition. Chemicals are also obtained through co-production with pulp mills.

A handbook on wood chemistry listing the chemical components of wood is now available. [116]

Further Information

[1] The relative yields obtained by the destructive distillation of different forms and species of hardwoods / L.F. Hawley, R.C. Palmer / Original communications, eighth international congress of applied chemistry Vol. 6 (1912)

[2] Wood turpentines: their analysis, refining, and composition, based upon experiments at the Forest Products Laboratory at Madison, Wis. / L.F. Hawley / USDA, Forest Service Technical bulletin 105 (1913)

[3] Improved practice in the production of naval stores / A.W. Schorger / Southern lumberman Vol. 77(1023) (May 1, 1914)

[4] Osage orange: its value as a commercial dyestuff / F.W. Kressmann / Journal of industrial and engineering chemistry Vol. 6(6) (June 1914)

[5] Preliminary experiments on the effect of temperature control on the yield of products in the destructive distillation of hardwoods / R.C Palmer / Journal of industrial and engineering chemistry Vol. 7(8) (Aug. 1915)

[6] The conifer leaf oil industry / A.W. Schorger / American lumberman (Apr. 29, 1916)

[7] Ethyl alcohol from wood: the process, its development and requirements / F.W. Kressmann / American lumberman No. 2148 (July 15, 1916)

[8] Study of wood structure in relation to the formation and yield of resin-producing tissue in southern pines: second progress report / Eloise Gerry / FPL Project No. 20–3 (1917)

[9] Use of wood pulp in the manufacture of nitrocellulose / Sidney D. Wells, Vance P. Edwardes / Paper, Vol. 23(23) (Feb. 12, 1919)

[10] Process for producing dense charcoal / Lee F. Hawley / U.S. Patent No. 1,369,428 (Feb. 22, 1921)

[11] Process for producing dense absorbent charcoal / Lee F. Hawley / U.S. Patent No. 1,385,826 (July 26, 1921)

[12] Process of producing highly-absorbent charcoal / Ernest Bateman / U.S Patent No. 1,573,509 (Feb. 16, 1926)

[13] Fifty years of wood distillation / Lee F. Hawley / Industrial and engineering chemistry Vol. 18(9) (Sept. 1926)

[14] Density of wood substance, adsorption by wood, and permeability of wood / Alfred J. Stamm / Journal of Physical Chemistry 33(3):398–414, (March 1929)

[15] Improvement in the production of oleoresin through lower chipping / Eloise Gerry / USDA, Forest Service Technical bulletin No. 262 (1931)

[16] Holocellulose, total carbohydrate fraction of extractive-free maple wood: its isolation and properties / G.J. Ritter, E.F. Kurth / FPL No. 1052 (1933)

[17] Structure of the cell wall of wood / G.J. Ritter / Paper industry 16(3):178–183 (June 1934)

[18] A naval stores handbook dealing with the production of pine gum or oleoresin / compiled by the Forest Service, the Bureau of Entomology and Plant Quarantine, and the Bureau of Plant Industry / E. Gerry / USDA Miscellaneous publication 209 (1935)

[19] Colloid chemistry of cellulosic materials / Alfred J. Stamm / USDA Miscellaneous Publication No. 240 (1936)

[20] Chemical composition of wood: determination of holocellulose / William G. Van Beckum, G.J. Ritter / Paper trade journal Vol. 104(19) (May 13, 1937)

[21] Forest Products Laboratory's work on plastics from wood lignin / FPL No. 1134 (1937)

[22] Minimizing wood shrinkage and swelling: effect of heating in various gases / Alfred J. Stamm, L.A. Hansen / Industrial & engineering chemistry Vol. 29(7):831–33 (July 1937)

[23] Some Forest Products Laboratory developments of particular significance to the South / FPL report 0–8 W73Sf (April 15, 1940)

[24] Hydrogenation of lignin / E.E. Harris / Paper trade journal Vol. 111(24) (Dec. 12, 1940)

[25] Hydrolysis of wood in a stationary digester by successive treatments with dilute sulfuric acid / Elwin E. Harris, E. Beglinger, G.J. Hajny, E.C. Sherrard / FPL No. 1455 (1944)

[26] Kinetics of wood saccharification / J.F. Saeman / Industrial and engineering chemistry Vol. 37(1) (Jan. 1945)

[27] The Madison wood-sugar process / E.E. Harris, Edward Beglinger / FPL No. 1617 (1946)

[28] Charcoal production / Edward Beglinger / FPL Report No. R 1666–11, (Revised Oct. 1947)

[29] Fodder yeast from wood hydrolyzates and still residues / Elwin E. Harris, J.F. Saeman, R.R. Marquardt, M.L. Hannan, S.C. Rogers / Industrial and engineering chemistry Vol. 40(7) (July 1948)

[30] Destructive distillation of solids in a liquid bath / Alfred J. Stamm / U.S. Patent No. 2,459,550 / (Jan. 18, 1949)

[31] The manufacture of rayon / FPL Technical note No. 217 (1949)

[32] The swelling and shrinking of wood, paper, and cotton textiles and their control / H. Tarkow / Tappi Vol. 32(5) (May 1949)

[33] Derived products and chemical utilization of wood waste / E.E. Harris / FPL Report No. R 1666–10 (June 1949)

[34] Forest Products Laboratory resin-treated, laminated, compressed wood (compreg) / Alfred J. Stamm, R.M. Seborg / FPL No. 1381 (1949)

[35] Lignin hydrogenation products / Elwin E. Harris, J.F. Saeman, C.B. Bergstrom / Industrial and engineering chemistry Vol. 41(9) (Sept. 1949)

[36] Bound water and hydration / Alfred J. Stamm / TAPPI Vol. 33(9) (Sept. 1950)

[37] Wood: an example of a colloid system / Harold Tarkow, Alfred J. Stamm / Colloid chemistry, New York, NY: Reinhold Publishing Company (1950)

[38] Stabilizing the dimensions of wood / Alfred J. Stamm / Crops in peace and war, USDA yearbook of agriculture, 1950–1951 Washington, DC (1951)

[39] Effect of formaldehyde treatments upon the dimensional stabilization of wood / Harold Tarkow, Alfred J. Stamm / Journal of the Forest Products Research Society (June 1953)

[40] Wood resources / E.G. Locke, K.G. Johnson / Industrial and engineering chemistry Vol. 46 (March 1954)

[41] Purified hardwood pulps for chemical conversion. II, Sweetgum prehydrolysis-sulphate pulps / F.A. Simmonds, R.M. Kingsbury, J.S. Martin / TAPPI Vol. 38(3) (Mar. 1955)

[42] Molecular properties of hemicellulose fractions / Merrill A. Millett, Alfred J. Stamm / FPL No. 1668 (1956)

[43] Thermal degradation of wood and cellulose / Alfred J. Stamm / Industrial and engineering chemistry Vol. 48(3) (Mar. 1956)

[44] Loblolly pine high alpha prehydrolysis-sulphate pulps / F.A. Simmonds, R.M. Kingsbury, J.S. Martin, R.L. Mitchell / TAPPI Vol. 39(9) (Sept. 1956)

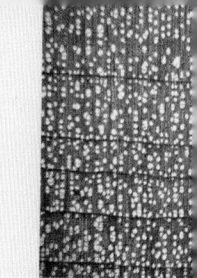

[45] Review of chemical utilization / E.G. Locke / Forest products journal Vol. 7(2) (Feb. 1957)

[46] Production of charcoal in a masonry block kiln: structure and operation / Compiled by Forest Products Laboratory / FPL No. 2084 (1957)

[47] Charcoal production in kilns / E. Beglinger / Forest products journal Vol. 7(11) (Nov. 1957)

[48] Dimensional stabilization of paper by crosslinking with formaldehyde / W.E. Cohen, A.J. Stamm, D.J. Fahey / TAPPI Vol. 42(12) (Dec. 1959)

[49] A method for determining the distribution, by sizes, of interstructured opening in wood / Alfred J. Stamm, Eugene Wagner / Forest products journal Vol. 11(3) (Mar. 1961)

[50] Economics of the production of furfural from xylose solutions: effects of variations in plant design and process variables on production costs / J.F. Harris, J.M. Smuk / Forest products journal Vol. 11(7) (July 1961)

[51] The biphenyl group in lignin / J.C. Pew / Nature Vol. 193(4812) (Jan. 20, 1962)

[52] Sugar units present: hydrolysis and quantitative paper chromatography / J.F. Saeman, W.E. Moore, M.A. Mille / Cellulose methods in carbohydrate chemistry Vol. 3, New York, NY: Academic Press, Inc. (1962)

[53] Gas-aerosol filter material / John R. Conlisk, Leonard A. Jonas, Alfred J. Stamm, Harold Tarkow, Richard C. Weatherwax / US Patent No. 3,034,947 (May 15,1962)

[54] Lignin / FPL Research note 079 (Superseded by Lignin and its uses / J.M. Harkin / FPL Research note 0206 (Jul 1969) (1964)

[55] Effect of wood on setting of Portland cement / R.C. Weatherwax, H. Tarkow / Forest products journal Vol. 14(12) (Dec. 1964)

[56] Modification of cellulose fine structure: effect of thermal and electron irradiation pretreatments / M.A. Millett, V.L. Goedken / TAPPI Vol. 48(6) (June 1965)

[57] Polyphenols of pine pollens: a survey / Mary Jane Strohl, Margaret K. Seikel / Phytochemistry Vol. 4 (1965)

[58] Interaction of wood with polymeric materials: penetration versus molecular size / Harold Tarkow, W.C. Feist, C.F. Southerland / Forest products journal Vol. 16(10) (Oct. 1966)

[59] Dimensional stability / Harold Tarkow / Encyclopedia of polymer science and technology Vol. 5 (New York, NY) John Wiley & Sons (1966)

[60] Quantitative gas-liquid chromatography of fatty and resin acid methyl esters / F.H. Max Nestler, D.F. Zinkel / Analytical chemistry Vol. 39(10) (Aug. 1967)

[61] New structures from the dehydrogenation of model compounds related to lignin / John C. Pew, W.J. Connors / Nature Vol. 25(5101) (Aug. 5, 1967)

[62] Procedures for the chemical analysis of wood and wood products (as used at the U.S. Forest Products Laboratory) / W.E. Moore, D.B. Johnson / Forest Products Laboratory (1967)

[63] Calorific values of the volatile pyrolysis products of wood / J.J. Brenden / Combustion and flame Vol. 11(5) (Oct. 1967)

[64] Extractives of jack pine bark: occurrence of cis- and trans-pinosylvin dimethyl ether and ferulic acid esters / J.W. Rowe, C.L. Bower, E.R. Wagner / Phytochemistry Vol. 8 (Jan. 1969)

[65] Furfural from spent sodium-base acid sulfite pulping liquor / L.L. Zoch, J.F. Harris, E.L. Springer / TAPPI Vol. 52(3) (Mar. 1969)

[66] A mechanism for improving the digestibility of lignocellulosic materials with dilute alkali and liquid ammonia / Harold Tarkow, William C. Feist / Cellulases and their applications. Washington, DC: American Chemical Society, 1969. Advances in Chemistry Series; Number 95. (1969)

[67] Mass spectra of diterpene resin acid methyl esters / Teh-Liang Chang, Thomas E. Mead, Duane F. Zinkel / Journal of the American Oil Chemists' Society Vol. 41 (Sept. 1971)

[68] Isolation and characterization of alpha-guaiaconic acid and the nature of guaiacum blue / J.F. Kratochvil, R.H. Burris, M.K. Seikel, J.M. Harkin / Phytochemistry Vol. 10(10) (Oct 1971)

[69] Diterpene resin acids: a compilation of infrared, mass, nuclear-magnetic resonance, ultraviolet spectra and gas chromatographic retention data (of the methyl esters) / Duane F. Zinkel, Lester C. Zank, Martin F. Wesolowski / (Unnumberd FPL report) (1971)

[70] A rapid analysis for total carbohydrate in wood or pulp: dehydrating to furans in concentrated sulfuric acid / Ralph W. Scott, Jesse Green / Tappi Vol. 55(7) (July 1972)

[71] Hydrolysis of wood and cellulose with cellulytic enzymes / Wayne E. Moore, Marilyn J. Effland, Merrill A. Millett / Journal of agricultural and food chemistry Vol. 20(6) (Nov./Dec. 1972)

[72] Effect of lignin on the in vitro digestibility of wood pulp / A.J. Baker / Journal of animal science Vol. 36(4) (Apr. 1973)

[73] Pretreatments to enhance chemical, enzymatic, and microbiological attack of cellulosic materials / Merrill A. Millett, Andrew J. Baker, Larry D. Satter / Cellulose as chemical and energy resource. New York: John Wiley & Sons, 1975. Biotechnology and Bioengineering Symposium No. 5 (1975)

[74] Chemicals from trees / Duane F. Zinkel / Chemtech Vol. 5(4) (Apr. 1975)

[75] Neutrals in southern pine tall oil / A.H. Conner, J.W. Rowe / Journal of the American Oil Chemists' Society Vol. 52(9) (Sept. 1975)

[76] Wood and wood-based residues in animal feeds / Baker, A.J. / Cellulose technology research / edited by A.F. Turbak. / Washington: American Chemical Society (1975)

[77] Naval stores: silvichemicals from pine / Duane F. Zinkel / Wood chemicals—a future challenge: proceedings of the eighth Cellulose Conference, Part I / edited by T.E. Timell. New York: Wiley, 1975. Applied polymer symposia No. 28 (1975)

[78] Microbial conversion of tall oil sterols to C19 steroids / Anthony H. Conner, Muneo Nagaoka, John W. Rowe, D. Perlman / Applied and Environmental Microbiology 32(2):310–311 (1976)

[79] Microscopic observations of paraquat-induced lightwood in slash pine / V.P. Miniutti / Wood science Vol. 9(3) (Jan 1977)

[80] A history of the charcoal industry in the U.S. / Andrew J. Baker / Text of talk presented at symposium on history of forest products, eleventh great lakes regional meeting American Chemical Society, Stevens Point, Wisconsin (June 6–8,1977)

[81] Wood as animal feed / Andrew J. Baker / McGraw–Hill Yearbook of science & technology (1977).

[82] Effects of paraquat treatment of northern and western conifers / A.H. Conner, M.A. Diehl, H. Wroblewska, J.W. Rowe / Lightwood research coordinating council: proceedings of the 1977 annual meeting. Asheville, North Carolina: Southeastern Forest Experiment Station (1977)

[83] Process alternatives for furfural production / J.F. Harris / Tappi Vol. 61(1) (Jan 1978)

[84] Chemistry of naval stores from pine lightwood: a critical review / D.F. Zinkel, C.R. McKibben / Lightwood research coordinating council, proceedings of 1978 annual meeting. Asheville, North Carolina: Southeastern Forest Experiment Station (1978)

[85] Modified cellulosics / R.M. Rowell, R.A. Young, eds. / Academic Press, New York, NY (1978)

[86] Extractives in eastern hardwoods: a review / John W. Rowe, Anthony H. Conner / FPL–GTR–18 (1979)

[87] Biological utilization of wood for production of chemicals and foodstuffs / George J. Hajny / FPL Research Paper 385 (1981)

[88] Biological decomposition of solid wood. Chemistry of solid wood / Roger M. Rowell, ed. / Washington: American Chemical Society, Advances in chemistry series 207 (1984)

[89] Prehydrolysis of hardwoods with dilute sulfuric acid / E.L. Springer / Industrial and engineering chemistry - product research and development Vol. 24 (1985)

[90] Effects of culture conditions on the fermentation of xylose to ethanol by *Candida shehatae*. Proceedings of 7th symposium on biotechnology for fuels and chemicals / Charles D. Scott, editor New York: Wiley, Biotechnology and bioengineering symposium No. 15 (1985)

[91] Two-stage, dilute sulfuric acid hydrolysis of wood: an investigation of fundamentals / John F. Harris, A.J. Baker, A.H. Conner, T.W. Jeffries, J.L. Minor, R.C. Pettersen, R.W. Scott, E.L. Springer, T.H. Wegner, J.I. Zerbe / FPL–GTR–45 (October 1985)

[92] The syringyl content of softwood lignin / John R. Obst, Lawrence L. Landucci. / Journal of wood chemistry and technology Vol. 6(3) (1986)

[93] Kinetic modeling of hardwood prehydrolysis. Part III, Water and dilute acetic acid prehydrolysis of southern red oak / Anthony H. Conner, Linda F. Lorenz / Wood and fiber science Vol. 18(2) (Apr. 1986)

[94] Naval stores research at the Forest Products Laboratory, past and present / Duane F. Zinkel / Presented at the 13th international naval stores meeting, New York, (Sept. 16, 1986)

[95] Cellulose: structure, modification, and degradation / R.A. Young, R.M. Rowell, eds. / John Wiley & Sons (1986)

[96] Naval stores: production, chemistry, utilization / Duane F. Zinkel, James Russell, editors. New York, NY: Pulp Chemicals Association (1989)

[97] Natural products of woody plants: chemicals extraneous to the lignocellulosic cell wall / John W. Rowe, ed. / Berlin; Springer–Verlag, 2 v. (xli, 1243 p.) (1989)

[98] Chemicals / Duane F. Zinkel, James F. Laundrie / FPL–GTR–67 (1991)

[99] Composition of American distilled tall oils / T.V. Magee, D.F. Zinkel / Journal of the American Oil Chemists Society Vol. 69(4) (April 1992)

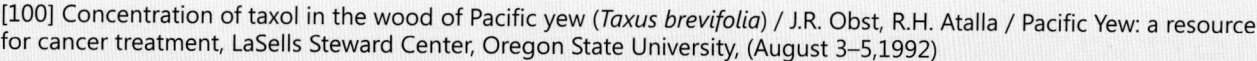
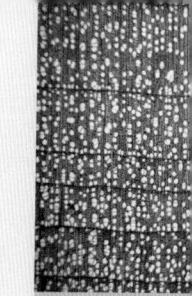

[100] Concentration of taxol in the wood of Pacific yew (*Taxus brevifolia*) / J.R. Obst, R.H. Atalla / Pacific Yew: a resource for cancer treatment, LaSells Steward Center, Oregon State University, (August 3–5,1992)

[101] Emerging technologies for materials and chemicals from biomass / R.M. Rowell, T. Schultz, R. Narayan, eds. / Symposium Series 476, American Chemical Society, Washington, D.C. (1992)

[102] Occurrence of taxol and related taxanes in the wood of Pacific yew / J.R. Obst / Yew (Taxus) conservation biology and interactions / S. Scher, B.S. Schwarzchild, eds. Proceedings international yew resources conference, U. of California, Berkeley p. 25, (1993)

[103] Isolation, separation, and characterization of organic acids / William V. Dashek, Jessie A. Micales / Methods in plant biochemistry and molecular biology. Chap. 9: 107–113, Boca Raton: CRC Press (1997)

[104] Chemical composition of fibers / James S. Han, Jeffrey S. Rowell, In: Rowell, R.M.; Young, R.A.; Rowell, J.K.; comp. ed. / Paper and composites from agro-based resources. Boca Raton, FL: CRC Lewis Publishers: 83–134. Chap. 5 (1997)

[105] A staining technique for evaluating the pore structure variations of microcrystallsine cellulose powders / Xiaochun Yu, Rajai H. Atalla / Powder technology Vol. 98: 135–138 (1998)

[106] 13C NMR characterization of guaiacyl, guaiacyl/syringyl and syringyl dehyrodgenation polymers / L.L. Landucci, S.A. Ralph, K.E. Hammel / Holzforschung Vol. 52:160–170 (1998)

[107] Fungal cellobiohydrolases and the degradation of crystalline cellulose / Tuula T. Teeri, Daniel Cullen, Christina Divne, Stuart Denman, Alwyn T. Jones, Anu Koivula, Markus Linder, JerrStåhlberg, Gerd Wohlfahrt, Ingemar von Ossowski, Jin-Yu. Zou / In: Genetics, biochemistry and ecology of cellulose degradation. Proceedings, Mie Bioforum 98, program and abstracts; September 7–11; Suzuka, Japan. Tse, Japan: Mie University: 25 (1998)

[108] Celluloses / Rajai H. Atalla / In: Barton, Sir Derek; Nakanishi, Koji; Meth-Cohn, Otto; eds. Comprehensive natural products chemistry. Pinto, B. Mario, Vol. ed. Vol. 3. Carbohydrates and their derivatives including tannins, cellulose, and related lignins. Oxford, UK: Elsevier Science Ltd: Chap. 16: 529–598 (1999)

[109] The individual structures of native celluloses: species specificity of secondary and tertiary structures and its implications with respect to processes of biogenesis / R.H. Atalla / Cellulose R & D, 6th Annual Meeting of the Cellulose Society of Japan, 1999: Vol. 1: 608–614. p. 9–12. (1999)

[110] Ethanol and thermo tolerance in the bioconversion of xylose by yeasts / T.W. Jeffries, Yong-Su Jin / Advances in applied microbiology 47: 221–268 (2000)

[111] Bacteria engineered for fuel ethanol production: current status / B.S. Dien, M.A. Cotta, T.W. Jeffries / Applied microbiology and biotechnology Vol. 63: 258–266. (2003)

[112] Characterization and reactions of a Salix extractive with a unique ring system / Lawrence Landucci, Sally Ralph, Kolby Hirth / 12th ISWPC international symposium on wood and pulping chemistry: Madison, Wisconsin, USA, June 9–12, 2003: Proceedings, Vol. II (2003)

[113] Enzymology and molecular biology of lignin degradation / D. Cullen, P.J. Kersten, In: R. Brambl and G.A. Marzluf, eds. The Mycota III. Biochemistry and Molecular Biology, 2nd Edition. Berlin–Heidelberg: Springer–Verlag: 13: 249–273 (2004)

[114] Top value added chemicals from biomass / T. Werpy, G. Petersen / Volume 1--Results of screening for potential candidates from sugars and synthesis gas / DOE/DOE_reports/35523/PNN–NREL–3552 (2004)

[115] Raman spectra of lignin model compounds / Umesh P. Agarwal, Richard S. Reiner, Ashok K. Pandey, Sally A. Ralph, Kolby C. Hirth, Rajai H. Atalla / In: Proceedings of the 59th APPITA Annual Conference and Exhibition incorporating the 13th ISWFPC (International Symposium on Wood, Fibre, and Pulping Chemistry) held in Auckland, New Zealand, May 16–19, 2005. Carlton, Victoria, Australia: APPITA: 1–8 (2005)

[116] Handbook of wood chemistry and wood composites / R.M. Rowell, ed. / CRC Press, New York, NY (2005)

Energy

◀ Electrical generating plant in Ashland, Wisconsin, that is partially fueled by wood. (2000s) {84G}

When the United States was settled, wood was a primary source of energy, especially for heating. Coal was later substituted because of its higher heat value in a ton of material; this was followed by oil, natural gas, and nuclear energy. With the continuing increase in world population and increasing standards of living, the need for increased energy is growing. New methods such as wind and solar power are growing, but there is a renewed interest in wood and other crop materials. The advantages of using wood for energy are that wood is renewable, and energy provides an outlet for low-quality wood that has no other market. Wood also is considered carbon neutral; that means the carbon dioxide that is released as it is burned is balanced by carbon dioxide that is used by growing trees. [9, 11, 12, 15, 18, 21, 22, 24, 25]

Challenge: How to meet the energy needs of the public and industry.

FPL's Contributions: FPL researchers increased the understanding of the use of wood for energy and production of ethanol, methanol, charcoal, and other chemicals from wood. [16]

Results: It is technically feasible to make transportation fuel energy from wood, but a major use of wood for energy is burning it and producing steam, which is converted to electrical energy and process heat used by industry. The primary producer of energy from wood is the pulp and paper industry. [18, 20, 22–24]

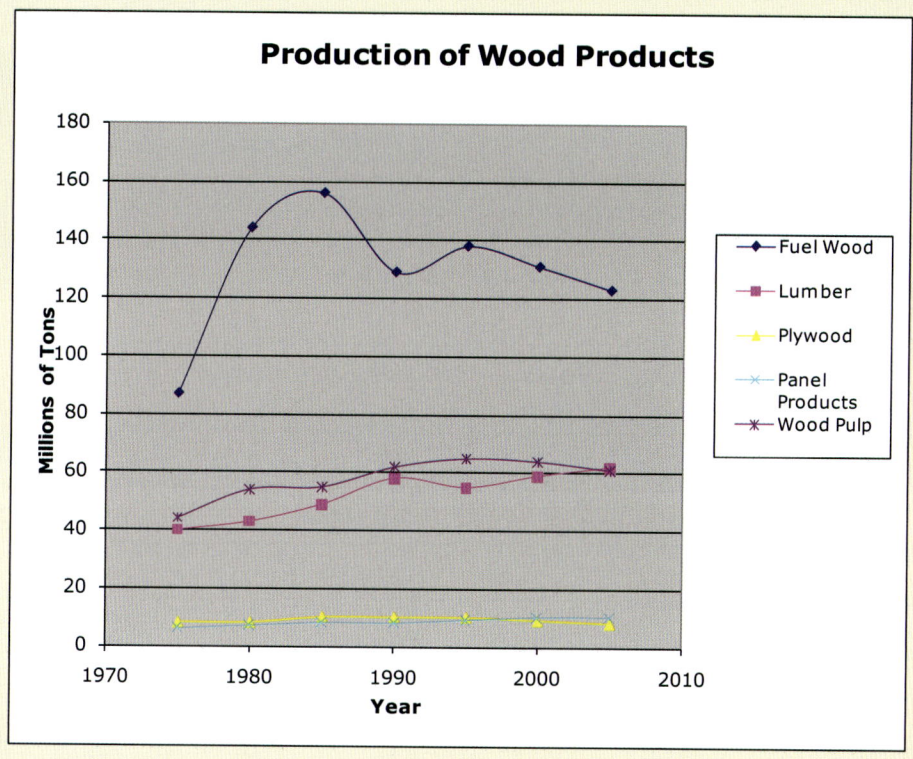

Production of Wood Products

Millions of Tons (y-axis: 0, 20, 40, 60, 80, 100, 120, 140, 160, 180)

Year (x-axis: 1970, 1980, 1990, 2000, 2010)

- Fuel Wood
- Lumber
- Plywood
- Panel Products
- Wood Pulp

◀ Tons of wood used to provide energy compared to that which is used to provide other wood products. The primary use of wood worldwide is for energy, primarily heating and cooking. [21, 28, 29]

Notes: Fuel wood is in ovendry tons and includes wood used by industry, residential, utilities, commercial, and pulp and paper, including black liquor. Other products are in air-dry (15% moisture) tons. Oriented strandboard is included in panel products since 1980.

Energy Efficiency in Light-Frame Construction

FPL researchers described aspects of design and landscape planning, engineering, and building code considerations as guides toward energy efficiency in light-frame wood construction. [13]

Wood-Gas-Generated Energy for Vehicles

◀ During WWII, FPL experimented with wood-gas-generated energy (gasogen) for powering vehicles. (1940s) {M 43359 F} [6]

Firewood

To help prevent the spread of diseases and insects, cut firewood should not be transported out of the local area or between states.

▶ Firewood cut and stacked ready for sale to the public. A big advantage of firewood is its availability. Burning wood not only provides heat, but for many, comfort and pleasure.

Ethanol

FPL has had a long history of research on liquid fuels such as ethanol. A brief time line of FPL efforts follows:

Brief History of FPL's Involvement with Ethyl Alcohol from Wood

1910s	FPL research on ethyl alcohol production from wood residues was started to provide an outlet for the vast amounts of wood residues at sawmills. Determined the yields of ethyl alcohol that could be obtained by dilute acid hydrolysis and fermentation; also the optimum cooking conditions for a single-cook process. This information helped in the successful operation of two commercial plants during WWI. [1]
1913	Commercial plant built in Georgetown, South Carolina (American single-stage, dilute acid, batch hydrolysis). (Yield: 20–22 gallons per ton of softwood.)
1914	Manufacture of ethyl alcohol from wood waste. [2]
1916	Commercial plant built in Fullerton, Louisianna (American single-stage, dilute acid, batch hydrolysis).
1917	Ethyl alcohol from wood waste: its possibilities as a motor fuel, and the process of manufacture [3]
1921	Manufacture of ethyl alcohol from wood waste [4]
1922	Kressmann conducted experiments at FPL on refining the single-stage, dilute acid, batch process and determined the alcohol yield from 24 wood species. [5]
1935	Cliffs Dow Chemical Co. of Marquette, Michigan, acquired rights to the Scholler process (dilute-acid percolation process) and built a pilot plant, but process was not commercialized. Pilot plant later moved to FPL. (Yield: 50–55 gallon per ton of wood.)
1940	Purpose of alcohol research was the need for ethyl alcohol in the synthetic rubber program.
1944	Installed the pilot plant from Cliffs Dow Chemical at FPL to study the Scholler process.
1944	Vulcan Copper & Supply Co. built a commercial size plant in Springfield, Oregon. Plant never went into production but one digester was run to confirm FPL pilot plant results. (Dilute acid, percolation process.)
1946	FPL modified the Scholler process to make the Madison wood-sugar process. (Madison process used continuous extraction of sugar solution.) (Yield: 54 to 68 gallons from softwoods and 35–48 gallons from hardwoods per ton.) [7]
1952	FPL studied the effects of high-energy cathode rays on cellulose. [8] TVA built a pilot plant at Wilson Dam to modify the Madison Process.
1970	Purpose of alcohol research was production of liquid fuel and chemicals as at least a partial replacement for those produced from petroleum.
1975	Katzen and Associates prepared a study on the economics of producing ethyl alcohol and other chemicals from wood waste. [10]
1979	Influence of fine grinding on the hydrolysis of cellulosic materials: acid vs. enzymatic was reported by Mike Millett, Marilyn Effland, and Dan Caulfield [14]
1980	Wood and the energy crisis [15]
1981	George Hajny summarized FPL's research in wood chemicals. [16]
1984	TVA built a pilot plant in Muscle Shoals, Alabama, for production of ethyl alcohol from wood and corn stover. (Concentrated acid, batch process.)
1985	A report on the two-stage, dilute sulfuric acid hydrolysis of wood was published, summarizing the fundamentals of the process steps in the production of ethanol. [20]
2007	Department of Energy selects six cellulosic ethanol plants for up to $385 million in federal funding. Department of Energy also established research centers in Tennessee, Wisconsin, and California to research new biofuel technologies with $125 million each over five years. [26, 27]
2008	J.Y. Zhu, X.J. Pan, G.S. Wang, and R. Gleisner developed a sulfite pretreatment to overcome recalcitrance of lignocellulose (SPORL) to convert softwood to its major components of cellulose, hemicellulose, and lignin. If successful in scale-up, this process helps remove a major barrier to converting wood to alcohol and other chemicals. [31]

Thermochemical Processing of Wood

Present research is on producing a clean gas that can then be used directly either in a turbine or internal combustion engine, with minimum gas clean-up, or in separation of gas components and reformulation of desired chemical products such as alcohol and diesel fuel. At an intermediate stage in thermochemical processing, flash pyrolysis may produce a liquid fuel that may be used as derived and refined or as an intermediate for further processing through gasification. [30]

◀ Mark Anderson (left), University of Wisconsin, and Mark Dietenberger (FPL) with the experimental flash pyrolysis equipment being used to evaluate the possibility of making liquid fuel from biomass. (2000s)

1910

Primary energy use of wood was burning it for heating.

2010

Primary energy use of wood is for heating and electrical generation. Research continues to economically make chemicals such as liquid fuels from biomass.

Further Information

[1] Preliminary report: manufacture of ethyl alcohol from wood waste / F.W. Kressman / Proj. 150 (May 15, 1911}

[2] Manufacture of ethyl alcohol from wood waste I, Preliminary experiments on the hydrolysis of white spruce / F.W. Kressmann / Journal of industrial and engineering chemistry Vol. 6(8) (Aug. 1914)

[3] Ethyl alcohol from wood waste: its possibilities as a motor fuel, and the process of manufacture / Rolf Thelen / The car owner (May 1917)

[4] The manufacture of ethyl alcohol from wood waste / E.C. Sherrard / Chemical age 29(2):76–79 (Feb. 1921)

[5] The manufacturing of ethyl alcohol from wood waste / F.W. Kressman / USDA Agricultural bulletin No. 983 (1922)

[6] Gasogens / R.H.P. Miller / FPL No. 1463 (1944)

[7] The Madison wood-sugar process / Elwin E. Harris, Edward Beglinger / Industrial and engineering chemistry, industrial edition Vol. 38: 890–895 (Sep. 1946}

[8] Effect of high-energy cathode rays on cellulose / Jerome F. Saeman, Merrill A. Millett, Elliott J. Lawton / Industrial and engineering chemistry Vol. 44(12) (Dec. 1952)

[9] Role of wood residue in the national energy picture / T.H. Ellis / Manuscript titled: Appraising wood residues for our nation's energy needs / Meeting held in Denver (Sep 3–5, 1975)

[10] Chemicals from wood waste / Allan E. Hokanson, V.B. Diebold, D.W. Bennett, R.P. Klier, S.A. Stein, W.W. Kline, D.C. Ferguson, R. Katzen / Forest Products Laboratory / Raphael Katzen Associates. (1975)

[11] Status and feasibility of utilizing forest residues for energy / John I. Zerbe / Presented at 8th World Forestry Congress, Jakarta, Indonesia, 16–28 Oct.1978. Special paper: forestry for quality of life, FQL/30–2 (1978)

[12] How to estimate recoverable heat energy in wood or bark fuels / Peter J. Ince / FPL–GTR–29 (1979)

[13] Energy efficiency in light-frame wood construction / Gerald E. Sherwood, Gunard E. Hans / Research paper FPL 317 (1979)

[14] Influence of fine grinding on the hydrolysis of cellulosic materials: acid vs. enzymatic / M.A. Millett, M.J. Effland, D.F. Caulfield / Hydrolysis of cellulose: mechanisms of enzymatic and acid catalysis / edited by Ross D. Brown Jr., Lubo Jarasek. Washington: American Chemical Society, Advances in chemistry series No. 181 (1979)

[15] Wood and the energy crisis / J.I. Zerbe / Forest farmer Vol. 39(5) (Mar 1980)

[16] Biological utilization of wood for production of chemicals and foodstuffs / George J. Hajny / Research paper FPL 385 (1981)

[17] Establishment of commercial-scale pilot plant for alcohol from biomass in developing countries / John I. Zerbe, Andrew J. Baker, Thomas W. Jeffries / FPL A 1402 (1981)

[18] Economic perspective on harvesting and physical constraints on utilizing small, dead lodgepole pine / Peter J. Ince / Forest products journal Vol. 32(11/12): 61–66. (Nov./Dec. 1982)

[19] Charcoal / A.J. Baker / Encyclopedia of American forest and conservation history Vol 1 / Richard C. Davis, ed. / New York: MacMillan (1983)

[20] Two-stage, dilute sulfuric acid hydrolysis of wood: An investigation of fundamentals. John F. Harris, Andrew J. Baker, Anthony H. Conner, Thomas W. Jeffries, James L. Minor, Roger C. Petterson, Ralph W. Scott, Edward L. Springer, Theodore H. Wegner, John I. Zerbe / General technical report FPL–45 (1985)

[21] Use of wood for energy in the United States—a threat or a challenge? / John W. Koning, Jr., Kenneth E. Skog / In: Energy from biomass and waste / Donald Klass / Institute of Gas Technology (1987)

[22] United States wood biomass for energy and chemicals: Possible changes in supply, end uses, and environmental impacts / Kenneth E. Skog, Howard N. Rosen / Forest products journal Vol. 47(2): 63–69. (1997)

[23] Use of wood energy for lumber drying and community heating in Southeast Alaskas / David L. Nicholls, John I. Zerbe, Richard D. Bergman, Peter M. Crimp / FPL–GTR–152 (2004)

[24] Fuel to burn: economics of converting forest thinnings to energy using Biomax in southern Oregon / E.M. (Ted) Bilek, K.E. Skog, J. Fried, G. Christensen / FPL–GTR–157 (2005)

[25] Thermal energy, electricity, and transportation fuels from wood / John I. Zerbe / Forest products journal Vol. 56(1) (January 2006)

[26] DOE selects six cellulosic ethanol plants for up to $385 million in federal funding / Craig Stevens / U.S. Department of Energy, Office of Public Affairs, (February 28, 2007)

[27] DOE announces bioenergy research center funding / Craig Stevens / U.S. Department of Energy, Office of Public Affairs (August 2, 2007)

[28] Energy Information Administration / Monthly Energy Review (September 2007)

[29] U.S. Timber Production, Trade, Consumption, and Price Statistics 1965 to 2005 / James L. Howard / FPL–RP–637 (2007)

[30] Vision of the U.S. biofuel future: a case for hydrogen-enriched biomass gasification / Mark A. Dietenberger, Mark Anderson / American Chemical Society, Published on web (Nov. 28, 2007)

[31] Sulfite pretreatment (SPORL) for robust enzymatic saccharification of spruce and red pine / J.Y. Zhu, X.J. Pan, G.S. Wang, R. Gleisner / Bioresourc. Technol.. (2008), doi:101016/j.biortech.2008.10.057.

Recycle/Reuse

◄ Self-sticking postage stamps. (1990s) {67 C}

One major advantage of wood-based products is that in many instances they can be used, reused, and recycled. A good example is the plain corrugated fiberboard box. First, it provides the consumer with a product, then it can be reused for storage, and then it can be recycled into another new box. Over 70% of corrugated boxes are recycled. However other products, such as self-sticking stamps, are more difficult to recycle. [16, 20, 25, 27, 29, 34]

Challenge: How can the recycling of wood products be increased to maximize the utilization of wood fiber, increase the timber supply, and decrease undesirable environmental impacts?

FPL's Contributions: Working with industry and the U.S. Postal Service, FPL developed methods to recycle self-sticking postage stamps. Working with the U.S. Bureau of Mines, City of Madison, Wisconsin, and industry, FPL developed a system for handling municipal solid waste. FPL evaluated the effects of recycling on paper products, developed cleaning systems for recycling paper, and developed panel products using recycled fibers. FPL scientists developed methods to dismantle buildings and reuse the wood.

Results: Reduction in the use of new wood (thus extending the fiber supply), reduction in landfill material, and improved wood products.

Paper Recycling

Recycling in the paper industry is not new. Paper that is off grade or is not sold often is shredded and fed back into the papermaking process

► Robert Kelley adding shredded printed paper into a hydropulper for recycling back into paper. (1970s) {84 P} [16]

◄ Bales of used corrugated fiberboard containers. Corrugated fiberboard is a particularly high grade paper, and containers are recovered for recycling. Many are reused and then recycled; some become long-term storage boxes and don't need to be recycled. Corrugated fiberboard is a premier paper product. It is made from a renewable resource, is light weight thus reducing shipping costs, and is low cost, reusable, and recyclable. (1970s) [9, 24, 32, 41]

▲ Bundles of old newspapers that will be further consolidated and either recycled in the United States or exported to other countries. (1970s) {81 A}

▲ Office waste was a challenge to recycle because of contaminates such as inks, toners, and adhesives. Here (left to right) are Dave Ketchum, Judy Gentry, and Mary Geske contributing to the in-house recycling program at FPL. With new methods of deinking, new adhesives, and new cleaning processes, more of this material is being recycled. (1980s) {18 H} [13]

◄ One of the most difficult recycling challenges is mixed household trash. (1970s) {14 B} [1]

Utilization of Urban Waste Fiber Resources

A 25-ton-per-day separation facility was constructed and operated by the City of Madison, Wisconsin, to recover paper and solid wood from municipal solid waste. Recovered materials studied included mixed wastepaper from household trash, discarded pallets and other dunnage, demolition wastes, and trees killed by Dutch elm disease. (1970s) [2, 4–6, 8, 12]

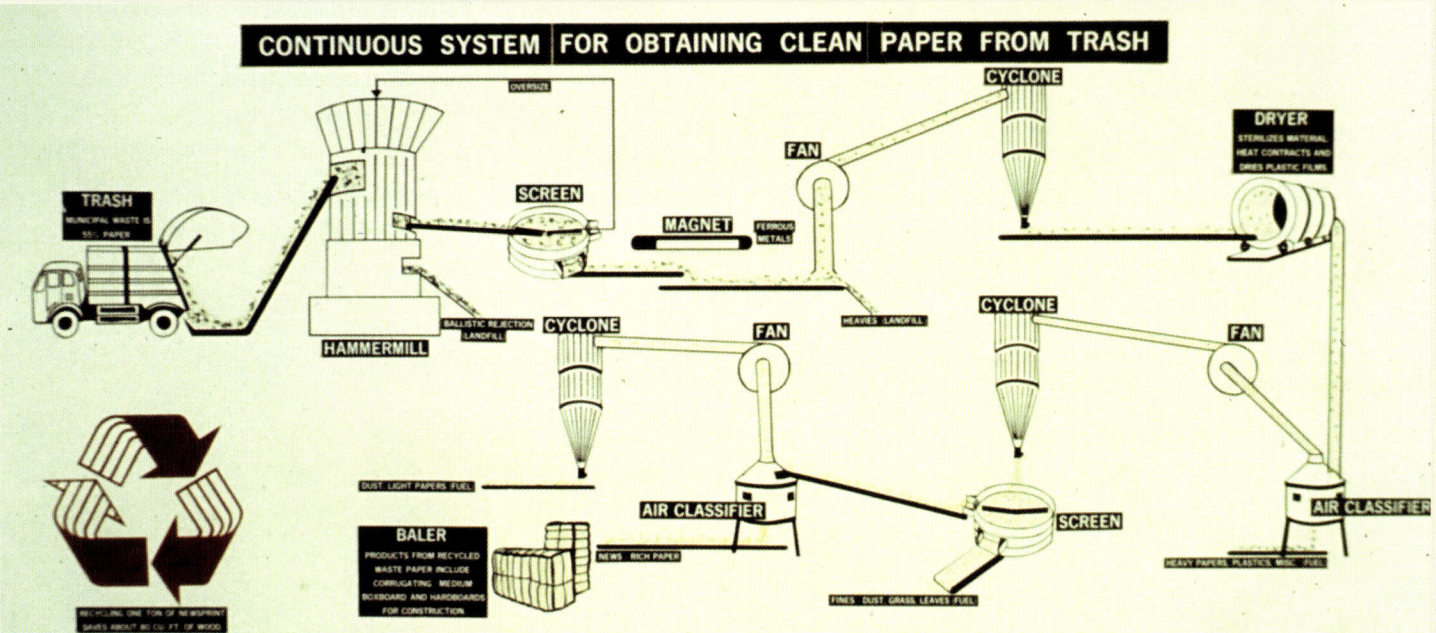

CONTINUOUS SYSTEM FOR OBTAINING CLEAN PAPER FROM TRASH

▲ FPL working with the U.S. Bureau of Mines, industry, and the City of Madison, Wisconsin, developed a successful process that was built and run by the City of Madison. (1970s)

▶ Shredded paper fraction being screened using the Madison process. Paul Steinmetz observing the screening operation. (1970s) {35 H}

Separation of Plastic from Recovered Paper in Recycling

Jim Laundrie developed two effective methods for separation of plastic from recovered paper in municipal solid waste: (1) utilizing the principle of thermoplastic films adhering to a heated surface, and (2) suspending mixtures of waste papers and thermoplastic films in a hot gas to contract the film to ball-like shapes. Both methods are practical, and the second method uses existing equipment. [4–6]

Removing Non-Wettable Contaminants from Wastepaper

◀ Spinning disk separator developed by FPL and the University of Wisconsin to separate contaminates from household trash fiber. (1980s) {129 L}

Working in cooperation with the University of Wisconsin, FPL invented a device, called a disk separator, that can remove non-wettable adhesive contaminants from wastepaper. The contaminants, called "stickies" by the papermakers, stick to the papermaking equipment or show up as undesirable spots in the recycled paper. The device is a smooth, metal disk with a wide lip that spins at high speeds. Because the pulp and the stickies have different degrees of wettability, they are separated as the disk spins. [10, 11]

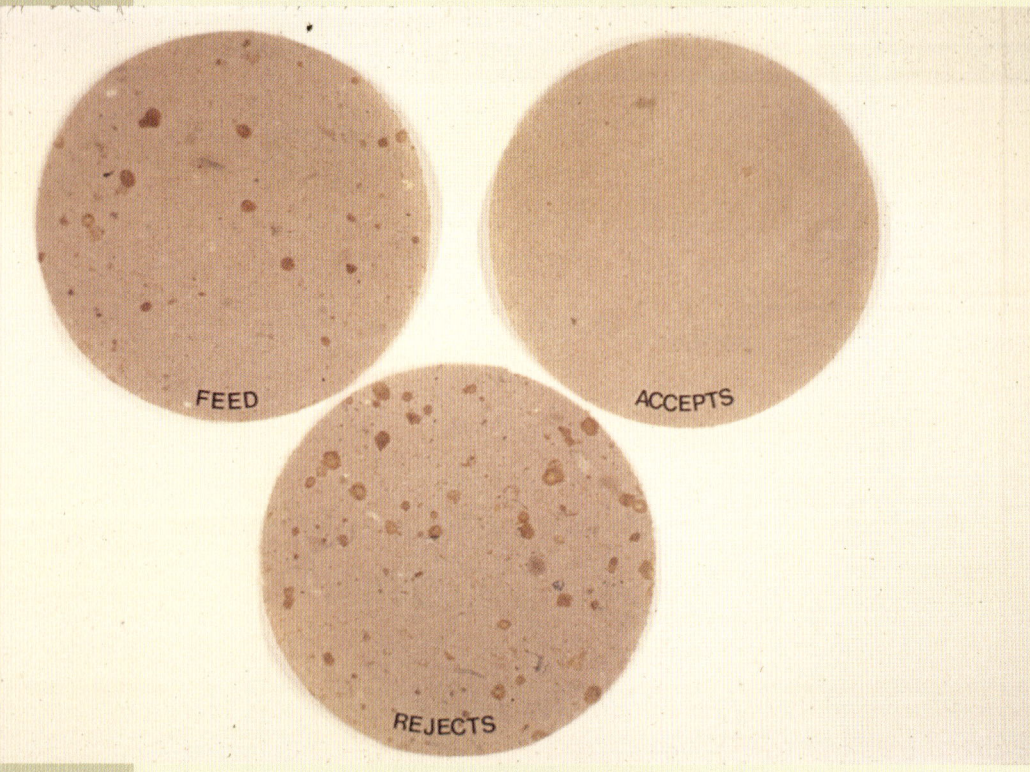

FEED

ACCEPTS

REJECTS

◄ Results of material going through the disk separator. The "feed" sample shows the concentration of mixed waste being fed to the disk. The "accepts" shows the effectiveness of the separation process and the "rejects" the concentration of the contaminants after passing over the disk separator. (1980s) {76 F}

Recycling Postage Stamp Adhesives

In the 1990s, Ted Wegner, Said Abubakr, Nancy Ross-Sutherland, Carl Houtman, and other FPL staff entered into an agreement with the U.S. Postal Service (USPS) to assist them in recycling stamps and other postal material. The FPL continues to work with the USPS to ensure that the extremely popular pressure-sensitive adhesive stamps pose no problems to recycling of postal materials. Working with companies using recycled fiber for papermaking, pressure-sensitive adhesive suppliers to the USPS, and recycling equipment suppliers, the FPL evaluated the recycling performance of current and new pressure-sensitive adhesive formulations, with a focus on making adhesive formulations that are compatible with papermill processing operations that use recycled paper. [21, 23, 27]

Recycling Postal Stamp Papers

Working with the U.S. Postal Service, the Laboratory evaluated postal stamp papers and ink and assisted the Postal Service in selecting materials that were easily removed in recycling and deinking processes. Subsequently this information was adopted as a government-wide standard that helps ensure that U.S. government agencies lead the way in using recycled paper and packaging. [18]

Recycling Postal Material from USPS Operations

Each year, the U.S. Postal Service must dispose of more than 440,000 tons of discarded lobby mail (DLM) and undeliverable bulk business mail (UBBM). Recycling this material typically was considered unfeasible because of the complex mixture of inks, dyes, adhesives, and coatings used on the papers. Working with USPS engineers and other industrial cooperators, FPL researchers overcame technical barriers to recycling DLM and UBBM. These waste papers are now being used by the paper industry to produce napkins, tissues, envelopes, and printing and writing papers, saving the Postal Service over $11,000,000 annually in disposal costs. [17]

Enzymatic Deinking

Paper from home and office recycling bins may end up as higher value products, thanks to FPL deinking research. These recycled papers usually contain mixed types and often are printed with inks and toners that are difficult to remove with conventional deinking chemicals, limiting their recyclability and potential market value. FPL research developed an enzymatic deinking method that is more effective and environmentally friendly than conventional deinking. The enzymatic process also improves subsequent fiber bleachability and helps remove residual contaminants, resulting in cleaner, brighter pulps. This process will permit increased recycling of mixed, low-quality wastepaper to higher value grades. [36, 37, 39, 40]

▶ Kathie Cropsey evaluating pulp that has been made from recovered office waste and exposed to deinking processes so it can be reused in paper. (1990s) {20 I} [15, 22, 26]

Corrugated Medium from Recycled Fiber

The semichemical pulping process, developed by FPL, successfully makes pulp for corrugating medium (the wavy fluting material in corrugated boxes). Today corrugating medium is one of the 12 major grades of paper and paperboard products. Currently, about 10 million tons of corrugating medium is produced in the United Stated, with about 36% coming from recycled fiber. However, corrugating medium produced from recycled fiber is lower in strength than that produced from virgin fiber and the resulting containers have lower strength. The FPL is working with an equipment manufacturer to improve the performance of corrugated fiberboard to permit greater amounts of corrugated medium produced from recycled fiber to be substituted for virgin fiber for use in corrugated shipping containers. [9]

Recycled Fibers in Panel Products

In recent years, with the advent of new adhesives, there has been increased interest in what types of composite panels can be made from various recoverable waste products. The following photos show some of the combinations. [35, 38]

◀ Hemlock fibers, recycled polyester, and adhesive. (2000s) {59 L}

◀ Hay straw, polyester, and adhesive. (2000s) {61 G}

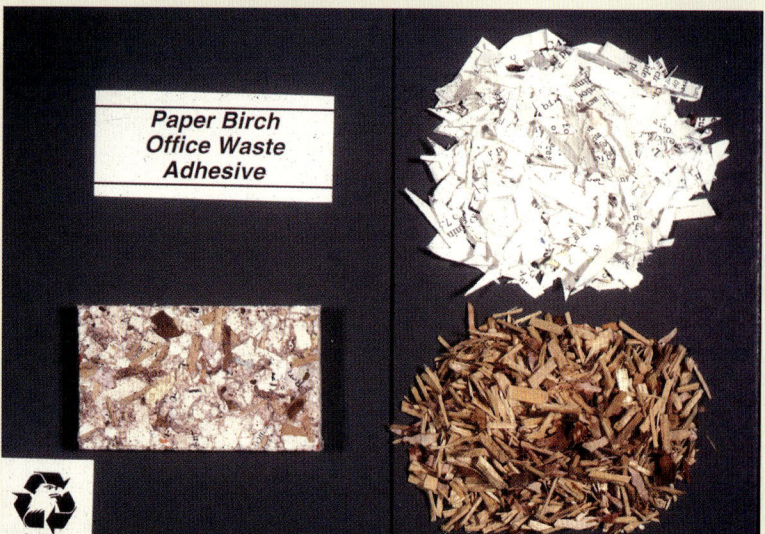

◀ Paper birch, office waste, and adhesive. (2000s) {59 M}

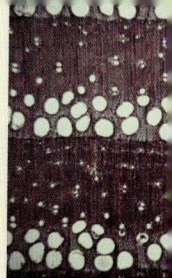

◀ White spruce, office waste, and adhesive. (2000s) {60 H}

◀ Kenaf fiber, office waste, and adhesive. (2000s) {59 Q}

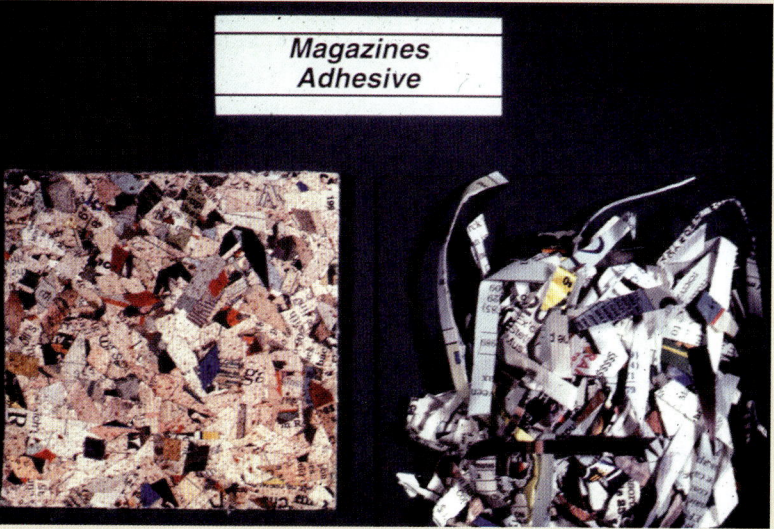

◀ Magazines and adhesives. (2000s) {61 N}

◀ Colored paper, envelopes, and adhesive. (2000s) {62 G}

Wood Fiber–Plastic Composite Blends

Working with industry, FPL scientists have developed new and improved wood fiber–plastics technologies that permit the manufacture of composite materials from blends of recovered wastepaper or waste wood and plastics. Technologies include compounding wastepaper and wood-waste-derived fiber–plastic blends at up to 90% fiber, devising ways to prevent thermal degradation of fiber in melt-blend operations, and evaluating and improving the material properties for the fiber–plastic composites. These technologies permit the recycling of large amounts of waste wood and wastepaper into value-added products. In addition, these waste wood and wastepaper fiber–plastic composite blends can in turn be recycled without serious loss in mechanical properties. [14, 35, 38]

◀ Recycled U.S. currency and polypropylene. (2000s) {61 K}

◀ Oak chips and fiberized waste plastics. (2000s) {59 E}

◀ Rice hulls and polypropylene. (2000s) {60 B}

Reuse of Lumber Reclaimed from Deconstruction of Buildings

Every year thousands of buildings are razed and the materials are sent to landfills. Recent research has been directed at methods to recover many of these building materials and recycle them. FPL has been involved with the challenges of recovering, refurbishing, and reusing wood products such as lumber from these buildings. [19, 28, 30, 31, 33, 42, 43]

◄ Unused munitions storage building at the U.S. Army's Badger Army Ammunitions Plant in Wisconsin. (1990s)

▶ Removal of the roof. (1990s)

◄ Top end of building removed. (1990s)

▼ Half the roof rafters removed. (1990s)

◄ Sides and end wall structure being dismantled. (1990s)

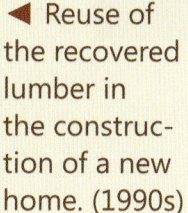

▶ Some of the lumber recovered from the munitions storage building. (1990s)

◄ Reuse of the recovered lumber in the construction of a new home. (1990s)

Remanufacturing of Wood Coated with Lead-Based Paint (LBP)

A significant barrier to salvaging wood materials found in buildings slated for demolition is lead-based paint (LBP). The lead metal found in the LBP that coats wood building materials poses a serious health hazard, and it is estimated that over 29 million pounds of lead go into our nation's landfills in the form of LBP each year from the demolition of single-family homes. Regulations and standards relative to LBP in buildings, LBP mitigation, and disposal of LBP-containing waste have been promulgated by many agencies. However, their applicability to recovering and reusing building materials is currently unclear. Guidelines for the remanufacture of wood coated with LBP are nonexistent.

In cooperation with the Department of Defense (DOD), FPL scientists have performed the first studies that demonstrate the feasibility of safely remanufacturing lumber coated with LBP. Results show not only that value-added wood products can be produced from the underlying wood, but also that remanufacturing can be performed safely while isolating the LBP from the environment. The U.S. Environmental Protection Agency (EPA) is currently incorporating the FPL technical data and regulatory and policy recommendations evolving out of this work to develop new regulations for LBP. [42]

▲ Siding used to study the handling of lead-based paint in the recovery of wood. (2000s)

◀ Bob Foss, Ted Smith, and Bob Falk (left to right) checking newly planed wood siding recovered from unbuilding the U.S. Army barracks building. Note the dust collection ducts in the background that isolate the planner dust and old lead-based paint for proper collection and disposal. (2000s)

▲ Recovered building siding before and after planing to remove the lead-based paint. (2000s)

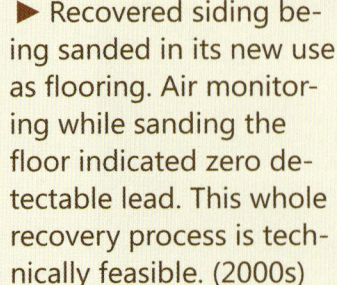

▶ Recovered siding being sanded in its new use as flooring. Air monitoring while sanding the floor indicated zero detectable lead. This whole recovery process is technically feasible. (2000s)

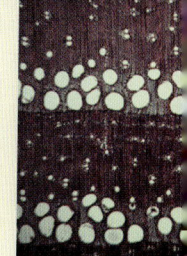

1910	**2010**
Minor concern in the United States.	Growing concern to extend the timber supply and decrease landfill disposal. Today, many commercial products are made with recycled and reused wood fibers.

Further Information

[1] What's in the wastepaper fiber collected from municipal trash / G.C. Myers / Paper trade journal Vol. 155(35) (Aug 30, 1971)

[2] FPL studies recovery of wood fiber from household refuse / G.C. Myers / Pulp and paper Vol. 46(11) (Oct 1972)

[3] Practical method for recycling wax-treated corrugated / A.A. Mohaupt, J.W. Koning, Jr. / Boxboard containers Vol. 79(6) (Jan 1972)

[4] Dry methods of separating plastic films from waste paper / J.F. Laundrie, J.H. Klungness / Paper trade journal Vol. 157(6) (Feb. 5, 1973)

[5] Effective dry methods of separating thermoplastic films from wastepapers / Research paper FPL 200 (1973)

[6] Separation of thermoplastic film and wastepaper /James F. Laundrie / United States Patent No. 3,814,240 (June 4, 1974)

[7] Hardboard: a potential outlet for waxed container waste / P.E. Steinmetz / Tappi Vol. 57(2) (Feb. 1974)

[8] Study indicates dry recovery of paper from shredded trash may be feasible / J.F. Laundrie / Paper trade journal Vol. 159(2) (Jan. 13, 1975)

[9] Repeated recycling of corrugated containers and its effect on strength properties / J.W. Koning, Jr. / Tappi Vol. 58(9) (Sep 1975)

[10] Disk separation: fiber recovery from recycled newsprint papermill tailings / John Klungness / TAPPI Pulping conference: Washington, D.C., Nov. 1–5, 1987: proceedings. Atlanta, Ga., TAPPI Press, (1987)

[11] Increased disk separator throughput /John H. Klungness, James W. Evans / Contaminant problems and strategies in wastepaper recycling seminar: 1989 April 24–26, Madison, WI: Atlanta, GA: TAPPI Press, TAPPI notes (1989)

[12] Materials interactions relevant to recycling of wood-based materials / R.M. Rowell, T.L. Laufenberg, J.K. Rowel / Materials Research Society symposium proceedings Vol. 266 (1992)

[13] Forest Service recycling research: new technology for wastepaper use / Thomas E. Hamilton, Theodore L. Laufenberg / Ride the wave of technology: Proceedings, TAPPI 1992 annual meeting, Atlanta, GA. Atlanta, GA: TAPPI Press (1992)

[14] Lignocellulosic–plastic composites from recycled materials / John Youngquist, George E. Myers, Teresa M. Harten / Emerging technologies for materials and chemicals from biomass. Washington, DC: American Chemical Society, ACS symposium series 476 (1992)

[15] Preliminary results of enzyme-enhanced versus conventional deinking of xerographic printed-paper / T.W. Jeffries, J.H. Klungness, M.S. Sykes, K.R. Rutledge-Cropsey / 1993 Recycling Symposium, New Orleans, Louisiana, February 28–March 4. Atlanta, GA: TAPPI Press Notes (Technical Association of the Pulp and Paper Industry) (1993)

[16] Expanding research horizons: USDA Forest Service initiative for developing recycled paper technology / T.L. Laufenberg, Said Abubakr / Proceeding of 3rd International symposium; 1994 September 13–14; Seattle, WA (1994)

[17] Deinking of business bulk mail and mixed papers by use of polymer scavengers / Yawen Chu, Mutombo Muvundamina, John Klungness / Progress in paper recycling. Vol. 5(1) (Nov. 1995)

[18] Recyclability of linerless PSA stamps / Said Abubakr, D.W. Bormett, G.M. Scott, N. Ross-Sutherland, A.A. Fatah, R. Kumar, T.V. Stagg, P.R. LaBrosse / 1995 Pulping Conference: Sheraton Chicago, Chicago, IL, October 1–5. Atlanta, GA: TAPPI Press (1995)

[19] Recycled lumber and timber / Robert H. Falk, D.W. Green, S.C. Lantz, M.R. Fix / Restructuring: America and beyond: proceedings of Structures Congress XIII, Boston, Massachusetts, April 2–5, 1995. New York, N.Y.: American Society of Civil Engineers (1995)

[20] A recursive linear programming analysis of the future of the pulp and paper industry in the United States: changes in supplies and demands, and the effects of recycling / Dali Zhang, Joseph Buongiorno, Peter J. Ince. / Annals of operations research Vol. 68 (1996)

[21] Repulping and cleaning of recovered paper: undeliverable and discarded mail / Said Abubakr, David W. Bormett, Marguerite S. Sykes, John Klungness, Nancy Ross Sutherland, Alim Fatah, Rajendra Kunar / In: Proceedings of the 1996 TAPPI pulping conference; October 27–31; Nashville, TN. Atlanta, GA: TAPPI Press: 410–417. Book 1. (1996)

[22] Deinking flexographic newsprint: using ultrafiltration to close the water loop / Bradley H. Upton, Gopal A. Krishnagopalan, Said Abubakr / Tappi J. 80(2): 155–164. (1996)

[23] Recycling evaluation of pressure sensitive adhesive using USPS–FPL protocol / Said Abubakr, David Bormett / Environmentally benign PSA for postal applications conference, July 7–9, 1997, Washington, D.C. (1997)

[24] Factors influencing the supply and demand of OCC / Howard James L. In: Abubakr, Said, ed. / Recycling. Chap. 6. Old Corrugated Containers. TAPPI Press anthology of published papers: 467–474 (1997)

[25] Wood recycling—opportunities for the wood waste resource / Bob Falk / Forest products journal Vol. 47(6) (1997)

[26] Enzymatic removal of stickie contaminants / Marguerite S. Sykes, John H. Klungness, Freya Tan, Said Abubakr / In: Proceedings of the 1997 TAPPI pulping conference; October 19–23; San Francisco, CA. Atlanta, GA: TAPPI Press: 687–691 (1997)

[27] Earth-friendly stamps on the roll. USDA news Vol. 56(8) (Oct./Nov. 1997)

[28] Feasibility of recycling timber from military industrial buildings / Scott F. Lantz, Robert H. Falk / In: Proceedings, The use of recycled wood and paper in building applications. Proceedings No. 7286. Madison, WI: Forest Products Society: 41–48 (1997)

[29] North American paper recycling situation and pulpwood market interactions / Peter J. Ince / In: USDA Forest Service, Forest Products Laboratory; Hillring, B., Dr., compilers. Recycling, energy, and market interactions. Proceedings, United Nations Economic Commission for Europe Timber Committee workshop; 1998 November 3–6; Istanbul, Turkey. Turkey: Ministry of Forestry: 61–72 (1998)

[30] Stress grading of recycled lumber and timber / Robert H. Falk, David Green / In: structures congress: structural engineering in the 21 century; 1999 April 18–21; New Orleans, LA. Reston, VA: American Society of Civil Engineers: 650–653 (1999)

[31] Evaluation of recycled timber members / Douglas R. Rammer / Materials and Construction: Exploring the Connection: Proceedings of the Fifth ASCE Materials Engineering Congress: May 10–12, 1999, Cincinnati, OH. Reston, VA: American Society of Civil Engineers 1999: 46–51 (1999)

[32] Strength and processing properties of wet-formed hardboards from recycled corrugated containers and commercial hardboard fibers / John F Hunt, Charles B. Vick / Forest products journal 49(5): 69–74 (1999)

[33] The properties of lumber and timber recycled from deconstructed buildings / Robert H. Falk / Proceedings of the pacific timber engineering conference, Rotorua, New Zealand (March 15–20, 1999)

[34] How woody residuals are recycled in the United States / David B. McKeever / BioCycle: journal of composting & recycling Vol. 40(12): 33–44 (Dec. 1999)

[35] Nonwoven and melt-blended composite panels from recycled plastics, wastepapers, and natural fibers / James H. Muehl, Andrezj Krzysik, John A. Youngquist / Forest Products Society 54th Annual Meeting: June 13–21, 2000, South Lake Tahoe, NV. Madison, WI: Forest Products Society (2000)

[36] Enzyme processes for pulp and paper: A review of recent developments / William R. Kenealy, Thomas W. Jeffries / In: Goodell, Barry; Nicholas, Darrel D.; Schultz, Tor P., eds. Wood deterioration and preservation: advances in our changing world. ACS symposium series 845. Proceedings, 221st national meeting of the American Chemical Society; 2001 April 1–April 5; San Diego, CA. (2003)

[37] Deinking selectivity (Z-factor): a new parameter to evaluate the performance of flotation deinking process / J.Y. Zhu, Freya Tan, Karen L. Scallon / 7th Research Forum on Recycling, September 27–29, 2004, Quebec City, QC, Canada, Montreal: PAPTAC, (2004)

[38] Composte panels made with biofiber or office wastepaper bonded with thermoplastic and/or thermosettng resin / J.H. Muehl, A.M. Krzysik, P. Chow / FPL–RN–0294 (2004)

[39] Drainage and fractionation of wood fibers in a flotation froth / J.Y. Zhu, Freya Tan / Progress in paper recycling Vol. 14(4) (Aug. 2005)

[40] On measurements of effective residual Ink concentration (ERIC) of deinked papers using Kubelka–Munk theory / D.W. Vahey, J.Y. Zhu, C.J Houtman / Progress in paper recycling Vol. 16(1) (Nov. 2006)

[41] Binderless fiberboard: comparison of fiber from recycled corrugated containers and refined small-diameter whole treetops / John F. Hunt, Karen Supa / Forest products journal. Vol. 56(7/8) (July/Aug. 2006)

[42] Investigation of Mechanical Processes for the Removal of Lead-Based Paint (LBP) From Wood Siding / R.H. Falk, J.J. Janowiak, R.G. Lampo, T.R. Napier, S. Cosper, S. Drodz, S. Larson, E. Smith / U.S. Army Corp Report ERDC/CERL 06–30, Construction Engineering Research Laboratory, September, 132 pg. (2006)

[43] Unbuilding: salvaging the architectural treasures of unwanted houses / Bob Falk, Brad Gay / Taunton Press, Newtown, CT (2007)

Brief History of Forest Products Utilization and Marketing Assistance

In recognition of the need to increase the use of FPL research by the wood industry and the general public who used wood, the Forest Utilization Service (FUS) was set up during World War II to facilitate the transfer of information. This area of focus is known as technology transfer, and it became quite an active area between 1946 and 1950.

The FUS units were staffed by subject-matter specialists located at the Forest Service Experimental Research Stations. These specialists would analyze the wood-using problems in their respective regions and determine the best approach to solve the problem. The outcomes could include presenting the problem to FPL researchers or to other qualified agencies for research, solving the problem based on existing research knowledge, or modifying existing practices to enhance efficiency and effectiveness.

In 1950, the Cooperative Forest Management Act established the Forest Service State and Private Forestry (S&PF) authorities to provide forest products utilization and marketing assistance and technology transfer. The mid-1960s saw the end of the Forest Service's Forest Utilization Service and the birth of the Forest Products Utilization (FPU) program. As authorized in the Cooperative Forest Management Act, the Chief of the Forest Service transferred the responsibility for technology transfer from Forest Service Research to State and Private Forestry. Thus, the Technology Marketing Unit was born to work with FPL, state utilization and marketing specialists, industry, universities, and the general public.

One of the more effective ways to transfer technology is through training workshops, which have been held at FPL and across the country. These workshops focused on subjects such as kiln drying, boxing and crating, sulphite cooking of wood pulp, sawing, and gluing of wood. The S&PF staffs would develop these workshops in partnership with FPL researchers.

Since the mid-1960s, S&PF has maintained liaison with researchers at the FPL. Eldon Estep was the first Utilization and Marketing person in this capacity. As a generalist, Eldon acted as an information broker between the Laboratory and people requesting information on virtually any subject related to wood and fiber products. Eldon would answer inquiries by either using the wealth of information he had stored in his files or going directly to the researcher. It was said that he maintained such an excellent filing system that he never had to research the same question twice.

In 1968, the Forest Service hired a contractor to evaluate the impact of the S&PF technology transfer program. The final report (published in 1973) found that on average, for every $1 invested in the FPU program, there was a return of $11 to the national economy within one year. Improved forest products conservation as a result of improved use of mill residues was also noted. In terms of rural economic development, communities benefitted with better economic stability and growth.

Around 1970, Paul Bois joined the S&PF unit at FPL as the National Lumber Drying Specialist. Paul had tremendous impact in improving drying technology in the forest products industry. He restarted the successful kiln drying short courses that are still used today.

In 1971, Stan Lundstrum joined the unit as the National Sawmill Specialist. Stan was charged with developing a national plan for improving sawmill conversions, based on FPL research by Hiram Hallock and Dave Lewis. The Sawmill Improvement Program (SIP) was created and helped improve the efficiency in 249 sawmills the first year.

This led to increasing demand for technologies that could improve sawmill conversion efficiencies. Thus, the Best Opening Face (BOF) system was developed. Researchers Hallock and Lewis developed a series of BOF algorithms (a mathematical sawing model) to maximize timber vol-

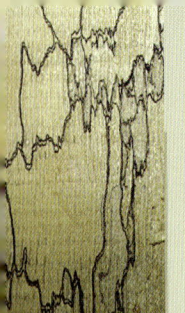

ume recovery from small logs. In the 1970s and 1980s, this FPL-developed technology aided the retooling and automation of the sawmill industry to successfully adapt to the shift from large old-growth trees to small-diameter second-growth timber supplies.

Success of BOF was accomplished by industry as equipment vendors started incorporating BOF algorithms with optical log scanners and computerized sawing systems to maximize recovery of softwood dimension lumber. Today, most softwood dimension lumber in the United States and around the world is manufactured using BOF-based sawing optimization technology pioneered at the FPL and brought to industry by the Sawmill Improvement Program.

In 1987, the Integrated Mill Production and Recovery Options for Value and Efficiency (IMPROVE) system was developed. These were more computer routines that were developed to further enhance recovery. Survey results showed that for those who used the routines, improvements in efficiency of 5% to 10% or more could be achieved in their operations.

In 1992, the Cooperative Forestry Utilization and Marketing Program changed focus to meet the needs of the Forest Service and rural communities hit hard by changing forest policies. The refocus concentrated on (1) providing technical and financial assistance to rural communities that relied on forests for their livelihood, (2) recycling and reusing wood, particularly in urban communities, and (3) creating opportunities for using smaller diameter trees and woody biomass that often just gets piled and burned in the forest.

Today, these emphasis areas are still relevant, although there is more focus on utilization of small-diameter trees, especially thinning material removed from hazardous fuel reduction activities. With increasing energy prices, more emphasis is being placed on using this waste woody material for heat, power, and biofuels. Susan LeVan-Green, Linda Richter, John (Rusty) Dramm, Mark Knaebe, John Zerbe, and Steve Yaddof continue to operate under the mission of helping to improve the conservation of our nation's resources by transferring appropriate technologies that extend the use of wood.

Technology Marketing Unit

◀ Structures like this were used as information kiosks at the 2002 Winter Olympic Games in Utah. They were built to demonstrate the use of small-diameter timber that had been removed from overstocked forests. Overstocked forests can lead to higher intensity wildfires.

The Technology Marketing Unit (TMU) is a part of the State and Private Forestry, Cooperative Forestry staff of the USDA Forest Service. The unit is located at the FPL and provides a broad scope of expertise in wood products utilization and marketing, technology transfer, and technical assistance.

The mission of the unit is to improve the use of wood by transferring technologies developed by the FPL, other Forest Service research installations, universities, and federal laboratories. The customers of the unit include citizens, local governments, rural communities, and forest industries, both large and small. The unit identifies opportunities and issues regarding wood products and helps develop solutions. The scope of the technology marketing work includes forest products conservation, processing, manufacturing efficiency, marketing, recycling, and wood for energy. A problem-solving approach with customers helps ensure sustainability of our natural resources.

The TMU accomplishes its mission through a number of activities, as summarized below.

Site visits: 62 site visits, technical assistance (2006)

Presentations: 60 presentations at workshops, conferences, seminars (2006)

Personal communications: More than 2,300 phone inquiries, letters, visitors (2006)

Publication distribution: More than 24,000 publications distributed (2006)

Grants: Provide funding in the form of competitive grants to assist communities and industries in the use of wood

Special reports: Prepare reports on specific subject areas such as utilization of small-diameter wood

- Primer on Wood Biomass for Energy, Richard Bergman and John Zerbe, revised January 2008
- 2006 National Forest Products Utilization and Marketing Personnel Directory, John Zerbe et al., 2006
- Small-Diameter Success Stories, I, II, III, Jean Livingston, May 2004, April 2006, April 2008
- Review of Log Sort Yards, John Dramm, October 2002

Technical notes: Prepare "Techlines" that focus on specific wood-related subjects

- Attractive, Durable Roofing Made From Recycled Plastic and Wood Fiber
- Biomass for Small-Scale Heat and Power
- Engineered Wood Fiber Surfaces Improve Accessibility for American with Disabilities
- Evaluating the Dead Yellow-Cedar Resource
- Fuel Treatment Evaluator 3.0
- Fuel Value Calculator
- Other Techlines can be accessed at www.fpl.fs.fed.us/document-lists/technlines.html

Special Initiatives: National Woody Biomass Utilization Program

One of the major challenges for forest managers is reducing the threat and ultimate damage that can be caused by wildfires, particularly those that occur where people have built in or adjacent to forested areas. This is often referred to as the wildland–urban interface. Because of a buildup of too many small, tightly spaced trees and underbrush, when a fire gets started it quickly can develop into a high-intensity raging inferno that spreads very fast, destroying forests, homes, and watersheds and costing taxpayers thousands or millions of dollars to extinguish. One method to help reduce this threat is to thin the forests of excess small trees and underbrush, thereby reducing the fire intensity.

▲ Wildland–urban interface project: White Mountain Stewardship contract before restoration. (2000s)

▲ Wildland–urban interface project: White Mountain Stewardship contract after restoration. (2000s)

This thinning process is very expensive and TMU works with communities and industries to utilize the removed material to help offset these costs, provide new products, and create local jobs. Following are a few examples of these successes.

Utilization of Small-Diameter Wood

▶ Sue LeVan-Green standing in the St. Marie's, Idaho, kiosk built using TMU's design. (2000s)

▲ Experimental system for producing electrical energy from wood. Chester Filipowicz, Richard Bergman, and Mark Knaebe discussing the infeed system. (2000s)

▼ Mark Knaebe and Chris Caldwell evaluating wood from harvested bundles of forest thinnings for their fuel value. (2000s)

◀ Getting research results out on the use of small-diameter wood. The registration booth with conference helpers (left to right) Adele Olstad, Eleanor Pape, unknown, Kathleen Walker, unknown, unknown, and Julie Lange.

▶ Interior of new Darby, Montana, library.

Friends of the Darby Library partnered with the TMU to help build a new 3,000-square-foot library that shows the use of small-diameter timbers. The timbers used were from fire-killed trees. TMU provided the funds and technical expertise for the architectural and engineering design for the building. Benefits of this project include demonstration of economical profits from forest fuel reduction activities, use of fire-killed material, expanded business opportunities for truss builders, and a beautiful new library for Darby's citizens. (2000s)

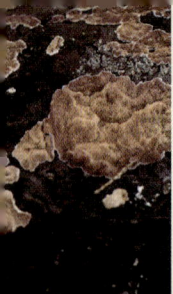

◄ The TMU provided financial and technical assistance to the town of Westcliffe, Colorado, to build a 32- by 64-foot park pavilion made from small-diameter roundwood that was removed from forest fuel reduction activities. This demonstration pavilion shows other communities one possible way of utilizing small-diameter wood. (2000s)

▲ Vehicular bridge cover manufactured from pre-assembled glued-laminated wood trusses. Small-diameter wood was used to make these glued-laminated trusses. (2000s)

▼ Larch flooring made from forest thinnings. (2000s)

◀ Mount Wachusett Community College's biomass power plant, Gardner, Massachusetts. (2000s) Wood residues from local sawmills are expected to meet the college's long-term supply needs. Using wood fuel for energy generation is another outlet for manufacturing wastes, forest thinnings, and other wood residues.

These are just a few examples of TMU activities. For details on these and many more, see www.fpl.fs.fed.us/tmu.

Further Information

Exploring the uses for small-diameter trees / Susan L. LeVan-Green, Jean Livingston / Forest products journal Vol. 51(9) (Sep. 2001)

Small-diameter success stories / Jean Livingston / FPL (2004)

Small-diameter success stories II / Jean Livingston / FPL–GTR–168 (2006)

Small-diameter success stories III / Jean Livingston / FPL–GTR–175) (2008)

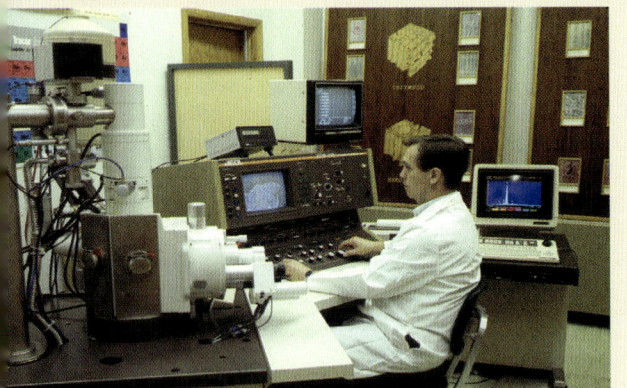 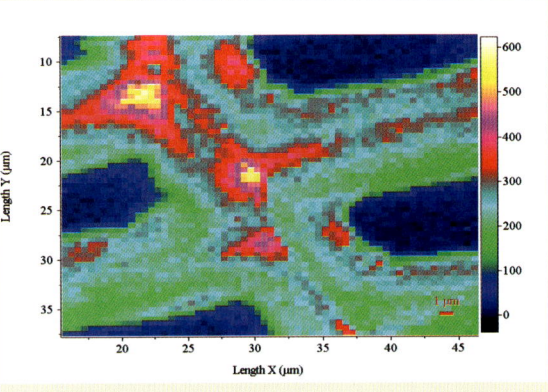 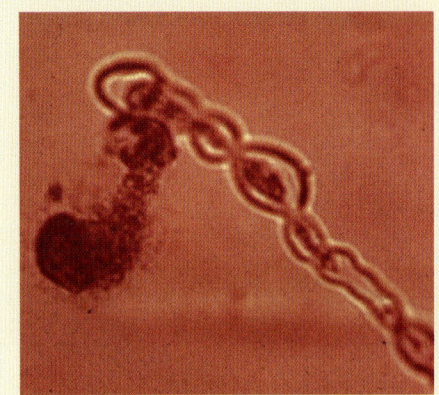

Summary and A New Century

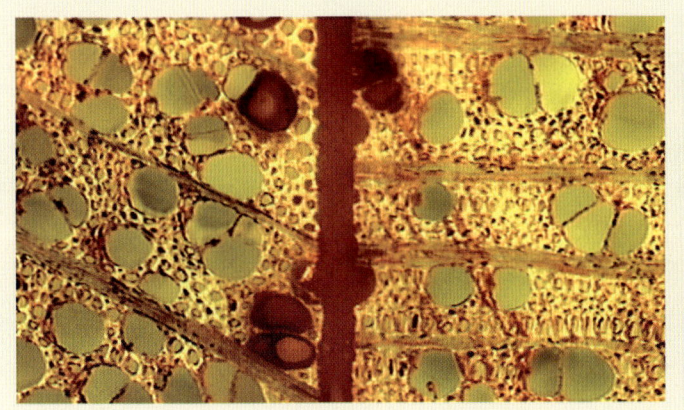  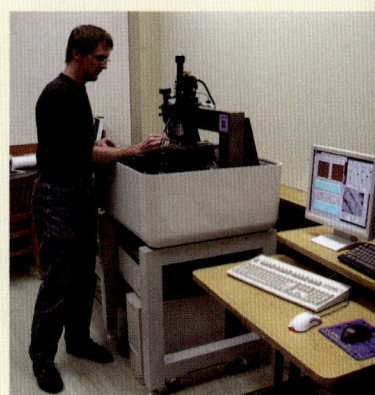

Summary

A brief summary of the material in this book covering the FPL's first 75 years was written by Robert J. Seidl, then President of Simpson Paper Company, San Francisco, California, and presented by him at the 75th Anniversary of the FPL. Though Robert worked at the Laboratory for a short time, he spent most of his time in private industry. Thus, this short summary of the FPL's accomplishments is from the point of view from someone in industry. [Quoted from "Progress through Wood Research: The Forest Products Laboratory Past, Present, and Future" / FPL 75th anniversary, general assembly presentations. (December 1985)]

From Robert Seidl's comments:

1. The first big mission was the orderly testing and identification of most of the world's wood and publication of findings in handbooks so that wood could be put to its best use. Nothing similar has been done before, and the Wood Handbook, first published in 1935, has long provided a basic reference.

2. Refinement of test methods for wood, in solid, fiber, or chemical form, which serve as a very important international language in the use of wood.

3. Studies of wood properties in relation to growth conditions as a guide to forest management, tree improvement, and genetic programs.

4. Fundamental studies of moisture movement in wood led to very useful charts showing the equilibrium moisture content reached by wood at various relative humidities. Forest products operations cannot ignore the fact that wood is always responding to its moisture environment. Observations on the need to remove water from wood gently to prevent checks, cracks, and warping led to the development of the internal fan dry kiln, a key early accomplishment.

5. Pioneering work in the preservation of wood against decay and fire, which was immensely important to its usefulness. Think of what this has meant in the building of railroads, bridges, and in general construction. The Laboratory is a center for mycology research, which deals with the collection, identification, and description of wood-rotting fungi.

6. A newer area is that of biotechnology, which involves the manipulation of microbes and enzymes to produce useful bioreactions. This promises to open entirely new vistas.

7. Fundamental work on paints and surface coatings for wood to extend it life and enhance its beauty. Beyond surface treatments, the Laboratory found methods to penetrate wood with resins, yielding durable products know as "Impreg" and "Compreg." Such products were used in floors, airplane propellers, aircraft carrier decking, and are often seen today in kitchen knife handles.

8. Understanding the complex chemistry of wood has been an ongoing effort, serving at various times in history the need for naval stores, chemical intermediates, extractives, explosives, rayon, flavorings, alcohols, fuel, and many other items. Beyond the yield of useful products, knowledge of wood in its molecular form is very helpful for other uses, such as for converting wood to pulp.

9. Studied the mechanism of bonding wood with adhesives and test methods for the products. These defended the integrity of the bond and made great contributions to exterior plywood, laminated beams, and more recently to strandboards.

10. Detailed studies of structural use of lumber, plywood, and laminated beams, which extended their use and guided code decisions. I believe the first large laminated beam in the United States was produced here.

11. Design criteria were developed for beams and columns, including allowable stresses, use of fasteners, span tables for various species and for construction that best resist earthquakes. In recent years this has naturally involved computer-aided models of structural behavior.

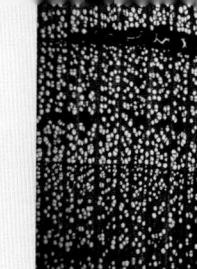

12. Methods were refined for testing the strength of lumber without breaking the piece under test, which is important for wood since it is variable in nature. This permits wood to be used in a true engineering manner.

13. Development of the first stress-skin glued prefabricated house, helping to pioneer the factory-built house. A much later development (1978 prototype) was the truss-framed system, which is a unitized element of roof, floor, and studs that saves time and lumber.

14. Methods of processing logs to get maximum volume of wood products were examined. You can imagine the complexity of obtaining the maximum yield of rectangular boards from logs that are round, tapered, and irregular. Other examples included the use of a power back-up roll in peeling logs for veneer, or in thick slicing of logs and gluing the slices into a lumber product. This eliminates sawdust and gives more uniform strength.

15. Early theory and recommendations on the use of vapor barriers in the walls of houses and how their abuse can cause decay and paint failures.

16. Theoretical mathematics and practical applications for stress-skin sandwich panels. These studies were sponsored by the military, but they opened vistas for a very efficient use of the resource.

17. Basic work on wood and fiber products in packaging, including necessary test methods. A very large percentage of all wood used in the form of lumber, plywood, and paper is dedicated to packaging.

18. Development of high-strength paper plastics, perhaps five times as strong as anything developed before, again done for the military, and used in glider floors, aircraft parts, and other products.

19. Overlays for plywood, both medium and high-density types, that emerged as a spin-off from the military structural plastics effort. This provided an interesting synergism on which I wish there were time to elaborate. I had the fun of helping this one into commercial use.

20. Semichemical pulping of hardwoods was a major development at Madison and led to an immense outlet for so-called "weed species," resulting in millions of tons of pulp production. The Laboratory helped to start the second commercial plant.

21. Hardwoods in corrugating board were stimulated by the mathematics of stress-skin panels, showing that stiffness rather than toughness of the corrugating medium was central to a good box. A small concept perhaps, but one with a vast commercial impact.

22. There was a general program on use of residues and otherwise unused species for pulp, including an ongoing search for higher yields and methods for bleaching such products. Many laboratories study wood, of course, but I know of no other that had such a determined and sustained focus on finding practical use for what many considered to be waste.

23. Pioneering work on pulping and especially bleaching of southern pine for packaging, printing papers, and newsprint, which helped to boost this gigantic industry.

24. Pulping of Douglas-fir, so important to the West, was intensively studied. I believe the Laboratory published about 20 technical reports on pulping this species before the industry began.

25. Press drying permitted short-fibered hardwood paperboards to have superior bonding and performance with less process energy needed. This study stems from fundamental understanding both of fiber bonding and engineering properties of the products. It is a creative idea, relatively new, and its commercial use is still ahead.

26. Pilot-plant studies, which are necessary to scale-up laboratory discoveries and also to demonstrate to the business community what might be achieved, are a continuing Laboratory activity. Such trials expose problems in large-scale operations and build conviction for commercial action. These have been a catalyst to a number of large commercial plants that I could identify.

27. Recycling of fiber has been the subject of other more recent studies. There is elemental logic in the assumption that reuse of paper reduces the need for new fibers. This enables us to tap the big city wastes, or the "asphalt forest," as someone has described this resource. It is also worth noting for the future that 1 ton of wastepaper upon combustion yields heat equivalent to about two barrels of oil.

28. As a central reference point for woods of the world, a reservoir known as the "Madison Collection," that now houses over 100,000 specimens. This is probably the largest in the world. Identification and retrieval is assisted by computer.

29. Some Laboratory studies are not for economic benefits. Such tasks as aiding law enforcement by analyzing wood found at crime sites, identifying the wood species and age of ribs from a long sunken ship, or helping to identify the wood in a family heirloom are undertaken as a public service.

Apart from specific items, one might condense Laboratory activities into the identification and understanding of wood in a macro and molecular sense, the relating of its properties to forestry conditions, the protection of wood, the bonding of wood in solid or subdivided form, and development of processes that increase its yield or usefulness in fiber or chemical form. The program of the Laboratory is often said to be justified as support of our big national forests, but in fact, the results of FPL research reach everyone that uses wood. Nature does not recognize ownership borders.

Though Robert Seidl covered FPL's accomplishment for the first 75 years, following are some of the major ones covering the past 25 years.

1. The FPL has continued to identify and evaluate the world's woods and updated the Wood Handbook in 2010.

2. Summarized years of acquired research on the chemical properties of wood and included them in the Handbook of Wood Chemistry and Wood Composites.

3. Continued to refine and develop new test methods and improve national and international test standards to provide meaningful and unbiased evaluations of wood in solid, fiber or chemical form.

4. Updated the Dry Kiln Operator's Manual and developed a solar kiln for wood drying.

5. Completed long-term evaluation of marine pilings, preserved using a dual treatment system, and found that the dual treatments were effective against a number of marine borers.

6. Biochemistry research has led to identification and understanding of the lignin-degrading enzymes that cause wood to decay, and fungal DNA studies provide information for new control approaches.

7. Fundamental work on the paint–wood surface interface continues with a growing understanding of the importance of the negative effects of exposure to sunlight of the wood, prior to painting, on paint performance.

8. A renewed interest in chemicals from wood has led to the concept of expanding chemical recovery from ongoing pulp mill operations.

9. Studies of wood surface–adhesive interactions on bonding properties of wood have resulted in the development of a wood coupling agent that increases the adhesive bond strength.

10. Use of computer modeling has allowed for a major change in the way wood connectors are used that has resulted in more efficient building design.

11. Non-destructive evaluation of wood has been expanded into automatic machine grading of lumber, evaluation of decay in wood structures and evaluation of logs and even standing trees.

12. Commercial development of structural insulated panels for building construction is one of the outcomes of the pioneering research done by the FPL on stress skin panels.

13. Development of new tests to evaluate paperboard and corrugated fiberboard containers (90%+ of all goods are shipped in corrugated fiberboard containers) as engineering materials and structures led to a major change in specifying corrugated in terms of performance.

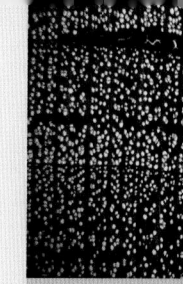

14. New pulping and bleaching methods that reduce energy consumption, improve paper strength properties, and eliminate the use of elemental chlorine were developed.

15. Recycling of wood fibers has continued with major accomplishment in enzyme deinking of printed paper and more efficient cleaning techniques. This has led to improved use of recycled fibers.

16. Identification of wood species for the American public and research scientists continues. The development of a wood anatomy web site to facilitate information transfer is successfully operating.

17. Modifying the properties of wood and fiberboard through acetylation has resulted in treated wood that is more dimensionally stable and resistant to decay.

18. New methods of removing the metals in preservative treated wood have allowed reuse of the fiber or safer disposal in landfills.

19. Developed techniques to evaluate the Nation's wood bridges and updated the Bridge Design Manual.

20. Evaluated carbon storage in wood products as part of the 2006 Guidelines for Greenhouse Gas Inventories of the United Nations Intergovernmental Panel on Climate Change. Kenneth Skog, one of the contributing scientists to this report, was one of the team of scientists that shared, with Vice-President Al Gore, the 2007 Nobel Peace Prize.

21. Evaluated small-diameter trees that are removed as part of the effort to reduce the severity of wildfires. Continued searching for uses for these thinnings that can offset the high cost to the American public for the removal and disposal of these trees.

22. Developed an environmentally friendly treatment to control termites.

23. Developed design information for wood poles to improve their safe use.

24. Evaluated various wood composites and structural designs for severe weather safe rooms for homes.

25. Evaluated formaldehyde emissions associated with the production of composite panels and how wood preservatives affect the environment where they are used.

26. Developed wood fiber–inorganic composites that are used along highways to control noise.

27. Developed heating schedules to eliminate invasive insects, such as the Asian longhorn beetle, from being shipped around the world in the wood used in pallets and crates.

28. Helped modified lumber standards based on machine grading.

29. Developed a whole family of new molded wood fiber products using recycled fibers.

30. Developed data and models required for fire safety engineering of forest products in a performance-based building code environment.

31. Successfully initiated the Shiitake mushroom industry in the United States.

32. Developed methods of successfully using wood from invasive species.

33. Evaluated the use of various wood species in wood fiber–plastic composite lumber.

34. Continued development of new method to utilized wood that in the past has been sent to landfills.

35. Developed a new soy-flour-based adhesive for use in bonding wood.

36. Initiated new research in nanotechnology.

It should be emphasized that as impressive as these accomplishments are, they really are the result of significant cooperative input from university, industry, associations, governments, and other professional people, and the continuing financial support of the American public. All contributors should take great pride in what was jointly accomplished.

It is hoped that you enjoyed reading about our shared accomplishments.

A New Century

Considering the next 100 years in wood and its uses, it is helpful to review our goals from the past 100 years and provide a look at our current plans for the future.

1. *Develop basic knowledge of wood and its use.*

In the future, a further understanding of wood fiber and its structure at the nanoscale and molecular level could lead to new ways of modifying wood growth characteristics to improve present forest products and create new uses for wood. Such understanding can be enhanced using research tools such as electron, Raman, and atomic force microscopy.

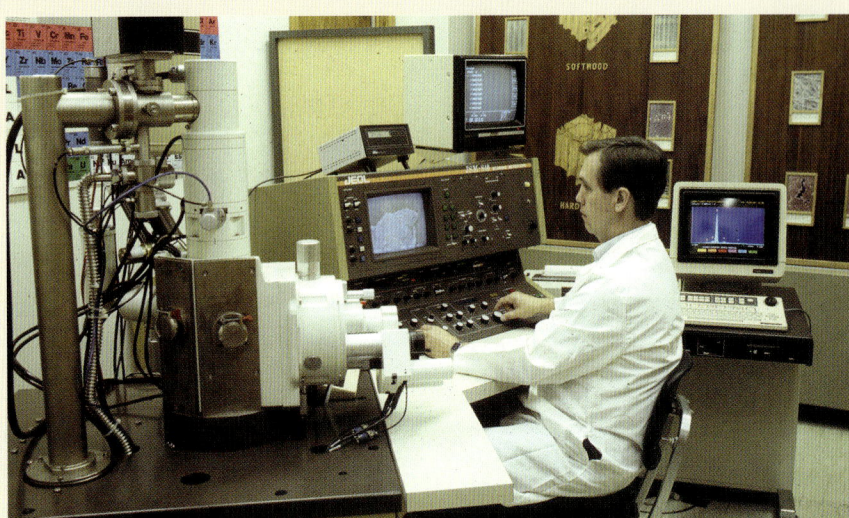

◄ Thomas Kuster exploring the structure of wood using an electron microscope. A whole different world.

► Raman microprobe analysis showing lignin distribution in a wood sample. Dark blue areas are the fiber lumen, red areas are the heavily lignified compound middle lamela.

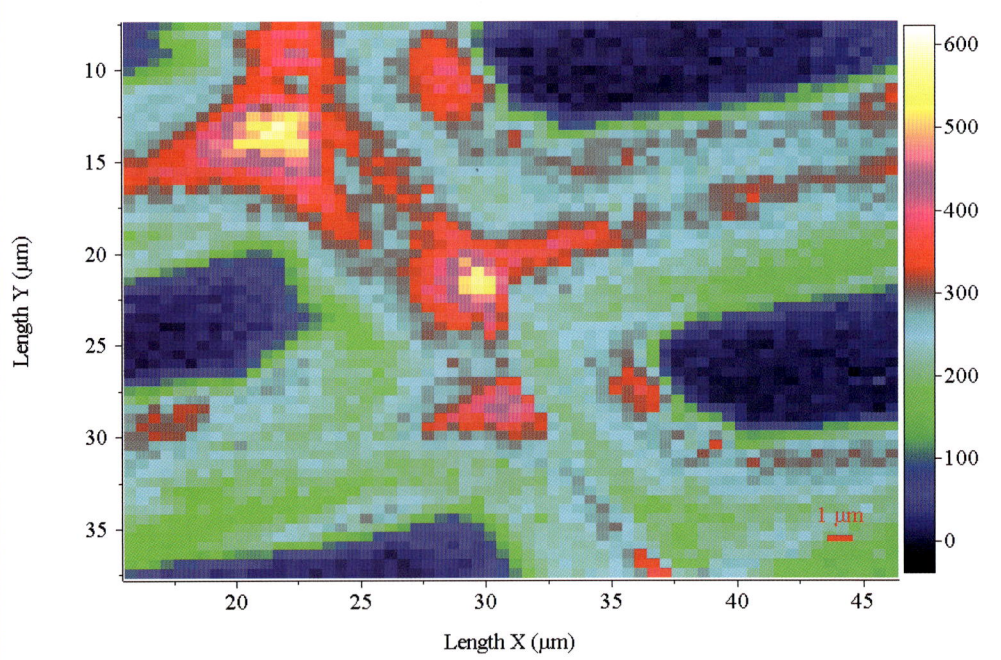

2. *Improve the processing of wood to increase yield and quality, and reduce costs and waste.*

Though it is possible to convert wood to pulp for paper, the processes are energy inefficient, require high capital costs, and some produce undesirable environmental impacts. The concept of creating a biorefinery, in which wood is processed in such a manner that it is converted into an adjustable array of usable chemicals, fuels, and fiber, is possible. Demonstration of this concept to determine its economic feasibility is necessary. The use of biological pretreatment for modifying wood is in its infancy. Improving the quality of wood chips going into new processes is essential. New methods to recycle used fiber, restore bonding strength, and upgrade recovered fiber from single-stream municipal collections need to be developed.

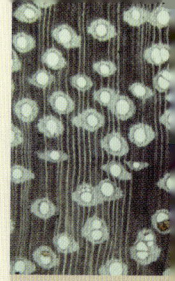

▼ Hypothetical process chart of multiple integrated biomass technologies that when used selectively and progressively may optimize economic, environmental, and social value for lignocellulosic biomass. [Integrated biomass technologies: a future vision for optimally using wood and biomass / J.E. Winandy, A.W. Rudie, R.S. Williams, T.H. Wegner / Forest products journal Vol. 58(6) (2008)]

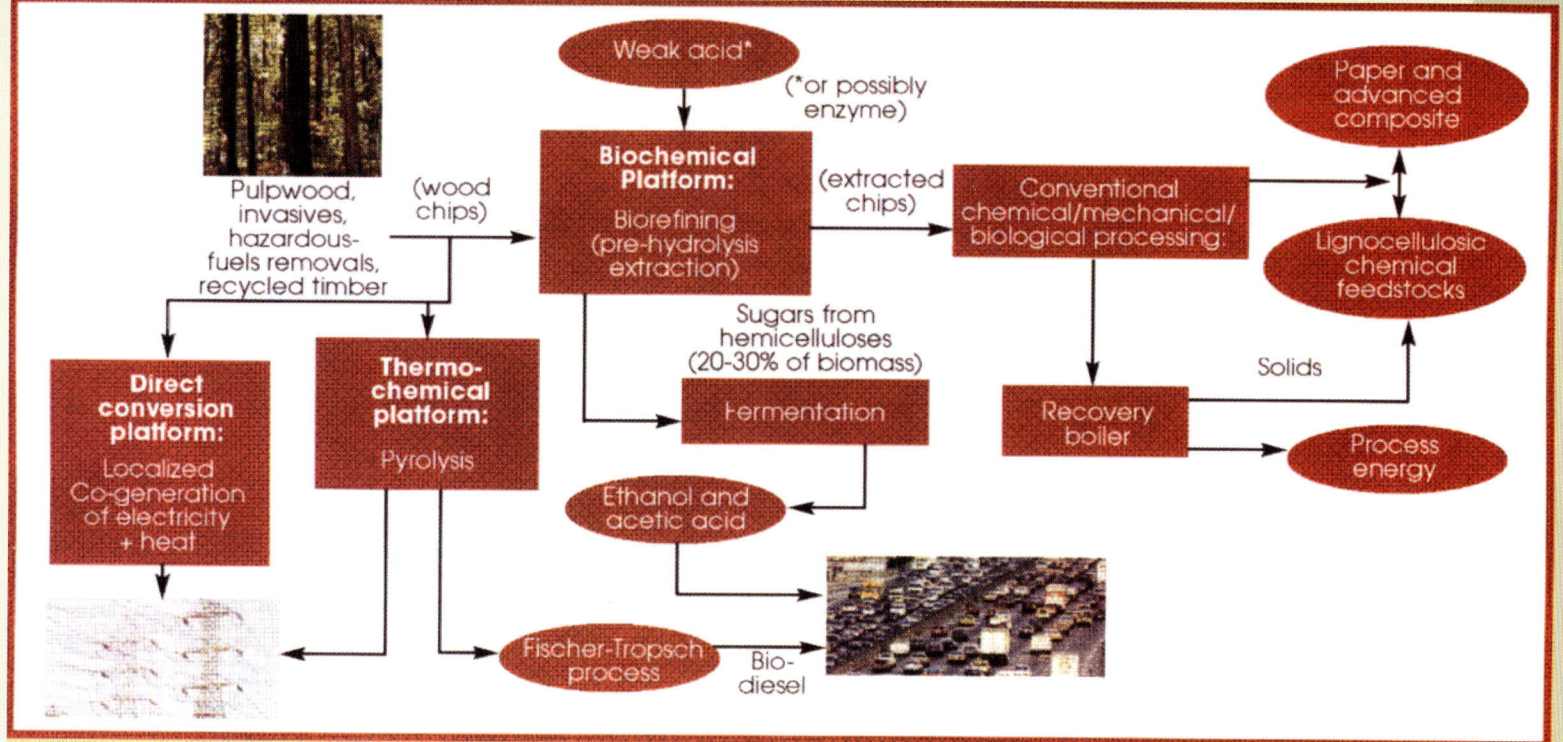

Sustainable biomass utilization may involve a series of sequential integrated processing approaches that may include an initial biorefinery stage to obtain biofuels, followed by production of bio-based products from the biorefinery residues, and then production of electrical bioenergy from bio-based residues. Such an integrated resource solution will also offer the optimum long-term solution to meeting both user needs and sustainable development.

3. *Extend the service life of wood products.*

We now know how to extend the service life of many wood products, but many of the treatments use undesirable broad-spectrum toxicants. A breakthrough in treatments that only kill or inhibit targeted wood-degrading organisms is needed. Making wood fiber less sensitive to moisture would also greatly extend its service life and increase its possible end uses.

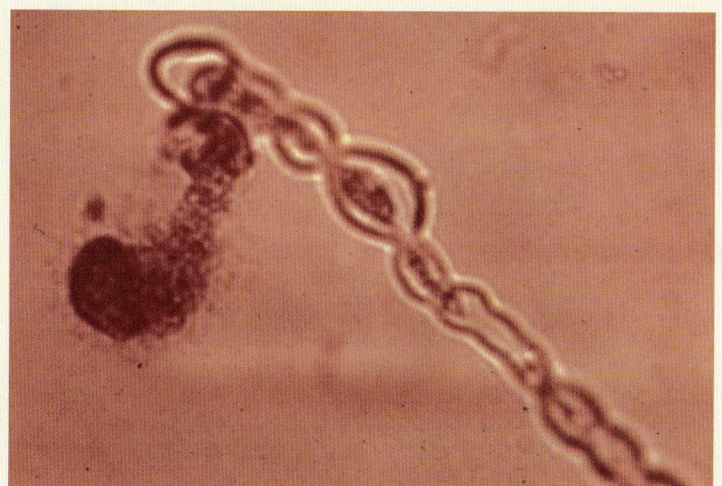

◄ Investigation of methods to inhibit chitin synthesis could lead to a much safer method of preventing decay, and perhaps termite and marine borer damage as well. The fungal hypha shown is beaded, rather than of uniform diameter, because synthesis of chitin was inhibited. It ruptured near the tip, causing cell contents to spill out and growth to stop.

4. *Utilize the various wood species and changing quality of wood.*

Removal and utilization of millions of acres of suppressed overgrown trees on the National Forests is essential to having healthy forests and decreasing the damage caused by wildfires. One of the keys to this utilization is in composite products, and they are dependent on adequate adhesives. Additionally, genetically modified commercial species grown on plantations or a short-rotation woody crop will be increasingly utilized.

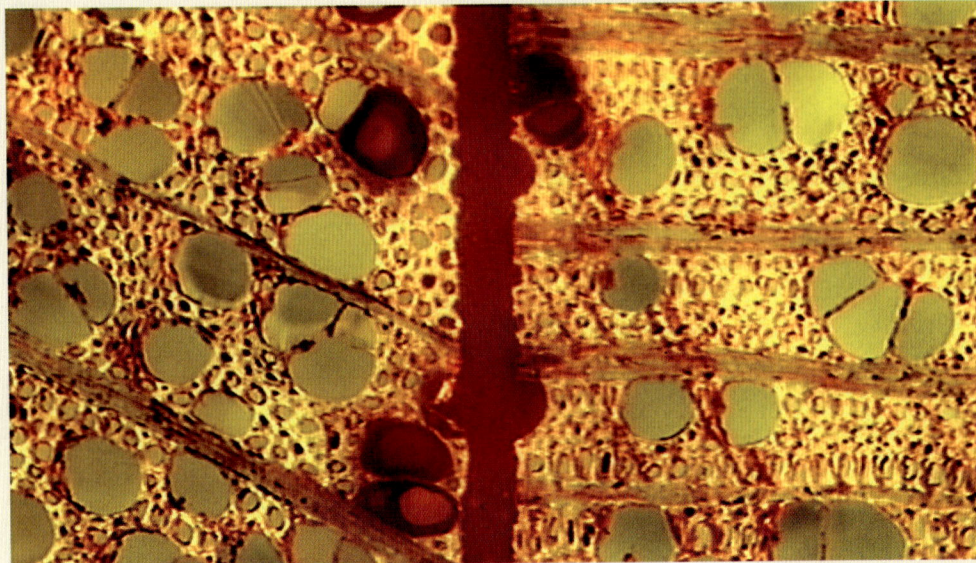

◄ Photomicrograph of an adhesive bond of two pieces of wood. The solid red areas are the adhesive, showing limited migration (wetting) into the wood structure.

5. *Improve the quality and usefulness of wood products.*

The application of nanotechnology to wood-based products and its potential to enhance quality and usefulness is in its infancy, with the potential of new products and jobs. One such product could be a paperboard composite that is waterproof that can be used to build low-cost shelters. Others could include biomedical uses.

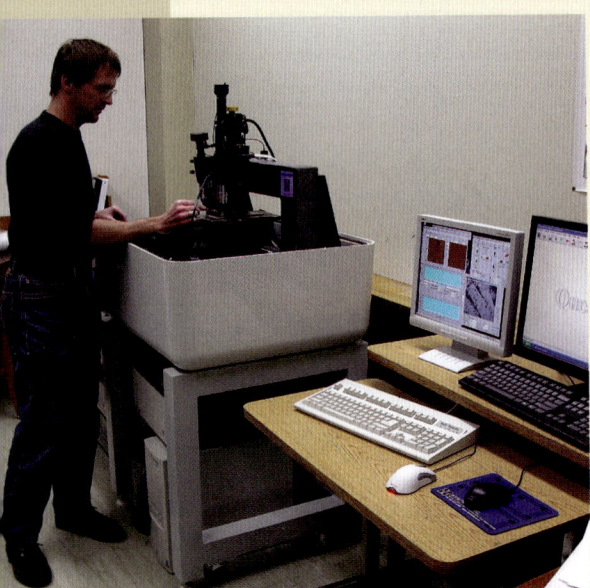

▲ Robert Moon examining a wood sample using a nanoindenter and atomic force microscope in order to study the mechanical properties and structure of cell walls.

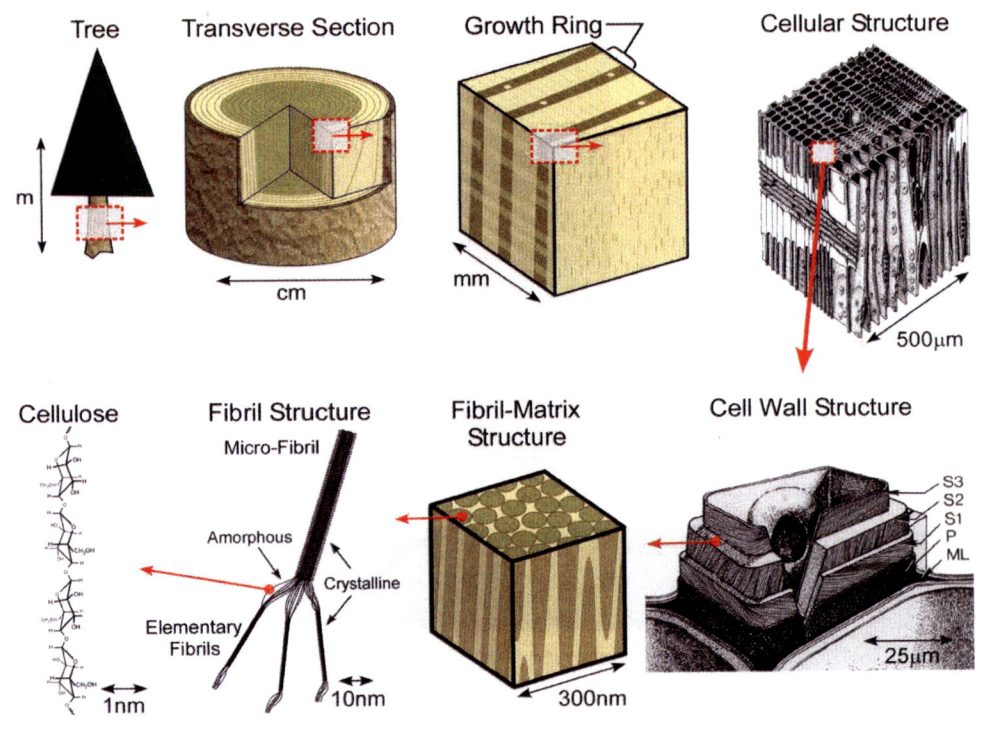

▲ The use of nanotechnology to better understand the cellular structure of wood.

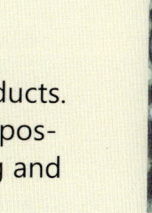

6. *Develop new products from wood.*

With the changing forest resource, new ways of utilizing the fiber will require new products. Engineered biocomposites are one of the opportunities to accomplish this. These composites can simultaneously meet the nation's diverse needs for high-performance building and commodity products while maximizing the sustainability of forest resources.

7. *Develop basic knowledge of the biological aspects of wood.*

The biochemistry of wood and its structure is just beginning to yield new insights into the complex nature of wood and potential for new chemicals.

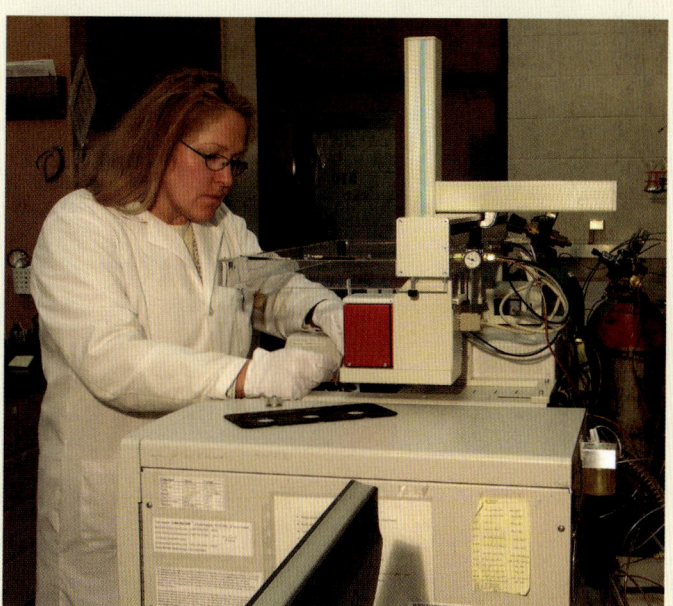

◀ Kolby Hirth maintains the gas chromatograph–ion trap mass spectrometer and uses the instrument to determine trace levels of degraded components of chlorinated lignin.

8. *Transfer the research developed both nationally and internationally.*

Though significant progress in technology transfer has been made in the past 100 years, getting needed technical information into the hands of users remains a challenge. With the advent of the internet, new methods of providing our research information to the national and international communities is now possible. New studies are needed to show what and where information can be the most helpful.

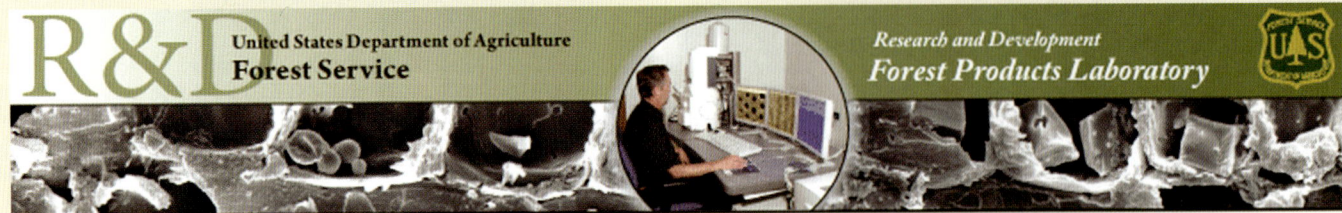

R&D United States Department of Agriculture **Forest Service** — Research and Development *Forest Products Laboratory*

Search

Search within FPL

Browse by Subject

Home

Research - Units, Facilities & Centers [+]

Information Products & Services [+]

Grants & Funding

People & Places [+]

News & Events [+]

Partnerships [+]

FAQs

FPL Library

About Forest Products Laboratory [+]

Contact Us

Contact Information

Forest Products Laboratory
One Gifford Pinchot Drive
Madison, WI 53726
Phone: (608) 231-9200
Fax: (608) 231-9592
Email: **mailroom_forest_ products_ laboratory@fs.fed.us**

You are here: FPL Home

Forest Products Laboratory

Our mission: to promote healthy forests and forest-based economies through the efficient, sustainable use of our wood resources.

News

FPL Centennial

100 Years of Service

Forest Products Laboratory
1910 – 2010

100 Years of Public Service...view»

Centennial Research Facility

The Centennial Research Facility (CRF), the newest addition to the FPL campus, is currently

Global Change

Climate, Biomass & Bioenergy

Areas of Focus

Advanced Composites

Advanced Structures

Upcoming Conferences & Events

Small Wood 2010 - Bridges, Business and Biomass Conference...view»

Grand Opening and Dedication of Centennial Research Facility...view»

Popular Items

2010 Woody Biomass Utilization and Grants...View»

Chapter 4 of the Wood Handbook - Wood as an Engineering Material ...View»

Paints and Finishes ...View»

Cone Calorimeter tests results - evaluation of the flammability of untreated and treated wood products ...view»

Conclusion

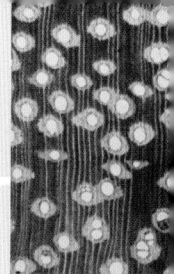

The American public has invested approximately $1,500,000,000 over 100 years, or an average of $15,000,000 per year (present dollars) in the FPL. In return the FPL has helped develop and sustain America's forests and a timber-based industry that according to the Annual Survey of Manufacturers employed 1,290,000 people and shipped products with a total value of $319,042,000,000 in 2006. (2006 Bureau of Labor Statistics)

This investment has returned significant benefits:

Helped develop test methods, codes, and standards that have resulted in safer buildings and homes and improved packaging, thus helping to protect consumers.

Contributed to improving the environment through safer wood preservative treatments, and increased reuse and recycling of wood and wood fiber.

Helped develop wood composites such as flakeboard and oriented strandboard, which have significantly reduced the waste disposal of wood residues that in the past were burned.

Reduced the demand for wood on America's forests from a growing population and consumption by significantly extending the useful life of wood in use.

Helped change the wood and paper industry from an art form to one that is based on scientific principles.

Finally, provided economic outlets for wood that in the past was not utilized, thus helping to cover the costs of forest management. This has helped ensure a sustainable supply of wood for the future and has helped improve the health and condition of America's forest lands.

We hope you will agree that FPL has made good use of your tax dollars to help protect and sustain your forests. Yet much remains to be done. With new tools such as nanotechnology and breakthroughs in chemical analysis we hope that you will continue or, if possible, increase your support.

"When the Forest Products Laboratory was started in 1910, it stood alone in the world. Today there are more than 90 research organizations that conduct research on forest products and related subjects, located in many countries. Details of the research organizations located in Australia, Austria, Canada, China, Finland, France, Germany, Ireland, Indonesia, Japan, Latvia, Malaysia, Netherlands, New Zealand, Norway, Philippines, Poland, Slovak, Republic South Africa, South Africa, Sweden, Switzerland, Taiwan, and United Kingdom were discussed in a recent Forest Products Journal article, in terms of their organization, governance, and measures of performance." [Forest Products Research and Development Organizations / P.V. Ellefson, M.A. Kilgore, K.E. Skog, C.D. Risbrudt / Forest products journal Vol. 57(10) (2007)]

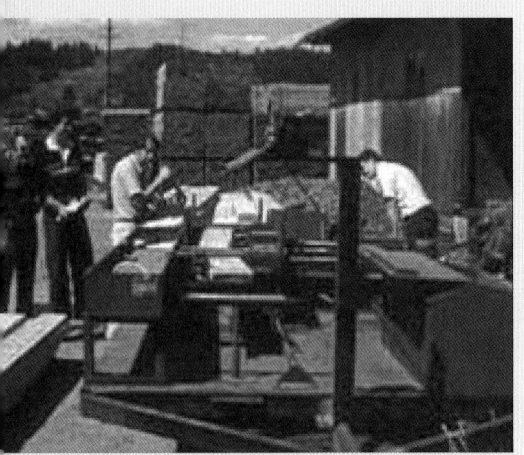

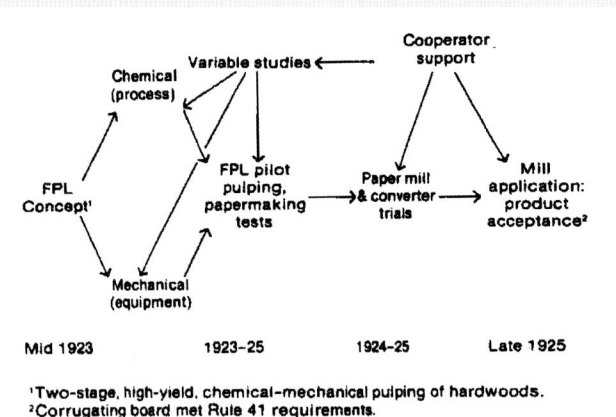

Chemical
(process)

Variable studies ←

Cooperator
support

FPL
Concept¹

FPL pilot
pulping,
papermaking
tests

Paper mill
& converter →
trials

Mill
application:
product
acceptance²

Mechanical
(equipment)

Mid 1923   1923–25   1924–25   Late 1925

¹Two-stage, high-yield, chemical–mechanical pulping of hardwoods.
²Corrugating board met Rule 41 requirements.

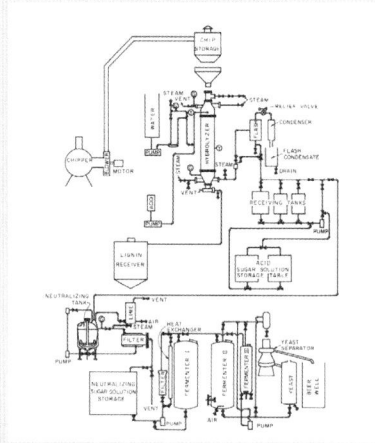

Related Articles

- **ASTM Committee D–7: Wood**
  Promoting Safety and Standardization for 100 Years

- **Development and Mill Transfer of the FPL Semichemical Pulping Process**

- **Biological Utilization of Wood for Production of Chemicals and Foodstuffs**

ASTM COMMITTEE D-7: WOOD

Promoting Safety and Standardization for 100 Years

By David W. Green, Robert L. Ethington,
Edward G. King, Bradley E. Shelley, and David S. Gromala
(2004)

In October 2004, Committee D-7 on Wood of the American Society for Testing and Materials (ASTM) is celebrating 100 years of contributions to the safe and efficient use of wood as a building material. Born during a period of rapid social, economic, and technological change, the Committee faced controversial issues and the challenge of a changing forest resource. This article highlights the technical and economic challenges we have faced over the years, discusses some of the controversial decisions, and speculates on future challenges.

The United States is the leading producer and consumer of wood products in the world. The majority of this wood resource is used to construct the approximately 1 million new single-family homes built each year, or to repair and remodel existing homes. This demand for construction material has led to better utilization of wood, the development of improved grading practices, and improved engineered wood products. Solid-sawn lumber continues to provide the bulk of structural lumber products used in construction. Mechanical grading procedures provide precise control of property assignments, but they have introduced an entire new set of "grades" to the marketplace. In addition, engineered wood products such as laminated veneer lumber and parallel strand lumber are being substituted for solid-sawn lumber. Large-diameter timbers are

increasingly difficult to obtain, and products from them are often replaced by prefabricated wood I-joists and glued laminated (glulam) beams. Plywood once replaced solid-sawn wood for sheathing material, and oriented strandboard has now largely replaced plywood. Compared to 100 years ago, the use of forest resources in the United States includes a more diverse array of species and trees with smaller diameters.

All of these 20th century changes have been a challenge to the building industry and have the potential to cause confusion and distrust in consumers. But similar challenges have been faced before. Wood was a useful material long before Europeans set foot in North America. Three centuries later, commerce in wood products had developed, in both sophistication and geographic reach, to a degree where standardization promised better consumer satisfaction and buyer/seller understanding. Shelley (1992) notes that standards are developed in response to one or more of at least five factors: 1) expansion of marketing areas; 2) advances in technology, both within and without the wood products sector; 3) new code or regulatory requirements; 4) competition from other products; and 5) changes in consumer demand.

Since the early 1900s, a system of codes and standards has been developed for lumber and wood

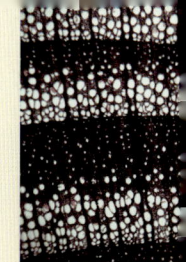

Officer	Term of Service	Length of Service (yr.)
Chair		
Hermann von Schrenk	1905 to 1948	44
L.J. Markwardt	1948 to 1964	16
Lyman W. Wood	1964 to 1966	2
W.A. Oliver	1966 to 1972	6
Wayne Lewis	1972 to 1974	2
F. Alan Tayelor	1974 to 1980	6
William L. Galligan	1980 to 1981	1
Al DeBonis	1981 to 1988	7
Ed Diekmann	1988 to 1992	4
Erwin Schaffer	1992 to 1994	2
Frank Beall	1994 to 1998	4
Robert Leichti	1998 to present	
Secretary		
William K. Hatt	1906 to 1915	10
John L. Newlin	1915 to 1943	28
L.J. Markwardt	1943 to 1948	5
L.W. Smith	1948 to 1958	10
W.A. Oliver	1958 to 1965	7
John Shope	1965 to 1972	7
L.L. Rappleyea	1972 to 1974	2
R.A. Hewett	1974 to 1979	5
Don Percival	1979 to 1984	5
R.E. Catchople	1984 to 1988	4
Chris K.A. Stieda	1988 to 1992	4
Donald Onysko	1992 to 1999	7
David Kretschmann	1999 to present	

Table 1. — Officers of ASTM Committee D-7: Wood (complete records could not be found for other offices).

products to assure quality and reliability (Green and Hernandez 2000). Committee D-7 on Wood has played a vital role in this system.

The Beginning: 1905 to 1929

The development and expansion of the railroads in the middle to late 1800s spurred interest in standardization with respect to wood utilization. Prior to 1895, grading procedures, lumber testing procedures, and methods for calculation of allowable properties were proprietary procedures developed by consulting engineers, university professors, and a few government scientists active in the field of timber engineering. The expansion of the railroads brought a need to standardize these procedures for the design of railway structures.

In 1895, a committee of the American International Association of Railway Superintendents of Bridges and Buildings reported on the strength of bridge and trestle timbers (Berg et al. 1907). The report presented 15 recommendations for timber engineering, factors used for converting breaking strengths from laboratory tests to allowable properties, and average safe design stresses for the principal species used for bridge and trestle timbers. In response to increased marketing opportunities provided by rail transportation, the lumber industry had begun to develop standardized sizes of lumber surfaced on two sides, primarily to limit shipping costs.

Organized in 1898, ASTM provided an opportunity for further standardization. At the 1903 annual meeting, W.K. Hatt provided a "preliminary report" on the timber test program of the Bureau of Forestry. At the 1904 meeting, the Executive Committee adopted a motion by Hatt for the formation of a committee on specifications for the grading of structural timber.

Committee Q

In the spring of 1905, a group of 12 men met to organize ASTM Committee Q on Standard Specifications for the Grading of Structural Lumber, the 7th technical committee to be formed (Fulweiler 1955). A report on the activities of Committee Q was presented by Herman von Schrenk at the 8th annual ASTM meeting in Atlantic City, New Jersey. In the report, von Schrenk listed himself as "Chairman pro tem," a post he held for 44 years (Table 1). The report proposed the following:

It is believed that the time has come for a comprehensive study and analysis of the grading of structural timbers, so as to arrive at a general understanding of the qualities of the various woods used for structural purposes, in order to standardize as far as possible, for the use of lumber manufacturers on the one hand, and architects and engineers on the other hand, the various grades and qualities of wood.

The subject was to be treated under three general headings:

1. Definition of structural timbers: define structural timber and indicate various uses for different species.

2. Standardization of trade names: establish a definitive list of common names of timbers used in this country.

In September of 1907 the first standard established by Committee Q, D 9-07,[1] the Standard Specification for Structural Timbers, was approved by a vote of the ASTM membership. The standard contained a definition of structural timber, definitions of standard defects, a list of standard names for structural timbers, and specifications for bridge and trestle timbers. Other topics proved more challenging and were to take several years to resolve. Among the problems to be addressed were the identification of longleaf and shortleaf southern pine, the effect of density and growth rings per inch on strength, and the variation in trade and botanical names.

Committee D-7

The year 1910 brought a name change for Committee Q to Committee D-7. A Standard Specification for the Grading of Yellow-Pine Bridge and Trestle Timbers was approved, and work on similar standards for Douglas-fir and western hemlock were in progress. Committee activities were changing to meet other challenges, especially with respect to preservative treatment of railroad ties and structural timbers. Some of the enduring standards approved over the next few years included D 25-15, which provided a standard specification on the physical characteristics of round timber piles; D 38-15, standard methods of sampling and analysis of creosote oil; D 52-18, wood paving blocks for exposed pavements; and D 93-21, determination of the flashpoint of hazardous liquids under laboratory conditions.

While the Committee had problems with some early specifications, they were more successful in methods of testing. For example, D 143-22 provided methods for sampling by species and procedures for mechanical testing of small clear specimens of wood, and D 198-24 addressed testing of lumber in structural sizes. These two standards were primarily developed at the USDA Forest Products Laboratory (FPL) in Madison, Wisconsin. Possible resolution of longstanding issues concerning lumber quality versus species and the development of allowable properties were suggested in USDA Circular 295, Basic Grading Rules and Working Stresses for Structural Timbers by Newlin and Johnson (1923). In 1926, these procedures were standardized with the adoption of D 245, Tentative Specifications for Structural Wood Joist and Planks, Beams and Stringers, and Posts and Timbers.

Wood products standardization in the early years of the 20th century was not the exclusive

Herman Von Schrenk, a pioneer of the science of forest pathology and Chair of ASTM Committee D-7 for 43 years. *Photo courtesy of the Hunt Institute for Botanical Documentation, Carnegie Mellon University, Pittsburgh, Pennsylvania.*

3. Grading: in establishing grading, define standard defects and their effect on strength and durability and establish specifications for both grading and grades.

The Proceedings of the 1906 meeting of ASTM stressed two distinct problems with respect to trade names. First, the rapid introduction of Pacific Coast species into the eastern states had brought about much confusion. For example, for over 50 years the term "white pine" had signified a soft white pine from Maine, Michigan, Wisconsin, and Minnesota. Now markets in the East were seeing "white pine" from Idaho and California. Similar problems were encountered with eastern and western hemlock, eastern and western larch, and the eastern and western spruces. The second problem was the use of "longleaf pine" for higher quality southern pine and "shortleaf pine" for lower quality southern pine rather than using these terms to indicate botanical species.

domain of ASTM. Rather, there was close cooperation between Committee D-7 and other organizations promoting standardization. Many members of D-7 were also members of these other organizations. Cooperation and joint meetings with the American Railway Engineering and Maintenance of Way Association were an early part of D-7 activities. Cooperation and coordination of standards between Committee D-7 and the American Wood Preservers Association (AWPA), founded in 1904, was of great benefit. Hermann von Schrenk, the longstanding Chair of D-7, was also a founding member of AWPA. World War I brought a great deal of attention to the benefits of standardization.

In 1919, the American Lumber Congress called for development of lumber standards for grades, nomenclature, forms, and molding, and in 1922, the Central Committee on Lumber Standards was established by the U.S. Department of Commerce (USDC 1924). In 1924 and 1925, there were extensive discussions with the Central Committee to coordinate lumber specifications recommended by the two organizations. Input from the American Lumber Standard Committee (ALSC), the current incarnation of the Central Committee, has always been important to Committee D-7 activities, and ALSC continues to be a prominent user of D-7 standards.

Great Depression and War: 1930 to 1954

The ASTM procedures provide that standards are dynamic, and they should be subject to review and reaffirmation on a periodic basis. As time passed, some standards were modified and improved; others disappeared altogether. The decade of the 1930s is probably most recognized for the Great Depression and the prelude to World War II. As the decade opened, Committee D-7 had sired four standards related to wood preservation, which are still in print, and five standards related to solid wood. The preservative standards dealt largely with creosote measurement and quality. The other standards dealt with wood terminology, testing small clear specimens, testing timbers in structural sizes, visual stress grading, and specifications for round timber piles. Ten standards were focused on solid wood in 1924. Thus, in 5 years, fully half of the extant standards had been allowed to lapse.

The research that underpinned the standards continued, perhaps at a reduced level, during this

Glulam arches under test at the USDA Forest Service, Forest Products Laboratory in 1935.

quarter century, and some research programs were concluded and published during the 1930s (e.g., Markwardt and Wilson 1935, Wilson 1934). But the slow business climate probably kept the industry component of D-7 mostly on the sidelines, and with that minimal participation no new ASTM wood standards were promulgated until 1949, well after World War II. The focus for D-7 standards throughout the war period was on solid-sawn lumber and wood preservation.

However, the extant standards were used increasingly during that period. It was possible to classify structural timbers into grades that had 'reasonable uniformity and fairly definite minimums of strength' (USDA 1935, p. 99). In turn, engineering principles could be used in design, although this prerogative was probably not practiced to any extent because most lumber was sold in 'yard grades' for which no allowable property claims were made. Under the auspices of the Central Committee, the lumber industry had created an American Lumber Standard in the 1920s (Simplified Practice Recommendation 16-24 (SPR 16), not an ASTM standard), the first national lumber standard in the United States. This standard relied heavily on the existing standards of D-7 and in turn provided guiding principles for published trade association grad-

ing rules. The first edition of the *Wood Handbook* (USDA 1935) gives a fair synopsis of how research knowledge, D-7 standards, and the industry standard SPR 16 were melded.

It is clear from the literature of the first quarter of the 20th century that FPL played a major role in determining the content of D-7 standards. In D 245, for example, the standard follows FPL reports in both style and content to a great extent. Thus, in that standard, much of the procedure and text can be seen to mirror Miscellaneous Publication 185 (Wilson 1934) and related FPL publications. The *Wood Handbook* (USDA 1935) in turn cites D 245 as the authority for substantially the same procedures and text. As Shelley (1992) notes:

> …from the time of publication of MP 185 in 1934 to 1949, the FPL moved from the position of a promulgator of lumber design stresses and grades to a more advisory role. At the same time the industry took on greater responsibility for the development of lumber grades and design values. By 1949 this shift of roles had progressed to the point that specific grade descriptions were removed from ASTM D 245 in favor of more generalized principles of stress grading.

World War II was a powerful motivator for changes that eventually translated into modification of standards. Demand for lumber products skyrocketed, and the War Production Board mandated changes in the practice of describing and using lumber as a way of extending precious supplies. Two significant mandates were that the principles in ASTM D 245 be extended to lower grades of lumber not heretofore embraced in the standard and that basic stresses for some properties in that standard be arbitrarily increased by 10 percent. These key changes in practice were imposed on both civilian and military use, although there was probably little civilian use during the war. But experience gained with the imposed changes over a half-dozen years demonstrated that the changes were acceptable, and after the war there was probably no compelling reason to return to earlier practices. So, strength ratio and basic stress tables in D 245 were changed, and SPR 16 (by now SPR 16-53) and industry grading rules were brought in line with D 245.

At the end of World War II, there was a pent-up demand for housing, the major use for lumber. In about the same timeframe, the Federal Housing Administration (FHA) came into being and began to impose requirements for lumber used in new construction supported by FHA financing. One of those requirements was that the wood must be American Standard Lumber. Thus, ipso facto, this immense quantity of lumber had to be graded by the principles set in D 245. Since homebuilders were likely to take advantage of FHA financing, in effect almost all new construction would require American Standard Lumber.

Growth and Controversy: 1955 to 1979

Standards on Mechanical Properties of Wood

Post-war demand fostered a rapid increase in lumber and plywood production, much of it from western forests. Coast Douglas-fir and larch had been the traditional western species of choice, but supply and demand dictated the use of many other species. The FHA required that products from these species have design properties consistent with D 245, which generally put them at a competitive disadvantage (i.e., lower design properties for similar grade descriptions). The industry unrest that followed included challenges that FPL-derived basic stresses might not be relevant to current production, were inconsistently assigned to various species groups, and did not treat all regions of the country equally, as well as a host of other issues.

As a result of these issues, in 1960 FPL announced that it would no longer recommend basic stresses for wood but would continue to provide clear wood mechanical property data for individual species. This announcement suggested that Committee D-7 provided the broad representation needed for establishing basic stresses that took into account all pertinent technical and marketing issues.

The foment about the way design properties for wood products were derived set in motion at least two decades of research and development, much of it focused on sampling methods. Snodgrass and Noskowiak (1968) investigated the collection of samples for testing small clear specimens from current lumber production, rather than the methods prescribed in D 143. In collaboration with the western softwood industry, FPL used a "double sampling" method to infer clear wood strength properties from extensive but less costly sampling of specific gravity (USDA 1965). Shortly thereafter, FPL investigated random sampling of small clear specimens from forests (Bendtsen et al. 1970).

Committee D-7 was the vector for codifying many of these advancements. In 1963, D-7 realigned subcommittee responsibilities to enable more effective handling of the controversial issues related to clear wood properties. A new subcommittee assumed responsibility for establishing and publishing clear wood strength properties for individual species or regions, and for establishing criteria for assigning clear wood strength properties to any combination of species. Under this restructuring, basic stresses were no longer to be published in D 245. Product standards on lumber, plywood, laminated timber, and round timber would be responsible for procedures for establishing stresses for product grades from information provided by the clear wood subcommittee, giving appropriate consideration to the technical and marketing factors particular to each product.

A new standard, Methods for Establishing Clear Wood Strength Properties (D 2555-66T), was developed during this period. This standard used historical clear wood property data from both the United States and Canada and new information from double sampling for some species and from random sampling for others. Specific criteria were given for calculating properties to assign to any combination of species grouped for marketing purposes. By 1970, voluntary product standards for softwood lumber and plywood used or referred to D 2555 as a basis for establishing species groupings.

A particularly significant new standard was D 2915-70T, Standard Practice for Evaluating Allowable Properties for Grades of Structural Lumber. This new practice introduced statistical methodology for sampling test material from a mill, lumberyard, geographical area, or entire species group such that authoritative judgments could be made about the properties of the population being evaluated. Especially noteworthy was that the practice introduced to D-7 standards the concept of assigning confidence levels to tolerance (content) limits developed from test results using either parametric (defined distribution) or nonparametric (distribution free) statistics. These concepts have been incorporated in many other D-7 standards.

In a major international effort in the mid-1970s, the North American In-Grade Testing Program was initiated to establish the key mechanical properties of dimension lumber by testing full-size pieces sampled from production (the Snodgrass and Noskowiak sampling concept), rather than by using clear wood properties.

Standards Development Takes Off

Standards D 2555 and D 2915 are just two of the 43 new standards developed by the committee after 1960, almost all of which continue to be published in updated form. This remarkable burst of activity occurred as regulatory agencies and user groups sought greater assurance that commodity wood products would provide satisfactory performance in service.

The 1970s saw the introduction of the first standards for deriving allowable properties for round timber piles (D 2899-70 and D 3200-74). In 1970, the glulam industry assumed responsibility for developing lamination grade and species combinations and related design values and so developed D 3737.

The ongoing need to assure regulatory agencies and users of the long-term decay and termite resistance of pressure-treated wood products continued to be reflected in the development of D-7 standards. Nineteen new specifications and test methods cover-

The Western Wood Density Survey provided information on nine species for formulating ASTM D 2555.

ing the content, analysis, evaluation, and treatment process for wood-preserving chemicals were introduced. Among these were the 1959 publication of specifications D 1624 and D 1625 and methods D 1627 and D 1628 for two waterborne salt treatments, ammoniacal copper arsenate (ACA) and chromated copper arsenate (CCA), marking a time of accelerating demand for these "clean" treatments in construction.

One of the most significant standards related to preservation issued in the history of Committee D-7 is D 1760-60, Standard Specification for Pressure Treatment of Timber Products. This standard is a virtual handbook of treating and penetration/retention information of importance to manufacturers and users of treated wood products. Included in each of seven tables is information on conditioning, the pressure treatment process, and results of treatment for each preservative solution (e.g., creosote, creosote-coal tar, pentachlorophenol, CCA, ACA) for each species or species group used for a particular timber product. The retention and penetration requirements provided for aboveground, ground contact, and coastal water uses are those required to assure long-term durability.

Two test methods related to the properties of fire-retardant chemicals were developed in the 1970s. One of these, D 2898-70T, provides a method for evaluating the durability of fire-retardant treatments of wood products under accelerated weathering. The impetus for this new method was the use of fire-retardant-treated wood shakes and shingles in the high-fire-risk brush areas of southern California. Standard D 2898 documents for regulatory agencies that the required flame spread rating of treated shakes and shingles as determined in the ASTM E 84 tunnel test is not changed by long-term exposure to the weather. The second fire treatment standard, D 3201-73T, was developed in response to chemical exudation and fastener corrosion problems associated with the use of interior fire-retardant-treated lumber in roof systems or other applications involving exposure to over 80 percent relative humidity.

Standard D 2017-62 involves evaluating the resistance of untreated wood to decay. It provides weight loss criteria for identifying decay-resistant species such as redwood and western redcedar for use in decks, sills, sleepers, and other aboveground applications.

The remaining 15 standards issued in this productive 25-year period deal with mechanical, physical, chemical analysis, and machining test methods; simulated service testing for flooring; and terminology. Seven standards involve test procedures for evaluating individual properties of plywood.

The strength of mechanical fasteners is a critical element in ensuring the satisfactory performance of wood structures. After World War II, university and industry research laboratories became increasingly involved in investigating the load-carrying capacity of different types of fasteners in both new and traditional wood products. Thus, the need for standard fastener test methods to assure reliable and comparable results was recognized. To meet this need, D 1761-60T was developed. This standard provides testing procedures for evaluating the withdrawal and lateral load resistance of nails, screws, bolts, and other types of connectors.

Since fastener design loads for mechanical connections are a function of species specific gravity and are classified on this basis, this overview of the third 25 years of Committee D-7 standardization activities concludes with reference to one of the Committee's most important fundamental test method standards, D 2395, Standard Test Methods for Specific Gravity of Wood and Wood-Based Materials, which was first issued in 1965. This method provides six different procedures for determining specific gravity, including volume by measurement, volume by water immersion, flotation, and increment core. Of particular significance are equations for converting specific gravity values to different moisture content bases, such as from a green volume basis as given in D 2555 to the ovendry volume basis used for fasteners in the National Design Specification (AF&PA 1991).

Completing the Century

The demand for standardization of wood and wood-based products that developed during the first 75 years of Committee D-7 continued to build as the Committee completed its first century. The standards developed during the last 25 years reflect the industry trend to define product sufficiency in terms of performance-based criteria. In some cases, these changes have resulted in new performance-based standards such as D 6570, Practice for Assigning Allowable Properties for Mechanically Graded Lumber. But even the new prescriptive standards such as D 1990, Practice for Establishing Allowable Properties for Visually-Graded Dimension Lumber From In-Grade Tests of Full-Size Specimens, reflect these trends by utilizing test results of full-size lumber specimens rather than relying on the traditional clear wood properties of D 2555.

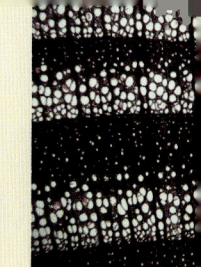

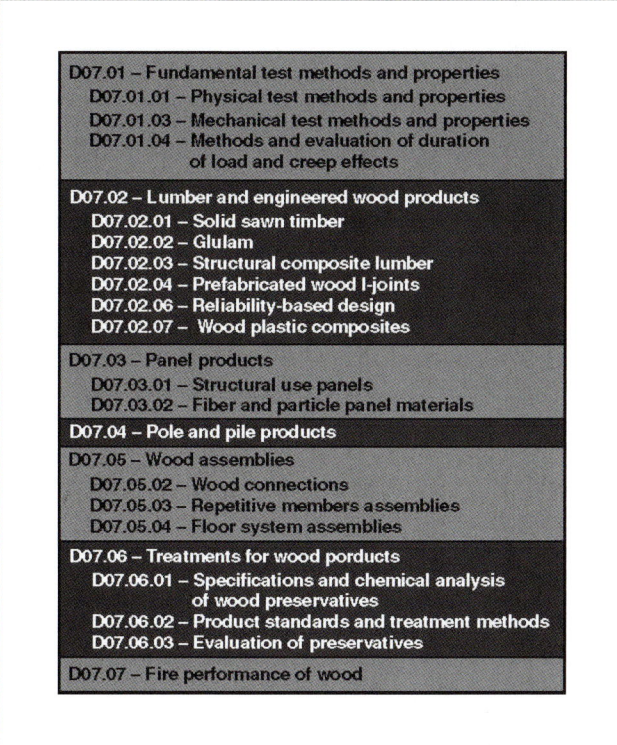

| D07.01 – Fundamental test methods and properties |
| D07.01.01 – Physical test methods and properties |
| D07.01.03 – Mechanical test methods and properties |
| D07.01.04 – Methods and evaluation of duration of load and creep effects |
| D07.02 – Lumber and engineered wood products |
| D07.02.01 – Solid sawn timber |
| D07.02.02 – Glulam |
| D07.02.03 – Structural composite lumber |
| D07.02.04 – Prefabricated wood I-joints |
| D07.02.06 – Reliability-based design |
| D07.02.07 – Wood plastic composites |
| D07.03 – Panel products |
| D07.03.01 – Structural use panels |
| D07.03.02 – Fiber and particle panel materials |
| D07.04 – Pole and pile products |
| D07.05 – Wood assemblies |
| D07.05.02 – Wood connections |
| D07.05.03 – Repetitive members assemblies |
| D07.05.04 – Floor system assemblies |
| D07.06 – Treatments for wood porducts |
| D07.06.01 – Specifications and chemical analysis of wood preservatives |
| D07.06.02 – Product standards and treatment methods |
| D07.06.03 – Evaluation of preservatives |
| D07.07 – Fire performance of wood |

Table 2. — D-7 subcommittees and sections.

This shift from the more traditional clear wood basis used in standards such as D 245 to an approach based more on product test results or performance as reflected in standards such as D 1990 and D 6570 was fostered by events in the wood products industry that had originated in the 1970s. The advent of ready access to testing equipment that could easily test full-size specimens of products such as lumber and glued laminated beams facilitated many research studies on these and other wood products. The resulting data called into question components of some prescriptive models used to establish allowable properties for wood and wood-based products.

The North American In-Grade Testing Program was the first major industry-sponsored program that attempted to characterize the structural properties of commercial lumber used in the United States by systematically sampling and testing full-size lumber specimens of several grades and species. This program is probably the most visible example of the changes occurring in the industry that were subsequently reflected in standards under the jurisdiction of Committee D-7 (Green et al. 1989). Many sampling procedures developed for the program were incorporated into D 2915. Test methods were incorporated into a new standard, D 4761, adopted in 1988. The analytical methodology used to derive allowable properties from the thou-

sands of data collected in the program was formalized into standard D 1990, adopted in 1991. This standard has become the de facto basis for the allowable properties used for all visually graded lumber used in the United States.

By the completion of the In-Grade Program in the mid-1980s, more than 70,000 full-size specimens of production lumber had been tested. In some cases, new standards were promulgated, such as D 1990; in other cases, standards were revised or expanded to provide for criteria based on tests of full-size components, such as D 3737 (glued laminated timber). While these changes in "traditional" standards were under development in D-7, the wood products industry was also developing new "engineered" products, such as parallel laminated veneer lumber, oriented strandboard, and structural I-joists. These products did not fit easily into the old clear wood prescriptive models for assigning allowable properties; they could be more easily accommodated by standards that set minimum product or test performance criteria. Standards such as D 5055 (I-joists) and D 5456 (composite lumber) were developed for these products.

As standards began to address these new products and data, it became clear that the structure of Committee D-7 did not provide adequate opportunity for review of the standards by representatives of other wood product lines. It also became clear that some needs for new standards did not fit well in the existing structure. The Committee was restructured in 1984 to address both of these concerns.

Historically, standards had been developed unique to specific products. D 2555 was the first standard to provide fundamental procedures that could be utilized by several product-specific standards. The new focus on product performance criteria resulting from new engineered products, and research on traditional products, also began to highlight other issues with broad product applicability. The reorganization of D-7 facilitated the Committee's ability to address these issues.

The Committee was divided into a number of subcommittees (Table 2). Fundamental test methods and properties were placed under the purview of subcommittee D07.01. All structural product standards that had been in separate product-specific subcommittees were consolidated into a single structural subcommittee (D07.02), with individual sections organized for each product line. This new structure ensured review of standards by all interested and affected committee members, and it also created a more streamlined and efficient process for standards development.

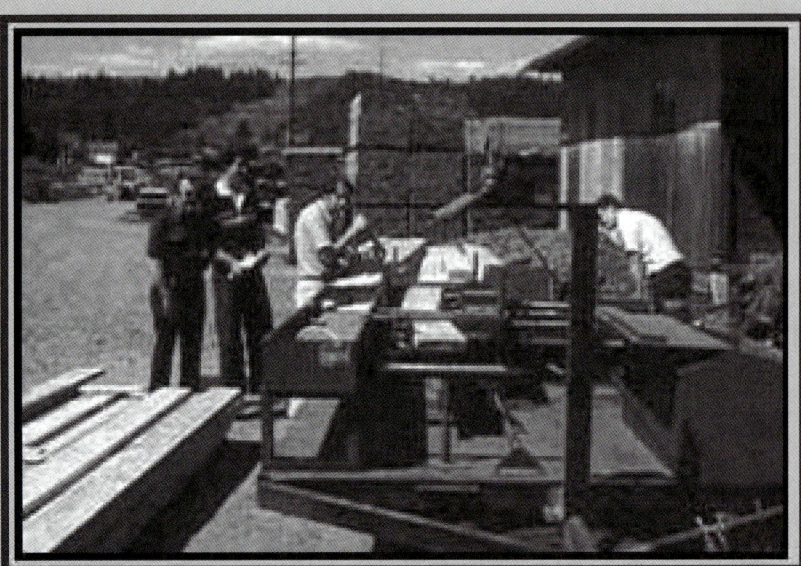

In-grade testing using portable equipment, 1980.

At times, Committee D-7 has been required to develop standards quickly in response to an urgent need. In the 1980s, some fire-retardant-treated lumber and panel products used in multi-family residential roof systems began to fail. Committee D-7 quickly began to develop standards to evaluate the effects of fire-retardant treatments on the structural properties of wood and wood-based products. In response to the urgency of the situation, standard ES 20-91 was first approved and published as an emergency standard. Emergency standards have been a seldom-used option in the long history of D-7, but were clearly appropriate in this circumstance. ES 20-91 was later balloted through the traditional committee process and approved as standard D 5516-94. A similar standard describing a test method for lumber, D 5664, was approved in 1995. Two additional standards, D 6305 and D 6513, were added in 1998 and 2000, respectively, to provide a standard practice for the development of appropriate fire-retardant-treatment adjustment factors.

The development of D 1990 is another example of the ability of Committee D-7 to quickly produce needed standards. This standard was written by a broad-based committee representing many segments of the wood products industry and other interested parties. And even though this standard broke new ground in many ways, it was completed in less than 4 years.

During the 1980s, the wood products industry also developed design methodologies that reflect the trend toward reliability based design concepts. The adopted protocol is referred to as Load Resistance Factor Design (LRFD) (AF&PA 1996) in the industry. In conjunction with the industry effort, Committee D-7 developed a new standard, D 5457, which provides a means for translating allowable properties developed according to the standards of D-7 to an appropriate value and format for use with LRFD.

Advances in testing and analysis of wood and wood-based products also prompted Committee D-7 to review existing standards in a new light, and some were revised in format to be more in line with the developing standards. Two examples of this effort were D 143 on the determination of clear wood properties and D 2016 on the determination of moisture content. In the case of D 143, the clear wood test methods were left intact in the standard, but the sampling procedures were removed and placed in a new standard, D 5536. In the case of D 2016, the revisions needed were such that D-7 felt it more appropriate to withdraw the standard and create two new standards, D 4442 and D 4444.

During this period, a number of new standards were also adopted relating to the performance of connectors used for wood, among these are D 5652 (bolted connections), D 5764 (metal shear plates), and D 5933 (dowel bearing). These standards were developed and adopted to reflect major changes in the design of fasteners and the determination of connector performance.

Meeting the Challenges of the Future

As Committee D-7 enters its second century, what are the prospects for the future? When asked this question, four trends were identified by people at the cutting edge of business, technology, and new product development: 1) a continuing trend toward globalization; 2) new products made of more than one type of material (composite materials); 3) an evolution from 'standard products' to 'product standards;' and 4) repercussions from advances in electronic technology.

Globalization

The direct implications of globalization for standards are fairly obvious. Multinational committees must assess the compatibility of standards from different countries. Differences in technical require-

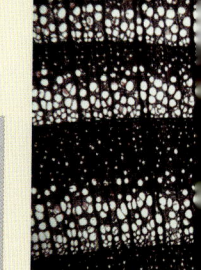

ments, expressed in different terms and different languages, must be reconciled. The indirect implications of globalization are more profound. Manufacturers will continue to grow and consolidate. The needs of internationally focused companies will drive changes in U.S. standards development. These companies generally employ more sophisticated quality management systems than do smaller companies with a purely domestic focus.

Composite Materials

When is a glulam beam no longer a glulam beam? ASTM D 3737 includes the embedded judgment of more than 100 years of collective technical knowledge and manufacturing experience in its provisions. However, even today, Committee D-7 is testing the boundaries of this knowledge and experience. Is a glulam beam still a glulam beam when the tension ply on the outer face shifts from sawn lumber to laminated veneer lumber? How about when the entire tension zone is replaced with fiber-reinforced laminations?

In a more general sense, are there hidden limitations within our basic testing standards that make them unsuitable for use with composite products? These questions are already in the pipeline for D-7. And, as the range of new products containing new combinations of materials continues to expand, we must be prepared to answer increasingly tough questions in this regard.

Standard Products vs. Product Standards

In the early days of Committee D-7, the development of standard designations for commercial grades of lumber was a key goal. In a time when lumber was produced by thousands of small mills, it was in the best interests of both manufacturers and consumers to have standard grades and sizes available to the marketplace. This concept was carried over into standards for glulam and wood structural panels. However, when the next generation of engineered wood products (EWPs) was presented to ASTM for standardization, companies were looking for something dramatically different. The companies introducing these products had invested heavily in their development. Patents typically covered many aspects of the products and their manufacturing techniques. Because of the many variables in manufacturing processes, it was generally accepted that each manufacturer should be responsible for evaluating the performance characteristics of its own production. This is in stark contrast to the approach used for visually graded lumber, in which fewer manufacturing variables impact structural performance.

The development of standards to cover EWPs presented D-7 with a unique challenge: achieving a balance in maintaining minimum safety levels and providing the manufacturer flexibility to optimize product performance. The difference between the old and new approaches is often contrasted as follows: the "old" system defines a standard product, while the "new" system establishes a product standard.

Electronic Technology

Why would advances in electronic technology affect Committee D-7? After all, our scope of products includes lumber, plywood, and EWPs – not satellite phones and digital cameras. Here's a list of possible areas in which electronic technology will change D-7 standards:

• *Laboratory testing and data acquisition.* — Test machines continue to evolve. Data acquisition systems are more sophisticated and cost less than ever before. Some current standards contain limitations related to limits in the technology available at the time the standards were developed. As the equipment evolves, standards must evolve to take advantage of the additional information available.

• *Advanced engineering analysis tools.* — Historically, some conservatism was embedded in traditional designs as a result of the limitations of hand-based engineering calculations. As software-based designs continue to become more sophisticated, many structures will be designed with less conservatism. What are the implications of this change relative to our standardized safety factors?

• *Electronic monitoring systems.* — As electronic devices shrink in size and cost while expanding their capabilities, embedded devices will be able to "report" about the condition of structures. "Smart materials" will know when they are over-stressed or exposed to conditions likely to cause degradation. When coupled with radio-frequency transmission capability, they will be able to "talk" to remote receiving equipment. These radio-frequency identification tags will also be able to track individual structural members to the underlying quality assurance data from their production run. What new standards can help us to leverage these advances?

Other Challenges

In addition to facing these long-term challenges, Committee D-7 will need to remain vigilant to subtle

"erosion" in the relevance of current and historical standards as products and their applications continue to evolve.

Products. — We will continue to support our traditional products while concurrently supporting "improved" versions of those products (e.g., machine-graded versus visually graded lumber).

Applications. — We must continue to match the requirements of the standard to the type of application. For example, while a purely statistical approach to developing design values is appropriate for sawn lumber, it is incomplete for engineered trusses, for which engineering stress analysis is a more important starting point.

Summary

Committee D-7 will continue to be challenged by new products, new materials, new technologies, and new regulations. Some future issues can be anticipated, and some will be thrust upon us. At stake in these issues will be public safety, economic vitality, and the competitive position of the U.S. economy. D-7 has successfully faced these types of challenges in the past, and it will successfully face them in the future.

Literature Cited

American Forest & Paper Association (AF&PA). 1991. National Design Specification for Wood Construction. AF&PA, Washington, DC.

_____. 1996. LRFD: Load Resistance Factor Design Manual for Engineered Wood Construction. AF&PA, Washington, DC.

Bendtsen, B.A., F. Freese, and R.L. Ethington. 1970. A forest sampling method for wood strength. Forest Prod. J. 20(11):38-47.

Berg, W.G., J.H. Cummin, J. Forman, and H.L. Fry. 1907. Strength of bridge and trestle timbers. *In*: Proc., Annual Convention of American Inter. Assoc. of Railway Superintendents of Bridges and Buildings, New Orleans, LA, October 16, 1895. pp. 14-63.

Fulweiler, W.H. 1955. 50 years of Committee D-7 on wood. ASTM Bulletin. No. 206. Am. Soc. for Testing and Materials, West Conshohocken, PA. pp. 40-41.

Green, D.W. and R. Hernandez. 2000. Codes and standards for structural wood products and their use in the United States. *In*: Proc, Forest Products Society Annual Meeting, June 23, 1998, Merida, Yucatan, Mexico. Forest Prod. Soc., Madison, WI . pp. 3-16.

_____, B.E. Shelley, and H.P. Vokey, eds. 1989. In-grade testing of structural lumber. *In*: Proc., ASTM workshop on North American In-Grade Testing Program. No. 47363. Forest Prod. Soc., Madison, WI. www.fpl.fs. fed.us/documnts/pdf2000/green00d.pdf.

Markwardt, L.J. and T.R.C. Wilson. 1935. Strength and related properties of woods grown in the United States. Bull. 479. U.S. Government Printing Office, Washington, DC.

Newlin, J.A. and R.P.A. Johnson. 1923. Basic grading rules and working stresses for structural timbers. USDA Circular 295. U.S. Government Printing Office, Washington, DC.

Shelley, B.E. 1992. Evolutionary standards development. *In*: Proc., Wood products for engineered structures: Issues affecting growth and acceptance of engineered wood products. No. 47329. Forest Products Soc., Madison, WI.

Snodgrass, J.D. and A.F. Noskowiak. 1968. Strength and related properties of Douglas-fir from mill samples. Bull. 10. Oregon State University, School of Forestry, Forest Research Laboratory, Corvallis, OR.

USDA Forest Service, Forest Products Laboratory (USDA). 1935. Wood Handbook. Superintendent of Documents. U.S. Government Printing Office, Washington, DC.

_____. 1965. Western wood density survey. Report. No. 1. Res. Pap. FPL-27. U.S. Department of Agriculture, Forest Service, Forest Products Laboratory, Madison, WI.

U.S. Dept. of Commerce (USDC). 1924. Simplified practice recommendation 16-24. American lumber standards for softwood lumber. Superintendent of Documents. U.S. Government Printing Office, Washington, DC.

Wilson, T.R.C. 1934. Guide to the grading of structural timbers and the determination of working stresses. Misc. Publ. MP-185. Superintendent of Documents, U.S. Government Printing Office, Washington, DC.

The authors are, Supervisory Research General Engineer, USDA Forest Serv., Forest Prod. Lab., Madison, WI 53705; Former Department Head and Professor (retired), Dept. of Wood Science and Engineering, Oregon State Univ., Corvallis, OR; President, Wood Construction Technologies, Inc., McLean, VA; Executive Vice President, West Coast Lumber Inspection Bureau, Portland, OR; and P.E., Director of Code and Product Acceptance, Weyerhaeuser Technology Center, Federal Way, WA.

Development and Mill Transfer of the FPL Semichemical Pulping Process

John N. McGovern Emeritus Professor, Department of Forestry, University of Wisconsin-Madison
Tappi Journal Vol. 67(6) (June 1984)

The technology of a high-yield pulping process for hardwoods, the so-called semichemical pulping process, was developed on experimental and pilot scales at the Forest Products Laboratory during the period of mid-1923 through 1925. The process featured many innovations in use today.

Semichemical pulp was first made in this country—first made anywhere—in late 1925 using a two-stage, chemical-mechanical, high-yield process developed at the Forest Products Laboratory (FPL) during the period 1923–1925. Progress in the development is shown chronologically in Table I, including the first two mills using the process. The process was described in detail in a landmark publication [11].

Date	Events	References
1923–1924	FPL sodium sulfite semichemical pulping process concept expressed: process variables investigated on laboratory and pilot scales. First fiberizing in a beater followed by ball mill and finally rod mill. Spent liquor recovery examined.	1
1924	Evaluation of extracted chestnut chips for standard chemical (unsatisfactory) and semichemical (promising) pulping and products. Successful mill trials of FPL chestnut semichemical pulp for corrugating medium and corrugated board.	2,3
1924–1925	Further process studies, including species (including black gum), reagent composition, ratios and additions, digester corrosion, rod mill variables, and product properties.	4,5
1925	Potential semichemical mill evaluations. First publication in sodium sulfite pulping series. Successful evaluation of a commercial disc refiner for fiberizing FPL semichemical chips.	6,7
1925–1926	Knoxville[1] mill (started up in late 1925) described including parallel rod and disc mill fiberizing lines.	8
1926	Black gum semichemical pulp made at Knoxville used for paper products trials at Hartsville[2]. Landmark publication on FPL process.	9,10,11

1927	Startup at Hartsville[2] (January 1927) of FPL process on black gum for butchers wrap (later corrugating). 1—Southern Extract Co. 2—Carolina Fiber Co.	12

Table I. Chronology of developments and mill applications of FPL semichemical pulping processes.

Semichemical pulp production has grown from about 7000 metric tons in the first mill in 1925-1926 to 3.5 million metric tons in the U.S. in 1981 [13]. The latter production was used to make 4.5 million tons of corrugating material, 80% of the total corrugated material made that year. The hardwood pulpwood consumption for making the semi-chemical pulp in 1981 was approximately 2.25 million cords.

This report documents the above development, including unpublished FPL reports. The report will describe the basic process concept and procedure, delineate the process variables investigated, outline the innovations in process procedure and equipment, and comment on the transfer of the technology into mill practice. The early industrial expansion of the FPL process will also be covered (Table II).

Startup date	Location	Daily capacity, tons	Wood species	Digester	Fiberizer	Product
1925	Knoxville, Tenn.	30	Extracted chestnut	Globe rotary	Rod mill[1]	Corrugating
1927	Hartsville, S.C.		Black gum	Globe rotary	Rod mill[2]	Butchers wrap[3]
1928	Sylva, N.C.	75	Extracted chestnut	Globe rotary	Rod mill	Corrugating
1928	Nashville, Tenn.	50	Extracted chestnut	Globe rotary	Rod mill	Corrugating
1929	Harriman, Tenn.	75	Extracted chestnut	Globe rotary	Rod mill	Corrugating
1929	Lynchburg, Va.	100	Extracted chestnut	Globe rotary	Rod mill	Liner[4]
1930	Bogalusa, La.	40	Pine slabs[5]	Globe[6] rotary	Disc attrition mill	Corrugating

1—An initial parallel installation of a disc mill discontinued
2—Hammer mill in 1929 and disc mill in 1930
3—Later corrugating board
4—30-40% kraft
5—Later oak and then gum
6—Later vertical digester

Table II. Early semichemical pulp mills [13]

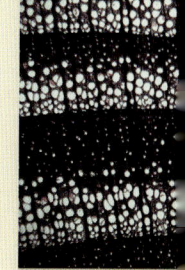

The FPL semichemical pulping concept and procedure

High pulp yields and increased utilization of hardwoods were important objectives at the FPL even in the early 1920s. Hardwoods made up only 10% of total pulpwood consumption in 1925, compared with 28% in 1982. A process for obtaining a high yield, over 70%, was conceived as comprising (a) a partial chemical pulping of wood chips with a carbohydrate-preserving pulping reagent, sodium sulfite, and (b) a mechanical disintegration of the softened chips into a papermaking pulp. The idea of a partial chemical pulping, using a bisulfite reagent, had been proposed by Mitscherlich in 1874, but there is no history of fiberizing the partially pulped chips. The use of sodium sulfite was suggested in 1880 by Cross, and the full sodium sulfite process of Bradley and McKee (Keebra) had short-lived success in the 1920s [*15*]. The combining and augmenting of the best features of past ideas, elaborating on them and developing a working pulping procedure were the contributions made by the FPL technologists.

The delignifying reagent in the FPL process was sodium sulfite, which is effective in lignin solution and at the same time causes the least solution of the carbohydrates of any of the common pulping reagents. In the early stages of pulping with this reagent, wood acids were formed that caused steel digester corrosion. This effect was reduced by the presence of sodium carbonate or bicarbonate, either added directly to a sodium sulfite solution or formed in the sulfiting of soda ash, the mill way of making the pulping reagent. It was also necessary to have thorough impregnation of the chips with the pulping solution in order to obtain uniform delignification and fiberizing for the highest pulp quality. This necessary penetration was achieved by (a) steaming of the chips at slightly above atmospheric pressure to obtain uniform moisture content, (b) submerging the conditioned chips in a concentrated pulping solution at slightly elevated temperature and pressure and (c) withdrawing (blowback) the excess free liquor for subsequent fortification and reuse (Fig.1).

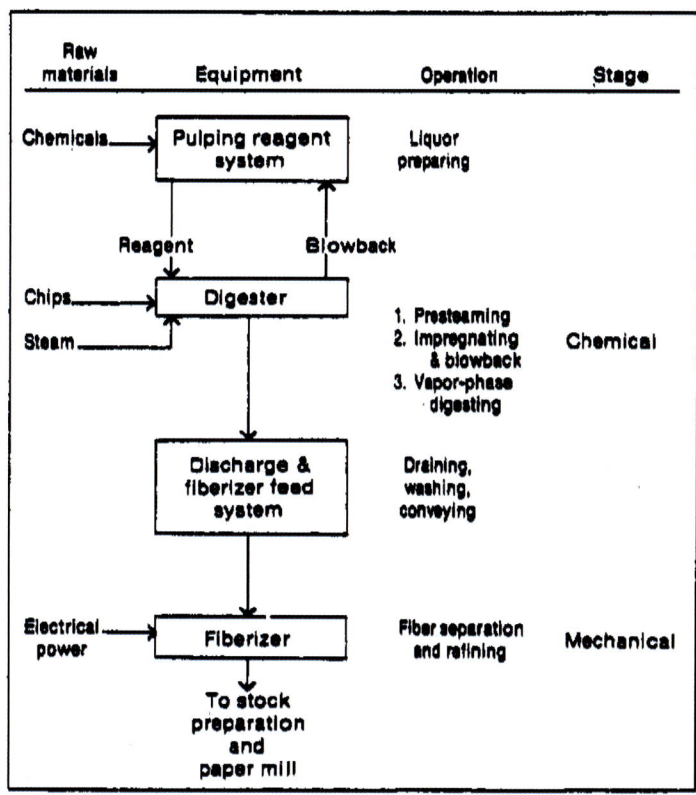

1. Outline of FPL semichemical pulping process.

The pulping was then conducted in the vapor phase at elevated temperature and pressure. This sequence of impregnation and vapor phase pulping is practiced in present-day wood chemical pulping systems. Typical chemical-stage conditions were for extracted chestnut made in 81% yield [11, 16] (Table III).

Pulping		
Chemical application	Impregnation	Pulping
Na_2SO_3, 8.0 lb/100 lb wood $NaHCO_3$, 3.2 lb/100 lb wood	Time, 0.5 h Temperature, 120-130°C	Time, 3 h Temperature, 150-160°C

Table III. Typical chemical-stage conditions.

The fiberizing stage of semichemical pulping is considered the heart of the process. The fiberizing of the softened chips was first performed in the laboratory in a papermill beater. This equipment gave poor fiber separation and required excessive energy. A pebble mill, while an improvement because it used less energy, was inadequate. A rod mill (adapted from the mining industry) had been installed at the FPL for general fiberizing investigations. This machine was especially suitable for fiberizing the semichemically softened chips. A comparison of fiberizers and fiberizing energy in the FPL process development is given in Table IV.

Wood species	Aspen, paper birch, jack pine, tannin-extracted chestnut, black gum, yellow poplar, loblolly pine, beech, effect of bark.
Pulping reagents	Na_2SO_3; Na_2SO_3+Na_2CO_3 or $NaHSO_3$; $NaOH$
Impregnation and pulping	Presteaming—Time, temperature, pressure, green and seasoned wood Impregnation Step—Time, temperature, pressure (steam hydraulic) vacuum, reagent composition, concentration and wood ratio Pulping Step—Time, temperature, pressure (relief of CO_2), spent liquor composition and concentration, and spent liquor recovery, liquor testing
Fiberizing and pulp processing	Equipment (beater, pebble mill, rod mill, disc attrition mill, conical (refiners); conditions (consistency, additives, temperature); throughput, energy consumption, pitch control
Pulp treatment	Bleach, washing, screening, riffling
Papermaking and mill products	Additives, alum, pH, news; wrapping, containerboard, corrugating board
Mill trials	Corrugating board, corrugating trials
Pulp and paper properties	Color, stiffness, opacity, burst, tear, fold, and tensile; basis weight
Miscellaneous	Bleaching evaluations

Table IV. Outline of semichemical pulping process variables investigated at the FPL

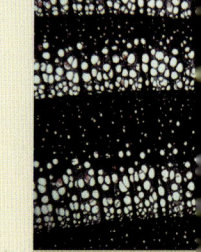

In the rod milling operation, there was a certain amount of pulp processing ensuing. This observation led to the use of the rod mill for chemical pulp refining or processing in the 1930s [17,18]. The final stock preparation for FPL and mill papermaking was done in conical refiners, jordans, and claflins.

The potential of the disc attrition mill, as adapted from the grain grinding industry, for fiberizing softened chips was recognized during the development of the FPL process.* In fact, the first semichemical pulpmill started up with two parallel fiberizing lines, one with a rod mill and the other with a disc mill. The higher energy demand and short plate life of the disc mill, at that time, were decisive in choosing the rod mill for fiberizing in the early semi-chemical mills (TableII). Improvements in disc mill design, especially precision and ruggedness and plate materials, and the disadvantages of limited capacity and metal contamination for the rod mills, reversed the choice above by the early 1930s.

Investigations of process variables and products
*The disc mill was also being investigated independently during the early second quarter of the 20th century for two of the important applications in the paper industry: (a) the Sutherland refiner for processing papermaking pulps as first introduced in 1928 [19] and (b) the Asplund defibrator for making a coarse-fibered softwood pulp for building and wallboard as first introduced in the U.S .in 1937 [20). The forerunner of the disc mill for pulp refining was the Kingsland machine of 1856, although it was unexploited commercially [19,21]. The Asplund process involved fiberizing wood chips in a disc mill under high temperatures and pressures. The Sutherland refiner was also used several decades later for fiberizing and refining high-yield (55-60%) softwood Kraft and sulfite pulps.

Extensive investigations of process variables of the FPL process were conducted during the period of its development, using a number of wood species, as summarized in Table IV. These variables were studied in conjunction with a mill cooperative investigation aimed at utilization of waste tannin-extracted chestnut chips. These chips had been evaluated for pulping by the standard processes-soda, sulfate, and sulfite-and were found to give very inferior pulps. Their suitability to make a promising semichemical pulp was established in the course of the variables studied. Blackgum, a little-used, copiously available southern hardwood, was also found at this time to make an excellent semichemical pulp (later, Kraft also).

An original product objective in the FPL process work was to make a bulky, light-colored pulp suitable for use in newsprint and wrapping paper, but the pulps tended to be dark, especially those from dark-colored woods, and un-bleachable. An exception was aspen semichemical pulp, which was used for a period several decades later in a Canadian newsprint mill. The exceptional stiffness (high hemicellulose) of board products from these pulps, including those from chestnut, focused product activity on corrugating board. Glassine-type papers, another future semichemical pulp product, were also made.

The recovery of chemicals from the spent liquor was studied on a laboratory scale, as acetic acid was a potential chemical for recovery.

Innovations associated with the development and application of the FPL process
The development of the FPL semichemical pulping process and its commercial implementation incorporated innovative features in wood fiber utilization, pulping reagents and procedures, fiberizing, and paper products.

Wood fiber utilization

The experimental work included investigations of northern and southern hardwoods, as mentioned previously. The suitability of the FPL process for hardwood pulping in general, and the waste chestnut chips and little-used black gum in particular, was advantageous for hardwood utilization, a problem then as now.

Two-stage chemical-mechanical pulping process

The two-stage FPL process comprising partial chemical pulping and mechanical fiberizing was new to the industry and included innovations both in the chemical pulping and mechanical fiberizing stages.

Chemical pulping stage

The employment of the little-used sodium sulfite for partial pulping provided minimum carbohydrate solution and maximum delignification for the highest yield. The digester corrosion from wood acids generated in sodium sulfite pulping was largely offset by adding or forming sodium carbonate or bicarbonate, a discovery of an earlier FPL investigation. (Hence, neutral sulfite semichemical or NSSC process.)

The unique impregnation procedure comprising (a) steaming the chips for moisture conditioning, (b) full submersion of the chips under pressure in a relatively concentrated pulping reagent, and (c) removal (blowback) of the excess or free liquor for complete penetration ensured uniform delignification and uniformly softened chips for fiberizing.

Pulping in the vapor phase using the chemicals within the chips made possible shorter pulping times at moderate temperatures with low steam consumption.

Mechanical fiberizing stage

The fiberizing of the softwood chips was first conducted in a laboratory beater; the disintegration was unsatisfactory. A pebble (mill) was an improvement, but adequate fiberizing was obtained in a small pilot-sized rod mill introduced at the FPL in 1924 for general fiberizing. (The rod mill was a standard ore grinder.) This machine, which was also found to impart strength development to pulps, as mentioned previously, was the standard fiberizing machine for the softened chip since early semichemical pulpmills.

In June 1925, under FPL sponsorship, a disc attrition mill was investigated for fiberizing. As mentioned earlier, a disc mill fiberizing line was installed in the first semichemical pulpmill. The disc mill eventually replaced the rod mill for fiberizing.

Paper products

Trials on the FPL paper machine of the papermaking potential of the new semichemical pulp led to the recognition that the inherently unbleachable, relatively low color, and high stiffness characteristics of the products made them suitable for the production of corrugating medium (nine-point) of superior quality in meeting industry use requirements [16].

Mill operation

The globe digesters of the early mills discharged into chests with false bottoms for draining the spent and wash liquors. These chests were fitted with leach casters, a device with revolving arms to sweep the chips toward an opening, as adapted from the chemical extraction industry. This device provided control of feed to the fiberizer.

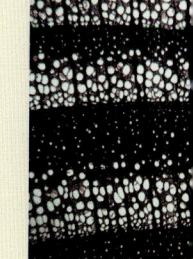

Progression of the FPL process concept to corrugating board production

The FPL two-stage, chemical-mechanical pulping process was apparently conceived sometime in mid-1923, and its mill start up occurred in December 1925, a period of about 2~years. The steps in the progression are given in Fig. 2.

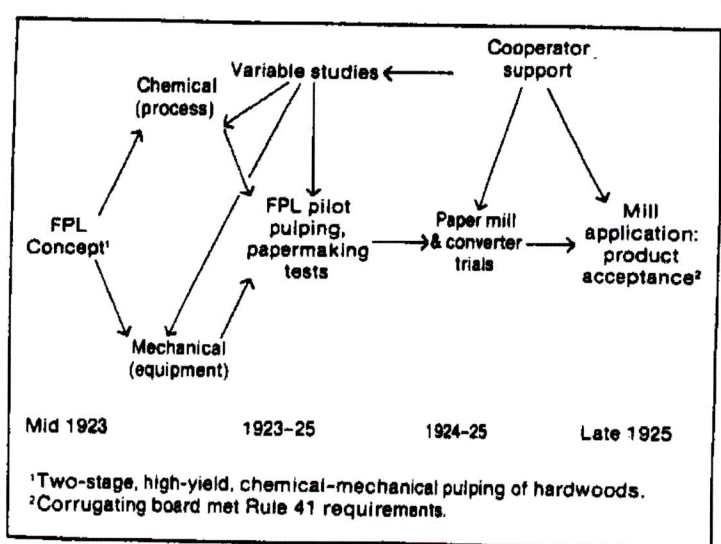

2. Evolution of semichemical corrugating board manufacture from FPL concept.

¹Two-stage, high-yield, chemical-mechanical pulping of hardwoods.
²Corrugating board met Rule 41 requirements.

This time of technology development and transfer is extraordinarily short in comparison with the 10 years being currently reported [22]. An important factor in this transfer of technology was the cooperative support of the Southern Extract Co. This firm had submitted to the FPL its problem of the utilization of waste or by-product chestnut chips from which the tannin had been extracted in a hot water process. The attack on the problem coincided with the development of the FPL process. The cooperator specifically supported variables studies, FPL pilot mill demonstrations, and mill corrugating board and corrugator trials. The importance of an industrial "receptor" in technology transfer has been reported [23]. In this case it was essentially a process development needing support. The situation is different with equipment development where the receptor may carry out the entire development of the concept, as in the case of a paper coater development [24].

A factor in the smooth transfer of the FPL process to the mill was the completeness with which the process variables, demonstrations, and trials had been conducted. These studies had been possible because, among other factors, complete pulping and papermaking facilities were available at the FPL, including the capability of producing the quantities of pulp and paper products necessary for substantive mill trials.

A key factor in the evolution of the FPL process was the unchanged adaptation of the rod mill for fiberizing in mid-1924. A small, commercial unit required no development in design and information on the variables of its operation, for fiberizing of different fibrous materials was readily obtained. Larger units were available "off the shelf" for mill operation. Another factor in facilitating the transfer was the essentially non-complicated nature of the process and equipment. Only simple instrumentation and control were needed. Other innovations which contributed to the application of the FPL process to industry have' already been discussed.

The quality of the FPL personnel involved in these investigations was necessarily of the highest caliber in creativity, ingenuity, chemical and engineering skills, and practicality.

CONCLUSIONS

The development of the FPL semichemical pulping process and its industrial application had the following results:

1. The utilization of northern and southern hardwoods, particularly in 1925, and later of by-product tannin-extracted chestnut chips

2. A two-stage process involving partial chemical pulping and mechanical fiberizing for producing a pulp, in a yield of 70% and higher, useful for making a stiff, marketable, corrugating board for container fabrication

3. A process incorporating numerous technological innovations, including (a) a carbohydrate-preserving and delignifying-efficient pulping reagent, sodium sulfite buffered with sodium carbonate or bicarbonate to neutralize wood acids, (b) a pulping procedure comprising impregnating the chips for uniform delignification by presteaming for moisture conditioning, pressure impregnation of concentrated reagent, removal of excess reagent, and pulping in the vapor phase, (c) a new type of chip fiberizer, the rod mill, for disintegrating the softened chips at low energy expenditure and producing a papermaking pulp and a recognition of the potential of a disc attrition mill for this purpose, and (d) a procedure for recovering the sodium acetate and other by-products from the spent pulping liquor.

LITERATURE CITED

[1] Rawling, F. G., "Pulping Wood by the Action of Sodium Sulfite and Subsequent Mechanical Disintegration, "Project 168-1, unpublished report, Forest Products Laboratory, April 21, 1924.

[2] Rawling, F. G., and Staidl, J. A., "Paper and Board From Extracted Chestnut Chips, "Project 1168-J14, unpublished report, Forest Products Laboratory, Aug. 6, 1924.

[3] Rawling, F. G., "A Mill Study of Board from Extracted Chestnut Waste, "Project 168-J14, unpublished report, Forest Products Laboratory, Feb. 17, 1925.

[4] Rawling, F. G., and Staidl, J.A., "Pulping Wood by Action of Sodium Sulfite and Carbonate and Subsequent Mechanical Disintegration,"Project168-1, unpublished report, Forest Products Laboratory, May 4, 1924, to January 1925.

[5] Rawling, F. A., and Staidl, J.A., "The Pulping of Extracted Chestnut Waste, "Project 2168-J14, unpublished report, Forest Products Laboratory, June 8, 1925.

[6] Rawling, F. A., and Staidl, J.A., Paper Industry 7(6):901 (1925).

[7] Edwardes, VinceP., "The Bauer Refiner for Disintegrating Semi-cooked Chestnut Chips, "unpublished report, Forest Products Laboratory, Oct. 2, 1925.

[8] Staidl, J. A., "Manufacture of Chestnut Board by the Semi-chemical Process, "Project 2168-J14, unpublished report, Forest Products Laboratory, Oct. 11, 1926.

[9] Rawling, F. A., and Staidl, J. A., "Mill Study of Pulping Black Gum, "Project 168-1, unpublished report, Forest Products Laboratory, March 10, 1926.

[10] Rue, John D., "Papermaking Characteristics of Semichemical Black Gum Pulp, "Project 168-1, unpublished report, Forest Products Laboratory, April 5, 1926.

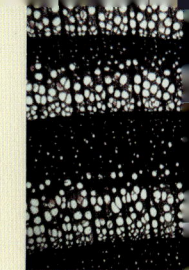

[11] Rue, J. D.,Wells, S. D., Rawling, F. G., and Staidl, J., Paper Trade. J. 83(13):50 (1926).

[12] Rue, J. D., "Installation of a Semichemical Pulping Plant at Hartsville, S. C., Project 1968-1 J32, unpublished report, Forest Products Laboratory, Sept. 16, 1927.

[13] American Paper Institute, Statistics of Paper, Paperboard and Wood Pulp, New York (1982).

[14] Durgin, A. G., and Small, Thaxter W., Jr., Semichemical Pulping in Modern Pulp and Paper Manufacture, John B. Calkin, Ed., Reinhold Publishing Corp., New York (1957).

[15] Bradley, L., and McKee, E., Paper Trade J. 88(8):72 (1929).

[16] Wells, S. D., Paper Trade J. 133(24):32 (1951).

[17] Gocke, W. A., Paper Trade J. 86(23):79 (1928).

[18] Rue, J. D., and Wells, S. D., Paper Trade J. 83(12):53 (1926).

[19] Clark, J.D' A, Pulp Technology and Treatment of Pulp, Miller Freeman Publications, San Francisco, 1978, p. 215.

[20] Lowgren, D., Paper Trade J. 113(11):29 (1941).

[21] Gilbert, H. S., Paper Trade J. 156(22):226 (1972).

[22] Abelson, P. H., Science 319(4582):243 (1983).

[23] Gomory, R. E., Science 220:576 (1983).

[24] Ketchen, W. A., Paper Age 19(5):21 (1983).

The writer is greatly indebted to the Library, Forest Products Laboratory, Madison, Wis. for the loan of unpublished research reports and other reference materials.
Received for review June 16. 1983. Accepted July 18, 1983.

Biological Utilization of Wood for Production of Chemicals and Foodstuffs

GEORGE J. HAJNY   Research Paper FPL 385 (**Abridged**)
March 1981

ABSTRACT

In the long term, mankind will have to depend on solar energy and photosynthetic processes rather than on fossil materials for energy and material needs. Cellulose and the hemicelluloses, which make up about 70 percent of the dry matter of trees and shrubs, are the most abundant, renewable, raw materials on earth. At present, the highest uses of wood are for structural material and as a source of fiber. There are, however, large quantities of wood residues produced during harvesting and manufacture that might be used. Intensive silviculture can greatly increase the supply of wood for all purposes. This paper reviews the work the U.S. Forest Products Laboratory has done over nearly 70 years to produce chemicals and feedstuffs from wood residues. Wood has been converted successfully to fermentation chemicals such as ethyl alcohol, glycerol, arabitol, erythritol, butanol, acetone, and 2,3-butylene glycol as well as to feedstuffs such as molasses and yeast, and to wood modified to make the polysaccharides digestible by ruminants. At present such use of wood is economically marginal, but is potentially economic in the future.

BACKGROUND

From prehistoric time to the late19th century, wood was mankind's principal source of energy for the production of heat and power. Even today, on a worldwide basis, one half of all wood harvested is used for fuel. In the developed countries it makes only a small contribution to the energy budget, but in many developing countries it is still an important source of energy.

In the United States, in colonial times and for many years after, wood was the energy source even though coal deposits were known at the time of the Revolution. It was not until about 1850 that the use of coal as an industrial fuel grew rapidly in the United States, but by 1885 it had surpassed wood.

In the early 19th century, there was no synthetic organic chemical industry. Organic chemicals were natural products either, extracted from the raw material, obtained as byproducts, or produced by fermentation.

The production of charcoal and tars is the oldest of all chemical wood processing methods. Charcoal was probably discovered by an early caveman when he found that the black remains of a previous fire burned with intense heat and little smoke. In all early civilizations charcoal braziers were used for heating and cooking. In the days of the wood sailing ships, naval stores–tar, pitch, and turpentine–were obtained by the destructive distillation of the pines. The softwoods were destructively distilled to obtain the volatiles, with charcoal being a byproduct; with hardwoods, the charcoal was the principal product.

In colonial times hardwood charcoal was produced as a fuel for blast furnaces for the production of pig iron, and as an ingredient in gunpowder. Collection of volatiles from hardwood distillation began in the early part of the 19th century. Pyroligneous acid was refined to produce methanol, acetone, acetic acid, and tars. These byproducts of the hardwood distillation industry were important commercial chemicals until the market was lost to the petroleum industry. In 1920 approximately 100 plants in the United States were recovering these materials from charcoal production; the last of these plants closed in 1969.

Intense and widespread study of coal chemistry began in 1856, when W. H. Perkins discovered mauve, the first coal tar dye. This discovery created the basis for the growth of industrial organic chemistry. A half-century later, industrial chemists turned their attention to petroleum chemistry with results that need no recounting. Wood chemistry has never received that kind of attention, although wood at one time was an important source of chemicals.

The soaring cost of petroleum products, shortages of natural gas in some areas during severe winters, long lines of cars at gasoline stations on days when they are open, have shown, as no amount of lecturing could do, that supplies of fossil fuels are indeed finite. Other incidents raise doubts about nuclear energy. Solar energy, in its broadest sense, offers a safe and dependable alternative, especially in regard to plant growth.

The use of solar energy can take many forms, but one of particular interest that is being evaluated is renewable photosynthetic materials or biomass for the production of chemicals. At the present time, there is great research activity throughout the country on biomass for the production of organic chemicals, especially ethyl alcohol. Wood is probably the most promising form of biomass. This paper will discuss the work of the U.S. Forest Products Laboratory in research on production of chemicals by fermentation of sugars derived from wood, and some other biological studies directed at increased utilization of wood.

RAW MATERIAL

The world's most abundant renewable organic raw materials are the carbohydrates, represented by the sugars, starches, hemicelluloses, and cellulose. Of these plant materials, cellulose is by far the most abundant. In 1819, Braconnot showed that cellulose or vegetable fiber could be converted into fermentable sugars by concentrated acids [13]. From Braconnot's time to the present, much work on the hydrolysis of cellulose has been directed primarily to production of sugars that could be fermented to ethyl alcohol. For example in 1854, two French patents, Nos. 1246 and 2281, had for their respective titles "Manufacture of Ligneous Alcohol" and "Obtaining Alcohol from Organic Substances" [74]. One hundred and twenty-five years later, the literature abounds in articles on the production of ethyl alcohol from biomass.

Chemically, wood consists of two groups of components. The major group comprises the structural components of wood and consists of cellulose, hemicellulose, and lignin. The minor group, designated as extraneous and extractive substances, varies greatly in the nature of the compounds and the quantities in the various species. In this group are found tannins, coloring matter, resins, essential oils, fats, waxes, sterols, pectins, gums, starch, organic acids, nitrogen compounds, and mineral matter. Wood contains 40 to 50 percent of cellulose, 20 to 30 percent hemicellulose, 15 to 30 percent lignin, and 2 to 15 percent of extraneous material [161].

The total carbohydrates of the cell wall of extractive-free wood have been named holocellulose by Ritter [124]. Holocellulose consists of alphacellulose and a heterogeneous mixture of polysaccharides called hemicellulose.

Lignin is a complex three dimensional polymer whose building blocks are phenyl-propane units. Lignin is a generic term as the name is applied to the material in plants as well as the material from pulping and hydrolytic processes and whose properties differ from each other. The lignin from softwoods contains coniferyl groups, while the lignin from hardwoods is made up of both coniferyl and syringyl groups [49]. Lignin is not hydrolyzed by acids as are cellulose and hemicellulose. Lignin appears to have little effect on the acid hydrolysis of wood, but it greatly inhibits hydrolysis of associated polysaccharides by the enzyme, cellulase [8, 128].

Softwoods, as a class, contain more mannose and less xylose than do the hardwoods. Galactose and arabinose are minor constituents in all the materials. The agricultural residues closely resemble the hardwoods in the distribution of the sugars. Softwoods contain more lignin than do the hardwoods, which in turn contain more lignin than do the agricultural residues.

Of the sugars produced by the saccharification {A process of breaking a complex carbohydrate (such as starch or cellulose) into its monosaccharide (simple sugar) components (fructose, glucose, ribose)} of wood, only glucose, mannose, and galactose are fermentable to ethyl alcohol by yeast [167]. A rapid method for determining the potential fermentable sugars for ethyl alcohol production from biomass is quantitative saccharification by strong acid followed by sorption of the fermentable sugars by *Saccharomyces cerevisiae* [90, 133, 134]

There are significant differences in yields of sugar from the various woods. Because of the lower lignin content, hardwoods tend to give higher yields of total reducing sugars (as glucose) than the softwoods. However, the fermentability of the sugars from the softwoods is higher than that of the hardwoods. The net result is that the potential fermentable sugar yield from softwoods is higher than from the hardwoods. Thus, for ethyl alcohol production, the softwoods would be the species of choice.

SACCHARIFICATION

Because only simple sugars can be used by yeasts in ethyl alcohol and most other fermentations, it is necessary to hydrolyze the wood polysaccharides. The hydrolysis of cellulosic materials to the monomer sugars appears to be a simple hydrolytic cleavage of glycosidic bonds. As such, one would expect the reaction to be simple, analogous to starch hydrolysis, and the manufacturing costs to be low. Such is not the case, for cellulose is extremely resistant to hydrolysis; this resistance is the most important factor in determining the cost, methods of production, and character of wood sugar solutions. The glycosidic bonds of solubilized cellulose can be readily hydrolyzed, but the crystalline organization of untreated cellulose results in low accessibility to the dilute acid commonly used as a catalyst [48]. As a result, the conditions of temperature and acid concentration necessary to carry out the reaction in a reasonable time cause serious degradation of the sugars produced, resulting in low yields and solutions which are difficult to ferment.

The kinetics {mechanism by which a physical or chemical change is effected} of batch saccharification of cellulose with dilute acid was first studied by Luers [82, 83]. This early work showed that the yield of sugar from cellulose followed a growth and decay curve that was well represented by consecutive first-order reactions. Changing reaction conditions were believed to affect the reaction rates, but not the maximum yield obtainable. Later work by Saeman [129] showed that the maximum yield does change with changes in the reaction conditions. The first-order rate constants for the hydrolysis of cellulose and for the decomposition of glucose were found to be of similar magnitude. The maximum yield of glucose is a function of the ratio of the rates of hydrolysis and decomposition. Saeman's work showed that increasing the temperature and acid concentration resulted in increased efficiency in the conversion of cellulose to reducing sugar [61].

Wood hydrolysis processes, which have been carried out on a commercial scale, fall into the following categories [48]:

1. A single-stage dilute-acid hydrolysis carried out at elevated temperatures without separation of the product as it is formed.

2. A percolation process using a dilute acid at elevated temperatures in which yields are higher than from the single-stage hydrolysis, because the sugars produced are continuously removed as they are formed.

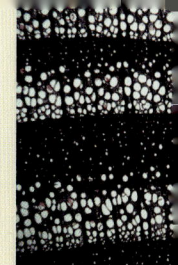

3. A concentrated acid process in which the crystalline nature of the cellulose is destroyed; the cellulose is solubilized, after which it is completely hydrolyzed by dilute acid.

As noted earlier, the two major deterrents to the utilization of lignocellulosic residues for chemical, enzymatic, or microbiological conversion processes are cellulose crystallinity and lignin. Lignin acts by restricting enzyme access to the cellulose, while crystallinity restricts the rate of all three modes of attack on the cellulose. Chemical and physical pretreatments have been investigated to overcome these problems. Tsao et al. [165] have shown that cellulose solvents, such as cadoxen or an iron sodium tartrate solution, can decrystallize and dissolve cellulose from biomass and that the precipitated cellulose is easily hydrolyzed enzymatically. Millett et al. [92] reviewed the physical and chemical methods for pretreatment of lignocellulosic materials that have been used to increase the reactivity of cellulose toward acid or enzymatic hydrolysis. Chemical methods which induce swelling or dissolution of cellulose, and physical methods such as vibratory ballmilling or electron radiation, can markedly increase the rate of cellulose hydrolysis. Commercial feasibility of any of the pretreatments remains in doubt until the costs of the pretreatment can be shown to be economically viable.

Research on ethyl alcohol production from wood residues has been carried on intermittently at the U.S. Forest Products Laboratory since 1910 with the main emphasis on dilute acid hydrolysis. The earliest work was done in response to the vast amounts of wood residues accumulated at sawmills in excess of their power requirements.

At the Forest Products Laboratory, Kressmann [74] conducted a long series of experiments on a pilot-plant scale in which he investigated: (1) the influence of temperature and pressure of digestion, (2) time of digestion, (3) ratio of water to sawdust, (4) concentration of sulfuric acid in water, (5) ratio of acid to dry sawdust, (6) size of wood particles, (7) yields of alcohol from different wood species, (8) fermentation variables, and (9) steam consumption. A porcelain-lined rotary digester with a capacity of 100 pounds of dry sawdust, heated by direct steam, was used for the experimental hydrolysis. The resulting sugars were extracted from the hydrolyzed wood in diffusion batteries, neutralized with lime, filtered, and prepared for fermentation. Maximum yields of fermentable sugars were obtained in 20 minutes at 7.5 atmospheres with a water-to-wood ratio of 1.25 to 1 and a sulfuric acid concentration of 2 percent. Twelve species of softwoods and 10 of hardwoods were hydrolyzed and gave about the same yield of reducing sugars. However, the sugars from softwoods were about 70 percent fermentable while those from hardwoods were 30 percent fermentable by yeast. This is understandable, since under the conditions of hydrolysis, most of the hemicellulose would be hydrolyzed and little of the true cellulose, so the hardwood hydrolyzate would consist chiefly of xylose. The single-stage batch hydrolysis, which became known as the "American Process" was applied on an industrial scale in a plant at Georgetown, S.C., in 1913 and then in one at Fullerton, La., in 1916. The Forest Products Laboratory assisted in the development and pilot-plant work of both these plants [161], and in fact Kressmann, formerly at the Forest Products Laboratory, became plant manager of the Fullerton plant [151]. Both plants operated successfully during World War I and for several years after. They produced from 5,000 to 7,000 gallons of ethyl alcohol a day from southern yellow pine sawdust and chips obtained from nearby sawmills. During the war the sawmills overcut their holdings so that, as a result of the curtailment of mill operations and a decrease in the price of blackstrap molasses, both hydrolysis plants were forced to close [151].

Batch Hydrolysis
Further experimental work with the rotary digester continued at the Forest Products Laboratory. Successive treatments with dilute sulfuric acid resulted in sugar yields of about 40 percent of the weight of the wood [150]. This, in effect, was a forerunner of the Scholler percolation process. Plow et al. [120] showed

that, by successive acid treatments with water wash between the acid cycles, yields of ethyl alcohol of over 50 gallons per ton of wood could be achieved. However, the involved operating schedule and time required for the completed run were considered to be serious disadvantages that could be overcome by use of the vertical stationary percolator.

Percolation Process

The rates of sugar production and destruction are of similar magnitudes in the single-stage hydrolysis of cellulose. Thus, the yields ordinarily are limited to about one-third of theoretical. To make the ratio of production to destruction more favorable, Scholler [142, 143] developed a dilute-acid percolation process using a vertical stationary digester. The Scholler process has been described by Luers [82, 83, 84], Fritzweiler and Karsch [36], and by Saeman et al. [135]. Basically, the process involves the hydrolysis of wood in a vertical cylindrical pressure vessel by a percolation process in which dilute acid is injected into the top of the vessel; acid addition is stopped and the hydrolyzate is withdrawn from the bottom of the vessel. This cycle of addition and withdrawal is repeated until the hydrolysis is complete, which requires from 14 to 20 cycles.

The Scholler plants constructed in Germany (1930s) contained six to eight digesters of 50 cubic meters capacity. The digester is loaded with chips, sawdust, or shavings from the top. The digester holds about 10 tons of wood, which is packed to a density of 12.5 pounds per cubic foot by steam shocking. The bottom of the digester is equipped with a filter cone, lines for removing the hydrolyzate, and a quick-opening valve through which the lignin is discharged. After the percolator is filled, the wood is heated with direct steam to 135° C. A charge of 1.4 percent sulfuric acid is injected at a temperature 5 to 10° C lower than the percolator contents. After the acid is added, the charge is brought back to temperature. The solution is then "pressed" from the percolator by applying steam to the top of the charge. This operation is then repeated the required number of times with the temperature gradually increased to a maximum of 185° C. The complete percolation cycle requires about 15 hours. Yields of 95 percent alcohol in a well-run plant amounted to about 53 gallons per ton of wood.

Madison Wood Sugar Process

The rights to the Scholler process in the United States were acquired by the Cliffs Dow Chemical Company of Marquette, Mich., in 1935. Pilot plant studies were made of a modified Scholler process but never commercialized.

Due to the need for ethyl alcohol in the synthetic rubber program, the War Production Board in 1943 requested the Forest Products Laboratory to study the Scholler process. Arrangements were made to use the pilot-plant at Marquette, Mich. but, after some preliminary work, the facilities were transferred to Madison, Wis. The Vulcan Copper and Supply Company was requested to follow the pilot-plant work and to prepare an engineering report which could serve as the basis for the design of a commercial plant suitable for the manufacture of alcohol from wood waste [28].

The pilot-plant equipment for the hydrolysis and fermentation is shown, diagrammatically, in figure 1. The original pilot-plant hydrolyzer was a 27-cubic-foot, 350-pound capacity, vertical digester, made of silicon bronze with a length of 11 feet and a diameter of 2 feet. A second hydrolyzer also used was a 60-cubic-foot, 600-pound-capacity digester made of Monel metal, with a length of 13 feet and a diameter of 2-1/2 feet.

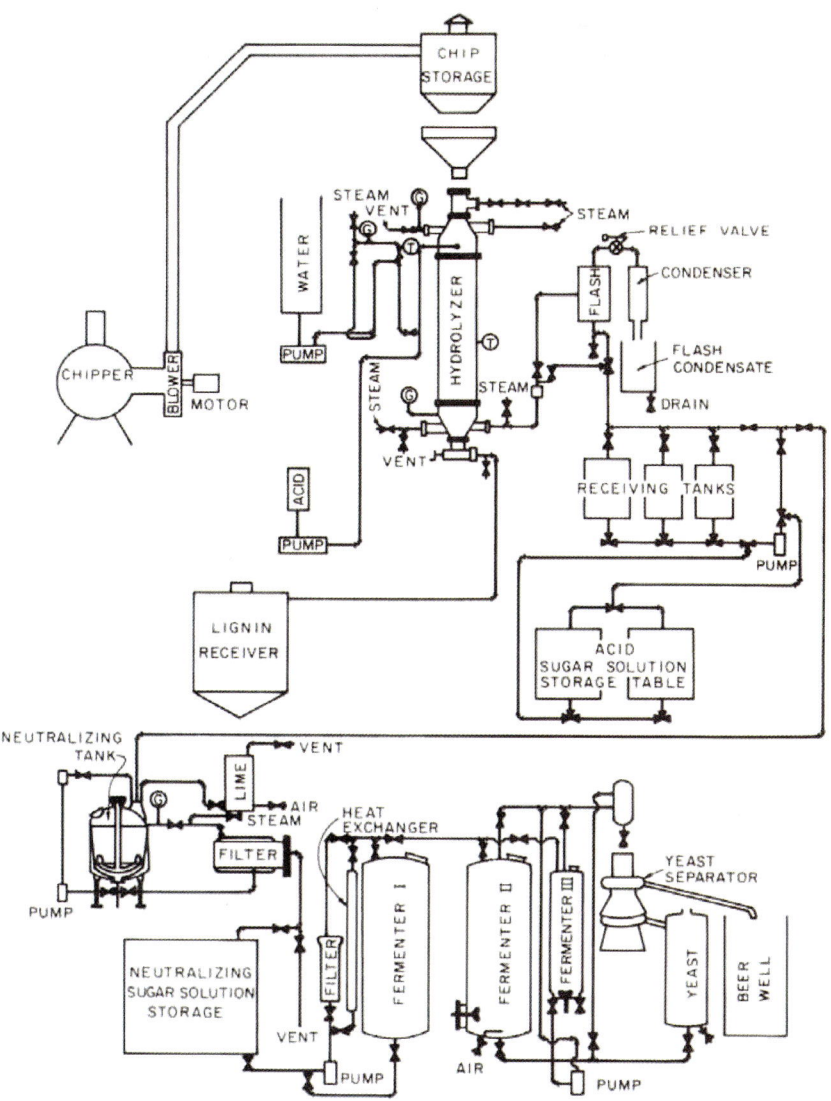

Figure 1.—Equipment for saccharification of wood and fermentation of wood sugar solutions.
(M 149 123)

Preliminary work with the pilot plant equipment was by the Scholler process [52]. Taking advantage of the kinetics of the process [129], modifications were soon introduced to reduce the time required and to increase the yield of reducing sugars [51]. A rapid percolation process, the Madison Wood-Sugar Process, was developed which is similar to the Scholler method of hydrolysis in that dilute sulfuric acid is percolated through a bed of wood chips in a stationary digester. The Madison Process differs in that, after an initial period of low-temperature hydrolysis, the acid solution is pumped in continuously at the top of the hydrolyzer and hydrolyzate is removed at the bottom with no interruptions until the hydrolysis is complete.

A description of the Madison Wood-Sugar Process follows [80]:

Madison Process
Wood waste in the form of shavings, sawdust, or chips is loaded into the percolator. When filled, the hydrolyzer is temporarily closed and steam is rapidly injected above the chip bed. This packs the bed and

permits more wood to be added. The procedure is repeated once or twice. With Douglas-fir, a density of about 14 pounds per cubic foot was found to be satisfactory. Dense hardwoods could be charged in amounts up to 20 pounds per cubic foot.

With the cover closed securely, steam is admitted to the digester and allowed to pass through the chip bed and out the bottom to remove air and heat the chips. When steam begins to flow rapidly from the bottom vent, this line is closed and steaming continued until the charge reaches a temperature of 145° C. When this temperature is reached, steaming is discontinued and dilute sulfuric acid at 145°C is injected. A calculated quantity of hydrolysis solution containing 2.5 percent acid is first injected to compensate for the diluting effect of the condensate from the steaming operation and the moisture content of the wood. The composite acid concentration in the digester is thus brought to 0.7 percent. The entering acid solution is then readjusted to a concentration of 0.7 percent and a constant flow rate is maintained to give a 2.5 to 1 liquor-to-wood ratio in the hydrolyzer in 45 minutes. The temperature of this entering stream is held at 145° C for 30 minutes and then taken to 170° C in the next 15 minutes.

After 45 minutes of elapsed time, the hydrolyzate control valve is opened at the bottom of the digester and the takeoff rate is adjusted so the liquor level coincides with the chip-bed surface. The hot acid liquor flow rate is adjusted to equal 0.08 times the weight of the dry charge 6 per minute. The temperature of the acid liquor is increased at the rate of 1° C per minute so that, after 1 hour's elapsed time, it is at a maximum of 185° C. At this time, most of the low temperature hydrolyzate has left the reactor, and hydrolysis of the resistant cellulose is underway in the upper part of the chip bed. The high through-put rate is continued until the sugar concentration in the hydrolyzate stream reaches about 3 percent. The flow rates are then halved and maintained until the sugar concentration drops below 2 percent, when addition of the acid is stopped. Hydrolyzate removal continues until the liquor level in the vessel is 10 to 12 inches below the surface of the chip bed. The blow-down valve at the bottom of the hydrolyzer is opened and lignin residue is discharged by steam pressure into the lignin receiver.

The sugar solution withdrawn from the hydrolyzer is passed to a flash tank where the pressure is reduced to 30 pounds per square inch. The solution is neutralized with lime and filtered at this pressure. Because of the temperature-solubility relationship of calcium sulfate, this procedure avoids subsequent scaling in the stills.

The condensed steam from the flash tank contains the following volatiles [52]:
Acetic acid - 1.50 grams per liter
Formic acid - 0.19 grams per liter
Methanol - 0.15 grams per liter
Acetone - 1.59 grams per liter
Furfural - 2.4 grams per liter
The lignin recovered in the cyclone has a moisture content of about 70 percent. On pressing to about 45 percent moisture content, sufficient sugar solution could be recovered to raise the total yield of reducing sugars by about 3 percent. The lignin residue from various runs amounted to 25 to 40 percent of the original wood. In representative runs on Douglas-fir, the lignin residue amounted to 29 percent of the original wood and contained 17.7 percent of cellulose. The heating value of the dried residue is approximately 10,800 Btu's per pound.

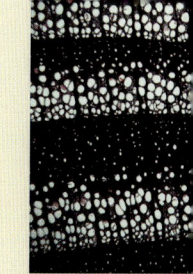

Yields of reducing sugars obtained from various residues and wood species by the Madison Wood-Sugar Process *[161]* are listed in table 5.

Type of wood product	Hydrolysis time Hours	Yield of sugar Percent	Average sugar concentration Percent
White spruce chips	3.1	54.2	5.1
Douglas-fir chips	3.0	52.5	5.3
Douglas-fir sawdust	3.1	44.7	4.9
Douglas-fir hog fuel	3.0	38.7	5.1
Douglas-fir bark	2.9	15.3	2.3
Southern yellow pine woods waste	3.3	50.0	4.8
Southern yellow pine sawdust	3.1	47.5	4.6
Ponderosa pine chips	3.0	51.5	5.35
Eastern white pine sawmill slabs	3.1	44,6	4.55
White fir chips	3.0	53.8	5.4
Western white pine chips	3.0	46.6	5.0
Sugar pine chips	3.0	46.9	4.0
Western hemlock chips	3.0	51.5	5.0
Western larch chips	3.0	54.0	4.9
Western larch sawmill slabs	3.0	42.0	4.9
Lodgepole pine chips	3.0	51.0	4.9
Spent turpentine chips, long-leaf pine stumps	3.1	40.0	4.8
Western redcedar chips	3.1	46.9	4.3
Redwood chips	3.1	42.6	4.0
Mixed southern oak shavings	3.0	51.0	5.0
Mixed southern oak sawdust	3.0	46.9	5.05
Mixed southern oak sawmill waste	3.1	42.9	4.7
Sugar maple sawmill waste	3.0	48.0	4.75
Aspen sawmill waste	3.2	50.2	5.0
Yellow birch sawmill waste	3.1	49.5	5.3
Hickory sawmill waste	3.2	42.8	4.9
Beech sawmill waste	3.1	46.5	5.04
Willow sawmill waste	3.2	50.0	5.0

Table 5.—Yield of sugar by hydrolysis of wood by the Madison wood-sugar process

Experimental Operation

In 1944, design and construction of a Government-financed plant at Springfield, Oreg., was begun by the Vulcan Copper and Supply Company, based on data obtained at Madison. A description of the plant is given by Saeman *[131]*. The plant was designed with a through-put of 220 tons of dry bark-free Douglas-fir wood residues per day, with a production of 10,700 gallons of 190° proof alcohol. The plant with its five percolators 40 feet high and 8 feet in diameter was not completed when World War II ended. One digester was put into experimental operation to determine conditions necessary for successful operation. The production of wood sugars presented some problems, none of which were considered fundamental to the process. The fermentation of the wood hydrolyzate was carried out successfully and confirmed the pilot plant results. Some 50,000 gallons of 190° proof alcohol were produced. Unfortunately, the plant operators were not financially able to put the plant into full operation so that data on costs were never obtained".

LITERATURE CITED (Abridged list)

8. Baker, A. J., A. A. Mohaupt, and D. F. Spino. 1973.J.AnimalSci.37(1):179-182.

13. Braconnot, H. 1819. Gilbert's Annalen der Physik 63:348.

28. Faith, W. L. 1945.Ind.Eng.Chem.37(1):9.

36. Fritzweiler, R., and W. Karsch. 1938.Z. Spiritusind. 61:207.

37. Frost, D. V., and H. R. Sandy. 1949. J. Nutrition 39:427.

48. Hail, J. A., J. F. Saeman, and J. F. Harris. 1956. Wood saccharification–A summary statement. Unasylva 10(1):7-16.

49. Harkin, J. M. 1966. Fortschr. Chem. Forsch. 6:101. 34

51. Harris, E. E., and E. Beglinger. 1946.Ind.Eng.Chem.38:890-895.

52. Harris, E. E., E. Beglinger, G. J. Hajny, and E. C. Sherrard. 1945.Ind.Eng.Chem.37(1):12-23.

61. Harris J. F., J. F. Saeman, and E. G. Locke. 1963. In The Chemistry of Wood, Chapter 11. B. L. Browning, ed. Interscience Publishers, New York, N.Y.

74. Kressman, F.W. 1922. U.S. Dep. Agric. Bull. 983.

80. Lloyd, R. A., and J. F. Harris. 1955. U.S. For. Prod. Lab. Rep. No.2029.

82. Lüers., H.Z. 1930.Angew.Chem.43:455.

83. Lüers, H. Z. 1932.Angew.Chem.45:369.

84. Lüers, H. Z. 1937. HolzRoh-und Werkstoff 1:35.

90. Menzinsky, G. 1942.SvenskPapperstidn.45:421.

92. Millett, M. A., A. J. Baker, and L. D. Satter. 1976. Biotechnol. and Bioeng. Symp. No. 6, p. 125-153.

120. Plow, R. H., J. F. Saeman, H. D. Turner,and E. C. Sherrard. 1945. Ind. Eng. Chem. 37(1):36-43.

124. Ritter, G. J., and E. F. Kurth. 1933.Ind.Eng.Chem.25:1250.

128. Saarinen, P., W. I. Jensen, and J. Alhojarvi. 1959. Acta. Agral. Fennica 94:41-62.

129. Saeman,J. F. 1945.Ind.Eng.Chem.37:43.

131. Saeman, J. F., and A. A. Andredson. 1954. *In* Industrial fermentation. L. A. Underhofler and R. J. Hickey, eds., Chemical Publishing Co., NewYork, N.Y.

133. Saeman, J. F., J. L. Bubl, and E. E. Harris. 1945. Ind. Eng.Chem.,anal.ed. 17:35.

134. Saeman, J. F., E. E. Harris, and A. A. Kline. 1945. Ind. Eng.Chem.,anal.ed. 17:95.

135. Saeman, J. F., E. G. Locke,andG. K. Dickerman. 1945. Production of wood sugar in Germany and its conversion to yeast and alcohol. U.S. Dep.Commerce,Off. Publ. Bd., Rep. No. 7736.

142. Scholler,H. 1935. Werkbl. Papierfabrik No. 5.

143. Scholler, H. 1939.Chem.Ztg.63:737.

150. Sherrard, E. C., and P. B. Davidson. 1927. Paper delivered at Am. Chem. Soc, meeting, Detroit, Mich.

151. Sherrard, E. C., and F. W. Kressmann. 1945. Ind. Eng. Chem. 37(1):5.

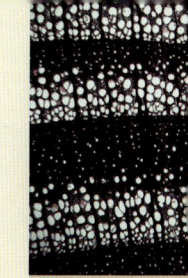

161. Stamm, A. J., and E. E. Harris. 1953. Chemical processing of wood. Chemical Publishing Co., New York, N.Y.

165. Tsao, G. T., M. Ladisch, C. Ladisch, T. A. HSu, B. Dale, and T. Chou. 1978. In Fermentation processes. D. Perlman, ed., Academic Press.

167. Underkofler, L. A., and R. J. Hickey. 1954. Industrial fermentations. Chemical Publishing Co., New York, N.Y.

Brief History Follow Up

Some of the details of the development of ethanol from wood (1914-1949) are discussed in the Energy section and the literature. [1 through 19]

After WWII interest in ethanol from wood waned. Research was continued on some of the chemistry, however, there was some renewed interest in the utilization of the hardwoods in the Tennessee Valley and in 1952 the Tennessee Valley Authority built a pilot plant at Wilson Dam to modify the Madison Process. [20]

During the 1960s research continued in universities and the FPL but again on a low key level, until the oil crisis occurred in 1973. Thus there was another major effort to produce liquid fuel and chemicals as at least a partial replacement for those produced from petroleum. This resulted with FPL contracting with Katzen and Associates to prepare a study on the economics of producing alcohol and other chemicals from wood waste. No plant was built. [23]

Also during the 1970s there was significant research done by others on the enzymatic hydrolysis of cellulose to glucose. A report of the Natick Program was prepared in 1981 / Natick Research and Development Command, U.S. Army / (September 1981) [24]

In 1979, Mike Millett, Marilyn Effland and Dan Caulfield conducted experiment on the influence of fine grinding on the hydrolysis of cellulosic materials: acid vs. enzymatic [25]

The oil price eased around 1976 but in 1979 there was another sharp increase in oil prices. However, additional oil had been discovered and in the early 1980s there was an oil glut that brought the price of oil down dramatically by 1986. Interest in ethanol waned again.

One of the major challenges with ethyl alcohol yield had been the problem with converting the large amount of pentose sugars found in hardwood to alcohol. FPL research had continued on a reduced scale but in 1981, Tom Jefferies developed a method for converting pentoses to ethanol. [27]

Also during 1981, John Zerbe, Andrew Baker, and Tom Jefferies prepared a short summary of the state of the art to help developing countries understand the options to producing alcohol from wood. [28]

In 1984 TVA, in cooperation with DOE, built a pilot plant in Muscle Shoals, AL, for production of ethyl alcohol from wood and corn stover. (Concentrated acid, batch process.)

As part of this cooperative TVA/DOE effort, TVA designed, built, and operated a 10 gal/hr pilot fuel alcohol facility. The facility was designed to be relatively energy efficient, relatively simple to construct and operate, and capable of producing 190-proof fuel alcohol. Information in this publication includes procedures and recommendations for building and operating small-scale fuel alcohol plants ranging in size from 6 to 50 gal/hr of 190-proof ethanol and are for plants using corn grain as the feedstock material for alcohol production. Plants may be built onsite using conventional tools and locally available components.

Experiences from design, construction, operation, and modification of the Muscle Shoals pilot facility were used as a basis for this information.

In 1985 an FPL report on the two-stage, dilute sulfuric acid hydrolysis of wood was published summarizing the fundamentals of the process steps in the production of ethanol and other chemicals. [30]

During the 1990s FPL researchers continued to find new or improved strains of yeast to increase the yield of ethanol from pentose and glucose sugars, and utilization of an enzymatic approach to wood hydrolysis.

In 2007, again because of a need for another source of liquid fuels, the Department of Energy selected six cellulosic ethanol plants for up to $385 million in federal funding. Department of Energy also established research centers in Tennessee, Wisconsin, and California to research new biofuel technologies with $125 million each over five years. [33] [34] [35]

In 2008, J.Y. Zhu, X.J. Pan, G.S. Wang, R. Gleisner, developed a sulfite pretreatment to overcome recalcitrance of lignocellulose (SPORL) to convert softwood to its major components of cellulose, hemicellulose, and lignin. This process helps remove a major barrier to converting wood to chemicals. [38].

In summary, the production of ethanol from cellulose is not only technically possible but was done commercially in 1913 and 1916 in the United States. (It has also been produced in Russia and Germany.) However, any of the processes are complex and when oil is cheap the processes are uneconomical. Researchers around the world continue to try to reduce the complexity and improve the yields of alcohol from wood and biomass to make the process cost competitive. [37]

Additional Literature

[1] Manufacture of ethyl alcohol from wood waste I, Preliminary experiments on the hydrolysis of white spruce / F.W. Kressmann / Journal of industrial and engineering chemistry Vol. 6(8) (Aug. 1914)

[2] Manufacture of ethyl alcohol from wood waste II, The hydrolysis of white spruce / F.W. Kressmann / Journal of industrial and engineering chemistry Vol. 7(11) (Nov. 1915)

[3] Manufacture of ethyl alcohol from wood waste III, Western larch as a raw material / F.W. Kressmann / Journal of industrial and engineering chemistry Vol. 7(11) (Nov. 1915)

[4] Ethyl alcohol from wood: the process, its development and requirements / F.W. Kressmann / American lumberman No. 2148 (July 15, 1916)

[5] Ethyl alcohol from wood waste: its possibilities as a motor fuel, and the process of manufacture / Rolf Thelen / The car owner (May 1917)

[6] Ethyl alcohol from waste sulphite liquor using an acclimated yeast / FPL Technical note / Forest Products Laboratory No. E3 (1919)

[7] The manufacture of ethyl alcohol from wood waste / E.C. Sherrard / Forest Products Laboratory No. 668 (1920)

[8] Yields of alcohol from wood waste / Technical note / FPL No. 120 (1920)

[9] The manufacturing of ethyl alcohol from wood waste / F.W. Kressman / U.S.D.A. Bulletin 983 (1922)

[10] Ethyl alcohol from western larch (Larix occidentalis, Nuttall) / E.C. Sherrard / Journal of industrial and engineering chemistry Vol. 14(10) (Oct. 1922)

[11] A study of factors affecting the hydrolysis of wood: progress report / Jerome F. Saeman / Forest Products Laboratory No. 1446A (Jan. 1944)

[12] Ethyl alcohol from waste wood by a modified Scholler process / W.L. Faith, J. A. Hall / Chemical and engineering news. Vol. 22 (Apr. 10, 1944)

[13] The quantitative saccharification of wood and cellulose / Jerome F. Saeman, Janet L. Bubl, Elwin E. Harris / FPL No. 1458 (1944)

[14] Fermentation of wood sugars to ethyl alcohol / Reid H. Leonard, George J. Hajny / FPL No. 1466 (1944)

[15] A proposed design of a plant for the production of ethyl alcohol by the hydrolysis of waste wood / William Lawrence Faith, T. O. Wentworth / War Production Board, Office of Production Research and Development (1944)

[16] Hydrolysis of wood / E.e. Harris, E. beglinger, G J. Hajny, E.C. Sharrard / Industrial and engineering Chemistry Vol 37(1) (January, 1945)

[17] Chemical utilization of wood : its opportunities and obstacles / Alfred J. Stamm / Journal of Forestry 44(4):258-265 (April 1946)

[18] The production of wood sugar in Germany and its conversion to yeast and alcohol / Jerome F. Saeman, Edward G. Locke, and G.K. Dickerman / Paper trade journal Vol. 123(12) (Sept. 19, 1946)

[19] Derived products and chemical utilization of wood waste / Elwin E. Harris / Forest Products Laboratory No. 1666-10 (1949)

[20] Hydrolysis of wood using dilute sulfuric acid / N. Gilbert, I. A. Hobbs, J. D. Levine / Tennessee Valley Authority, Wilson Dam / Industrial and engineering chemistry Vol 44(7) (July 1952)

[21] Sixth meeting of the FAO Technical Panel on Wood Chemistry: Stockholm 1953 / Technical Panel on Wood Chemistry Meeting / Food and Agriculture Organization of the United Nations, Rome, Italy / (March 1954)

[22] Separation and utilization of hardwood hemicelluloses / J. F. Harris, J. F. Saeman, Edward G. Locke / Forest products journal Vol. 8(9) (Sept. 1955)

[23] Chemicals from wood waste / Allan E. Hokanson, V. B. Diebold, D. W. Bennett, R. P. Klier, S. A. Stein, W. W. Kline, D. C. Ferguson, R. Katzen / Forest Products Laboratory / Raphael Katzen Associates. (1975)

[24] Production of ethanol from biomass: A bibliography / M. C. Nason / Tennessee Valley Authority TVA/PUB-86/8 (OSTI ID: 6314429; DE86900545) Muscle Shoals, AL (Jan. 1979)

[25] Influence of fine grinding on the hydrolysis of cellulosic materials: acid vs. enzymatic / Millett, M. A. Effland, M. J., Caulfield, D. F./ Hydrolysis of cellulose: mechanisms of enzymatic and acid catalysis / edited by Ross D Brown Jr. and Lubo Jarasek. Washington : American Chemical Society, 1979. American Chemical Society, Advances in Chemistry series; no. 181 (1979)

[26] Biological utilization of wood for production of chemicals and foodstuffs / George J. Hajny / FPL Research Paper FPL 385 (1981)

[27] Conversion of xylose to ethanol under aerobic conditions by Candida tropicalis / Jeffries, T. W. / Biotechnology letters Vol. 3(5) (May 1981)

[28] Establishment of commercial-scale pilot plant for alcohol from biomass in developing countries / John I. Zerbe, Andrew J. Baker and Thomas W. T. Jeffries / Forest Products Laboratory, WC Z5E (1981)

[29] Construction and operation of small-scale fuel alcohol plants / P. C. Badger, R. S. Pile, C. E. Madewell / Tennessee Valley Authority, Muscle Shoals, AL / TVA/PUB-86/16 / OSTI ID: 5989218; DE86901139 (Mar 1, 1984)

[30] Two-stage, dilute sulfuric acid hydrolysis of wood: An investigation of fundamentals. John F. Harris, Andrew J. Baker, Anthony H. Conner, Thomas W. Jeffries, James L. Minor, Roger C. Petterson, Ralph W. Scott, Edward L. Springer, Theodore, H Wegner, John I. Zerbe / FPL Gen. Tech. Rep. FPL-45. (1985)

[31] United States wood biomass for energy and chemicals: Possible changes in supply, end uses, and environmental impacts / Kenneth E. Skog, Howard N. Rosen, / Forest products J. 47(2): 63-69. (1997)

[32] Thermal energy, electricity, and transportation fuels from wood / John I. Zerbe / Forest Products Journal Vol. 56(1) (January 2006)

[33] DOE selects six cellulosic ethanol plants for up to $385 million in federal funding / Craig Stevens / U.S. Department of Energy, Office of Public Affairs, (February 28, 2007)

[34] DOE announces bioenergy research center funding. Craig Stevens / U.S. Department of Energy, Office of Public Affairs, (August 2, 2007)

[35] Energy Information Administration / September 2007 Monthly Energy Review

[36] U.S. timber production, trade, consumption, and price statistics 1965 to 2005 / James. L. Howard / FPL Research Paper FPL-RP-637 (2007)

[37] Integrated biomass technologies: a future vision for optimally using wood and biomass / J. E. Winandy, A.W. Rudie, R. S. Williams, T. H. Wegner / Forest products journal Vol. 58(6 (2008)

[38] Sulfite pretreatment (SPORL) for robust enzymatic saccharification of spruce and red pine / J.Y. Zhu, X. J. Pan, G. S. Wang, R. Gleisner / Bioresourc. Technol.. (2008), doi:101016/j.biortech.2008.10.057.

FPL People

Partial list of people who worked at the Forest Products Laboratory.

Following is a partial list of people who have worked at the Forest Products Laboratory since its beginning in 1910. The names were compiled from past employee directories, phone lists, and photographs. Most of those listed were permanent employees of the Laboratory, but some were visiting scientists, temporary employees, students, or volunteers. Unfortunately, some names are missing. The major gap in records was 1911 to 1917; however, also missing were directories and phone lists from 1948, 1951, 1956, and 1977. A few people worked at the Laboratory for such a short time that they were not included in the available sources and thus not included in this list. My apologies for any person who was inadvertently not included or whose name is misspelled. Also, a few people may be listed twice (under a birth name and married name).

Abaly, Henry
Abel, Joseph
Abitz, Bethsheba
Abitz, Peter
Abo-Sharkh, Basel
Abrams, Julian E.
Abubakr, Said
Acker, Rachel
Acree, S.F.
Acres S.F.
Adam, Martin A.
Adams, Arnel B.
Adams, Bill
Adams, Bonnie J.
Adams, James A.
Adams, Marion F.
Adams, Mary D.
Adams, Robert G., Jr.
Adank, Harvey L.
Aderman, O. Darrell
Adkins, Zanie B.
Adler, Genevieve, E.
Adriano, Reynaldo A.
Aflenzer, FriedrichA.
Agard, Mallory
Agarwal, Umesh P.
Ahmed, Aziz
Ahn, Jae Hong
Ahrens, Helen L.
Ahrens, Otto G.
Ahrens-Sather, Judy M.
Ainsworth, Henry B.
Aiyar, S. S.
Akhtar, Masood
Akins, Virginia
Albers, Arline C.
Albers, William B.
Albrecht, Estrella M.
Albright, Bernice
Alden, Harry
Alderman, Matthew
Alexander, Mike
Alexander, Russell M.
Alfke, Douglas
Alford, Cora M.
Ali, Mohammed Omar
Alig, Joanne
Allen, Anthony
Allen, Avis
Allen, Claudia M.
Allen, Michael W.
Allen, Shirley W.
Allin, John W.
Allinger, Ida H.
Allsop, Simon
Almusin, Oscar C.
Alonte, Antonio D.
Alsteens, Clarence E.
Alston, Linda K.
Alt, Karen E.
Alwin, Herman F.
Amacher, Robert A.
Amartey, Samuel
Ambrose, Patricia L.
Amundson, Jeffrey J.
Ancona, Edward P.
Anderegg, Robert J.
Anderson, Albert
Anderson, Allan E.
Anderson, Allan F.
Anderson, Alvin A.
Anderson, Arthur B.
Anderson, Cynthia L.

Anderson, Douglas A.
Anderson, Elsie D.
Anderson, Eric A.
Anderson, Evert A. F.
Anderson, Francis R.
Anderson, Fredolf T.
Anderson, Irsk
Anderson, James R.
Anderson, L. Andy
Anderson, Lauris A.
Anderson, Lawrence A.
Anderson, LeRoy O.
Anderson, Malchin T.
Anderson, Malcomb T.
Anderson, Marjorie
Anderson, Mary Ann
Anderson, Maurice M.
Anderson, Maybert E.
Anderson, Norman A.
Anderson, Philip C.
Anderson, Ralph L.
Anderson,Tonnes N.
Andrews, Leslie H.
Andrews, Troy M.
Andrews,Thomas A.
Andruszkiewicz, Michal F.
Angus, Marlin
Anthony, Charles E.
Anthony, Walter B.
Antonioni, David T.
Arango, Rachel A.
Aratson, Anna
Aravamadhan, Shrinivas
Arivamahdan, Shrinivas
Arcoren, Tony A.
Ard, Patricia J.
Ardito, Elizabeth
Argus, Marlin
Arimoto, K.
Armbrecht, Helen
Armbrecht, Prescott
Armstrong, A.K.
Armstrong, James C.
Armstrong, Robert L.
Arneson, Ellizabeth C.
Arneson, Gus N.
Arnold, Kathleen M.
Arntson, Anna
Arrivee David A.
Arthur, Charles M.
Arvela, Marianna
Ashley, Terry
Asleson, OscarT.
Atalla, Rajai H.
Attridge, Mike
Atwood, Robert
Aubey, Rolland A.
Auchter, Richard J.
Ault, Melvin E.
Austerman, Fredericks
Austin, Monte M.
Austin, Susan R.
Austin, Thomas
Auth, Hak
Avanzado, Melecio B.
Avanzado, Norma A.
Avila, Marcela
Ayers, James E.
Babcock, Virginia L.
Babcock, Wesley W.
Bachelder, Calvin L.
Bachelder, Roy H.

Bachhuber, John G.
Bae, JuYun (ju-li-an)
Baechler, Adela
Baechler, Roy H.
Baerwolf, Sheila K.
Bahr, Herbert
Baille, Cecil C.
Baird, Eva F. G.
Baird, Parker K.
Baker, Andrew J.
Baker, Evelyn J.
Baker, Gloria J.
Baker, Hazel. B.
Baker, Nicholas D.
Baker, O.
Baker, William J.
Bakken, Eunice F.
Bakken, H. June
Bakken, Lance
Bakulin, Virginia L.
Balczewski, Joseph D.
Bales, Anna M.
Ballistreri, Michael
Ballmer, Doris M.
Ballweg, Bonnie K.
Ballweg, Jane M.
Balsamo, Virginia L.
Baltes, Bertha Z.
Baltes, Kenneth J.
Baltes, Roman C.
Baltzer, Barbara R.
Bambery, James E.
Bancroft, Ann
Bancroft, Russell H.
Bangel, James. M.
Banick, Mark
Banik, Mark T.
Banks, Shirley M.
Bao, Wuli
Barbosa, A.
Barbour, Bert S.
Barden, Edward W.
Bareis, Marie
Barkalow, David
Bar-Lev, S.
Barley, Leanne A.
Barlow, Darlyne J.
Barman, Sandra R.
Barnes, Charles F.
Barnes, Doris
Barnes, Grace C.
Barnes, Vernon S.
Barnett, John
Barnum, Charles T.
Barry, C.
Barry, Richard
Barsness, Gloria
Bartholomew, I.
Bartholomew, Marguerite
Bartle, Helen A.J.
Bassett, James L.
Bassett, Jennifer
Bassett, Leo C.
Bast, Adelbert C.
Bast, Victor W.
Basthemer, Janet
Bateman, Ernest
Bates, Carlos G.
Battaglia, James
Baudendistel, Martin E.
Bauer, Elsie C.
Bauer, H.E.

Bauer, Leonard W.
Bauer, Richard B.
Bauernfiend, Arlene A.
Baughman, Elba
Baughman, Jas. W.
Baumann, Melissa G.
Baumgartner, Anne M.
Baumgartner, Bertha M.
Baxter, Laura O.
Bayer, Robert J.
Beachem, Ivy J.
Beall, Francis C.
Beatty, Janet L.
Beaumont Rhett
Beaumont, John
Beauverd, H. A.
Beaver, Christi
Bechle, Nathan
Bechtel, Stanley C.
Beck, Lillian M.
Becker, C. Dennis
Becker, John
Becker, LaVergne E.
Beckwith, Chauncey, Jr.
Beckwith, Chauncey
Beckwith, Frances L.
Beckwith, Howard W.
Beecher, James F.
Beerly, John
Begel, Marshall M.
Beglinger, Edward
Behling, Donald E.
Behr, Eldon A.
Behrend, Steve
Behrnd, Henry A.
Beireis, Richard L.
Beitz, William C.
Beitzel, Robert H.
Belknapp, Della M.
Bell, Arthur I.
Bell, Claude C.
Bell, Earl R.
Bell, Fern
Bell, John L.
Bello, Emmanuel
Bellosillo, Simplicio B.
Beloungy, Albert J.
Belovsky, Mike (John M.)
Belz, Brian
Bender, Donald
Bendtsen, B. Alan
Benedict, Chuck
Benedict, Reginald R.
Benisch, JoAnn H.
Bennett, Mary Jo
Bennett, Wyman E.
Bennin, John W.
Beno, Louvain A.
Benoy, Allen R.
Bensend, Dwight A.
Benson, Arnold O.
Benson, Barbara
Benson, Roy E.
Bentley, Richard J.
Bently, Louis
Benz, Gilbert H.
Berg, Betty M.
Berg, George C.
Berg, Harold P.
Berg, Hobart R.
Berg, Lucile H.
Berge, Lila G.
Bergen, Lilyan L.

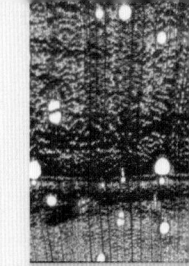

Berger, Richard W.
Bergman, Cindy R.
Bergman, Richard D.
Bergseng, Margaret S
Bergstrom, Clarence G.
Bergum, Bertha J.
Berkeypile, Arlene F.
Berna, Kathleen F.
Bernard, Amy
Bernard, Randy W.
Berthy, Howard Phillip
Best, Frank R.
Betlach, E. Claire
Bettermann, Alan
Betts, Charles N.
Betts, Clifford A., Jr.
Betts, Harold S.
Bewick, Grace B.
Bewick, Wayne E.
Beyer, Frank K.
Beyers, Amanda
Bickelhaupt, Jill
Biddison, Henry C.
Bienfait, Jacques L.
Bienfang, Arnold J.
Biermann, Ruth M.
Biersach, Herbert F.
Biesanz, Oakley
Big, Edward J.
Bigger, Roger F.
Bigham, Donald R.
Bigham, R.
Bilbrey, Dan
Bilek, Ted
Bilkey, Peter C.
Billington, Florence L.
Billington, Paul S.
Bilse, Cora S.
Bingham, Donald R.
Birchmeier, Matt
Bird, Carroll D.
Birkeland, Mike
Birnstengel, Robert L.
Bishop, Aaron
Bishop, Rosemary
Bishop, Shelley D.
Bissen, Steven P.
Bitner, Carol A.
Bittinger, Kristen
Bittner, Adela M.
Biwer, Marvin A.
Bixby, James O.
Bjorke, Solveig
Black Spotted Horse, Kenny P.
Black, John M.
Blackmon, Ann-Louise
Blair, Ron
Blanchard, Shirley M.
Blanco, Galo. W.
Blandino, George
Blankenburg, Julie J.
Blatterman, Wm. H.
Blau, Robert O.
Blauwkamp, Jennifer
Blew, J. Oscar, Jr.
Blied, Doris M.
Bliefernicht, Enid J.
Bliss, Dorothy E.
Bliss, Hugh P.
Bliss, P.
Block, Mary A.
Blodgett, Amy
Blodgett, George V.

Blomdahl, Mary L.
Blomgren
Blomquist, Richard F.
Blucher, Eva
Blucher, Frank
Bluhm, Edgar J.
Blum, Gordon
Boardman, Charles R.
Bocanegra-Severson, Apolonia
Bodell, Elaine
Boehmer, H.R.
Boelsing, Emilie A.
Boelsing, Rudolph G.
Boelsing, Zelda J.
Bogan, William
Bogart, Kris R.
Bohannan, Billy
Bohn, Oscar E.
Bohnsack, Albert J.
Bohnsack, Robert H.
Bois, Paul J.
Bokina, Lorraine
Boland, Leon B.
Bolden, Stephen
Bolduc, Jennifer L.
Bolen, Richard L.
Boling, Chas. W.
Boller, Kenneth H.
Bolles, Steven
Bollman, Raymond E.
Bollman, Raymond H.
Bolzt, Wm. O.
Bond, Francis M.
Bond, James S.
Bonnarme, Pascal
Bonner, Frank E.
Bonner, Nancy L.
Bonser, William E.
Bonura, Mike
Booker, J.D.
Boone, R. Sidney
Booth, Charles W.
Booth, Jean S.
Borchers, Bartel B.
Bordeaux, Damita
Borkenhagen, Edward H., Jr.
Bormett, David W.
Boro, Judith M.
Bosch, Erin
Bosch, Joanne M.
Bossenberry, Dale B.
Botten, Gale W.
Bouchard, Charles
Bourdeau, Napoleon J.
Bourne, Keith
Boushea, George W.
Bowar, Roman M.
Bower, Carol L.
Bowlby, David R.
Bowler, Geraldine M.
Boyd, Carol P.
Boyd, Paul D.
Boyes, Florence E.
Braden, Ann
Bradford, Anna Mae
Bradley, Clarence E.
Bradley, H.
Bradley, John H.
Bradley, Rosemary R.
Brager, Susan J.
Bram, Martin
Brandel, Eva P.

Brandon, Rishawn-Sneed
Brann, Betty Lou
Brantmeyer, Josephine M.
Brashi, George
Brault, Gilbert
Braun, Michael O.
Braund, Duane
Bray, Mark W.
Bray, Stuart
Brayton, Shirley C.
Bredendick, Kenneth E.
Breeden, Powhatan, III
Breezee, Mike L.
Breitenbach, Hazel H.
Brenden, John J.
Brendler, Iva E. M.
Brenner, Helen L.
Brenner, Joseph
Brewer, William B.
Brewer, William S.
Brewster, Donald R.
Brey, Franklin R.
Brickl, Elaine A.
Brickl, Jessica L.
Bridge, Bert B.
Bridger, Karen E. H.
Bridwell, James J.
Briggs, Edgar V.
Brigham, Robert H.
Brink, Barbara K.
Brinkerhoff, Mary W.
Brinkman, Edward W.
Bristol, H. Stanley
Britt, William T.
Britten, Stephen S.
Brndel, Eva P.
Brockert, Claude O.
Brockert, David J.
Brockmann, Marian O.
Broecker, Viole E.
Broge, Judith A.
Broich, Ryan
Brokaw, Max P.
Brooks, James K.
Broome, W.S.
Brouse, Don
Brown, Edna R.
Brown, Betty C.
Brown, Claude A.
Brown, Gordon H.
Brown, Helen M.
Brown, Kenton J.
Brown, LaNita J.
Brown, Olive M.
Brown, Robert A.
Brown, Robert R.
Brown, Samuel T.
Brown, Susanne M.
Browne, Frederick L.
Browne, Robert E.
Browne, Timothy P.
Brownell, Lori
Brownlee, James. L.
Brownlow, Wilfred J.
Bruak, Jean
Bruce, Herbert D.
Bruce, Ivy J.
Bruce, Marion L.
Bruegger, Myrtle M.
Brumm, Jerome L.
Brumm, Patricia A.
Brumm, Roy F.
Brunette, Fern L.

Brunker, Edward.
Brunner, Martin
Brunner, Pat W.
Brunner, Timothy A.
Bruno, Joe A.
Bruns, Henry A.
Brunsell, Linda
Brush, W.D.
Brussow, Albert R.
Bubl, Janet L.
Bubrick, Erin
Buchanan, C. Eleanor
Buchanan, John S.
Buchi, Henry H.
Bucina, Mariann J.
Buck, David L.
Buckley, James A.
Buckley, Odin L.
Buckmaster, Martha L.
Buckstrup, Judy M.
Buczak, Kirston
Buehler, Adeline K.
Bujanovic, Biljana
Bulgrin, Erwin H.
Bull, Carl F.
Bullock, A. A.
Bultman, August C.
Bultman, Helen H.
Bumkyoo, Choi
Burden, H. P.
Burdsall, Harold H., Jr.
Burgess, Horace T.
Burke, Floyd H.
Burke, Loretta
Burkhart, Ambrose
Burkhart, Jennifer
Burkhead, Eric M.
Burkholder, Susan K.
Burmeister, Ervin L.
Burns, Darleen M.
Burns, Florence J.I.
Burns, Roger K. Jr.
Burns, Walter J.
Buroker, Belinda K.
Burr, Horace K.
Burroughs, James
Burroughs, Laurence R.
Burwitz, Richard J.
Busacca, Basil
Busby, Julie A.
Busby, Lynn J.
Buscemi, Sam
Busch, Mary
Busch, Simon H.
Bush, Gayle
Bushor, Joseph M.
Buskov, Richard J.
Busse, Margaret D.
Busse, Marilyn J.
Busse, Warren F.
Butler, Lyle E.
Butler, Ovid M.
Butler, Russell, J.
Butler, Walter, T.
Butterman, Samuel
Button, Betty
Butts, Emily V.
Butzler, George J.
Byrd, Pamela J.
Byrd, Von L.
Byrd, Willie
Byrns, Daniel D.
Caballero, Mike

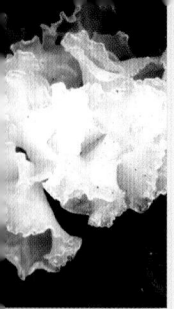

Cable, Donald E.
Cadwell, James J.
Cahil, Carlito
Cai, Zhiyong
Cain, Orison J.
Caine, George I.
Caire, Gretchen
Calderwood, H. N., Jr.
Caldwell, Christopher M.
Caldwell, Comazel
Calil, Carlito
Call, Elizabeth Z.
Callahan, M. Patricia
Callahan, Marian B.
Callahan, Mary P.
Callahan, Patrick J.
Caluwaert, Edward C.
Cameron, Charles B.
Cameron, Kimberly
Campbell, Diann L.
Campbell, Dorothy O.
Campbell, Gerald L.
Campbell, Kevin H.
Campbell, Robert N.
Campbell, William
Cannady, Jonathan E.
Cannon, Anita
Canto, Vilma E.
Capossela, Marjorie R.
Caravella, Mike, M.
Caravello, Mariano
Caravello, Vincenzo
Carden, John E. III
Cardoza, Yasmin
Carey, Dudley C.
Carey, Jeannette M.
Cargen, Isabella B.
Carll, Charles G.
Carlson, Charles O.
Carlson, Gladys E.
Carlson, Harold R.
Carlson, Louis A.
Carlson, Thorwald A.
Carnes, Diane E.
Caronna, Jennie G.
Carpenter, Betty
Carpenter, Lois E.
Carpenter, Lynn A.
Carr, Everett T.
Carr, Wayne F.
Carroll, Geraldine M.
Carroll, Grace M.
Carson, Clair C.
Carter, Leslie
Carter, Veda V.
Cascio, Giuseppe
Cash, E.
Cassens, Daniel L.
Cassity, Nathan
Catalino, Angelo
Cator, Russell, H.
Caulfield, Daniel F.
Caygill, Elizabeth R.
Cedar, Ellenore
Cefalu, Salvatore
Celimene, Catheine C.
Ceragioli, Giorgio.
Cerda,Gina
Chadwick, Katherine A.
Chakmakas, John J.
Chaline, Georges R.
Chamberlain, R.
Chamberlin

Chambers, Ira R.
Champion, Connie
Champion, Francis J.
Chan, Mike
Chandler, Jermal
Chandra, Andy
Chandra, Alexander
Chang, Ying-pe
Chapman, Dale
Chapman, Pierre
Chappelle, Jake
Charles, Eva
Charles, L..
Charlton, Terri M.
Chastang, Rita C.
Chaudhry, Gulzar
Chavera, Raymond R.
Check, Kevin
Chen, Chen-Loung
Chen, Dezhao
Chen, George C.
Chen, Tsung H.
Cheng, Shun
Cheng, Sun
Chern, Joseph
Cherry, Jerome F.
Chesley, Kenneth G.
Chi, Yu Jie
Chidester, Gardner H.
Chidester, Mae S.
Chilson, Warren A.
Cho, Choong Yun
Cho, Jae-yong
Christensen, Al
Christensen, Clyde M.
Christensen, Donna J.
Christensen,Carl A.
Christenson, Alfred C.
Christenson, E.
Christenson, John .
Christesen, Robert J.
Christiansen, Alfred W., Jr.
Christison, Frances L.
Christofferson, Ruth
Christopher, Kaye
Christopherson, Andrew
Christus, Daniel N.
Chu, Boyer B.
Chudnoff, Martin
Chun, Se-Kyoung
Chung,
Church, Beverly J.
Churchill, Guy W.
Cieciwa, James A.
Ciula, Mary C.
Clampitt, Shirley D.
Clark, Anna M.
Clark, George W.
Clark, Ira T.
Clark, Joe W.
Clark, Marion J.
Clark, Marlyn E.
Clark, Marriott A.
Clark, Paul A.
Clark, Richard J.
Clark, William L.
Clarke, Anne
Clarke, Edward H.
Clarke, Wilbur C.
Clausen, Carol A.
Clausen, Roger L.
Clay, Harvey
Clay, Jethro

Cleggett, Paula M.
Clemens, Arthur G.
Clemens, Harold L.
Clementi, Lucille D.
Clemons, Craig M.
Cleven, Thea O.
Clevenger, C. H.
Clevenger, Koren
Clifcorn, Edward F.
Clifford, Ray R.
Cline, McGarvey
Closs, John O.
Cloukey, Homer
Clouser, William S.
Cnare, Edward G.
Coates, Joyce J.
Cockrell, Robert A.
Cockrum, Tim
Cocroft, Mark
Coens, Charles, L.
Coffey, Daniel J.
Coffman, Wilbur L.
Cohen, Roni
Cohen, Wilby E.
Cohn, Frieda S.
Cohodas, Leah E.
Cole, Flecia
Coleman, Donald G.
Coleman, Robert
Colgan, Richard A.
Collet, Mike
Collett, Mary
Collette, Jos. W.
Colletti, Vincent
Colley R.M.
Colley, Reginald H.
Collins, Earl J.
Collins, Emmons
Collins, James M.
Collins, Kelly
Collins, Lawrence J.
Colon, Harry C.
Colson, Jeanine N.
Compton, Kenneth C.
Compton, Lake F.
Comstock, Gilbert L.
Conde, Martin Y.
Cone, Joyce L.
Conlin, Vern W.
Conner, Anthony H.
Conner, Thomas R.
Conners, Helene A.
Connolly, Claire A.
Connor, Eugene P.
Connors, Ethel V.
Connors, William J.
Conohan, James E.
Conrad, Sara
Considine, John M.
Contezac, Michel C.
Conway, Hilma U.
Cook, Arthur M.
Cook, Gerald R.
Cook, James A.
Cook, John W.
Cook, Leo R.
Cook, Mildred I.
Cooney, William N.
Cooper, Eugene R.
Cooper, George, P.
Cooper, Jose D.
Cooper, Robert E.
Cooper, Russell L.

Cooperrider, Chas. K.
Copeland, Marie A.
Coppock, Kathleen
Corbin, Kathy
Corcoran, Michael T.
Cork, E.H.
Cork, W.
Cornwell, Edward
Correa, Laciene
Corriveau, Allen
Corscot, Jennie M.
Corscot, Marian R.
Coster, Sarah A.
Cottingham, Willard A.
Cottinghan, Willard S.
Cottrell, Mildred A.
Couch, George
Coughlin, John F.
Cover, L. E.
Covert, Sarah
Covert, Edgar A.
Covier, Iona
Cowan, William C.
Cowen, Patrick J.
Cowgill, Nancy E.
Cowling, Ellis B.
Cox, John E.
Cox, John R. Jr.
Cox, Roger G.
Coye, Clarence W.
Coyne, Barbara
Craig, Ollison
Craig, Robert Jr.
Cram, Sally A.
Cramer, Calvin O.
Cramer, George F.
Crane, Rufus
Cranston, Annis R.
Crapp, Beverly J.
Cratsenberg, E.
Crawford, Douglas M.
Crawford, Harold W.
Cree, Elisa
Creft, Henry C.
Cretney, Mary J.
Cripps, Aileen M.
Cripps, Guy
Crispin, Norman D.
Croan, Suki C.
Croft, Henry C.
Crollard, Harriet A.
Crook, Dorothy
Crooks, Casey
Crooks, Merritt E.
Cropsey, Kathleen R.
Crosby, Clifford E., Jr.
Crounse, Dennis D.
Crow, Gerald. W.
Crow, Mary A.
Cruger, Mildred F.
Cruz, Jose
Cuadra, Jose A.
Cuccia, Nick
Cuevas, Diego
Cuevas, Jose
Cull, Irene M.
Cullen, Daniel
Culp, Joseph
Culp, John
Cummins, John E.
Cunningham, Karla J.
Cuno, John B.
Cunzenheim, Marie A.

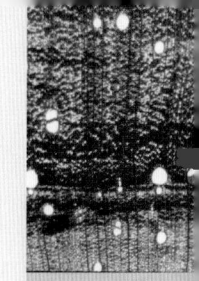

Curley, Jeanne W.
Curling, Simon
Curran, Carleton E.
Curran, George W.
Curran, S.
Currie, Jane W.
Currier, Lloyd A.
Currry, Jean L.
Cusick, Mae P.
Custer, Alma
Custer, Frank
Custer, Glen S.
Cutrell, James E.
Cyr, Sylvio J.
Cyte, Virginia A.
Czederpiltz, Daniel L.
Dadswell, Herbert E.
Dafoe, Percy L.
Dahl, Clarence J.
Dahl, John C.
Dahle, Ella
Dahlgren, Daniel L.
Dahlke, Jeanne L.
Dahn, Kris
Dailey, Evelyn G.
Dain, Bryant D.
Daines, Linda O.
Dalebroux, Zachary
Daley, Walter C.
Dally, Brice N.
Dalton, Louise
Daly, Frank J.
Damron, Leatha
Danca, Vincent J.
Daniel, Charles
Daniels, Albert F.
Daniels, David H.
Danielson, Betty Ann
Danielson, Jeanne D.
Danielson, Priscilla J.
Danna, Anne M.
Danna, Pamela J.
Darling, Eric H.
Darrin, Marc
Das, Dilip Kumar
Dashek, William
Datta, Asit
Daugherty, Jane R.
Daugherty, Martha A.
Davenport, Eleanor L.
Daverin, Edward
Davidson, Harley L.
Davidson, Lloyd
Davidson, Max A.
Davidson, Paul B.
Davie, Lisa
Davies, Idris I.
Davila, Antonio
Davis, Edward M.
Davis, George K.
Davis, James H.
Davis, Mark W.
Davis, Martha E.
Davis, Peter N.
Davis, Susan A.
Davis, William A.
Davis, Willis P.
Davister, Michael D.
Dawley, Earle R.
Day, Howard E.
Day, Judith A.
De Chant, Connie L.
Dean, Charles E.

Dean, M.
Dean, Ruth C.
DeCampa, David
DeChant, Connie L.
Dedolph, Carter
Dedow, Donald R.
DeFlon, Leland L.
Dega, Kris
DeGroot, Rodney C.
DeHaven, Chas. D.
DeHoeger, Iris
Deihl, Joseph D.
Dekker, Grant
deKoker, Theo
DeLa Portilla, Elizabeth
DeLa Portilla, Thomas
Delaney, Helen A.
Delaney, Mary Ellen
Delfosse, David
Della-Moretta, Leonard B.
Delong, Chang
DeMaster, Robert J.
DeMaster, Sandra L.
Demmer, Mary J.
Dempsey, Angeline M.
Demro, Mary Jane
Denes, Ferenca
Dengel, Philip J.
Denicke, Lowell R.
Denk, Joseph E.
Dennis, C.E. Jr.
Dennis, E. M. Pearl
Dennison, Henry C.
DeRidder, Charles G.
Derks, Richard J.
DeRosia, John J.
DeSmidt, Arthur J.
Destree, Brian
Destree, Joseph C.
Detra, Lloyd A.
Deutsch, Margaret F.
Deverall, Flora J.
Devitt, Martin T.
Devitt, Mary K.
Devlin, Edward P.
DeVoe, Roy H.
DeWane, Harold J.
Dewey, Elizabeth A.
Dewey, Ruth E.
Dexter, A. Kendall
Diaz, Lucy
Dickerhoff, Edward
Dickerson, Michael
Dickert, Marie
Dickinson, Fred E.
Diebold, Ferne D.
Diebold, Sylvester J.
Diederich, Marie C.
Diehl, Marilyn A.
Diemer, Melvin E.
Dieruf, Louetta M.
Dietenberger, Mark A.
Dietrich, Diane M.
Dietz, Albert G. H.
Dietzman, Debra J. S.
Dill, Iris L.
Dillion, Barbara
Dillman, Sandra L.
Dilworth, Emily J.
Dingledy, Fred
DiPiazza, Andrew
Disrud, Chris
Dittmer, Margaret

Divers, Alice R.
Dix, George E.
Dix, Judith A.
Djergou, Solange
Dlugopolski, Joseph E.
Dobbels, Carol Jean
Dobbins, Linda S.
Dobson, Dulcie L.
Dockin, George M.
Dockry, Michael J.
Doelle, Klaus
Doering, Ronald J.
Doggett, R. Janet
Dohr, Alfred W.
Dohr, Herbert L.
Dohse, Emilie C.
Doke, Howard B.
Doke, Howard C.
Dokken, Edith F.
Dokken, Merle K.
Dolan, Francis J.
Dold, Agnes K.
Dolinky, Melvyn A.
Doll, Edna C.
Doll, Sandra M.
Doll, Vincent W.
Dolphin, Anthony P.
Dominguez, Roberto
Domini, Joseph B.
Domini, Roy
Donagricho, Anton
Donaldson, Elwin C.
Donny, Joe
Donovan, Margaret
Doraiswamy, L. K.
Doran, Mary C.
Dorau, Benjamin
Dorn, Donald E.
Dorn, Joann B.
Dorworth, Elizabeth
Dos Santos, Gracielza
Dottle, Helen
Doty, Edmund E.
Doubleday, Albert E.
Douds, David D.
Doughty, Randall H.
Douros, Mary T.
Dove, Roger D.
Downes, Frances H.
Downs, James J.
Downs, Leslie, E.
Doyle, Cecilia M.
Doyle, Donald V.
Doyle, James E.
Doyle, Leo J.
Drake, Herman L.
Drake, Shirley M.
Dramm, John R.(Rusty)
Draper, Helen M.
Dreher, Geraldine E.
Drenning, Robin.
Dresen, Dorothy
Drewes, Curt P.
Drewry, Mary E.
Drives, Janet A.
Drow, John T.
DuBois, R. Clifton
DuBois, Roy C.
Duch, Tracey
DuCharme, Arlene
DuCharme, Jacqueline L.
Dudgeon, Robert
Duff, Andrew C.

Duff, John E.
Dufresne, Paul R.
Dugan, Bernard
Duke, Atina
Dukes, Peter
Dulin, Stanley S.
Dumphy, Marie C.
Duncan, Catherine G.
Dunlap, Fredrick
Dunlap, Matthew E.
Dunn, Ben
Dunn, Virginia A.
Dunphy, Mary E.
Duquaine, Camille J.
Durand, Delphine M.
Durbak, Irene
Durfey, Forest E.
Durkee, Earle
Durkin, Ursula
Durrani, Sardar
Dusch, Chris
Dussling, Anton A.
Duyser, Ardith R.
Dyer, Lenny J.
Dyer, Melvina M.
Ealey, Jackie
Eannelli, Katherine J.
Earle, Lydle J.
Eastwood, Paul R.
Eaton, David
Eberline, Richard K.
Ebert, Kristin
Ebewele, Robert O.
Ebisch, Lucy
Eboh, Dan. O.
Echevarria, Rene
Eckbo, Nils B.
Ecker, Charles
Eckerle, Gilbert L.
Edd, Herbert A.
Eden, Philip
Edlund, Robert P.
Edmonds, Martha Ann
Edwardes, Vance P.
Edwards, Ann E.
Edwards, B. A.
Edwards, Blithe
Edwards, Harriet Y.
Edwards, Joan
Edwards, John C.
Edwards, Tameka L.
Edwards, William R.
Effland, Marilyn J.
Efird, Frank K., Jr.
Egan, Thomas H.
Egbert, Kathryn E.
Egbert, William C.
Eggleston, Edgar A.
Ehret, Beverly S.
Ehrhardt, Priscilla D.
Eickner, Herbert W.
Eid, Hebert A.
Eide, Bernice T.
Eierman, Stanley L.
Einerson, Joanne
Eisert, Leonard N.
Ekum, Walter R.
Ekvall, Leonard E.
Elegir, Graziano
Elert, Ervin E.
Ellenbecker, Margaret E.
Ellickson, Samuel C.
Elliott, David D.

Elliott, Howard
Elliott, Leroy P.
Ellis, Cecil
Ellis, Nona E.
Ellis, Pierce G.
Ellis, Thes.
Ellis, Thomas H.
Ellis, Walter Dale
Ellison, C. William, Jr.
Ellison, J.
Ellwood, Chas. D.
Elmendorf, Armin
Elmer, Thomas S.
Elrod, Helen M.
Elsom, Dorothy H.
Elton, Robert L.
Elver, Kenneth E.
Elvert, Marcella M.
Elvord, Shelly D.
El-Wakil, Tatiana
Ely, Alexander W.
Emasitti, Chao
Emden, Dennis R.
Emens, Gregg
Emery, Judith D.
Emig, Ray M.
Emmons, Verdie I.
Emory, Madge I.
Empey, Leroy W.
Endress, L.
Engel, Alfred H.
Engelke, Beverly J.
Englebert, Doug
Engler, Catherine C.
Englerth, George H.
English, Brent
Entwistle, Gregory C.
Erbe, Bernard J.
Erbes, Dorothy J.
Ericksen, Leyden N.
Ericksen, Wilhelm S.
Erickson, Donald W.
Erickson, Edwin C.O.
Erickson, John R.
Erickson, Lucille L.
Erlandson, Eunice M.
Ermeling, Helen V.
Ernst, George C.
Ersland, Beatrice L.
Ersland, Leif J.
Ervin, Harold O.
Escolano, Jaime O.
Esenther, Glenn R.
Eslyn, Wallace E.
Espen, Gerhard A.
Espenas, Leif D.
Esse, Mildred S.
Esser, Margaret
Esser, Noreen M.
Estep, Eldon M.
Esther, M. Lotta
Ethington, Robert L.
Eubanks, James O.
Euhardy, Steven J.
Eusebio, Mario A.
Eustice, David R.
Evan, Jessie M.
Evans, Bernice M.
Evans, Faye J.
Evans, Frank W.
Evans, James T.
Evans, James W.
Evans, Jane K.

Evans, Kathy
Evans, Paul G.
Evans, Russell H.
Evans, Virginia
Eveland, Burrell D.
Ever, Nancy A.
Everett, T.
Everhart, Helen E.
Evert, Linda L.
Every, Beulah M.
Every, Donovan R.
Every, Roderick D.
Eves, Florence H.
Evins, Anthony L.
Faga, Karen J.
Fagan, Mathias E.
Fahey, Donald J.
Fahey, Ellen J.
Fahey, Julia M.
Fahlberg, Ernest D.
Fahlberg, The' A.
Fahlstrom, George B.
Faison, Brenda
Falk, Robert H.
Fanega, Salvador M.
Farkasch, Ruth
Farmer, Eldon L., II
Farmer, Janis E.
Farnsworth, June I.
Farnum, Jean C.
Farnum, Richard J.
Farrell, John H.
Farrell, Josephine M.
Fason, Gwendolyn E.
Faulkes, William F., Jr.
Faust, Gilbert
Faustino, Dominador G.
Faville, Harry C.
Fay, Verllyn J.
Feather, Milton S.
Feezel, Jenelle H.
Fehling, Clarence C.
Feist, Colleen F.
Feist, William C.
Felby, Claus
Felder, John H.
Feldman, Linda
Feldman, Samuel S.
Felker, Michael C.
Fell, Ambrose M.
Fellers, Nils C.
Felt, Earl, J.
Felton, Colin
Fender, Albert S.
Feng, Xu
Fennell, Francis L.
Fenner, Ellen R.
Fenton, Sidney E.
Ferdon, Clara
Ferge, Leslie A.
Ferger, William T.
Fergus, David A.
Ferguson, Betty J.
Ferguson, Clarence N.
Ferguson, Josephine
Ferguson, W. Rex
Fernandez, Luis
Fett, Ralph R.
Fetty, Hugh J.
Field, Eunice M.
Filipowicz, Chester J.
Finch, Carroll, H.
Fink, Delmar A.

Finn, Mary G.
Finnegan, Kathie J.
Finnie, Edward J.
Fisher, Dolores N.
Fisher, Donald H.
Fisher, Jeanette R.
Fisher, Joan E.
Fisher, Joan, R.
Fisher, Nellie
Fishwild, Sara J.
Fitschen, Fred A.
Fitzgerald, Charlotte
Fitzgerald, Lawrence C.
Fix, Jacob M.
Flach, Dwight David
Flach, Scott A.
Flamme, Wayne H.
Fleck, Louis C.
Fleischer, Herbert O.
Fleischmann, Michael
Fleming, Archie J.
Fleming, Cathleen
Fleming, Michael
Fleming, Ruth M.
Fletcher, T. Lloyd
Floeter, Lester H.
Florence, John N.
Flournoy, Douglas
Fluck, Paul G.
Flynn, Lucy B.
Fobes, Eugene W.
Foellmi, Barbara G.
Foellmi, Donald R.
Foerst, Ruth L.
Follensbee, Robert
Follick, Jack H.
Fondren, Kirsten
Fong, Samuel T.
Font, Javier E.
Foote, Kevin
Foote, Nelle M.
Ford, Charlotte M.
Ford, Daniel P.
Ford, E.H.
Ford, J.T.
Formiller, Peter M.
Fornasiere, John
Forrest, Helen M.
Forrest, Wm.
Forrester, Ian
Forsythe, Reno H.
Foss, Robert E.
Fossen, Violet I.
Foster, Adrian
Foster, Daniel O.
Foster, Glenn W.
Foster, Linda M.
Foster, Russell J.
Foulger, Albert N.
Fox, Genevieve P.
Fox, Jessica
Fox, John M.
Fox, Marion A.
Fracheboud, Michel G.
Francis, Copeland Y.
Frank, Arthur R.
Frank, Helen M.
Frank, Miriam S.
Frank, Warren A.
Frary, Hobart D.
Frater, Norman K.
Frear, Jenness B.
Freas, Alan D.

Fredenberg, David G.
Frederici, Gertrude F.
Frederick, James R.
Frederick, William J., Jr.
Freedland, Cassia
Freeman, Trula B.
Freese, Frank
Freidig, Lylas L.
French, George E.
Freyburger, Edwin
Frias, Janice
Friedl, Alois J.
Friedl, Marcia C.
Friedrich, Katherine A.
Fries, Beverly J.
Frihart, Charles R.
Frisque, Alvin J.
Fritz, Jeanette S.
Froding, Daniel R.
Froehlich, Sandra K.
Froehlke, Adolph
Froh, David
Fromm, Henry J. III
Fronczak, Frank J.
Frothingham, Chester
Frye, Charles L.
Frye, Dan W.
Fuchs, Wm. R.
Fukushima, Yachi
Fuller, Anne M.
Fuller, Collins F.
Fuller, James J.
Fuller, Keith L.
Fuller, Willard L.
Fumusa, Saverio
Fumuso, Judy A.
Funmaker, Mary A.
Gabbert, Carl H.
Gabler, Nathan
Gabriel, Arthur E.
Gadzik,
Gaffney, George P.
Gahagan, John M.
Gaidula, Kathryn Z.
Galiger, Lynn J.
Gallagher, Ann M.
Gallagher, Cora M.
Galligan, Robert R.
Galligan, William L.
Gambaro, Anthony
Ganeson, Kulasekaram
Gangstad, John E.
Ganshert, Peter J.
Garcia, Kevin
Gard, Andrea H.
Gard, Harold K.
Gardner, Betty G.
Gardner, Jo Ellen
Gardner, Karen O.
Garfoot, Harvey P.
Garrett, George A.
Garrett, Helen E.
Garrison, Lillie R.
Garrow, Thomas S.
Garske, Michael L.
Garthwaite, Myra J.
Garver, Raymond D.
Garvoille, Dian I.
Garvoille, Wm. F.
Gary, Joseph J.
Gary, Mary Frances
Gaskell, Jill A.
Gasner, Phyllis A.

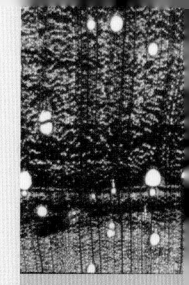

Gastrow, Frederick E.
Gatchell, Charles J.
Gates, Mary M.
Gatz, William A.
Gauer, Ferdinand J.
Gauger, William H.
Gawlik, LaVern J.
Gay, Ralph, C.
Gaynor, George J.
Geer, W. C.
Geetha, Holalkere V.
Gehlhoff, Sarah
Gehred, Ruth J.
Gehrke, Doug
Geier, Milton G.
Geier, Richard K.
Geier, Steven E.
Geiger, Clarence C.
Geiger, Eddneata
Geiger, Linda K.
Geimer, Robert L.
Genn, Agnes
Genna, Hallie
Gentry, Judy I.
George, Clarence P.
George, Gene R.
Gerard, Eugene L.
Gerbitz, Richard M.
Gerhards, Charles C.
Gerhardt, Daniel
Gerhardt, Terry D.
Gerrow, Mary
Gerry, Eloise
Gerry, Everett C.
Gertjejansen, Roland O.
Geske, Earl A.
Geske, Mary K.
Gesme, Gordon A..
Gettens, Rebecca H.
Geurkink, Deborah K.
Ghani, Qudrat
Ghosh, Sujata
Gia, Russo Don P.
Gibson, Sandra L.
Giese, Jennie C.
Giese, Pamela J.
Gigot, Scott
Gilardi, Bert W.
Gilbertson, James T.
Gilbertson, Lucille D.
Gilbertson, Merrill T.
Giles, Eber T.
Giles, Esther
Giles, Kenneth A.
Gilkey, Barbara J.
Gilkey, Claudia O.
Gillespie, Robert H.
Gillette, Lucille I.
Gillingham, Linda J.
Gilson, Jane
Ginter, John F.
Ginter, Paul
Girard, Arthur G.
Gish, Marvin
Gjestson, Herman J., Jr.
Gjinolli, Agron
Gjovik, Lee R.
Glaeser, Jessie A
Glass, George C.
Glass, Samuel V.
Gleisner, Grant
Gleisner, Roland L.,Jr.
Glenz, Emma L.

Glover, Nathan
Godell, Horace R.
Godfrey, James R.
Godoy, Pablo A.
Godsey, Doris
Godsey, Jean E.
Godshall, W. Duncan
Goeden, Lorraine M.
Goers, Mary P.
Goikovich, Nick
Gojraty, Sattar
Goldberg, Esther H.
Goldgruber, J.
Goldschmidt, Mona R.
Goldsworthy, George J.
Gombar, Francis S.
Gonzalez Carlos
Gooch, George W.
Goodell, Horace R.
Goodman, Alicia M.
Goodman, Joshua L.
Goodman, Maurice E.
Googins, Joseph Q.
Gordon, Dennis
Gordon, Florence W.
Gorgis, Donald J.
Gorman, Thomas
Gorsege, Steve
Gorski, Susan B.
Goslin, Robert A.
Gosline, Mark.
Goss, John R.
Goss, O. P. M.
Gosz, Jordan
Gotter, Elroy E.
Gottfredsen, James E.
Gough, Tivoli C.
Goul, Ida
Gould, Clark W.
Gounaris, George
Govier, Iona A.
Govinden, Roshini
Grabarczyk, Randolph
Grabel, Virginia M.
Grabow, Rudolph H.
Grady, Frank T.
Grady, James, B.
Graf, John C.
Graham, Barbara L.
Graika, Thomas
Grambsch, Michael J.
Granados, Luis Alberto
Granberg, Leonard A.
Grancis,
Grandy, Juanita K.
Grant, Dorothy F.
Grantham, John B.
Gratz, Janet K.
Graves, Fay D.
Graves, Glyde E.
Graves, Robert M.
Gray, Hazel, E.
Gray, Jacalyn
Gray, Rebekah S.
Grebel, Robert L.
Green, Beatrice L.
Green, Charles B.
Green, David W.
Green, Ellen F.
Green, Frederick
Green, Gertrude L.
Green, Heather
Green, James D.

Green, Jesse O.
Green, Phyllis
Green, Yvonne L.
Greene, George E.
Greenman, Paul R.
Greer, Robert W. III
Grefe, D.
Gregerson, Lois M.
Greinert, Matthew
Gremminger, Dennis
Grenoble, Herbert S.
Gribbins, Neil R.
Gribble, Mary Jo
Griffee, Willet E.
Griffin, Gertrude J.
Griffin, Gregory L.
Griffin, Janet L.
Griffiths, Bryn A.
Grimes, Henry J.
Grimes, James J.
Grinsteiner, Timothy J.
Griswold, James F., Jr.
Gromala, David S.
Gross, Richard
Gross, Suzanne
Grosse, Raymond
Groth, Alvin L.
Gruen, Daniel H.
Grundahl, Gloria J.
Grundahl, Vickie
Grundman,
Gu, Hongmei
Guelzow, Marjorie A.
Guelzow, Midge
Guerrero, Dorothy M.
Guimaraes, J.
Gulseth, Darrell M.
Gunderman, Howard L.
Gunderson, Dennis E.
Gunderson, Harvey S.
Gundlach, Leslie M.
Gundlach, M. Josephine
Gunkel, Grace M.
Gunter, Chandrika
Gurrie, Dennis T.
Guss, Cyrus O.
Gust, Dolores J.
Gustman, Doris E. Kolstad
Guthaug, Jos. G.
Guthneck, Isidore H.
Gutzmer, David I.
Guy, Charles U.
Gwynne, Chas. S.
Gyte, Henry
Gyte, Virginia A.
Haas, Rex A.
Haberkern, Kirsten
Habermann, Hermann
Habich, Adolph C.
Hackney, John M.
Haft, Everett E.
Hagan, Marian J.
Hagen, Mark
Hagengruber, Virginia
Hagerty, John
Hagge, Wallace J.
Hahm, Eldona L.
Hahne, Howard
Haight, John
Haima, Robert A.
Haines, Margie A.
Hajny, George J.
Halbrook, Andronike

Hale, Harry M.
Haley, Thomas F.
Hall, Douglas, R.
Hall, J. Alfred
Hall, Maybert E.
Hall, Mike
Hall, Robert T.
Hall, Stan S.
Hall, Vicki L.
Hall, William G.
Hall, William L.
Hallauer, Frank J.
Hallock, Hiram Y.
Halma, E.
Halperin, Bessie
Halperin, Harry
Halpin, Corey
Halverson, Dorothy L.
Halverson, Martin
Halverson, Steven A.
Halvorsen, JoAnn M.
Ham, Victor H.
Hamann, Shirley L.
Hambrecht, Walter L.
Hamill, Dennis W.
Hamilton, Frank J.
Hamilton, Harry L.
Hamilton, Thomas E.
Hammel, Kenneth E.
Hammel, Scott
Hammes, H. Clyde
Hammes, Harlow C.
Hammes, Harlow E.
Hammill, Dennis W.
Hammond, Rodney A.
Hammonds, Charlie L.
Hampson, Laura R.
Hamre, Robert H.
Han, James S.
Han, Yousoo
Hand, Silas C.
Handell, Thomas A.
Hanger, J. Dennis
Hanifin, Ruth S.
Hankel, Steven G.
Hankel, Virginia O.
Hankinson, Hazel
Hann, Robert A.
Hannan, Kerry
Hannan, Marian M.
Hans, Gunard E.
Hansard, Vera A.
Hansbrough, J. R.
Hansen, Carl
Hansen, Erica
Hansen, Helen M.
Hansen, Jeshua
Hanshaw, Larry C.
Hanson, Cynthia A.
Hanson, Dorothy
Hanson, Einar
Hanson, Fred G.
Hanson, Hans
Hanson, Hilma
Hanson, Lester A.
Hanson, Lester M.
Hanson, Louis W.
Hanson, Mary M.
Hanson, Robert L.
Hanson, Selma
Hantke, Richard W.
Harding, John P.
Hardy, Katherine

Hare, Clarence, M.
Hargons, Frederick G.
Hargons, Madree K.
Hargons, Valya N.
Harker, Betty A.
Harker, David L.
Harkin, John M.
Harloff, James A.
Harman, Rebecca
Harmon, Lisa M.
Harmon, Sandy
Harper, David
Harper, Zenor
Harpole, George B.
Harrington, Floyd
Harris, Brad R.
Harris, Elwin E.
Harris, John F.
Harris, Johnny
Harrison, Albert C.
Harrison, Charles A.
Harrison, Harold M.
Harrison, Robert S.
Harrison, Sandra E.
Harrison, Willie Jr.
Hart, Jean M.
Hart, Jo Ramon
Hart, Narcisse P.
Hart, Pierre R.
Hart, Ruth M.
Hart, Suzanne V.
Hart, Victor D.
Hartenberg, Richard S.
Hartley, Julius A.
Hartley, Norman J.
Hartman, John
Hartman, Marvin F.
Haskell, Arthur J.
Haskell, Henry H.
Haskell, James H.
Haskins, Byron O.
Haskins, Mary L.
Hassan, Mohammad
Hasselkus, Jane
Hassell, Arthur J.
Hatfield, Cherilyn A.
Hatfield, Ira
Hatfield, W.
Hathaway, Avery
Hatt, W.K.
Haubrich, Kurt E.
Haukereid Mary K.
Haukereid, Morris W.
Hausman, Eleanor, H.
Hausman, Loran
Hausman, Millie L.
Havilik, Frank
Hawk, Emilie J.
Hawkins, Andrew B.
Hawkins, Margaret E.
Hawley, Lee F.
Hawthorne, Gladys R.
Hay, Harold R.
Hayes, Charlie
Hayes, Elmer L.
Hayes, Mark V.
Haynes, Judith M.
Hays, Eddy L.
Hays, Logan C.
Haywood, Kay R.
Haywood, Samuel L.
Head, Dorothy
Headman, Jenifer

Healy, Andrew F.
Heck, George E.
Hedquest, Julia A.
Heebink, Bruce G.
Heebink, Thomas B.
Hefty, Francis V.
Hegel, Henry J.
Hegge, Arthur P.
Hegge, Wallace J.
Heick, Ramon J.
Heider, Lenny J.
Heil, Lawrence
Heilman, Lura F.
Heim, Arthur L.
Heimann, Elmer A.
Heimann, Grace S.
Heine, Joan I.
Heine, Robert, W.
Heinig, Melburn
Heinrichs, Arline A.
Heinrichs, J. Frank
Heinrichs, Leroy C.
Heintz, Robert A.
Heinzelman, Chad
Heitepriem, Arthur L.
Helgoe, Willard L.
Hellenbrand, Susan C.
Heller, James T.
Heller, Leslie C.
Hemphill, Ray L.
Hendershot, William F.
Henderson, Benjamin A.
Henderson, George W.
Henderson, Willie C.
Hendrick, Jim
Hendricks, Berry E.
Hendricks, LewisT.
Hendricksen, Laura
Hendrickson, Gary L.
Hendrickson, Rex R.
Hendry, Allan P.
Henke, Milton J.
Henke, Robert D.
Henkel, Steven G.
Henkel, Willard G.
Henley, John W.
Henning, S.B.
Henningsen, Carleton
Henrichs, J. F.
Hensel, Rose Marie
Henseler, Gerald R.
Hensen, Jeshua
Henshue, Emily E.
Hensley, Janelle L.
Hentzen, Erwin A.
Hentzen, Herbert D.
Hepp, Amiel L.
Herian, Victoria L.
Hering, Fred C.
Heritage, Clark C.
Hermann, Albert
Hermanson, Gene
Hermanson, John C.
Hernandez, Roland
Heronemus, William S.
Herreid, Sylfest J.
Herrick, David E.
Hertel, Stella M.
Hertzler, Richard A.
Hess, Lawerence
Heuser, C.W.
Heuser, Peter E.
Heyer, Otto C.

Hickman, Richard A.
Hicks, James R.
Hicks, Jean C.
Hicks, P.R.
Hicks, Richard D.
Hidayat, Syarif
Higgins, Albert C.
Higgins, Beverly J.
Highley, Terry L.
Hilbrand, Howard C.
Hilbrich Lee, Paula D.
Hill, Clifford H.
Hill, J.
Hill, Melissa
Hill, William E.
Hillard, Robert J.
Hillary, Charles J.
Hillcoat, Jim
Hiller, Charlotte H.
Hillestad, James B.
Hilliker, Bruce V.
Hillis, John L.
Hillman, Wade L.
Hilton, Robert D.
Hinderlie, M. E.
Hinderlie, Martha E.
Hinman, Dorothy C.
Hinn,William H.
Hinz, Paul N.
Hirth, Kolby C.
Hislop, Lola E.
Hittmeier, Michael E.
Hiziroglu, S.
Hjort, Mary
Hjortland, Carl J.
Hobert, Victor C.
Hobson, Alice E.
Hockett, Maria E.
Hodges, Mildred A.
Hoebel, Pauline V.
Hoernemann, Cyreal G.
Hoff, Brian
Hoffland, Michael J.
Hoffman, Anne J.
Hoffman, Arthur M.
Hoffman, Ellen A.
Hoffman, Frank J.
Hoffman, Louis G.
Hoft, John P.
Hogan, Alice C.
Hogan, Barbara
Hohensee, Greg
Hohf, John Paul
Hoiberg, Lester Hans
Holbrook, Herbert T.
Holfinger, Michael
Holland, Donald A.
Holland, William T.
Holler, Karen J.
Holley, James S.
Hollis, Keith
Holm, Dave
Holm, Mary Ann
Holman, Sylvia B.
Holmen, Paul W.
Holmes, Carlton A.
Holmes, George B.
Holmes, Leonard G.
Holmes, Mary
Holmes, Paul N.
Holmes, Start
Holmquist, Earl W.
Holst, H. Eugene C.

Holstein, Gladys M.
Holum, Fay F.E.
Holway, Ethel
Holzbower, Joe
Holzem Martin H.
Holzem, Jerilyn
Holzman, Sylvia B.
Holzwarth, Lloyd R.
Hookanson, Mildred L.
Hoover, William L.
Hopper, Glenna M.
Horn, Eric
Horn, Eugene F.
Horn, Richard A.
Horst, Arthur J.
Hossain, S. M.
Hostettler, Frances D.
Hostman, M.H.
Hotle, Brad
Hougan, Bette J.
House, John P.
Houston, M. Lloyd
Houtman, Carl J.
Hovey, Elizabeth J.
Hovey, Mark, H.
Howard, James L.
Howard, Marilyn T.
Howard, N.O.
Howard, O.
Howard, Rachel Ann
Howard, Richard A.
Howards, Susan E.
Howden, Charles J.
Howden, Margaret H.
Howe, Margaret A.
Howland, George H.
Hoyt, Jean D.
Hozeny, Walter J.
Hrubesky, Clarence E.
Hubanks, Dorothy M.
Hubbard, Ralph B.
Hubbard, Steven
Hubenthal, Homer H.
Hubert, Ernest E.
Huberty, Sarah L.
Huegel, Lillian
Huffman, Susan A.
Hughes, Bernice J.
Hughes, Cevia E.
Hughes, Charles F.
Hughes, Michael E.
Hullerman, Madeline F.
Humbert, Lori
Humphrey, John.
Humphrey, Lucile E.
Humphreys, C. J.
Hunt, Christopher G.
Hunt, Edd M.
Hunt, George M.
Hunt, H. Wayne
Hunt, John F.
Hunt, Paul J.
Hunter, Charles J.
Hunter, Ethel C.
Huntley, H.W.
Hunzicker, Dean L.
Hurd, David C.
Husain, M. Sarwat
Huss, Mabel
Hussa, Robert O.
Hustad, Janet E.
Hustad, Judith M.
Hustad, Milton B.

Hustad, William F.
Hustad, William L.
Huston, Harry H.
Hutchins, Martha E.
Hutchins, Wayne F.
Hutton, Arthur
Hutton, Floyd B.
Hvamb, Gullik.
Hyatt, Joan R.
Hying, Thomas A.
Hyland, Janet K.
Hyttinen, Axel
Hyvarinen, Matti
Ibach, Rebecca E.
Ibenthal, Eunice J.
Illman, Barbara L.
Imel, Clifford J.
Imhoff, Michael G.
Imrie, James E.
Ince, Jane A.
Ince, Peter J.
Ingersoll, L.R.
Ingraham, B.
Ingram, C. Denise
Innes, Beulah L.
Ireland, William H.
Iron Shooter, Darrell G.
Irwin, Charles E.
Irwin, Don R.
Irwin, Gladys G.
Irwin, William
Isogai, Akira
Itoh, Tomoyuki
I-uinones, Hugo
Ivory, Edward P.
Jackelen, Nicole
Jackson, Craig A.
Jackson, Gary
Jackson, Gene A.
Jackson, Gerry
Jackson, Harry T.
Jackson, Marjorie
Jackson, Travis
Jacobs, Marion F.
Jacobs, Tom S.
Jacobsen, Adolph L.
Jacobson, James
Jacobson, Rodney E.
Jacobson, Theo. B.
Jacobson, Thomas M.
Jacobson, Virginia C.
Jaech, Harold W.
Jaeger, Deborah
Jaeger, Edward H.
Jaeger, Michael J.
Jahnke, Walter E.
Jahr, Dorothy M.
Jakes, Joseph E.
James, Shantelle N.
James, William L.
Jameson, Patricia N.
Jang, Min
Janse, Bernard
Janssen, Dirk W.
Jaquish E. Del Mar
Jaranilla, Emelio
Jardine, James B.
Jardine, James L.
Jarvis, James A.
Jeffers, Brian
Jeffries, Thomas W.
Jelinski, Martha
Jenkins, Bernie H.

Jenkins, Katie M.
Jenkins, Sarah
Jenkinson, Paul M.
Jensen, Estelle
Jensen, Harland B.
Jensen, J. Norman
Jensen, James P.
Jensen, Kenneth A.
Jensen, M. C. Tr.
Jensen, Mogens C.
Jensen, Ruth E.
Jenson, Joseph A.
Jeremission, Jerry R.
Jess, John A.
Jewell, James W.
Jewett, Douglas M.
Jewson, Robert G.
Jhansale, Shaum
Jiang, Hongquan
Jiang, Shu Yong
Jillson, Jack
Jimenez, Alberto
Jin, Yong-Su
Joachim, Lila B.
Joe, Janet, M.
Johannson, Tomas
Johlin, Jacob M. Jr.
John, Barbara
John, Sarah
Johns, Donald
Johns, Michael L.
Johnson, Alfred N.
Johnson, Bruce R.
Johnson, Caryl G.
Johnson, Charles A.
Johnson, Cheryl A.
Johnson, Clarence M.
Johnson, Curtis L.
Johnson, Dale M.
Johnson, David B.
Johnson, Dean S.
Johnson, Dewey W.
Johnson, Don P.
Johnson, Donald R.
Johnson, Donna M.
Johnson, Doris A.
Johnson, Edna May
Johnson, Edward L.
Johnson, Eleanor F.
Johnson, Elsa
Johnson, Ernest P.
Johnson, Ethel M.
Johnson, Eugene L.
Johnson, Frank
Johnson, Henry C.
Johnson, Jimmie Jr.
Johnson, Julia Agnes
Johnson, Kathleen L.
Johnson, Kenneth G.
Johnson, LaVerne F.
Johnson, Marjorie H.
Johnson, Martha C.
Johnson, Martha, L.
Johnson, Marvin J.
Johnson, Mary E.
Johnson, Melvin L.
Johnson, Milton A.
Johnson, Nancy C.
Johnson, Nancy S.
Johnson, Oral M.
Johnson, Portia E.
Johnson, Robert K.
Johnson, Robert P.A.

Johnson, Robert S.
Johnson, Seymour J.
Johnston, Bonita A.
Johnston, Don P.
Johnston, Ralph E.
Jokerst, Ronald W.
Jonas, George
Jones, Alma C.
Jones, Arthur T.
Jones, David
Jones, Don Carl
Jones, Hiawatha
Jones, Jeanne
Jones, Jenne
Jones, John L., Jr.
Jones, Joseph C.
Jones, Norma E.
Jones, Paula G.
Jones, Robert E.
Jones, Terry P.
Jones, Wanda, B.
Jones, Wesley W.
Jones, Wilbur L.
Joppa, Neal N.
Joppa, Roe E.
Jordan, Clarence A.
Jordan, Freda
Jordan, J. Robert
Jordan, Janet E.
Jordan, Richard L.
Jordens, Harry W.
Jotter, Ernst V.
Joyal, Marc R.
Joyce, Edward A.
Ju, Li Wei
Juckem. Charles P.
Judd, Harold, L.
Judd, Roy C.
Julson, Loral M.
Junck, Leonard A.
Jung, Joseph
Jungmann, Edward C.
Juntunen, David D.
Juola, Arne W.
Justiliano, Vito L.
Jutte, Susanna M.
JuVette, Darrell D.
Kaap, Norma M.
Kachelhoffer, Edith F.
Kaether, Helen B.
Kaether, James E.
Kaether, John P.
Kahl, Karla
Kahl, Wanda M.
Kailin, Clarence
Kainz, James A.
Kaiser, Daniel R.
Kaiserlik, Joseph M.
Kaland, Michael
Kalar, Wm. L.
Kalinosky, Steven R.
Kalnins, Martin A.
Kaltenbach, John
Kaltenberg, Wm. E.
Kamasudirdja, Soeparmar
Kamp, Holly F.
Kane, Beatrice
Kane, Kenneth J.
Kaneshige, Harry M.
Kang, Min Hyung
Kanvik, A.
Kanvik, Elizabeth A.
Kanvik, Karl J.

Kanz, James
Kapitch, Alexander
Karch, Lillian M.
Karlsen, Ole
Karmaker, Ajit
Karpe, William S.
Kartal, Nami
Karthikeyan, K.G.
Kaspszak, Michael J.
Kass, Andrew J.
Kass, Marilyn J.
Katovich, Kerry
Katsfuros, Kathryn Z.
Katz, Carole
Katzenberger, Claire
Kaufert, Frank H.
Keen, Nancy
Keenan, Margaret
Keliher, Ruth D.
Kelleher, John M.
Keller, Eugene L.
Keller, Joyce E.
Kelley, Robert E.
Kellicutt, Keith Q.
Kellner, Magdalene J.
Kellor, John G.
Kellor, John K.
Kelly, Betty Lee
Kelly, Brian
Kelly, Clayton
Kelly, Dewey
Kelly, George T.
Kelly, Janice K.
Kelly, Myron W.
Kelso, T.
Kemmeter, James A.
Kempen, Robert
Kemper, Louise E.
Kempfer, Willam H.
Kempka, Mary Lea
Kenealy, William R.
Kenison, Fred C.
Kenison, Roy G.
Kennedy, Donald C.
Kennedy, Gerald L., Jr.
Kennedy, Janet
Kennedy, Theodore W.
Kent, Judy A.
Kent, Samuel M.
Kepke, Clara M.
Kerem, Zohar
Kerl, Herman R.
Kern, Jeffrey A.
Kern, John R.
Kerr, Jean
Kershasky, Michael J.
Kerst, Ashley A.
Kersten, Philip J.
Kerwin, Anastasia
Kerwin, Joseph J.
Kerwin, Stasia
Kessenich, Judy H.
Kessler, Amy J.
Ketcham, Henry H.
Ketcham, Marie M.
Kewin, Jennifer
Key, Henry J.
Keyes, Benjamin J.
Khatri, Chandra
Khurana, Sangot
Kiatgrajai Prffcha
Kieckhefer, Herbert H.C.
Kiefer, Donald A.

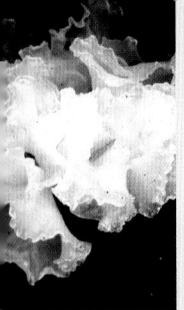

Kiefer, Ella
Kietzke, Fred O.
Kiffer, Donald
Kihle, Lowell E.
Kikkert, Donna R.
Kile, Todd
Kilgore, Dante T.
Killebrew, John C.
Kills Plenty, Theodore N.
Kim, Jong Man
Kim, Juyoung
Kim, Sumg Ryul
Kim, Won Sup
Kimball, Frank B.
Kimball, Kenneth E.
Kimel, William R.
Kindschi, John A.
King, H. Theodore
King, Helen E.
King, Kevin A.
King, Margaret P.
King, Ted
King, Victoria G.
King, W.T.
King, Wayne M.
Kingsbury, Alan P.
Kingsbury, Harold N.
Kingsbury, Ralph M.
Kingsland, George E.
Kingston, Donald J.
Kinney, Edmund F. Jr.
Kinney, James L.
Kinney, Richard E.
Kinney, Robert L.
Kinney, Sadis D.
Kinney, Sarah D.
Kinney, Wilfred E.
Kinney, Will A.
Kipp, Hildegard P.
Kirachner, Marian B.
Kirchhoff, Astrid
Kirchner, Jerry A.
Kirk, Celeste H.
Kirk, Douglas K.
Kirk, T. Kent
Kirker, Grant T.
Kirsch, Simon
Kirsten, Gordon
Kirwin, Richard
Kirwin, Rose
Kisker, David W.
Kissane, Leo F.
Kistler, Samuel E.
Kitchen, Joseph H.
Kitchin, P.C.
Kittell, Fredrick J.
Kjelland, Alan R.
Kjelland, Sharon R.
Kjervik, Andrew O.
Kjichefski, Anita L.
Klade, Richard J.
Klainert, Barthold E.
Klassy, Romaine K.
Klebesadel, Many Jane
Klee, Eugene H.
Klein, Mary Jane D.
Klein, Rosemary
Kleinheinz, Anton F.
Kleinheinz, George F.
Klem, Gustav G.
Kleman-Leyer, Karen
Klepp, Lawrence S.
Klimm, Josef

Kline, Albert A.
Klinger, Robert J.
Klongland, Lydia M.
Klopf, Leonard W.
Kluever, Maria R.
Kluever, Thyra S.
Klug, Donald A.
Klug, Lucy
Kluga, Edmund R.
Kluge, Annette S.
Kluge, Atcharaphan
Klun, Cheryl K.
Klungness, John H.
Kmiecik, Paul A.
Kmiotek, Alice M.
Knaebe, Mark T.
Knauss, Archibald C.
Knauss, Evangeline E.
Knechtges, Richard G.
Kneebone, John U.
Knepper, Jeanne G.
Knight, Megan
Knight, Peter C.
Knispel, Ronald O.
Knorr, Steven G.
Knowlton, Harry N.
Knox, Elizabeth M.
Knuth, David T.
Knutson, Oscar G.
Knutson, Torkel
Koberstein, Adeline H.
Koch, Arthur W.
Kocs, Andrew M.
Koehler, Arthur
Koehler, Jean E.
Koellen, Adolph P.
Koelsch, Roman J.
Koepoke, Elinor W.
Koeppen, Robert C.
Kohl, Lloyd R.
Kohl, Walter H.
Kohlman, Jane D.
Kohlmann, Robert F.
Koivisto, Janet M.
Kojis, Daniel D.
Kolacke, Robert F.
Koleske, Joseph V.
Kollberg, Paul G.
Kolosick, Paul
Kolstad, Doris E.
Kolstad, Jalmar
Kolstad, Sever A.
Koltz, Shirley M.
Kommers, Jesse B.
Kommers, William J.
Koning, John W., Jr.
Konkol, Margaret R.
Kopp, Frank M.
Kopper, Brian
Korb, Ralph E.
Korenkiewicz, Bartley L.
Kosowicz, James S.
Koster, Arthur L.
Koster, Doris V.
Kovacsvolgyi, Gabor
Kowalke, Otto L.
Kozak, Robert J.
Kozak, Sharon L.
Kraege, Carl A.
Krafft, Gertrude
Kramer, Robert M.
Kramer, Wendy H.
Kramp, Andrew

Krane, Arnold
Kratochvil, John F.
Kraus, Denita A.
Krause, Robert L.
Krause, William L.
Kregel, William J.
Kress, Otto
Kressman, Frederick W.
Kretschmann, David E.
Kreuger, Gordon P.
Kreul, William C.
Kreutzer, Mark A.
Kriel, Robert
Krin, J.
Kroening, Ada B.
Kroll, David
Kromar, Gene F.
Kronberg, Barbara
Krone, Robert H.
Krouse, Rachel
Krueger, Beverly R.
Krueger, Catherine E.
Krueger, Esther
Krueger, Gordon P.
Krueger, James H.
Krueger, Kurt J.
Krueger, Mark T.
Krueger, Norman T.
Kruel, William C.
Kruger, Kenneth W.
Krugman, Isabell E.
Kruser, Mark
Kryn, Jeannette M.
Krzysik, Andrzej M.
Kubitschek, Catherine
Kubitz, Ernest C.
Kubitz, Marlin A.
Kubler, Hans
Kuehl, James F.
Kuehn, Catherine
Kuelling, Herbert J.
Kuenzi, Edward W., Jr.
Kuester, Arland W.
Kuettel, Glen M.
Kuhlman, Evelyn M.
Kuhnhenn, Margaret A.
Kukachka, B. Francis
Kulik, Henry J.
Kulp, John W.
Kunishi, Alice T.
Kunzweiler, John J.
Kupfer, Carl A.
Kurtenacker, Robert S.
Kurth, Ervin F.
Kurth, Perter
Kuskowski, Steven J.
Kuster, Thomas A.
Kutscha, Dieter F. A.
Kutscha, Norman P.
Kutzback, Victor R.
Kuzma, James
Kwon, Jin Heon
Kyrkjeeide, Per Arne
Kyung-Park, Yoon
LaBarca, Irene
Labissoniere, Tim
Lacher, Steven
Lacina, Shannon
Lacina, Shawn
Lacy, Geraldine R.
Lafarga, Suzanne P.
Lafontaine, Sara
Lagraves, Robert G.

Laing, John J.
Lake, Clarence L.
Lake, Robert M.
Lal, Raj
Lallier, Chas. J.
Lamar, Richard T.
Lamb, Jerrold B.
Lamberson, Winifred
Lambrecht, Michael
Lamont, Nancy R.
LaMore, William
LaMotte, F.L.
Lancer, Emilio A., Jr.
Landis, Merrill W.
Landis, Orin
Landon, Joan S.
Landry, Robert W., II
Landt, Eugene F.
Landucci, Lawrence L.
Lane, Nancy L.
Lang, Gary K.
Lang, Jake P.
Lang, Lloyd N.
Lange, Albert G.
Lange, Irene E.
Lange, Jane R.
Lange, Julie
Lange, Sandra E.
Lange, William J.
Langhammer, David
Langrehr, Richard J.
Langsdorf, Jon H.
Lanier, Bryan
Lanier, Emilio A., Jr.
Lanphier, Ira B.
LaPidus, Ann G.
Lapinski, T.
Laplaza, Jose M.
Lappley, Melva
Larmore, Frederick Dale
Larrondo, Luis
Larsen, Michael J.
Larsen, Ruben A.
Larson, Robert L.
Larson, Camen C.
Larson, Gary R.
Larson, Gertrude N.
Larson, Jennifer
Larson, Kathy M.
Larson, Oliver
Larson, Robert L.
Larson, Rodney G.
Larson, Sandra
Larsson, Pia
Lasisi Rasadki Afolabi
Lassen, Lawrence E.
Lassen, Leon
Lasson, Elizabeth L.
Last, Walter F.
Latham, E.O.
Lathrop, Benjamin A.
Lathrop, Mary B.
Latif, Farooq
Latif, M. Abdul
Latip, Rozaida
Laue, Edmund J.
Laue, J.
Laufenberg, Theodore L.
Laughnan, Don F.
Laughren, Thomas P.
Laundrie, James F.
Lauret, Barbara
Laux, Daniel R.

LaVaque, Theodore J.
Lawrence, Christine A.
Lawrence, Darlene
Lawrence, George L.
Lawrence, James C.
Lawson, Beulah
Lay, David
Layden, John
Lea, Gerald W.
Leadabrand, Joseph A.
Leadley, Warren D.
Lean, Holly J.
Leao, Alcides
Leatham, Gary
Lebo, Robert B.
Lebow, Patricia K.
Lebow, Stan T.
Lecher, Sylvia M.
LeCount, Fred L.
Lee, Agnes
Lee, Beom-Goo
Lee, Hong-Lin
Lee, Jae-won
Lee, John J.
Lee, Karen
Lee, Mary A.
Lee, Michele
Lee, Sung K.
Leeck, Charles J.
Lefebvre, Ellen M.
Left Hand Bull, Cleo P.
Left Hand Bull, Paul L.
Lehman, Andrew J.
Lehmann, Jeffrey
Lehmann, William F.
Lehtola, Wm. E.
Leigh, C. T.
Leistra, Fred
Leitner, Edward V.
Lemanski, Barbara A.
LeMay, David J.
LeMay, Maurice W.
LeMieux, Rita W.
Lemon, Ethel M.
Lenahan, Nellie
Lengel, Joan E.
Lennon, Erin
Lentz, Michael
Lenz, Bernard L.
Leo, David W.
Leonard, Reid H.
Leopold, Aldo
LePage, Marjorie H.
Leslie, Harry C. III
Lessig, Margaret B.
Lestan, Domen
Lestan, Marjeta
Letcher, S.L.
Leuterio, Leuber
LeVan-Green, Susan L.
Levenhagen, Paul
Levey, Barbara J.
Levine, Howard
Lewicki, Leonard S.
Lewis, David W.
Lewis, Earle S.
Lewis, Ethelyn E.
Lewis, Kevin
Lewis, Raymond
Lewis, Wayne C.
Li, Kenning
Li, Li
Li Xian Jun

Li, Yongxiang
Li, Yue Xiang
Liang, Shu-chi
Libert, Ryan S.
Libkie, Kimball A.
Lichtenberg, Gary J.
Liddle, Albert H.
Liddle, Ruth M.
Liebl, Mary A.
Lien, Gunnar
Lighthall, Marshall B.
Lillesand, Lynn
Lillie, Richard A.
Limbach, John P.
Lin, Amy
Lin, Bernice C.
Lin, Chyong Huey
Lincoln, Blaine D.
Lind, Sidney R.
Lindauer, Alfred C.
Lindeborg, Richard A.
Lindell, Gary R.
Linden, David J.
Lindgren, Ralph M.
Lindley, Lynnette
Lindner, Daniel
Lindsay, John A.
Lindstrom, Sandra A.
Liner, David
Lingle, Mary J.
Link, Carol L.
Link, Charles
Link, John
Link, Leo P.
Link, Roman A.
Lins, Christian
Lins, Terrence J.
Lintner, Marilyn A.
Linton, Joe
Linwalter
Lipert, Robert J.
Lipski, Jerome J.
Liscum, Maynard E.
Liska, Joseph A.
Little, Charles A.
Little, James E.
Little, James K.
Little, Joseph
Little, Kevin
Littlel, Edward
Littlel, Joseph
Litwin, Dan
Liu, Hao
Liu, Jen Y.
Livesey, Bonnie M.
Livingston, Jean M.
Livingston, Philip
Lloyd, Omar D.
Lloyd, Raymond T.
Lloyd, Roger A.
Lloyd, Thomas R.
Locher, Ted J.
Lochner, Helen C.
Lochner, Jean L.
Locke, Edward G.
Loder, Janice E.
Lodge, D. Jean
Loehnertz, Stephen P.
Loeprich, Henry
Loftus, Theresa
Logan, Gordon D.
Logan, Joseph
Lollis, Vivian R.

Lombard, Frances F.
Lomblot, Sandra L.
Long, Tanya
Longfield, Sara R.
Loos, Emil
Lorberter, Geraldine M.
Lord, Richard E.
Lord, Ted J.
Lord, Timothy D.
Lorenz, Linda F.
Lorrigan, Anna M.
Lott, Cynthia L.
Lott, Louis Y.
Lotterer, David W.
Lough, Charles H.
Loughborough, Dwight L.
Loughborough, W. Karl
Loughead, Harvey J.
Love, James B.
Lowe, Frederick N.
Lowell, Herbert M.
Lowery, David P.
Lowrey, Elsie B.
Lu, Chenfeng
Lu, Ke Yang
Lu, Ping
Lucas, Marilyn J.
Luding, William A.
Ludwig, William F.
Luebke, Robert W.
Luedeke, Michael C.
Luetscher, Harold M.
Luhman, Dale C.
Lujan, Brenda L.
Luksich, Helena
Lulling, Robert M.
Lunak, S. E.
Lund, Elmer A.
Lund, H. Beatrice
Lund, J. Edward
Lund, James E.
Lund, John A.
Lund, Richard J.
Lunde, C. Eleanor
Lundgren, David C.
Lundgren, James F.
Lundin, Robert
Lundin, Thomas J. L.
Lunstrum, Stanford J.
Lunte, Louis
Luther, Andrew
Luther, Mattie A.
Lutz, Joann E.
Lutz, John F.
Luxford, Ronald F.
Luze, Deana
Lyles, Deandrea
Lynch, Mary C.
Lynch, Maybelle
Lynes, Percy F.
Lyngaas, Victor A.
Lynwalther, Rose
Lyon, Charles W.
Lyons, Anastesia S.
Lyons, D.
Lyons, Thomas
Lysager, Alvin S.
Lysager, Rachel A.
Lythjohan, Bruce L.
Ma, Agnes
MacArthur, Marcia C.
MacDonald, Fred A.
MacDonald, Storm I.

Mace, Terry
Machnik, Diane S.
Mack, David J.
Mackay, Elizabeth
MacKenzie, Andrew H.
Mackesey, Stephen A.
Mackin, George, E.
MacLean, James D.
MacLeod, Etta R.
MacMurray, M.
MacNaughton, Leslie
Macon, John W.
MacRavey, F.
Maedje, Carl W.
Maeglin, Robert R.
Magann, Edith
Magnuson, Linda F.
Mahaffey, Margaret V.
Mahan, Sarah
Mahmoud, Tayeh
Mahood, Samuel A.
Main, Laudra M.
Maine, F. Leland
Mair, Mary Jane S.
Maki, A. Carl
Malandri, Nicole
Malas, Jackie
Malcom, Fredrick B.
Malinauskas, Vyto
Malison, Jeff
Malone, Philip V.
Maloney, Jerome F.
Maloney, Mark
Maloney, Wm. J.
Malseed, Dorothy L.
Mamauag, Aurora S.
Mamlatdarna, Ramlal
Maney, Patricia A.
Manion, Paul E.
Mankie, Earl K.
Mann, Doreen H.
Mann, Richard A.
Mannering, Sidney
Manske, Brian
Mansoorabadi, Karen
Manthe, Alfred F.
Mantz, Raymond A.
Mapson, Norman
March, Herman W.
March, James G.
Marcin, Thomas C.
Margerum, Stanley D.
Marin, L.
Marinack, Robert J.
Mariotte, Walter W.
Marking, Betty B.
Marking, Conrad A.
Marks, Jere A.
Marks, Lillian F.
Marks, Nellie
Markwardt, Lorraine J.
Marquardt, Ralph R.
Marr, David J.
Marra, George G.
Mars, Ruben
Marschner, Charles F.
Marshall, B.
Marshall, Robert E.
Marston, Doris V.
Marti, Ruben A.
Martin, Adriana
Martin, Dorothy E.
Martin, Irvin B.

Martin, J. Stanley
Martin, John S.
Martin, Jonathan D.
Martin, Kay C.
Martin, Larrry
Martin, Rebecca E.
Martin, Theodore J.
Martinez, Albert J.
Martinez-Roacho, Hugo
Martinson, Karen L.
Martinson, Leon A.
Martinson, Norman C.
Martley, J. F.
Marts, Ralph O.
Marty, Helen M.
Marty, Wanda M.
Marty, Wilbert H.
Marx, Catherine M.
Masino, Samuel
Masko, Ann M.
Mason, H. M. Jr.
Mason, Lola
Matalya, Margaret L.
Mathens
Mathews, Ashley
Mathews, Dolores O.
Mathewson, James S.
Mathewson, Nancy P.
Mathison, Morris
Matson, Bertha R.
Matson, Gerald B.
Matson, Lillian H.
Matt, Barbara J.
Matt, Frederick J.
Matthews, Etta L.
Matthewson, J.
Mattson, Paul M.
Mauch, Steven A.
Maudlin, C.V.
Maul, Derek
Maurer Edward R.
Maurer, Phyllis
Maves, Nancy
May, Adrain
May, Mike
Mayer, Susan J.
Mayo, H.M.
Mays, Doris L.
McAlister, Robert H.
McAteer, James R.
McBeath, Lida W.
McBurney, Robert S.
McCaffrey, Howard
McCall, Ann H.
McCall, Russell V.
McCann, Alice
McCarthy, Joanne J.
McCarthy, Monica
McCarthy, Thomas W.
McCartney, Wm. M.
McConnell, Lyle M.
McConnell, Lyle W.
McCord, Orel E.
McCormick, James P.
McCormick, Norbert G.
McCourt, Joseph E.
McCoy, Arthur R.
McCoy, Jenette M.
McCulloch, Melissa
McCullough, Kevin
McCutcheon, William J.
McDonald, Dwight
McDonald, John K.

McDonald, Kent A.
McDonald, Merle, R.
McDonald, Nathan
McDuffie, Arthur F.
McFarland, John
McFee, John
McGahey, C. Elizabeth
McGee, Robert T.
McGee, Tom
McGilligan, Thomas J.
McGlynn,William J.
McGovern, John N.
McGraw, Dianne
McGregor, George H.
McHenry, Neva Jean
McKean, Herbert B.
McKee, Aasta
McKeever, Chris
McKeever, David B.
McKeever, Tim
McKenna, Lucille T.
McKenzie, Andrew H.
McKenzie, H.E.
McKenzie, Malcom
McKeough, Patrick J.
McKibbins, Samuel W.
McKinnen, Tom
McKinnon, Paul F.
McKoy, Cheryl L.
McLain,Thomas E.
McLean, James D.
McLean, Stewart M.
McLeod, Arthur M.
McManus, William
McMaster, Michael D.
McMillen, John M.
McMillin, Casey R.
McNatt, J. Dobbin
McNaughton, George C.
McPherson, J. A.
McRoberts, Donald E.
McRoberts, Ronald E.
McSherry, Eleanor
McSherry, Eugene F.
McSwain, George A.
McSweeny, James D.
McWatty, Della M.
Meadowcroft, Kent B.
Meassinger, Adam
Medcraft, Clyde
Medd, Tim
Medicielo, Leodulo
Medicielo, Teodulo J.
Medley, James W.
Meer, Robert
Megraw, Robert A.
Mehlberg, Ryan
Mehlig, Joseph F.
Meicher, Gregory A.
Meier, Myrtle L.
Meindl, Frank J.
Meinecke, Eberhard A.
Meisnest, Marilee
Meixelsperger, Anita C.
Meixelsperger, Diane E.
Melcher, Fred
Mell, C.D.
Mellin, Alice M. A.
Meloche, Villiers W.
Melrose, Clayton A.
Melsen, Josephine K.
Melvin, Judith D.
Mendes, Alfredo

Menzel, Carl A.
Mergen, Alfred F.
Mergen, Alois M.
Mergen, Herman L.
Merhemic, Irene
Merklein, William R.
Merle, Raymond K.
Merli, Angelo
Merrel, Edward C.
Mersberger, Mary R.
Meseck, Diana D.
Messerschmidt, Ralph O.
Metcalf, Robert L.
Mettel, Mark C.
Metz, Carl C.
Metzler, Allen
Metzler, Benjamin S.
Meuer, William J.
Meuler, David
Meyer, Edmund L.
Meyer, Gary W.
Meyer, Herbert R.
Meyer, Kathy
Meyer, M. Noreen
Meyer, Maria S.
Meyer, Mary.
Meyer, Russell, E.
Mianowski, Theodore
Micklewright, James T.
Middlecamp, Catherine B.
Middleswart, Eugene L.
Middleton, Leonore B.
Middleton, Oral L.
Midthun, Kari
Midthun, Oliver A.
Mikulski, Ariana
Militzer, Walter E.
Millar, Wentworth A.
Miller, Ann C.
Miller, Brenna
Miller, Burton F.
Miller, Carl E.
Miller, Carol M.
Miller, Charles D.
Miller, Doris, M.
Miller, Edward H.
Miller, Henrietta E.
Miller, Keith S.
Miller, Kim S.
Miller, Leona M.
Miller, Louise P.
Miller, Mary Anne
Miller, Mary S.
Miller, R.
Miller, Raymond H.P.
Miller, Regis B.
Miller, Roland N.
Miller, Steven D.
Millett, Kenneth
Millett, Merrill A.
Milliman, Frank C.
Milota, Michael R.
Milton, Estrella R.
Min, Soo-Hong
Miner, Oliver H.
Miniutti, Victor P.
Minor, James L.
Minotte, Flo
Miro, Carlos J.
Mishra Chittra
Mitchell, Angela
Mitchell, Dorothy
Mitchell, Harold L.

Mitchell, John A.
Mitchell, Kenneth G.
Mitchell, Raymond L.
Mitchell, Shirley M.
Mitchell, William W.
Mitsunaga, Tohru
Miyashita, Ernie S.
Moe, Kenneth V.
Moede, Michele
Moehlman, Julius F.
Moehlman, Leroy
Moehlman, Louis
Moehlman, William
Moen, Abbie M.
Moen, Jean M.
Moen, Jerrold N.
Moen, Kenneth A.
Moen, Virginia A.
Moerke, Ida
Mohaupt, Alvin A.
Mohr, Harvey W.
Mohrhauser, Carol C.
Moiseye, Alexander
Moldenhauer, Ida E.
Moldenhauer, Judith A.
Moldenhauer, Vivian G.
Moldes, Ana
Monahan, Anna
Monfried, Clara
Monsalud, Manuel R.
Monsson, William H.
Montana, Estevan
Montrey, Henry M. III
Monum, Alvie W.
Moody, Russell C.
Moon, Marion W.
Moon, Robert J.
Moore, Dennis R.
Moore, Lorraine A.
Moore, Pearl K.
Moore, Robert W.
Moore, Roderquita
Moore, Thomas F.
Moore, Wayne E.
Moore, William G.
Morbeck, George, C.
Moreell, Samuel
Morehauser, Jos.
Morehouse, Fred E.
Morgan, Lois G.
Morgan, Lucille E.
Morgan, Sandra L.
Morgan, Steve A.
Mori, Scott A.
Mork, Donald
Morman, Frank C.
Moroney, Frances
Moroney, Robert
Morrell, Samuel
Morrill, Majorie E.
Morris, William W.
Morris, Leslie R.
Morris, Meade M.
Morris, Nathan A.
Morrison, Gerald E.
Morrissey, Leone J.
Morrissey, Lillian
Morrissey, Thomas L.
Mortensen, Betty G.
Morton, Hudson T.
Mosdal, Bob
Moses, Clayton S.
Mosey, Linda D.

Mosher, Dorothy H.
Mosley, Thomas J.
Moss, Peter J.
Mossing, Susan L.
Motelet, Alphonse J.
Mott, Wallace C.
Moubry, Robert J.
Moyer, Arthur M.
Mozuch, Michael D.
Mraz, Edward A.
Mu, Yi
Muehl, James H.
Mueller, Herman R..
Mueller, Kathleen R.
Mueller, LaVern H.
Mueller, Lincoln A.
Mueller, Scott A.
Muench, Paul
Muender, Olga M.
Mukavitz, Hilarie
Mulcahy, Kim M.
Mulchaey, Rupert J.
Muller, Carl E.
Muller, Christine E.
Muller, Debra J.
Munson, Robert A.
Munson, Spencer M.
Munthe, Bert P.
Murmanis, Lidija L.
Murphy, Blanche R.
Murphy, Harriette P.
Murphy, Joseph F.
Murphy, Marion R.
Murphy, Stephen
Murray, Jean F.
Murray, John T.
Murray, Jorene M.
Murray, Michael
Murugesan, Goulavanan
Musil, Margaret
Myashita, Ernie
Myers, Earl C.
Myers, Gary C.
Myers, George E.
Myers, Kathy
Myren, Andrew O.
Myren, Krista M.
Myrland, Isabel
Mytton, Patricia A.
Nagaoka, Munfo
Nagasampagi, B. A.
Nagel, Robert
Nagler, Anton W.
Nahar, Shamsum
Nair, Gangadharan
Nairn, Linda S.
Nakasone, Karen K.
Napp, Charles P.
Naprstek, John
Nash, Casper
Nash, James
Nash, Richard M.
Nash, T. Romale
Nasierowski, Adam
Nastke, Heinz F.
Natera, Rodolfo V.
Natwick, John W.
Navarro, Andrew
Neal, Claudius A.
Neal, Clyde U.
Neal, Mary E.
Neckar, Jeanette M.
Nee, Michael

Nellis, Fay F.
Nelson, Benjamin L.
Nelson, Brenton
Nelson, Charles A.
Nelson, Gladys L.
Nelson, Grace M.
Nelson, Gregory O.
Nelson, Herbert R.
Nelson, John W.
Nelson, Karen K.
Nelson, Mary O.
Nelson, Neil D.
Nelson, Nona F.
Nelson, Oliver J.
Nelson, Ralph
Nelson, Russell, G.
Nelson, Shawn S.
Nelson, Stanley C.
Nelson, Timothy C.
Nelson, William J.
Nelson, William R.
Nerdrum, Christ
Ness, Joseph
Nestler, F. H. Max
Neubauer, Agnes
Neugent, Mable T.
Nevalainen, Tony
Newel, Anton
Newel, Frances
Newel, Peter
Newlin, John A.
Newman, Leah
Newman, Ralph E.
Newman, Stephanie
Ni, Haiying
Nibbelnk, Karl
Nicholls, Herbert C.
Nicholson, Geraldine
Nickles, William, C.
Nickolls, Steven K.
Niedziela, Terrance F.
Nielsen, Max H.
Nielsen, Tor
Niemann, G. J.
Nigbor, William P.
Nightingale, Lee
Nilles, Darrell, F.
Ninedorf, Jacalyn K.
Nishijima, Kate A.
Niskanen, Kaarlo
Noe, Ralph W.
Nondahl, Peter J.
Nondorff, Marie A.
Nonn, Albert
Noordewier, Michiel J.
Nordbrock, Earl E.
Nordenson, David E.
Nordgren, Fred C.
Nordgren, John E.
Nordgren, John L.
Noren, David T.
Norris, Charles B.
Norton, Kyah L.
Norton, Lee
Norton, Newell A.
Novotny, John J.
Nowell, Sidney M.
Nusinoff, Max S.
Nutter, Peggy L.
O'Brien, Elizabeth L.
O'Brien, Harold A.
O'Brien, John F.
O'Brien, M. Lee

O'Connell, J. A.
O'Connell, Michael
O'Connell, Paul F.
O'Connor, Bernette
O'Connor, Mary H.
O'Connor, Ralph D.
O'Connor, Therese R.
O'Dea, Catherine M.
O'Dea, Thomas A.
O'Dell, Jane L.
O'Gara, Patrick W.
O'Hera, Agnes
O'Hora, Regina
O'Leary, Charles
O'Leary, Clyde W.
O'Malley, Frances M.
O'Neil, Mary K.
O'Neill, Bessie
O'Neill, Eric
O'Neill, Raymond J.
O'Neill, Robert J.
Oakey, James J.
Oakey, Madeleine N.
Oberg, Tena
Obst, John R.
Ochs, Donald E.
Oen, Gladys I.
Ofanne, L. I.
Ogden, Janet C.
Oh, MiYoung
Ohba, Toshio
Ohm, Allan O.
Okkonen, E. Arnold
Oldham, Stanley
Olexer, David L.
Olietti, Mike J.
Oliver, Charles A.
Oliver, Karl K.
Oliversen, Heidi J.
Olsen, Bernetta A.
Olsen, Christopher P.
Olsen, Ingman
Olsen, Mary C.
Olson, Alma L.
Olson, Bruce
Olson, David
Olson, Douglas E.
Olson, Enid J.
Olson, Frank R.
Olson, Gladys
Olson, Lewis O.
Olson, Lynn C.
Olson, Olaf A..
Olson, Paul H.
Olson, Ronald R.
Olson, Violet L.
Olson, Warren Z.
Olstad, Adele M.
Onan, Robert W.
Ondahl, Peter
Ondal, Arthur I.
Onsrud, Harold M.
Oostdyk, Theresa
Orloff, Daniel L.
Orne, Nels, H.
Orosz, Ivan
Orozco, Jose B.
Orr, D.
Ortiz-Bermudez, Patricia
Orvis, Gary W.
Osborne, Kenneth A.
Osius, Virginia A.
Ossmann, Erward C.

Osteraas, John D.
Ostorero, Andrew J.
Ostrander, Nora
Otiz-Santana, Beatriz
Ott, James R.
Otterson, Maybelle
Otto, Mary S.
Ouraipryvan, Piya
Oviatt, Alfred E. Jr.
Owen, Nina
Oxman, Lynn
Ozburn, Kathryn E.
Ozburn, Robert D.
Ozburn, W.
Pablo, Arturo
Padfield, Hammond H.
Padgham, Lorenzo J.
Padley, Eunice
Paepke, Benjamin
Page, Donovan W.
Page, Rita A.
Pagel, Bruce A.
Paine, Paul M.
Paine, William
Paksys, Hilda L..
Paley, Anne
Palkovic, Gloria M.
Palms, Jerome
Palma, Joseph
Palmer, Antron
Palmer, Herbert P.
Palmer, John G.
Palmer, John T.
Palmer, Lora
Palmer, Patricia A.
Palmer, Phillip M.
Palmer, Robert C.
Palmquist, Ronald W.
Paltz. N.
Palubicki, Lonny
Palzkill, Mary M.
Palzkill, Mildred
Pan, Chang-Pi
Panabaker, Betty J.
Panek, Edward
Panshin, Alexis J.
Pantazes, Edna
Panzer, Harry R.
Pape, Eleanor M.
Pappas, Richard
Pappe, Herbert
Parisi, Angeline C.
Parisi, Susan D.
Park, John S.
Park, Youngki
Parker, Ethel
Parker, John
Parkinson, Kathryn E.
Parkyn, Betty C.
Parr, J. L.
Parr, Olive M.
Parrish, John R.
Parson, Roger K.
Parthun, Veronica R.
Paschall, Wells, H.
Paskin, Bessie
Patch, Richard
Patenaude, Judy A.
Patrick, Lillian L.
Patrie, Carmen R.
Patterson, Frederic B.
Patterson, Maree
Patton-Mallory, Marcia

Patzer, Robert A.
Patzer, Walter E.
Pauhs, L.C. Elsie
Paul, Benson H.
Paul, Jennifer
Pauley, Susan
Paulos, James P.
Paulson, Arlene C.
Paulson, George C.
Paulson, George M.
Paulson, Susan K.
Paulson, Thelma
Payne, Juanita
Payne, Kay E.
Payne, Ruth G.
Paynter, Richard V.
Peakes, Lawson V., Jr.
Pearson, Clarence D.
Pearson, Richard J.
Pechan, Majorie I.
Pecher, Elizabeth H.
Pechie, William J.
Peck, Annabelle
Peck, Edward Charles
Pecker, Elizabeth H.
Peckham, Cynthia M.
Peckham, Karen J.
Peckham, Venor
Peckman, Lori A.
Pedder, Russel E.
Pedracine, Katherine S.
Peffer, Roland J.
Peirce, Benjamin
Pelech, Dennis
Pelton, Iva E.M.
Pelton, Marion H.
Penewell, John B.
Pennie, Laverne
Perdue, James H.
Pericchi, Sixto Jose
Perkins, Richard
Perkl, Virginia L.
Perlman, Charles M.
Pertzborn, Carol T.
Pestalozzi, James H.
Peter, Herbert M.
Peter, Ralph K.
Peters, Avis M.
Peters, Curtis C.
Peters, E.W.
Petersen, Axel J.
Peterson, Beulah L.
Peterson, Amos L.
Peterson, Clifford E.
Peterson, Donald E.
Peterson, Eleanor L.
Peterson, Gary M.
Peterson, John
Peterson, June
Peterson, Kenneth R.
Peterson, Larry B.
Peterson, Oscar
Peterson, Richard R.
Peterson, Theodore A.
Peterson, William H.
Pettersen, Roger C.
Pettibone, Heman N.
Petz, Earl W.,Jr.
Pew, John C.
Peyer, Ben
Pfeifer, David
Phalen, Edward J.
Pharo, Kenneth H.

Phelps, Fred W.
Phelps, John F.
Phelps, Rose Marie
Phillips, Alexander
Phillips, Dorothy
Phillips, Jacob
Phillips, Lucia S.
Phillips, Rufus S.
Phillips, Wilbur H.
Phillips, Wm. H.
Philumalee, Joe G.
Phipps, Kenneth L.
Piao, Cheng
Piazza, Russell
Pickett, Gerald
Piediscalzzi, James
Pieh, Myrtle M.
Pien, Roy W.
Pieper, E. J.
Pieper, Thyra S.
Pierce, David S.
Pierce, Maurice C.
Pieringer, Jo Anne
Pierstorff, Otto L.
Pike, Joseph P.
Pillow, Maxon Y.
Pilon, Crystal L.
Pingel, Clareen H.
Pingel, Eldona L.
Piper,Wilmer A.
Pirola, Alexander J.
Pittman, Paul
Plantinga, Pamela L.
Plaskett, Clyde A.
Plath, Wesley C.
Platkowski, Steven J.
Plechaty, John R.
Ploog, Blanche
Plow, Richard H.
Pluff, Gary
Plumb, Valworth R.
Podlipec, Mark R.
Polley, Charles W.
Pomeroy, Leslie K.
Pongratz, Eileen M.
Pongratz, Herman A.
Ponti, M.
Popp, Janet
Poppy, Gertrude, I.
Porter, Julius T.
Porter, Margaret C.
Porter, Robert E.
Porter, Ryan
Porter, Shielda
Porter, Wichaune C.
Porterfield, Dorothy H.
Post, Charles
Post, Walter A.
Postler, John R.
Potter, Philip A.
Potter, William E.
Poulle, Leigh Ann
Poulos, Christ G.
Powelson, Robert C.
Powers, Don A.
Poylos, Chris
Poynor, Fred R.
Prausa, Robert L.
Precourt, Priscilla
Predith, Patricia S.
Prenger, David B.
Prestemon, Dean
Preston, Richard J., Jr.

Price, M.H.
Price, Priscilla D.
Price, Raymond L.
Prideaux, Julie A.
Priestley, Herbert S.
Prince, Robert E.
Prindiville, James B.
Pritchard, John B.
Privett, Walter H.
Promboon, Warot
Promenschenkel, Jim
Pronin, Dimitri
Propp, Patricia L.
Protheroe, Susan L.
Prothfrof, Susan
Prston, Betty J.
Pugh, Wanda Lee M.
Pullara, John J.
Pullara, Joseph A.
Pulvermacher, Judith E.
Purcell, Mark T.
Putnam, Grace
Putnam, Susan L.
Pyle, Russell W.
Pyne, Clyde R.
Qazi, Aasma
Quam, Jean E.
Quam, Karon K.
Quam, LaVerne D.
Quarles, Denise L.
Quillin, Dan
Quilty, John M.
Quimby, Oscar T.
Quincy, R. B.
Quinn, Don L.
Quinn, Kenneth J.
Quinn, Lois M.
Quinn, Raymond J.
Quinn, Veronica R.
Quinn, William L.
Quirk, John T.
Raabe, Mark E.
Rabe, Allen E.
Rabyor, Mary E.
Rachmat, Zake F.
Rader, Elizabeth B.
Radske, Betty
Radue, B.
Rahman, Adeeb A.
Rahman, Moe O.
Raimond, John J.
Raktiprakara, Chote
Ralph, Sally A.
Ralston, Brian
Ralston, Michael
Ramaker, Terry J.
Ramanathan, Ayalur
Rambadt, Kenneth L.
Ramirez, Sally T.
Ramiro, Mariano P.
Ramlet, Nellie
Rammer, Douglas R
Ramos, Agustin N., Jr.
Ramos, Robert C.
Ramos, Rudolfo H.
Randall, Marilyn R.
Randall, Theo
Rankin, Charles, C.
Rankin, Michael K.
Ranney, Richard L.
Rapson, Ira J., Jr.
Raschein, Pamela R.
Rasmussen, Edmund F.

Rasmussen, Florence V.
Rasmussen, Howard R.
Rasmussen, Jere
Rasmussen, Sherrie
Ratell, Jesse C.
Rattner, Fred
Rau, Herman
Rauch, August H.
Rauch, Dorothy T.
Raville, Milton E.
Rawling, Francis G.
Ray, Brian H.
Rayford, Annette
Razzaque, M.A.
Reda, Pasquale
Reddick, Melvin L.
Redell, Ritchie R.
Redemann, William W.
Redmond, Elizabeth
Reed, Connie
Reed, Gladys M.
Reeves, Ronald E.
Regan, Ann A.
Rehm, Leo F.
Reich, Elsa
Reichel, Steven E.
Reid, William H.
Reidtan, Edith
Reif, Donald, J.
Reif, Walter J.
Reigle, Bruce A.
Reimann, Doris E.
Reindl, Jackie
Reineke, Lester H.
Reinemann, Doug
Reiner, Richard S.
Reinholtz, Ann C.
Reinke, Jeffrey S.
Reinke, Ronald F.
Reinsvold, Robert H.
Reis, Joseph J.
Reisdorf, Louan
Reiss, William L.
Rentmeester, Rita M.
Replingen, Mildred
Rettew, D. William
ReTua, E.M.
Reuter, Mary H.
Revels, Charles E.
Reyes-Chapman, Sanja
Reyhner, Theordore O.
Reynel, Carlos
Reynolds, David T.
Reynolds, Hugh W.
Reynolds, James E.
Reynolds, Jean E.
Reynolds, Milo B.
Reynolds, Terry P.
Rhoda, David
Rhude, Maurice
Rhyne, Regina R.
Rhyner, C.
Riba-Ramirez, Ramon
Ricard, Jacques
Rice, James W.
Rice, Theresa
Rich, J. Harry
Richard, Thomas C.
Richards, C. Audrey
Richards, William G.
Richardson, Charlotte H.
Richardson, Thomas J.
Richgels, Leone M.

Richgels, Patricia A.
Richmond, Daniel W.
Richter, Adrian L.
Richter, Klaus
Richter, Linda K.
Rickard, Thomas C.
Rieck, Kenneth W.
Rieder, John B.
Rieder, John R.
Riek, Donald H.
Riese, Bonnie F.
Rietz, Raymond C.
Rigney, Margaret E.
Riker, John L.
Riley, Carl J.
Riley, Mae M.
Rinder, Corey M.
Ringelstetter, Leo A.
Ringer, James R.
Ringwood, James B.
Ripp, Ardith M.
Ripp, Gerhard M.
Ripp, Patricia A.
Risbrudt, Christopher D.
Ritter, George J.
Ritter, Michael A.
Rivas-Torres, Beatrix
River, Bryan H.
Rizzo, Dave
Roa, Aicardo
Roach, Barbara
Roark, Raymond J.
Robar, Diane, J.
Roberson, P. J.
Roberts, Alvin S.
Roberts, David L.
Roberts, Gaige S.
Roberts, Marie
Roberts, Merle W.
Robertson, George M.
Robertts, Randal L.
Robidoux, Raymond J.
Robinson, Donna
Robinson, Evelyn A.
Robinson, Jeffrey D.
Rocca, Ruth K.
Roche, Edward H.
Rockwood, Donald L
Rockwood, Fred G.
Rodgers, Clarence C.
Rodrigues, Rita
Rodriguez, Apolonio T.
Rodriguez, Joseph F.
Rodriguez, Robert
Rodriquez, C.
Rodriquez, Ruby
Roe, Charles W.
Roecker, Lawrence F.
Roemer, Bernard H.
Roeser, Marion J.
Roesler, Betty A.
Roessler, Chester G.
Rogers, Alice L.
Rogers, Ethel M.
Rogers, Rachel
Rogers, Sedgwick C.
Rogers, William R.
Rogers, Willis H.
Rogers, William R. E.
Rogge, Joan E.
Rohovetz, Clarence J.
Rohr, Robert Jeffrey
Rohr, Roscoe C.

Rohr, Thomas R.
Rollings, F.
Roloff, Rita K.
Rolston, Franklin E.
Rommelfanger, J.
Romstad, Karl M.
Ronald, Robert C.
Rooney, Marion L.
Rooney, Walter E.
Root, Donald F.
Rork, Wesley L.
Rosales, Augusto
Rosas, Jesus R.
Rose, Dennis C.
Roseboro, Phyllis M.
Rosenbaum, Charles F.
Rosenbaum, Colleen J.
Rosenbaum, Michael A.
Rosenberg, Betsy
Rosenberger, Keil
Rosenberry, Howard J.
Rosendahl, Russell O.
Rosenfeld, Stuart A.
Rosenthal, Harry J.
Rosenthal, Michael R.
Roshong, Richard L.
Rosien, Herbert H.
Rosier, Harlan E.
Rosin, Brian
Ross, Robert J.
Ross, Stephen
Ross, Wallace R.
Ross Sutherland, Nancy
Rostal, Phyllis J.
Roth, Albert C.
Roth, Eleanor G.
Roth, Henry G.
Roth, Jaina
Rothe, Knute L.
Roullier, Napoleon
Rounds, C. Roy
Rovsek, Frank J.
Rowe, John W.
Rowell, Jeffrey S.
Rowell, Roger M.
Rowlands, Robert
Rowley, Lorene
Royston, Calvin T.
Royston, Sharon
Rubadeau, Kenneth W.
Ruby, Evelyn E.
Ruby, Isabelle E.
Rudie, Alan W.
Rue, John D.
Ruess, Robert M.
Ruhland, Grace I.
Ruiz, Ricardo
Rumbold, CarolineT.
Rundell, Jeanette
Rundell, Milton C.
Runge, Charles D.
Rupp, Claude E.
Rusch, Frances D.
Russell, Edna F.
Russell, Marie B.
Russell, Roger B.
Rust, Stanley F.
Ruttimann-Johnson, Carmen
Ryan, Doris B.
Ryan, Edward J.
Ryan, George P.
Ryan, James
Ryan, John L.

Ryan, Lovedy M.
Ryan, Madge I.
Ryan, Menola
Ryan, Sally A.
Ryan, Willard T.
Ryan, Zach
Rybarczyk, Thomas M.
Rybicki, Jacquelynn L.
Ryu, Jae San
Sabean, Inez M.
Sablovitch, Bonnie J.
Sabo, Ronald C.
Sachs, Irving B.
Sachs, Leo
Sachtschale, Mary K.
Sacket, Walter H.
Sackett, H.S.
Sadowski, Thomas J.
Saeman, Jerome F.
Saeman, Marie C.
Sage, Robbie C.
Sagoe, John
Sagrado, Maximo A.
Salehuddin, A.B.M.
Saliklis, Edmond P.
Saliklis, Edward L.
Salzberg, Harold K.
Salzwedel, Edwin H.
Samp, Robert J.
Sanadi, Anand
Sanborn, Gertrude
Sanborn, William A.
Sandberg, Carl L.
Sander, Gustave H.
Sanders, Beau
Sanderson, William J.
Sandoval, Antonio
Sands, Waldo M.
Sanna, Charlene A.
Santaholma, Matti
Santana, Marcos
Sanyer, Necmi
Sarfo, James Sah
Sargent, Ronald A.
Sartin, James E.
Sato, Masatoshi
Satori, Holly
Saue, Reginald T.
Sauk, Leo
Sauve, Gregg
Saviano, Michael J.
Savonne, Charlotte M.
Saxton, John D.
Sayre, Ralph
Scallon, Jill
Scallon, Karen
Scarlett, Barbara E.
Schaefer, Dorothy L.
Schaeffer, Ralph
Schafer, Earl, R.
Schaffer, Erwin L.
Schalch, Heidi
Schalla, Mildred L.
Scharch, Marilyn A.
Scharfetter, Hans O.
Scharmer, Roger, C.
Scharnke, LeRoy O.
Schaub, Mary R.
Scheele, Rebecca E.
Scheele, Todd C.
Scheffer, Theodore C.
Schell, Sandra L.
Schemenauer, Jarrod

Scherer, Louis
Scheutz, Chas.
Schiavone, Dom
Schied, Eugene F.
Schiefelbein, Chas. J.A.
Schieg, Margaret H.
Schiller, Mathias
Schillinger, Wailliam J.
Schimming, Milo H.
Schindler, Gina
Schindler, Rita A.
Schiro, A.
Schlaak, August R.
Schlaefer, Genevieve A.
Schlagheck, Mary E.
Schlitt, Brian
Schmeding, Pamela
Schmelter, Sandra L.
Schmid, Christian
Schmidd, George C.
Schmidt, Elizabeth
Schmidt, Eugene C.
Schmidt, George P.
Schmidt, Russell A.
Schmieding, Stephen A.
Schmitt, Robert A.
Schmitz, John R.
Schmoldt, Dan
Schneider, Ione R.
Schneider, Jim
Schneider, Nan F.
Schneider, Wilfred N.
Schoenike, Barry
Schoenwetter, Otto C.
Schohl, James J.
Scholten, John A.
Schooley, Wilson F.
Schorger, Arlie W.
Schorger, M.
Schott, Hans
Schowalter, Wilbert E.
Schrader, O. Harry
Schroder, Mabel G.
Schroeder, Chris J.
Schroeder, David P.
Schroeder, Henry R.
Schroeder, Henry W.
Schroeder, Herbert A.
Schroeder, Lori
Schroeder, Mary B..
Schroedl, Francis A.
Schroeter, Joel
Schubert, Eileen K.
Schubert, Hazel M.
Schubring, Irene M.
Schueneman, Gregory
Schuete, Mike A.
Schuette, Henry A..
Schuette, John F.
Schuetz, Chas. J.
Schulenburg, Debra S.
Schuler, Julie A.
Schultheis, A. Elizabeth
Schultz, Dason
Schultz, David
Schultz, Hildegrade M.
Schultz, Jane S.
Schultz, John R.
Schultz, Mabel
Schultz, Nancy J.
Schultz, Paul W.
Schultz, Rufus C.
Schultz, Vicki L.

Schulz, Georg(e)
Schulze, Mary K.
Schumann, David P.
Schumann, David R.
Schumann, Rebecca L.
Schumann, Sharon L.
Schumann, Thomas H.
Schuna, Arthur A.
Schunning, Ann K.
Schutz, Virginia
Schwandt, Virgil, H.
Schwartz, Florence
Schwartz, Gail A.
Schwartz, Harry
Schwartz, Samuel
Schwartz, Sidney L.
Schwarz, Oscar R.
Schwebs, Arthur C.
Schwegler, Kathryn M.
Schwenn, Grace S.
Schwerdtfeger, Edith J.
Schwoegler, Joseph
Schwoegler, Leonard J.
Scordia, Danilo
Scott, Gary
Scott, Kenneth
Scott, LeRoy W.
Scott, Lucy M.
Scott, Ralph W.
Scott, C.Tim
Scotton, Gail L.
Scotton, Teresa E.
Scroggins, Jonnell H.
Seamonson, Beverly J.
Seaquist, Robert W.
Seborg, Clarence O.
Seborg, Raymond M.
Sechler, Stephanie
Sedbrook, Tod A.
Seder, Willard J.
Seeley, Grace
Seeley, Jeanette
Seeliger, Ilse M.
Seelinger, Edgar E.
Seely, David F.
Seffrood, Wendy
Seidl, Michael G.
Seidl, Robert J.
Seikel, Margaret K.
Seilbold, Frederick C. Jr.
Seiler, Dorothy R.
Selbo, Magnus L.
Sell, Jurgen
Selz, John P.
Serrano, Mayda
Sestan, Domen
Setterholm, Vance C.
Seutter, Louis
Sever, David J.
Severin, Geoffrey A
Severson, Claire G.
Severt, D. J.
Sexton, Joseph
Sexton, Teresa I.
Seybold, Barbara R.
Shade, Christina A.
Shafer, Clifford B.
Shafi, Tariq A.
Shanke, E. Chat
Sharkey, Bernard E.
Sharp, Joseph J.
Shary, Semarjit
Shea, Rodney M.

Sheahan, Mary B.
Shebesta, Kevin
Shelton, Alicia
Sher, Carol R.
Sherman, Elaine A.
Sherrard, Earl C.
Sherrer, Robert E.
Sherwood, Gerald E.
Shi, Nian-qing
Shields, Gordon
Shilts, Richard W.
Shim, Kug-bo
Shin, Eun Woo
Shinker, William M.
Shinn, Joseph H.
Shirley, Clemmie N.
Shisler, Everett W.
Shivers, Stanley M.
Shore, Herman, R.
Shorger, A. W.
Showalter, Janel M.
Shower, Regina R.
Showers, Susan I.
Shufelt, Marion G.
Shuga, Beverly J.
Shumway, Mark H.
Siebert, Edward J.
Sielaff, Edna C.
Siemering, Alice M.
Sietmann, Caroline
Sievers, Walter A.
Sigala, Constanza
Siggers, Paul V.
Silitonga, Toga
Silva, Rodrigo A.
Silver, Francis M.
Silver, Frank V.
Simangunsong, Bintang
Simmonds, Forrest A.
Simmons, Laurie A.
Simms, David H.
Simon, Reuben C.
Simons, William
Simonsen, John M.
Simonsen, Rita M.
Simonson, Herbert C.
Simonson, Roselyn L.
Simonson, Steven W.
Simpson, David K.
Simpson, Emil J.
Simpson, William T.
Sims, Jeffrey M.
Sinclair, Florence L.
Sinden, Robert L.
Singh, Rajinder
Singler, Donald
Sirit, Mollie
Sixel, Lois C.
Skaaland, Albert R.
Skaar, Christen
Skidmore, Ferne D.
Skidmore, Henry
Skidmore, Kenneth E.
Skidmore, Stanley C.
Skog, Kenneth E.
Skolas, Christine
Skolas, Margaret
Skrenes, Susan.
Slack, Anne W.
Slaughter, Steven G.
Slayton, Elwood C.
Sliker, Alan W.
Sloan, A.

Sloan, Daniel J.
Small, Jennie M.
Smerda, August Jr.
Smesrud, Isabel T.
Smiley, Vern, N.
Smith, Alice M.
Smith, Bliss H.
Smith, Bruce C.
Smith, Bruce J.
Smith, C. Bassel
Smith, C. E.
Smith, C. S.
Smith, Cynthia M.
Smith, David C.
Smith, Diana M.
Smith, Gary D.
Smith, Gordon E.
Smith, Greg
Smith, Harvey H.
Smith, James P.
Smith, James H.
Smith, Jeffrey L.
Smith, John
Smith, John M.
Smith, Lyle K.
Smith, Marie, L.
Smith, Matthew
Smith, Paul M.
Smith, Percy T.
Smith, Philip C.
Smith, R.
Smith, R. K.
Smith, Reuben W.
Smith, Robert N.
Smith, Sue
Smith, Susan K.
Smith, Theordore S.
Smith, Tim
Smith, Virginia E.
Smith, Walton R.
Smith, William E.
Smuk, John M.
Smyrski, Rose M.
Smyth, Bertha
Smythe, Richard D.
Snell, Mary O.
Snell, Walter
Snider, Maurice W.
Snorek, Teresa M.
Snow, Ron
Snyder, Wm. A.
Sobek, Joyce E.
Soerens, Natalie
Sohn, Richard F.
Sokel, Corry
Solari, Robert J.
Solberg, Lucy
Solbrig, Charles L., Jr.
Soltis, Lawrence A.
Somboonna, Naraporn
Somers, Peter A.
Somerset, Henry B.
Sommers, Florence H.
Sommers, Leo E.
Sommerville, Stephen W.
Sondern, Clarence W.
Song, Hong-Keun
Sonnen, Dan
Soper, Vernon R.
Sotos, Peter G.
Soto-Villa, Luis J.
Southard, Conway A.
Southerland, Carole F.

Sowunmi, Samuel
Spaanem, Cora E.
Spaeni, John H.
Spalding Brian P.
Spartz, James
Sparver, E.C.
Spear, Mary M.
Spelter, Henry N.
Spence, J. Robert
Spencer, Marshall, E.
Sperry, Jerome
Sperry, Michael R.
Spilde, Grace L.
Spinar, Frank J.
Spink, Donn R.
Spinti, Ernest P.
Spinti, Isabella C.
Spoerl, Edward L.
Sponsler, O.L.
Spotted War Bonnett,
   Nathaniel
Spradling, Mae
Spray, Harry M.
Spreen, Candice A.
Spring, Barbara K.
Springer, Edward L.
Sprowles, Bruce H.
Srebotnik, Ewald
Sreenath, Hassan K.
Sreenath, Hiki
Srinivasan, Vsha
Srivastava, Kailash
St. John, Franz J.
St. John, Macel R.
Staat, Norman E.
Stadelmann, Otto R.
Stadler, George
Staebell, Margaret E.
Stahl, Bernhard
Staidl, Joseph A.
Staie, Ione V.
Stamm, Alfred J.
Stamper, Fred
Stanfill-McMillan, Kim
Stangl, Josh
Stanley, Jeranice E.
Stark, Anna L.
Stark, Nicole M.
Starr, Charles
Stassi, Nicola
Staton, Donald E.
Statz, Winifred I.
Stauffacher, Jeff
Steckel, Vera
Steeb, Harry J.
Steel, Joseph I.
Steele, Hazel. L.
Steele, Richard H.
Steele, Shirley A.
Steele, Stanley D.
Steele, Susan V.
Steelink, Cornelius
Steffen, Irving D.
Steffes, David
Steffes, R. Florence
Steffes, Ruth A.
Stehr, M.
Stehr, Raymond
Stehr, Reginald D.
Stehr, Virginia M.
Steidemann, Robert W.
Stein, James, W.
Steinkopff, Cassandra T.

Steinman, Marilyn A.	Sukhov, Dmitry	Teesdale, Jerald T.	Tolo, Kenneth
Steinmetz, Paul E.	Suleski, Jane C.	Teesdale, Laurence V.	Tolonen, Yrjo
Steiss, Patricia F.	Sullivan, Beverly M.	Telford, Clarence J.	Tolson, Lawrence S.
Stempke, J. A.	Sullivan, Brady	Tellman, Stephen J.	Tomczyk, Gail R.
Stenjem, David W.	Sullivan, Cindy	Temple, Darrel M.	Tomei, Evalexa
Stenseth, LaVerne B.	Sullivan, John D.	Templeton, Hugh L.	Toole, Roland E.
Stephen, Roy J.	Sullivan, Timothy B.	Tennant, Walter J.	Top, Marjorie M.
Stephens, Aimee	Sumi, Hiroshi	TenWolde, Anton	Topp, Donald J.
Stern, Abigail R.	Summers, Bertrand S.	Terashima, Noritsugu	Torgersen, Anna S.
Stern, Robert K.	Sund, Matilda E.	Terrell, Bessie Z	Torgeson, Benjamin S.
Steuber, Bill	Sund, Walter F.	Terry, Elizabeth E.	Torgeson, Oscar W.
Steuber, William F.	Sundling, Walter	Terry, Joseph M.	Torkelson, Michelle
Steuck, Many Ann	Superfsky, Michael J.	Tessman, Norman A.	Tormey, Harold J.
Stevens, Gordon H.	Surface, Henry E.	Tessmann, John W.	Tormey, John J.
Stevens, Robert R.	Sutermeister, Edwin	Tetzlaff, Herbert E.	Tormey, Regina
Stewart, Linn A.	Sutherland, Bruce	Teude, Lester F.	Torres, Juana Perez
Stewart, Marilyn S.	Sutherland, Jeff	Textor, Clinton K.	Torresani, Peter
Stewart, Philip	Sutherland, O. M.	Theis, Micheal P.	Totolin, Vladimir
Stidgen, George	Sutter, Errin	Thelen, Rolf	Tourtellot, Edward B.
Stidgen, Laura B.	Sutter, Nancy E.	Thessin, Dorothy E.	Town, George G.
Stiefvater, Stacey L	Sutton, Dorothy L.	Thickens, John H.	Towne, Agnes B.
Stillson, Ferril S.	Sutton, Stephen S.	Thieding, Sharon	Townley, Margaret
Stimke, J.	Suzuki, Melissa	Thiel, Gordon	Townley, Marie
Stitgen, Carlos H.	Suzuki, Shuzo	Thielmann, Ann	Townsend, Gerald J.
Stivarius, Joan M.	Sveum, Richard S.	Thier, Dorothy E.	Townsend, Thomas G.
Stluka, Robert T.	Svoboda, Kay E.	Thomas, Charles R..	Tracy, Elwood
Stockhausen, Janet	Swanson, Kari	Thomas, Duane	Trainor, Philip
Stockman, Willy L.	Swanson, Lillian M.	Thomas, Lorraine C.	Trainor, Ronald J.
Stockmann, Volker E.	Swanson, Shawn	Thompson, Arlene H.	Trameri, Charles C.
Stoeber, Mary L.	Swanson, Terry B.	Thompson, Charles, J.	Tran, Hao
Stofen, Carl H.	Swanson, Walter H.	Thompson, David C.	Tran, Minh, T.
Stojanovich, Donna C.	Swartout, James T.	Thompson, Elaine T.	Traska, Amy
Stoker, Denise L.	Swed, Frieda S.	Thompson, Frances W.	Traver, Roy H.
Stoltenberg, T.R.	Sweet, Carroll V.	Thompson, Garace	Trayer, George W.
Stoltz, Benjamin F.	Swenson, Mildred K.	Thompson, Gladys C.	Trevorrow, Gloria J.
Stondall, Steiner	Swiggum, Clifford M.	Thompson, Grace V.B.	Triatik, Frederick J.
Stone, Robert N.	Swiler,	Thompson, Herbert O.	Triggs, Pamela, R.
Stoneman, David M.	Switzky, Ben C.	Thompson, John M.	Trivedi, Sarawatichandra A.
Stonis, John A.	Switzky, Saul S.	Thompson, John P.	Troia, Natale
Storkson, Jo Ann M.	Swndson, Norman L.	Thompson, Joyce	Trottman, Charles H.
Stormzand, Martin J.	Sydney, Kate	Thompson, Patricia M.	Troyer, Rodney K.
Stotzheim, Mary K.	Syftestad, Bertha	Thompson, Russ	Truax, Thomas R.
Straka, Thomas J.	Syftestad, Carl E.	Thompson, Sadie E.	Trumble, Patricia L.
Strang, Anna E.	Sykes, Marguerite S.	Thompson, William E.	Trumbo, Robert L.
Strasbaugh, Richard M.	Sylvester, Corwin L.	Thorkelson, Elias B.	Tsai, Pang-yen
Stratton, John W.	Sylvester, William W.	Thornsen, Michael	Tschernitz, John L.
Stream, Mark, E.	Syndergaard, Larry	Thorson, Lowell M.	Tshabalala, Eunice B.
Streitfeld, Marsha S.	Syska, Arthur D.	Thousand, John A.	Tshabalala, Mandla A.
Strenge, Frederick A.	Szymczak, Eugene J.	Thubauville, Arvella	Tu, Xin
Strickland, D.	Tabata, Haruro	Thurman, Robert E.	Tucker, Doug
Strickler, Bruce C.	Tait, Ina B.	Thurner, George	Tuomi, Roger L.
Strickler, Irene I.	Tajvidi, Mehdi	Tiedeman, Stanley H.	Turan, Tank
Strohl, Mary Jane	Talati, Viththal R.	Tiedemann, Henry F.	Turk, Christopher G.
Stroika, Matthew	Talbot, William G.	Tiedemann, Stuart	Turk, Lester J.
Strong, Milt L.	Taliaferro, Elizabeth	Tiemann, Harry D.	Turk, Virginia H.
Strutt, Maxine L.	Tamulevich, Stanley R.	Tien, Ming	Turnbull, Raymond F.
Struve, Janine A.	Tan, Freya	Tierney, Mary E.	Turnell, Pamela J.
Stuart, Sherry J.	Tan, Ka Hin	Tilleman, Johann	Turner, Elizabeth
Stucki, Gottlieb	Tang, Walter K.	Tilton, Thomas A.	Turner, H. Dale
Studee, Ernest E.	Tang, Xiaoyan	Tinker, Arthur M.	Turner, Horace G.
Studley, James D.	Taniwiryono, Darmono	Tipple, Robert	Turng, Tom
Stumbo, Donald A.	Taras, Michael A.	Tirumalai, V.C.	Tuschen, Roberta A.
Stump, Reva N.	Tarkow, Harold	Tirumalai, Vedakpat	Tustison, Chas. H.
Stump, William G.	Tartarian, C.S.	Titsworth, Kenneth M.	Tustison, Pearl K.
Stuntz, Beverly A.	Tasker, Marilyn A.	Tobey, Wayland E.	Tuttle, Clifton
Sturm, Henry J.	Taylor, Dorothy	Toda, Jeanette K.	Tuttle, Fordyce E.
Su, Yi-Kai	Taylor, Fred D.	Todd, Helen	Tuttle, Ray C.
Suddarth, Stanley K.	Taylor, James A.	Todd, Thomas F.	Twidwell, V.
Suder, Frank	Taylor, Katherine	Toepelman, Alfred F.	Twining, Nancy A.
Sudo, Syoji	Taylor, Minnie W.	Toepfer, Beth E.	Tyler, Robert R.
Suetter, Louis	Taylor, William G.	Togstad, Dorothy	Tylk, Helen J.
Suhm, Clarence F.	Techtman, Mae A.	Tollefson, Halvor W.	Tyner,Howard D.
Sukartana, Paimin	Teesdale, Clyde H.	Tollner, William	Tyrell, Gladys D.

Ujang, Salmiah
Ulrey, Walter R.
Ulvestad, Lorraine E.
Underbrink, Charles D.
Underwood, Rosemary
Upperman, William C.
Upson, Arthur
Upton, Mary P.
Urban, Eunice R.
Urbanik, Thomas J.
Usher, Florence B.
Usher, Frederick R.
Vader, Eugene M.
Vahey, David W.
Valcarcel, Angela M.
Valdes, Geoffrey
Valencia, Pedro
Valente, Enrique G.
Valesh, Helen A.
Vallim, Marcello
Valyocsik, Ernest W.
Van Arsdel, Rose P.
Van Den Heuvel, Zachary
Van Dyke, Florence M.
Van Egeren, James L.
Van Hagan, Charles E.
Van Handel, Alice C.
Van Kleeck, Arthur
Van Kleeck, Florence
Van Sickle, Esther
Van Vleet, Jennifer
Van Wagnen, Eltravis H.
Van Winter, Carl
Vance, Kathryn
Vanden Wymelenberg, Amber
Vanderhoef, Elijah R.
Vanney, Sheila
Van Vleet, Jennifer
Vargo, James R.
Varney, Vern V., Jr.
Vasaitio, Anthony J.
Vaughan, Coleman L.
Veerhusen, M.
Venden, Norman L.
Vergara, Esther Z.
Verrill, Steve P.
Vetter, Esther A.
Viar
Vick, Charles B.
Vickers, Rita R.
Vilbrandt, Karen L.
Vilker, John P.
Villarreal, Alberto
Villarreal, Jesse C.
Villarreal, Rolando
Vincent, Leon M.
Vindedahl, Harlow L.
Viray, Fernando
Virnig, Rita M.
Vital, R.
Vitale, Frank A.
Vitale, James Jr.
Vivas, Dorothy G.
Viviani, Richard L.
Voelker, Joseph G.
Voelker, Timothy
Vojacek, Emily H.
Volk, Thomas J.
Volkert, Robert M.
Vollmert, Dale
Vondrachek, Diann M.
Voorhies, Glenn
Vos, Dan J.

Voss, Arnold W.
Voytek, Christine M.
Vroman, J.A.
Vultaggio, Charles
Vultaggio, Frank
Wacker, James P.
Wacker, John
Wackman, Charline M..
Waege, Madge
Waggener, Thomas R.
Wagner, Donald
Wagner, Eugene R.
Wagner, Leo J.
Wagner, Michael
Wagner, Neil
Wagner, Regina D.
Wagner, Russel O.
Wahl, George C.
Wahl, John R.
Wahlgren, Harold E.
Wais, William D.
Wakefield, Shirley L.
Wakeman, James A.
Wald, Helen
Waldvogel, Jessica
Waldvogel, John A.
Walker, Craig
Walker, George
Walker, Kathleen C.
Walker, Ralph P.
Wall, Donald L.
Wall, Josephine O.
Wall, Mary Beth
Wallace, Harland E.
Wallace, Kathryn
Wallace, Rebecca M.
Wallace, Richard M.
Walling, Raymond C.
Walls, Clarence
Walls, Johnny J.
Walrath, Martin H., Jr.
Walsh, Bernard.
Walsh, Isabel C.
Walsh, Patrick M.
Walsh, Philip J.
Walter, Thomas J.
Walz, Ken
Wang, Dongmei
Wang, Gaosheng
Wang, Xiping
Wang, Zhitong
Wang, Ziping
Wangaard, Frederick F.
Wanta, Jody
Ward, Deidre L.
Ward, James C.
Ward, James W.
Ward, Michael K.
Ward, Sterling A.
Ward, Thomas P.
Wardell, Vernon
Warner, Adolph H.
Wartiovaara, Ilkka
Wasley, William L.
Wasser, Richard B.
Wastian, Robert M.
Wateman, Steve
Watkins, James R.
Watkins, John
Watkins, Leslie A.
Watson, B.
Watterson, Irene
Way, William J.

Weatherwax, Richard C.
Webb, Stella A.
Webb, William C.
Webber, Dorothy S.
Webber, Merton L.
Weber, Diane E.
Weber, O. L. E.
Weber, Peter J.
Weber, Roman F.
Weber, Wallace W.
Weber, Walter H.
Wegner, Judy V.
Wegner, Theodore H.
Weidner, Harold L.
Wei, Dongsheng
Weil, J.
Weiman, Fred
Weinauer, Dietrich E.
Weinbaum, Florence E.
Weinman, Fred
Weinstein, Arthur I.
Weinstock, Ira A.
Weir, William F.
Weise, Anne M.
Weiss, Howard F.
Welch, Charles W.
Welch, Chas. R.
Welch, Dehn S.
Welch, John R.
Welch, Michael L.
Welhoefer, Thomas
Welker, Howard N.
Wells, Joseph E.
Wells, Sarah
Wells, Sidney D.
Welo, Lars A.
Welsch, Heidi.
Welsh, Stephen L.
Wendt, Gerhard F.
Wendt, Kenneth W.
Wendt, Kurt F.
Wendt, Marla
Wendt, Robert E.
Wenger, Douglas M.
Wengert, Eugene M.
Wennesheimer, Jill M.
Wentz, Vicki
Wergin, Gary J.
Wermerskirchen, Jacob D.
Werner, Janet C.
Werner, Kenneth W.
Wernsted, Lage
Werren, Fred
Wescott, Erin
Wescott, James
Wesolowski, Betty P.
Wesolowski, Donald
Wesolowski, Martin F.
Wesphal, Colleen S.
West, John D.
Westbury, Edgar L.
Westby, Rebecca M.
Westenhaver, H. L.
Western, Loren J.
Westover, Bert J.
Wetzel, Harry J.
Weum, Martin
Weum, Olin L.
Weyls, Barney
Weyls, Harry
Weyna, Philip L.
Weynand, Lucille R.
Whalen, John F.

Wheat, Dan
Wheeler, Douglas R.
Wheeler, Herbert A.
Whitcombe, Marion H.
White, A.
White, David G.
White, E.G.
White, Edgar F.
White, Ethel M.
White, James A.
White, John
White, Portia E.
White, Robert H.
White, Ronald H.
White, Rosa
White, Veulah
White, Virginia D.
Whiting, Dennis
Whitmore, Kelly
Whiton, Arthur L.
Whitt, Anna
Wichmann, John F.
Wick, Charles H.
Wieben, Anella
Wiedabach, Jack R.
Wiedeback, Marian E.
Wiedenbeck, Diane
Wiedenhoeft, Alex C.
Wiedholz, W. G.
Wiedholz, William G.
Wieloch, Linda
Wiemann, Michael C.
Wiepking, Christopher A.
Wiertelak, Jan
Wiese, Howard E.
Wiese, John D.
Wiese, Julie A.
Wightman, George
Wikler, Karen
Wilcox, Larry C.
Wilcox, Roy W.
Wilcox, W. Wayne
Wild, Peggy S.
Wilder, Beulah M.
Wilder, W.S.
Wiley, Martha S.
Wilhelm, Fred J.
Wilkas, Garth.
Wilkie, George R.
Wilkinson, Thomas L.
Willgrubs, George M.
Willgrubs, George W.
Willgrubs, Roy T.
Williams, Anne E.
Williams, Darrell
Williams, Ingeborg
Williams, Judith C.
Williams, Karen F.
Williams, Llewelyn
Williams, R. Sam
Willis, Anita J.
Wilmont, Colin
Wilsie, Mary C.
Wilson, Gordon C.
Wilson, Willie W.
Wilson, Thomas R. C.
Wilson, Tracy
Winandy, Jerrold E.
Winburn, Samuel C., Jr.
Winch, Hazel, M.
Winch, Marvin D.
Wind, Violet E.
Wingender, Ronald J.

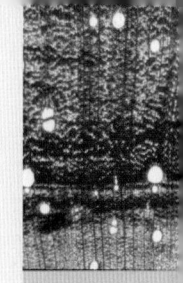

Winistorfer, Robin M.
Winistorfer, Steve
Winrich, K. M.
Winskill, Marianne L.
Winslow, Carlile P.
Winters,W. R.
Wipperforth, Herman J.
Wipperman, Jeanne M.
Wipprecht, James G.
Wirka, Raymond M.
Wirtala, Matthew
Wise, James L.
Wise, Madelon M.
Wisniewski, Joseph
Witherow, Boyd M.
Witherspoon, Gregory A.
Withey, Morton O.
Withey, Norman H.
Witt, Louis
Witt, Peter J., Jr.
Witting, Richard H.
Wittwer, Alice H.
Witz, Shirley
Wlodarczyk, Wencil W.
Wodzinski, Claudia G.
Wohlferd, Franklyn J.
Wold, Oscar E.
Wold, Oscar M.
Wolf, Geneva
Wolfe, Barbara
Wolfe, E.
Wolfe, Kathleen
Wolfe, Ronald W.
Wolfe, Stella V.
Wolferman, Chauncey G.
Wolff, Margaret E.
Wolff, Minnie
Wolff, Phillip S.
Wolff, Wilhelm
Wolkert, Robert M.
Wollin, Arno C.
Wolter, Karl E.
Wondergem, Jennifer
Wong, Jenny
Wong, Michael
Wongcharupan, Metha
Woo Lee, P.
Wood, Barry
Wood, Edgar M.
Wood, James, E., Jr.
Wood, Lyman W.
Woodburn, James G.
Woodfin, Richard O., Jr.
Woods, JoAnn M.
Woodward, Bessie E.
Woodzicka, William F.
Wooster, Charles R.
Wooten, Lillian B.
Worrell, Arthur F.
Worrill, Gloria A.
Worsham, Marie
Worthington, Oliver J.
Wright, Daniel J.
Wright, Dorothy L.
Wright, Geraldine
Wright, Gilbert M.
Wright, Paul L.
Wright, Rebecca H.
Wright, Robert E.
Wruck, Randall A.
Wuilleumier, Blanche
Wulff, Dorothy M.
Wurth, Charles

Xiong, Pangki
Xu, Dan-Ping
Yaddof, Steve C.
Yager, Stephanie
Yandle, David O.
Yang,Vina W.
Yanker, Frank J.
Yarborough, Linda
Yaun, Mingjun
Yeary, Lon
Yelle, Daniel J.
Yenn, Agnes
Yeong, Kim
Yin, Menping
Yoder, Ralph
Yokota, Tokuo
Yolton, Leslie A.
Yonjing, Li
Yort, Svend
Young, Arthur J.
Young, Edith, L.
Young, Betty J.
Young, Eugene P.
Young, Gertrude E.
Young, P. Robert
Young, Ruth A.
Young, Teresa L.
Youngquist, John A.
Youngquist, Waldemar G.
Youngs, Barbara V.
Youngs, Robert L.
Yu, Xiaochun
Yucebay, Feliz
Yule, James B.
Yunis, Luis
Zabel, James M.
Zaccone, Santo L.
Zahn, John J.
Zander, Florence M.
Zank, Lester C.
Zapata, Emily
Zarnke, Randall L.
Zastrow, Joyce J.
Zatko, JoAnn C.
Zauscher, Stefan
Zehner, Larry J.
Zehrt, William H.
Zeldin, Eric L.
Zeldin, Paul E.
Zeldin, Robin, J.
Zelinka, Samuel L.
Zenk, Robert R.
Zenz, David H.
Zenz, Rosemary
Zerbe, John I.
Zevnik, Ruth E.
Zeyher, Susan
Zhang, Houjiang
Zhang, Li
Zhanqian, Song
Zhu, JunYong (JY)
Zhu, Rongvian
Zhu, Wenyuan
Ziebarth, Bertha H.
Ziemann, Erich
Zimbric, Mary A.
Zimdars, Vivian J.
Zimmerman, James E.
Zimmerman, Wayne
Zimmermann, Marshall A.
Zingale, Olga S.
Zingg, Henry
Zinkel, Duane F.

Zinnsmeister
Zirngible, Candice A.
Zischke, Douglas A.
Zitzer, Jacob
Zoch, Lawrence L., Jr.
Zuelsdorf, Annette L.
Zuelsdorf, Richard J.
Zuther, Michelle M.
Zwerg, Gustave W.
Zwettler, Petra

A

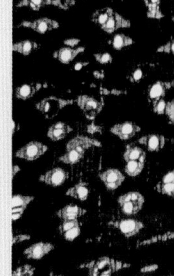

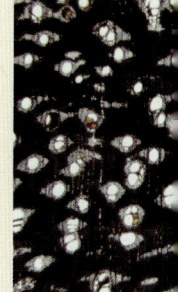